ENCYCLOPEDIA OF

Mathematics and Society

ENCYCLOPEDIA OF

Mathematics and Society

Sarah J. Greenwald
Jill E. Thomley
Appalachian State University

VOLUME 3

Salem Press

The paper used in these volumes conforms to the American National Standard for Permanence of Paper for Printed Library Materials, X39.48-1992 (R1997).

LIBRARY OF CONGRESS CATALOGING-IN-PUBLICATION DATA

Encyclopedia of mathematics and society / Sarah J. Greenwald , Jill E. Thomley, general Editors.
    p. cm.
Includes bibliographical references and index.
ISBN 978-1-58765-844-0 (set : alk. paper) -- ISBN 978-1-58765-845-7 (v. 1 : alk. paper) -- ISBN 978-1-58765-846-4 (v. 2 : alk. paper) -- ISBN 978-1-58765-847-1 (v. 3 : alk. paper)
   1. Mathematics--Social aspects. I. Greenwald, Sarah J. II. Thomley, Jill E.
   QA10.7.E53 2012
   303.48'3--dc23

2011021856

First Printing

PRINTED IN THE UNITED STATES OF AMERICA

Produced by Golson Media
President and Editor J. Geoffrey Golson
Senior Layout Editor Mary Jo Scibetta
Author Manager Joseph K. Golson
Copy Editors Carl Atwood, Kenneth Heller, Holli Fort
Proofreader Lee A. Young
Indexer J S Editorial

Contents

Volume 1

Publisher's Note   *vi*
About the Editors   *viii*
Introduction   *ix*
List of Articles   *xiii*
Topic Finder   *xxi*
List of Contributors   *xxvii*
Articles A to E   *1–376*

Volume 2

List of Articles   *vii*
Articles F to O   *377–744*

Volume 3

List of Articles   *vii*
Articles P to Z   *745–1090*
Chronology   *1091*
Resource Guide   *1109*
Glossary   *1113*
Index   *1127*
Photo Credits   *1191*

List of Articles

A

Accident Reconstruction 1
Accounting 2
Acrostics, Word Squares, and Crosswords 5
Actors, *see Writers, Producers, and Actors* 1081
Addition and Subtraction 7
Advertising 10
Africa, Central 13
Africa, Eastern 15
Africa, North 17
Africa, Southern 19
Africa, West 20
African Mathematics 23
AIDS, *see HIV/AIDS* 478
Aircraft Design 25
Airplanes/Flight 28
Algebra and Algebra Education 31
Algebra in Society 36
Analytic Geometry, *see Coordinate Geometry* 247
Anesthesia 42
Animals 43
Animation and CGI 49
Apgar Scores 51
Arabic/Islamic Mathematics 53
Archery 55
Archimedes 57
Arenas, Sports 61
Artillery 63

Asia, Central and Northern 66
Asia, Eastern 68
Asia, Southeastern 70
Asia, Southern 72
Asia, Western 74
Astronomy 76
Atomic Bomb (Manhattan Project) 79
Auto Racing 81
Axiomatic Systems 84

B

Babylonian Mathematics 87
Ballet 90
Ballroom Dancing 91
Bankruptcy, Business 92
Bankruptcy, Personal 94
Bar Codes 96
Baseball 97
Basketball 99
Basketry 102
Bees 103
Betting and Fairness 105
Bicycles 107
Billiards 110
Binomial Theorem 111
Birthday Problem 113
Black Holes 115

Blackmun, Harry A. 117
Blackwell, David 118
Board Games 119
Body Mass Index 122
Brain 124
Bridges 129
Budgeting 130
Burns, Ursula 132
Bus Scheduling 133

C

Calculators in Classrooms 137
Calculators in Society 139
Calculus and Calculus Education 142
Calculus in Society 148
Cameras, *see Digital Cameras* 304
Calendars 153
Canals 155
Carbon Dating 157
Carbon Footprint 159
Careers 162
Caribbean America 166
Carpentry 167
Castillo-Chávez, Carlos 169
Castles 171
Caves and Caverns 172
Cell Phone Networks 174
Census 176
Central America 178
Cerf, Vinton 180
Cheerleading 181
Chemotherapy 183
Chinese Mathematics 184
Circumference, *see Perimeter
    and Circumference* 761
City Planning 188
Civil War, U.S. 191
Climate Change 194
Climbing 200
Clocks 201
Closed-Box Collecting 204
Clouds 206
Clubs and Honor
    Societies 207
Cochlear Implants 209
Cocktail Party Problem 210
Coding and Encryption 212
Cold War 214

Combinations, *see Permutations
    and Combinations* 763
Comic Strips 218
Communication in Society 219
Comparison Shopping 225
Competitions and Contests 227
Composing 229
Computer-Generated Imagery (CGI),
    see Animation and CGI 49
Congressional Representation 231
Conic Sections 235
Connections in Society 238
Continuity, *see Limits and Continuity* 552
Contra and Square Dancing 243
Cooking 244
Coordinate Geometry 247
Coral Reefs 250
Counterintelligence,
    see Intelligence and Counterintelligence 508
Coupons and Rebates 252
Credit Cards 253
Crime Scene Investigation 255
Crochet and Knitting 257
Crosswords,
    see Acrostics, Word Squares, and Crosswords 5
Crystallography 259
Cubes and Cube Roots 260
Currency Exchange 262
Curricula, International 264
Curriculum, College 267
Curriculum, K–12 274
Curves 280

D

Dams 283
Data Analysis and Probability in Society 284
Data Mining 290
Daubechies, Ingrid 292
Deep Submergence Vehicles 294
Deforestation 295
Deming, W. Edwards 298
Diagnostic Testing 299
Dice Games 301
Digital Book Readers 303
Digital Cameras 304
Digital Images 306
Digital Storage 308
Disease Survival Rates 311

Diseases, Tracking Infectious	312
Division, *see Multiplication and Division*	685
Domes	315
Doppler Radar	316
Drug Dosing	317
DVR Devices	319

E

Earthquakes	323
Educational Manipulatives	324
Educational Testing	326
EEG/EKG	329
Egyptian Mathematics	330
Einstein, Albert	333
Elections	335
Electricity	340
Elementary Particles	342
Elevation	344
Elevators	346
Encryption, *see Coding and Encryption*	212
Energy	348
Energy, Geothermal, *see Geothermal Energy*	441
Engineering Design	351
Equations, Polar	353
Escher, M.C.	354
Ethics	356
Europe, Eastern	358
Europe, Northern	361
Europe, Southern	363
Europe, Western	365
Expected Values	368
Exponentials and Logarithms	370
Extinction	372
Extreme Sports	373

F

Fantasy Sports Leagues	377
Farming	379
Fax Machines	382
Fertility	384
Fibonacci Tuning, *see Pythagorean and Fibonacci Tuning*	823
FICO Score	386
File Downloading and Sharing	387
Fingerprints	388
Firearms	390
Fireworks	392
Fishing	394

Floods	395
Football	398
Forecasting	400
Forecasting, Weather, *see Weather Forecasting*	1052
Forest Fires	402
Fuel Consumption	404
Function Rate of Change	405
Functions	408
Functions, Recursive	410

G

Game Theory	413
Games, *see Board Games; Video Games*	119, 1032
Garfield, Richard	415
Genealogy	417
Genetics	419
Geometry and Geometry Education	422
Geometry in Society	427
Geometry of Music	433
Geometry of the Universe	436
Geothermal Energy	441
Gerrymandering	443
Global Warming, *see Climate Change*	194
Golden Ratio	445
Government and State Legislation	448
GPS	450
Graham, Fan Chung	453
Graphs	454
Gravity	456
Greek Mathematics	458
Green Design	461
Green Mathematics	463
Gross Domestic Product (GDP)	466
Growth Charts	468
Guns, *see Firearms*	390
Gymnastics	469

H

Harmonics	471
Hawking, Stephen	473
Helicopters	475
Highways	476
Hitting a Home Run	477
HIV/AIDS	478
Hockey	480
Home Buying	482
Houses of Worship	485
HOV Lane Management	488

Hunt, Fern   489
Hurricanes and Tornadoes   490

I

Incan and Mayan Mathematics   493
Income Tax   496
Individual Retirement Accounts (IRAs),
    see Pensions, IRAs, and Social Security   757
Industrial Revolution   499
Infantry (Aerial and Ground Movements)   501
Infinity   504
Insurance   506
Intelligence and Counterintelligence   508
Intelligence Quotients   511
Interdisciplinary Mathematics Research,
    see Mathematics Research, Interdisciplinary   632
Interior Design   514
Internet   515
Interplanetary Travel   520
Inventory Models   523
Investments, see Mutual Funds   691
Irrational Numbers, see Numbers, Rational
    and Irrational   724
Islamic Mathematics, see Arabic/Islamic
    Mathematics   53

J

Jackson, Shirley Ann   525
Joints   526

K

Kicking a Field Goal   529
King, Ada (Countess of Lovelace),
    see Lovelace, Ada   565
Knitting, see Crochet and Knitting   257
Knots   531

L

Landscape Design   533
LD50/Median Lethal Dose   535
Learning Exceptionalities   536
Learning Models and Trajectories   540
Legislation, see Government
    and State Legislation   448
Levers   544
Life Expectancy   545
Light   547
Light Bulbs   549

Lightning   550
Limits and Continuity   552
Linear Concepts   553
Literature   556
Loans   561
Logarithms, see Exponentials and Logarithms   370
Lotteries   563
Lovelace, Ada   565

M

Magic   567
Mapping Coastlines   570
Maps   571
Marine Navigation   574
Market Research   578
Marriage   580
Martial Arts   582
Math Gene   584
Mathematical Certainty   586
Mathematical Friendships and Romances   588
Mathematical Modeling   589
Mathematical Puzzles   593
Mathematician Defined   595
Mathematicians, Amateur   597
Mathematicians, Religious   600
Mathematics, African, see African Mathematics   23
Mathematics, Applied   603
Mathematics, Arabic/Islamic,
    see Arabic/Islamic Mathematics   53
Mathematics, Babylonian,
    see Babylonian Mathematics   87
Mathematics, Defined   608
Mathematics, Chinese, see Chinese Mathematics   184
Mathematics, Egyptian,
    see Egyptian Mathematics   330
Mathematics, Elegant   610
Mathematics, Greek, see Greek Mathematics   458
Mathematics, Green, see Green Mathematics   463
Mathematics, Incan and Mayan,
    see Incan and Mayan Mathematics   493
Mathematics, Native American,
    see Native American Mathematics   697
Mathematics, Roman, see Roman Mathematics   878
Mathematics, Theoretical   613
Mathematics, Utility of   618
Mathematics, Vedic, see Vedic Mathematics   1029
Mathematics: Discovery or Invention   620
Mathematics and Religion   622

Mathematics Genealogy Project 628
Mathematics Literacy and Civil Rights 630
Mathematics Research, Interdisciplinary 632
Mathematics Software,
    see Software, Mathematics 926
Matrices 634
Mattresses 636
Mayan Mathematics,
    see Incan and Mayan Mathematics 493
Measurement, Systems of 637
Measurement in Society 640
Measurements, Area 645
Measurements, Length 647
Measurements, Volume 651
Measures of Center 653
Measuring Time 655
Measuring Tools 657
Medical Imaging 659
Medical Simulations 660
Microwave Ovens 661
Middle Ages 663
Military Draft 665
Minorities 667
Missiles 671
Molecular Structure 672
Money 674
Moon 677
Movies, Making of 679
Movies, Mathematics in 681
MP3 Players 684
Multiplication and Division 685
Music, Geometry of,
    see Geometry of Music 433
Music, Popular, *see Popular Music* 786
Musical Theater 689
Mutual Funds 691

N
Nanotechnology 693
National Debt 695
Native American Mathematics 697
Nervous System 700
Neural Networks 701
Newman, Ryan 703
Nielsen Ratings 704
Normal Distribution 706
North America 708
Number and Operations 710

Number and Operations in Society 714
Number Theory 719
Numbers, Complex 721
Numbers, Rational and Irrational 724
Numbers, Real 727
Numbers and God 729
Nutrition 731

O
Ocean Tides and Waves, *see Tides and Waves* 993
Oceania, Australia and New Zealand 735
Oceania, Pacific Islands 737
Operations, *see Number and Operations;*
    Number and Operations in Society 714
Optical Illusions 739
Orbits, Planetary, *see Planetary Orbits* 771
Origami 741

P
Pacemakers 745
Packing Problems 746
Painting 748
Parallel Postulate 750
Parallel Processing 752
Payroll 754
Pearl Harbor, Attack on 755
Pensions, IRAs, and Social Security 757
Percussion Instruments 760
Perimeter and Circumference 761
Permutations and Combinations 763
Perry, William J. 766
Personal Computers 767
Pi 770
Planetary Orbits 771
Plate Tectonics 773
Plays 774
Poetry 777
Polygons 779
Polyhedra 782
Polynomials 784
Popular Music 786
Predator–Prey Models 788
Predicting Attacks 790
Predicting Divorce 792
Predicting Preferences 793
Pregnancy 796
Prehistory 798
Probability 800

Probability in Society,
  see Data Analysis and Probability in Society   284
Problem Solving in Society   804
Producers, see Writers, Producers, and Actors   1081
Professional Associations   809
Proof   812
Proof in Society,
  see Reasoning and Proof in Society   845
Psychological Testing   815
Pulleys   818
Puzzles   819
Puzzles, Mathematical, see Mathematical Puzzles   593
Pythagorean and Fibonacci Tuning   823
Pythagorean School   825
Pythagorean Theorem   827

Q
Quality Control   831
Quilting   833

R
Racquet Games   835
Radar, see Doppler Radar   316
Radiation   836
Radio   838
Raghavan, Prabhakar   840
Randomness   841
Rankings   843
Rational Numbers, see Numbers,
  Rational and Irrational   724
Reasoning and Proof in Society   845
Recycling   850
Relativity   853
Religion, Mathematics and,
  see Mathematics and Religion   622
Religious Mathematicians
  see Mathematicians, Religious   600
Religious Symbolism   855
Religious Writings   857
Renaissance   860
Representations in Society   863
Revolutionary War, U.S.   868
Ride, Sally   870
Risk Management   872
Robots   874
Roller Coasters   877
Roman Mathematics   878
Ross, Mary G.   880

Ruler and Compass Constructions   881

S
Sacred Geometry   885
Sales Tax and Shipping Fees   886
Sample Surveys   888
Satellites   890
Scales   892
Scatterplots   894
Scheduling   896
Schools   897
Science Fiction   899
Sculpture   903
Search Engines   905
Segway   906
Sequences and Series   908
Servers   910
Shipping   912
Similarity   914
Six Degrees of Kevin Bacon   915
Skating, Figure   916
Skydiving   918
Skyscrapers   919
SMART Board   920
Smart Cars   922
Soccer   923
Social Networks   924
Social Security, see Pensions, IRAs,
  and Social Security   757
Software, Mathematics   926
Solar Panels   930
South America   931
Space Travel, see Interplanetary Travel   520
Spaceships   933
Spam Filters   935
Sport Handicapping   936
Sports Arenas, see Arenas, Sports   61
Squares and Square Roots   938
Stalactites and Stalagmites   940
State Legislation, see Government
  and State Legislation   448
Statistics Education   941
Step and Tap Dancing   946
Stethoscopes   947
Stock Market Indices   949
Strategy and Tactics   951
Street Maintenance   954
String Instruments   956

Stylometry   957
Submarines, *see Deep Submergence Vehicles*   294
Succeeding in Mathematics   958
Sudoku   962
Sunspots   963
Subtraction, *see Addition and Subtraction*   7
Surfaces   964
Surgery   966
Swimming   969
Symmetry   970
Synchrony and Spontaneous Order   972

T
Tao, Terence   975
Tax, *see Income Tax; Sales Tax*
    and Shipping Fees   496, 886
Telephones   976
Telescopes   978
Television, Mathematics in   981
Televisions   985
Temperature   987
Textiles   989
Thermostat   990
Tic-Tac-Toe   992
Tides and Waves   993
Time, Measuring, *see Measuring Time*   655
Time Signatures   995
Toilets   997
Tools, Measuring, *see Measuring Tools*   657
Tornadoes, *see Hurricanes and Tornadoes*   490
Tournaments   998
Traffic   1000
Trains   1001
Trajectories, *see Learning Models and Trajectories*   540
Transformations   1004
Transplantation   1006
Travel Planning   1007
Traveling Salesman Problem   1009
Trigonometry   1010
Tunnels   1014

U
Ultrasound   1017
Unemployment, Estimating   1018

Units of Area   1020
Units of Length   1021
Units of Mass   1023
Units of Volume   1024
Universal Constants   1026
Universal Language   1027

V
Vectors   1029
Vedic Mathematics   1031
Vending Machines   1033
Video Games   1034
Vietnam War   1037
Viruses   1038
Vision Correction   1039
Visualization   1041
Volcanoes   1044
Volleyball   1046
Voting, *see Elections*   335
Voting Methods   1047

W
Water Distribution   1051
Water Quality   1053
Waves, *see Tides and Waves*   993
Weather Forecasting   1054
Weather Scales   1057
Weightless Flight   1059
Wheel   1060
Wiles, Andrew   1061
Wind and Wind Power   1063
Wind Instruments   1065
Windmills   1066
Wireless Communication   1068
Women   1069
World War I   1073
World War II   1075
Wright, Frank Lloyd   1080
Writers, Producers, and
    Actors   1081

Z
Zero   1087

P

Pacemakers

Category: Medicine and Health.
Fields of Study: Algebra; Geometry.
Summary: Artificial pacemakers send a signal to the heart to keep it pumping and mathematicians develop models to determine when and how often to do so.

While a pacemaker is often thought of as a regulator for the heart, a variety of natural pacemakers are responsible for regulating numerous bodily functions including circadian rhythms and menstruation. The actions of natural pacemakers can be modeled as coupled oscillators, where, for example, the behavior of the natural pacemaker influences the function of the heart and vice versa. Square waves or sine waves are often useful in understanding the theory of coupled oscillators, which dates back to 1665 when Christiaan Huygens noticed synchronization in pendulum clocks.

Scientists and mathematicians have shown that chaotic oscillation or amplitude death can also occur in coupled scenarios. A change in the rhythm or in the way they are coupled can result in a change in function, such as in irregular menstrual periods or menopause. Dynamical systems model the interactions between coupled oscillators and allow for theoretical predictions. Using these models, mathematicians, biologists, and medical professionals have made significant advances in understanding natural pacemakers and in designing effective artificial pacemakers. Some of the related mathematical theory is taught to undergraduate mathematics students.

Heart Rhythms and Pacemakers

The sinoatrial node (SA node) is thought to act as the heart's natural pacemaker via electrical impulses. The typical rate for a resting heart is 60 to 70 beats per minute. The pacemaker cells keep the heart pumping at a steady rate, but medical problems can lead to chaotic behavior and cardiac arrest.

Defibrillation may reset the rhythm in some cases but an artificial cardiac pacemaker may be required if the rhythm remains chaotic. Wavelet transforms have been used to effectively model cardiac signals but implementation is difficult because of high power consumption. Australian anesthesiologist Mark Lidwell and physicist Edgar Booth are believed to have designed the first artificial pacemaker in 1928.

American physiologist Albert Hyman also developed an early pacemaker. Many designers of artificial pacemakers have assumed that regular impulses from a pacemaker should be used to stabilize the heartbeat. However, a periodic signal may lead to chaos in some mathematical models, so scientists are developing pacemakers that send impulses based on chaos control theory.

Body Clock and Jet Lag

Jet lag is thought to to result from a desynchronization of the suprachiasmatic nucleus (SCN) pacemaker cells in the hypothalamus of mammals. Experimental studies suggest that the SCN may synchronize within one week. Scientists and mathematicians have mathematically modeled the system as a network with connections between the cells, which are called *nodes* in the language of graph theory.

For example, mathematicians Channa Navaratna and Menaka Navaratna have adapted a model of neuroscientist Peter Achermann and bioinformaticist Hanspeter Kunz. The hypothalamus is thought to have 16,000 pacemaker cells, so they analyzed computer data from a model with this many pacemaker cells and found that the number of long-distance connections in the network determined the synchronicity time. They examined the types of network connections that are needed between the nodes in order to make the model synchronize in a week, and they designed a model that consistently synchronized in close to seven days.

Scientists and mathematicians have also studied many other issues related to pacemakers, such as interference and power issues. There is controversy and conflicting evidence on whether devices such as cell phones or iPods affect pacemakers. Many medical professionals presume an association until clearer evidence to the contrary is found and recommend keeping the devices at least a few inches away from a pacemaker to err on the side of caution. Scientists have developed what some call "origami batteries" made of carbon nanotubes and cellulose that may power the next generation of pacemakers. The batteries can be cut into many shapes.

Further Reading

Barold, S. Serge, et al. *Cardiac Pacemakers Step by Step: An Illustrated Guide.* Hoboken, NJ: Wiley, 2003.

Glantz, Stanton. *Mathematics for Biomedical Applications.* Berkeley: University of California Press, 1979.

Strogatz, Steven. *Nonlinear Dynamics and Chaos: With Applications to Physics, Biology, Chemistry, and Engineering.* Boulder, CO: Westview Press, 2001.

SARAH J. GREENWALD
JILL E. THOMLEY

See Also: Clocks; EEG/EKG; Function Rate of Change; Medical Simulations; Origami.

Packing Problems

Category: Architecture and Engineering.
Fields of Study: Algebra; Data Analysis and Probability; Geometry.
Summary: Packing problems challenge the solver to optimally fill a given space or determine how many objects of some type may fit in a space.

The name "packing problems" has been given to a variety of mathematical problems in both "serious" and "recreational" mathematics. Packing problems are mainly geometric but the term is sometimes also applied to certain numerical problems important to computer science. The distinguishing feature of a geometric packing problem is the objective to position a family of shapes with no overlap and a minimal amount of leftover space.

If a baker has rolled out a certain sheet of dough and has a certain size and shape of cookie cutter, how should the shapes be cut to leave as little wasted dough as possible? How many tennis balls will fit in a very large box? These are packing problems of the most fundamental type. The most thoroughly studied case is that of packing identical spheres (circles in the two-dimensional case or hyperspheres in the four-dimensional—or more—space) in Euclidean space. The most efficient way to pack circles in the plane is to surround each circle with six others in a honeycomb formation, filling about 90.7% of the plane. It was conjectured by Johannes Kepler that a similarly symmetric arrangement of spheres (filling about 74% of space) is optimal in three-dimensions. This conjecture was generally considered to have been proved by Thomas Hales in 1998. Hales's proof of Kepler's conjecture relies on large-scale computer calculation and represents an important example of a well-studied mathematical problem that is "solved," but for which no hand-checked proof is known.

There is much potential for generalization. There is already much unknown for hyperspheres in four dimensions; other important variations include considering spheres of varying sizes, considering non-Euclidean geometry, and using different shapes instead of spheres. There is significant, active research along all of these lines.

Packing Puzzles

The above type of problem has been studied both by professional mathematicians and recreational mathematicians, but there is another category of packing problem particular to the recreational mathematician (and to the puzzle enthusiast). In these packing puzzles, the solver tries to fit a collection of shapes into a larger shape; typically the pieces are of a sort that fits together exactly (for example, packing rectangles into a rectangle rather than packing circles into a square). An old puzzle is to determine, for each n, how large a square is needed to accommodate a 1×1 square, a 2×2 square, a 3×3 square, . . . , and an $n \times n$ square. The oldest known problem of this type is the Tangram puzzle that originated in ancient China, a set of seven simple shapes that can be rearranged to perfectly fill a square (and many other shapes). In the twentieth century, a wide range of packing puzzles involving polyominos (shapes made by gluing together unit squares along their edges) have enjoyed considerable popularity. Packing puzzles can also be posed in three dimensions. Three particularly popular and interesting examples involving blocks illustrate the concept well. The Slothouber–Graatsma puzzle, named after architects Jan Slothouber and William Graatsma, is to pack a $3 \times 3 \times 3$ cube with six $1 \times 2 \times 2$ blocks (leaving three $1 \times 1 \times 1$ holes). The Conway puzzle, named after mathematician John Conway, is to fill a $5 \times 5 \times 5$ cube with 13 $1 \times 2 \times 4$ blocks, one $2 \times 2 \times 2$ block, one $1 \times 2 \times 2$ block, and three $1 \times 1 \times 3$ blocks. A harder puzzle is to pack 41 $1 \times 2 \times 4$ blocks into a $7 \times 7 \times 7$ cube (leaving 15 $1 \times 1 \times 1$ holes).

Covering Problems

A class of problems closely related to the first type of packing problem discussed are so-called covering problems. Covering problems are "dual" to packing problems; instead of positioning non-overlapping copies of a shape in a region with minimal leftover space (as in a packing problem), the solver positions overlapping copies of a shape so that they completely cover a region with a minimal overlap. For example, how many circles of radius 1 does the solver need to completely cover a circle of radius 10? Covering problems have historically received less attention than packing problems, perhaps because packing problems correspond more obviously to physical situations. Nonetheless, covering problems have applications: for example, to designing satellite or cellular networks. Covering a large region as efficiently as possible with circles corresponds to placing security guards in a large area as efficiently as possible so that each point is within a fixed distance of at least one guard.

Packing Problems in Computer Science

Another class of packing problem, the so-called knapsack problems, is numerical (rather than geometric) in nature. A typical example is sometimes called the "Aladdin's saddlebag problem."

Aladdin is in a cave full of a variety of treasures: gold, silver, rubies, diamonds, rare books, and other valuable objects. Each type of object takes up a certain amount of space in Aladdin's saddlebags, weighs a certain amount, and has a certain value. The problem is to decide how to get the most valuable hoard possible, if there is a limited amount of space and if Aladdin's mule can carry only a limited amount of weight. If the solver thinks of the quantities as continuous (if it makes sense in context to take exactly as much gold as is wanted), then this is a classical instance of linear programming, a powerful and efficient technique in applied mathematics.

On the other hand, if the quantities are discrete (for example, if the gold is in large bars, and Aladdin cannot take more than two bars but less than three), then the problem is in general very difficult. Indeed, the simplest version of the discrete knapsack problem is already believed to be computationally quite hard. In this problem, a list of integers is given as well as a large target integer. The goal is to achieve the target as a sum of integer multiples of the given numbers. Any progress toward finding more efficient solving methods or toward showing that the current methods are optimal would be extremely significant in the field of computer science.

Further Reading

Friedman, Erich. "Erich's Packing Center." http://www2 .stetson.edu/~efriedma/packing.html.

Golomb, Solomon. *Polyominoes: Puzzles, Patterns, Problems, and Packings.* Princeton, NJ: Princeton University Press, 1994.

Kellerer, Hans, Ulrich Pferschy, and David Pisinger. *Knapsack Problems.* Berlin: Springer, 2004.

Szpiro, George. *Kepler's Conjecture: How Some of the Greatest Minds in History Helped Solve One of the Oldest Math Problems in the World.* Hoboken, NJ: Wiley, 2003.

Michael "Cap" Khoury

See Also: Polygons; Polyhedra; Puzzles; Shipping.

Painting

Category: Arts, Music, and Entertainment.
Fields of Study: Geometry; Representations.
Summary: Painting incorporates many mathematical concepts, and mathematics is also used to analyze paintings.

Human beings strive to comprehend their reality in a number of ways, including artistic expression and mathematics. Examples can be found in many cultures, such as the long history of interesting mathematical patterns in Islamic art and in the cave paintings of Paleolithic people. Many artists throughout history also have been mathematicians, such as fifteenth-century painter Piero della Francesca. Modern painter Michael Schultheis also worked as a software engineer. He and Mary Lesser, a painter and printmaker, both explicitly include mathematical elements like numbers, equations, and geometric objects in their work. Mathematical concepts, especially geometry, are embedded throughout the art of painting. Some that are most commonly used for analyzing paintings involve symmetry, perspective, golden ratios and rectangles, and fractals, as well as fundamental geometric forms, shapes, fractals, and abstraction. Mathematicians and scientists also use mathematical methods to determine whether or not unidentified paintings belong to a particular artist.

Symmetry

M.C. Escher, a graphic artist, used transformational geometry to create a variety of works that explored symmetry. His classic work *Day and Night*, a 1938 woodcut, transforms rectangular fields into flying geese and uses a black and white color scheme to emphasize the transition of a setting from day to night. While many artists explore symmetry and transformational geometry, Escher took it further by exploring and emphasizing mathematical concepts including *Convex and Concave*, a 1955 lithograph, *Two Intersecting Planes*, a 1952 woodcut, and *Moebius Strip II*, a 1963 woodcut. The use of symmetry as the catalyst for transforming the plane is one of the more pleasing aspects of his work. Navajo sand painting also offers many good examples of various types of symmetry. Four-fold symmetry is widely found in Native-American painting and other art forms, and it plays a role in some spiritual and healing ceremonies.

Perspective

Early paintings did not use perspective to show a three-dimensional world on a two-dimensional canvas. Giotto di Bondone, a thirteenth-century painter began to develop depth of field in some of his work; but the first artist credited with a correct representation of linear perspective is Filippo Brunelleschi (1377–1446), who was able to devise a method using a single vanishing point. An architect and sculptor, he shared his method with fellow artist Battista Alberti, who wrote about the mechanics of mathematical perspective in painting. Leonardo da Vinci used perspective in his paintings and explored artificial, natural, and compound perspective in his work. He examined how the viewer's observation point changed the perspective, and how the perspective could be perceived by changing where the viewer was observing the painting. Notably, while perspective and the illusion of depth were widely used in Western painting from the 1300s onward, it was not universal. Painters from India rarely used this technique; rather, they tended to focus more on patterns and geometric relationships.

Golden Ratio and Golden Rectangles

Consider a rectangle with short side a and a long side that is $a+b$. A golden rectangle would be where the ratio a/b is equal to

$$\frac{a+b}{a}.$$

In other words, the large rectangle is proportional to the smaller rectangle formed by side b and side a—this is the golden ratio. Some claim that this proportion influenced many artists and early Greek architecture, while others note the variability of picking points in a painting to have golden rectangles superimposed. It is, however, a way of considering the proportionality of a work.

Fundamental Geometric Forms or Shapes

Geometric forms and shapes are the basis for drawing and painting. For example, Piet Mondrian (1872–1944) explored cubism in his work from black and white lines and blocks of primary colors that divided the plane. Other cubists, such as Pablo Picasso, broke with the Renaissance use of perspective to provide an alternative conception of form. Cubists made it possible for the viewer to see multiple points of view simultaneously.

Paul Cézanne ignored perspective in some of his work to construct color on the two-dimensional surface. Pointillism was used by Georges Seurat (1859–1891) to create *Sunday Afternoon on the Island of La Grande Jatte*. In pointillism, a series of small, distinct points of color are used to create a painting that relies on the viewer's eye to blend them into a cohesive form. The brain uses the dots to create a solid space. The primary colors are used to create secondary colors for shading and create the impression of a rich palate of secondary colors.

Art deco is characterized by the use of strong geometric forms that are symmetrical. This style of painting was popular in the 1920s and 1930s.

Abstraction and Fractals

Abstraction is an important tenet of mathematics. In mathematical abstraction, the underlying essence of a mathematical concept is removed from dependence on any specific, real-world object and generalized so that it has wider applications. In abstract expressionism, the artist is expressing purely through color and form, with no explicit representation intended. However, that does not mean that abstract art is entirely unstructured. Fractals are one tool used to quantitatively analyze and explain what makes some paintings more pleasing than others. The argument is that, even in an apparently random abstract work, there is an underlying logic or structure that the human brain recognizes as fractal patterns and that it inherently prefers over other works that do not have these patterns. This preference is perhaps because such works are more reflective of the geometry of naturally occurring spaces. For example, physicists Richard Taylor, Adam Micolich, and David Jonas analyzed Jackson Pollock's paintings and found two different fractal dimensions in his work that are mathematically and structurally similar to naturally occurring phenomena, like snow-covered vegetation and forest canopies. In addition to the application of fractals, mathematical concepts like open and closed sets have been used to compare and contrast the work of abstract expressionist artists like Pollock and Wassily Kandinsky to artists like Joseph Turner and Vincent van Gogh, whose works are among those credited with inspiring the expressionist movement.

Mathematical Analysis to Determine Authenticity

Sometimes, the painter of a particular artwork is unknown or disputed, which affects the study of art and the monetary valuation of paintings. Hany Farid and his team created a computer program that uses wavelets to analyze digital images of paintings and map the stroke patterns—some too small to be seen with the naked eye—that characterize an artist's unique style. In one case, known drawings by Pieter Bruegel the Elder were compared to five drawings originally attributed to him. The analysis determined that the five drawings were different from the original eight and also from each other, suggesting multiple creators. Chinese ink paintings are an example in which brush strokes are critical to identification, since they do not have colors or tones to distinguish style. One successful method, tested on the work of some of China's most renowned artists, used a mixture of stochastic models. In another case, fractal geometry was used to question the authenticity of some newly discovered Pollock works, based

French painter Georges Seurat used the painting technique of pointillism to create Sunday Afternoon on the Island of La Grande Jatte.

on his earlier patterns. Radioactive scans and X-ray analysis help to authenticate works by well-known and highly valued masters, such as Johannes Vermeer.

Additional Parallels in Painting and Mathematics

There are many natural parallels in the work of painters and mathematicians. In the same way that painters of different traditions and schools may represent the same scene in drastically different ways, mathematicians may approach the same problem from a variety of disciplines or perspectives. There are also varying degrees of connection to reality in both mathematics and painting. Applied mathematicians and realist painters may be primarily concerned with detailed and faithful representations of the real world in their work, while abstract painters and theoretical mathematicians often work in ways that are logically coherent and consistent, but that do not immediately or obviously connect to the real world. As with art, there is also subjective appreciation of the beauty of mathematics and arguments over what is or is not mathematically valid. Artist Michael Schultheis reported that he was often inspired by mathematical and scientific writing on whiteboards from his days as an engineer, and said, "I constantly revise equations with the Japanese calligraphy brush, rubbing out an area and thus creating a window into the equations. I draw and re-draw new ideas. All of these ideas are analytical. But they also live in the realm of beauty."

Further Reading

Field, J. V. *Piero della Francesca: A Mathematician's Art.* New Haven, CT: Yale University Press, 2005.

Jensen, Henrik. "Mathematics and Painting." *Interdisciplinary Science Reviews* 27, no. 1 (2002).

Robbin, Tony. *Shadows of Reality: The Fourth Dimension in Relativity, Cubism, and Modern Thought.* New Haven, CT: Yale University Press, 2006.

Taft, W. Stanley, and James Mayer. *The Science of Paintings.* New York: Springer, 2000.

Talasek, J. D. "Curator's Essay—Blending the Languages of Mathematics and Painting: The Work of Michael Schultheis." National Academy of Sciences. http://www.michaelschultheis.com/publications/talasek_essay.pdf.

LINDA HUTCHISON

See Also: Escher, M.C.; Geometry in Society; Golden Ratio; Greek Mathematics; Renaissance; Sculpture; Symmetry; Transformations.

Parallel Postulate

Category: History and Development of Curricular Concepts.
Fields of Study: Communication; Connections; Geometry.
Summary: The parallel postulate led to thousands of years of investigation and debate.

One of humanity's greatest intellectual achievements occurred in approximately 300 B.C.E. when the axiomatic method was born. The classic text *Elements*, written by the great Greek geometer Euclid of Alexandria, is a work that shaped the nature of mathematics and stands to this day as an example of the beauty and elegance of reasoning and proof.

Euclid was among the first people to understand that abstract mathematics is based on reasoning, from assumptions to general conclusions. From a very modest set of assumptions—his five postulates (called "axioms")—Euclid set out to argue the truth of a large number of propositions (called "theorems") in geometry.

The first four of Euclid's postulates appear reasonable enough: (1) any two points determine a unique line; (2) any line segment can be extended to an infinite line; (3) given any center and radius, a circle can be constructed; and (4) all right angles are congruent. But the fifth postulate stands out for its comparative complexity:

If a straight line falling on two straight lines makes the interior angles on the same side less than two right angles, the two straight lines, if produced indefinitely, meet on that side on which are the angles less than the two right angles.

This fifth postulate has come to be known as the "parallel postulate," in part for its very content, but also for the key role it plays in proving certain propositions about parallel lines.

From an historical perspective, Euclid himself seemed a bit uncomfortable with his fifth postulate. This discomfort is evidenced by the order of his work in Book I of *Elements*, where, on his way to eventually proving 48 propositions, he waited until proposition 29 to use the parallel postulate. The first 28 results rely only on the first four postulates and theorems that can be proven using those assumptions.

Attempts to Prove the Parallel Postulate as a Theorem

As subsequent mathematicians studied the *Elements*, most were troubled in some way by the parallel postulate. Because of its complexity, as well as its "if-then" format, it struck most mathematicians that Euclid's fifth postulate really ought to be a theorem. In other words, the parallel postulate ought to be a consequence of the first four postulates, and this fact ought to be provable, using only those four postulates and any theorems that could be derived from them.

Thus, many mathematicians set out to prove the parallel postulate as a theorem. It is one of the great tales of the history of mathematics that every single mathematician who attempted to prove the parallel postulate failed. Early on, many of these esteemed intellects made a common error that the rules of logic forbid—they assumed precisely what they were attempting to prove. Clearly, if the goal is to prove a statement *S*, one should never be allowed to simply assume that *S* is true. While certainly no mathematician was so dull as to say, "To prove the parallel postulate, I will assume the parallel postulate," many people did make the mistake of making the assumption that certain "obvious" statements were true. For example, they may have assumed statements such as the following:

- Parallel lines are everywhere equidistant.
- The sum of the measures of the interior angles of a triangle is 180 degrees.
- If a line intersects one of two parallel lines, then it must also intersect the other.
- There exists a rectangle (a quadrilateral having four right angles).

Remarkably, each of the above statements (along with many others) is equivalent to the parallel postulate. Said differently, if one of the above statements is called *P* and the statement of the parallel postulate is called *S*, then it turns out that *P* is true if and only if *S* is true—the truth of one implies the truth of the other, and vice versa.

Hence, when a mathematician said, "Using the fact that any triangle's angle sum is 180 degrees," and then went on to "prove" the parallel postulate, this argument was like saying "the parallel postulate is true because the parallel postulate is true." These errors came to be well understood by the end of the eighteenth century, perhaps most prominently in G. S. Klugel's 1763 doctoral dissertation in which he debunked 43 flawed "proofs" of the parallel postulate.

Girolamo Saccheri's Developments

Of course, even though nobody had found a valid proof of the parallel postulate did not mean that one could not be found, and many continued the search. Around the turn of the eighteenth century, a Jesuit priest named Girolamo Saccheri (1677–1733) made a lasting contribution to the study of the parallel postulate in particular, and to the history of mathematics in general. Saccheri considered the unthinkable, as part of his effort to prove the parallel postulate through a contradiction argument: what if the parallel postulate is false?

It was well understood by Saccheri's time that an equivalent statement of the parallel postulate was Playfair's Postulate, which states that

For any line *l* and any point *P* not on *l*, there exists a unique line through *P* parallel to *l*.

A contradiction argument works by assuming that the statement one wants to prove true is actually false and showing that some contradiction follows. Thus, it is natural to consider Playfair's Postulate and suppose that there is not be a unique line through *P* parallel to *l*. That is, one would assume that either there is not *any* line through *P* parallel to *l*, or there is *more than one* line through *P* parallel to *l*. Saccheri considered a similar scenario where he had transformed the problem about parallels to an equivalent one about quadrilaterals (now called "Saccheri quadrilaterals") in which the quadrilateral has two congruent sides perpendicular to the base. Fundamentally, Saccheri was trying to prove that a rectangle existed by showing that the summit angles of his quadrilateral were also right angles. After proving that the summit angles were congruent, he realized that there were three possibilities: the summit angles were each right angles, each was less than a right angle, or each was more than a right angle.

While Saccheri was able to rule out the possibility that the summit angles were obtuse by assuming that they were obtuse and finding a contradiction, when he

Modern Conceptions

Today, mathematicians understand a great deal about the role of Euclid's parallel postulate. Euclid's parallel postulate really is an axiom, and not a theorem. The parallel postulate is independent of the first four postulates. One can assume that the parallel postulate is true, or one can assume that the parallel postulate is false. Either leads to a perfectly valid geometry, with the truth of the parallel postulate leading to Euclidean geometry. Considering Playfair's postulate, named for John Playfair, if one assumes there are no parallel lines through a point *P* not on a line *l*, then this leads to so-called elliptic geometry, which is like the geometry of the sphere. If instead one assumes that there is more than one parallel line through a point *P* not on *l*, then this leads to "hyperbolic geometry," a geometry that some believe may help describe the shape of the universe.

It took approximately 2000 years for humankind to fully appreciate the work of Euclid and to reconcile the fact that Euclid was right—his fifth postulate really is an axiom, and not a theorem that can be derived. More than this, the parallel postulate is like a door that opens the world to one geometry—Euclidean—while there are other similar postulates that open doors to different universes, those of elliptic and hyperbolic geometries.

assumed that the summit angles were acute, he could not find a contradiction. From this assumption, he went on to prove many strange and unusual theorems. Unknowingly, Saccheri had discovered a whole new geometry, one that another mathematician named Janos Bolyai would call "a strange, new universe" in his own investigations. What both of these mathematicians, along with others such as Carl Gauss, started to realize is that there actually exists a geometry in which there is more than one line through a point *P* not on line *l* such that each is parallel to *l*. This realization stands as one of the greatest accidental discoveries in the history of the human intellect: Saccheri did not find what he set out to prove, but instead developed a collection of ideas that would radically change mathematics.

Further Reading

Dunham, Douglas. "A Tale Both Shocking and Hyperbolic." *Math Horizons* 10 (April 2003).

Greenberg M. *Euclidean and Non-Euclidean Geometries: Development and History*. New York: W. H. Freeman and Co., 2007.

Socrates Bardi, Jason. *The Fifth Postulate: How Unraveling A Two Thousand Year Old Mystery Unraveled the Universe*. Hoboken, NJ: Wiley, 2008.

Matt Boelkins

See Also: Axiomatic Systems; Geometry of the Universe; Proof.

Parallel Processing

Category: Communication and Computers.
Fields of Study: Algebra; Number and Operations.
Summary: Parallel processing speeds up the run-time of computing through the use of mathematical algorithms.

In computing, parallel processing is the action of performing multiple operations or tasks simultaneously by two or more processing cores. Ideally, this arrangement reduces the overall run-time of a computer program because the workload is shared among a number of engines—central processing units (CPUs) or cores. In practice, it is often difficult to distribute the instructions of a program in such a way that each CPU core operates continuously and efficiently, and without interfering with other cores. It should be noted that parallel processing differs from multitasking, in which a single CPU core provides the effect of simultaneously executing instructions from several different programs by rapidly switching between them, or interleaving their instructions. Modern computers typically include multi-core processor chips with two or four cores. The most advanced supercomputers in the early twenty-first century may have thousands of multi-core CPU nodes organized as a cluster of single processor computers and connected using a special-purpose, high-speed, fiber communication network. Although it is also possible to perform parallel processing by connecting computers together using a local area network, or even across the Internet, this type of parallel processing requires the individual processing elements to work predominantly in isolation because of the comparatively slow communication between nodes. Parallel processing requires data to be shared among processors and thus leads to the concept of "shared memory" where multiple processing cores work with the same physical memory. In large computer clusters, the memory is usually distributed across the nodes, with each node storing its own part of the full problem. Data are exchanged between nodes using message-passing software, such as Message Passing Interface (MPI).

Amdahl's Law and Gustafson's Law

The speed-up gained through parallelization of a program would ideally be linear; for example, doubling the number of processing elements should halve the runtime. However, very few parallel algorithms achieve this target. The majority of parallel programs attain a near-linear speed-up for small numbers of processing elements but for large numbers of processors the addition of further cores provides negligible benefits.

The potential speed-up of an algorithm on a parallel computing platform is given by Amdahl's law, originally formulated by Gene Amdahl in the 1960s. A large mathematical or engineering problem will typically consist of several parallelizable parts and several non-parallelizable parts. The overall speed-up attainable through parallelization is proportional to the size of the non-parallelizable portion of the program and is given by the equation

$$S = \frac{1}{1 - P}$$

where S is the speed-up of the program (as a factor of its original sequential runtime), and P is the fraction that is parallelizable. Amdahl's law assumes the size of the problem is fixed and that the relative proportion of the sequential section is independent of the number of processors. For example, if the sequential portion of a program is 10% of the run-time ($P = 0.9$), no more

than a 10-times speed-up could be obtained, regardless of how many processors are added. This characteristic puts an upper limit on the usefulness of adding more parallel execution units.

Gustafson's law is closely related to Amdahl's law, but is not so restrictive on the assumptions made about the problem. It can be formulated algebraically as

$$S(P) = P - a(P - 1)$$

where P is the number of processors, S is the speed-up, and a is the non-parallelizable proportion of the process.

Applications

Parallel computing is used in a broad range of fields, including mathematics, engineering, meteorology, bioinformatics, economics, and finance. However, all of these applications usually involve performing one or more of a small set of highly parallelizable operations, such as sparse or dense linear algebra, spectral methods, n-body problems, or Monte Carlo simulations. Frequently, the first step to exploiting the power of parallel processing is to express a problem in terms of these basic parallelizable building blocks.

Parallel processing plays a large part in many aspects of everyday life, such as weather prediction, stock market prediction, and the design of cars and aircraft. As parallel computers become larger and faster, it becomes feasible to solve larger problems that previously took too long to run on a single computer.

Further Reading

Barney, Blaise. "Introduction to Parallel Computing." *Lawrence Livermore National Laboratory*, 2007. https://computing.llnl.gov/tutorials/parallel_comp/.

Gupta, A., A. Grama, G. Karypis, and V. Kumar. *An Introduction to Parallel Computing: Design and Analysis of Algorithms*. Reading, MA: Addison Wesley, 2003.

Jordan, Harry F., and Gita Alaghband. *Fundamentals of Parallel Processing*. Upper Saddle River, NJ: Prentice Hall, 2002.

Chris D. Cantwell

See Also: Mathematical Modeling; Mathematics Research, Interdisciplinary; Software, Mathematics; Weather Forecasting.

Payroll

Category: Business, Economics, and Marketing.
Fields of Study: Algebra; Number and Operations.
Summary: Various payroll systems employ different mathematical calculations.

A variety of pay practices date back to ancient times, including compensation for services in the form of food, commodities, land, or livestock. Payroll systems are connected with the history of bookkeeping, which can be traced back to 4000 B.C.E. Paymasters were responsible for paying workers. Governments kept financial records called "pipe rolls" at least as early as the eleventh century. In 1494, Franciscan friar and mathematician Luca Pacioli published the book *Summa de Arithmetica, Geometria, Proportioni et Proportionalita*, which contained double-entry bookkeeping. The term *payroll* dates back to the seventeenth century, and compensation gradually changed from goods to money. In the mid-twentieth century, mathematician Grace Murray Hopper developed a compiler, later known as the FLOW-MATIC, which could be used for payroll calculations. When the U.S. Navy could not develop a working payroll plan, they called Hopper back to active duty. In the early twenty-first century, a payroll specialist is listed by some schools as a career option for mathematics majors. Accountants and actuaries calculate quantitative measures and predictions based on historic payroll information and salary increases. For example, the pensionable payroll is calculated as an integral that takes salary increases into account. In payroll analysis, the impact of changing salary expenses is compared to other factors, such as sales or profit.

Frequency

Some employees are paid each day they work; however, in many cases, an employer will withhold daily earnings and pay the cumulative amount earned at a later time as a lump sum. Common payroll frequencies include weekly, bi-weekly (every other week), semimonthly (twice a month), and monthly. Each of these frequencies would correspond to receiving 52, 26, 24, and 12 paychecks each year, respectively, assuming a full year of work. Some seasonal jobs pay only for part of the year, but still use the standard payroll frequencies. For example, teachers often receive pay for only nine months. Some schools offer for that pay to be spread over a full year to guarantee consistent income during the summer months when teachers are not actually working.

On payday, the employee will receive earned wages for the previous pay period. Rather than receiving cash, sometimes an employee will receive a check that can be exchanged for an equivalent amount of cash. Other times, an employee will receive income as a "direct deposit" where the income is automatically deposited into the employee's checking or savings account.

Earning Money

Some employees work for an hourly wage—for every hour of work they perform, they get paid a specified amount of money. Suppose that a worker had an hourly wage of $10 and worked for 20 hours. To find the total amount of the paycheck, the worker would multiply the hourly wage by the number of hours worked. For example, $10 \times 20 = \$200$.

Sometimes, contracts or laws dictate the number of hours a person can work per week and—should they work more than that amount—his or her income increases. For example, in the United States, 40 hours is a common workweek. A person working over 40 hours often gets paid "time and a half" or "wage and a half" for the number of hours over 40 that he or she works (called "overtime"). Again, assuming an hourly wage of $10, an employee who worked 48 hours in one week would earn $10 \times 40 = \$400$ for the first 40 hours they worked. The eight hours he or she worked beyond 40 hours would earn him or her extra money. If the employee earns "time and a half," the time would be multiplied by 1.5 before being multiplied by his or her hourly wage. If he or she earns "wage and a half," the wage would be multiplied by 1.5 before being multiplied by the number of hours worked beyond 40. In reality, the method of calculating overtime earnings is irrelevant since multiplication is associative. Time and a half would be calculated as $\$10 \times (1.5 \times 8) = \$10 \times 12 = \$120$, and wage and a half would be calculated as $(\$10 \times 1.5) \times 8 = \$15 \times 8 = \$120$. The total earnings for that week would be found by taking the sum of these wages: $\$400 + \$120 = \$520$.

Another method for earning money is a salary. Unlike the hourly wage, a salary is a predetermined amount of money that the worker earns regardless of how long (or how short) it takes the worker to accomplish those tasks. Often, salary is determined based on how much a person will make over a year's time. However, rarely does a person only receive one paycheck a year. The amount of money earned on each paycheck is calculated by taking the salary and dividing it by the number of pay periods in a year. That number will vary depending on how often a person gets paid. Suppose an employee agreed to work for a salary of $31,200 each year. Looking at the common pay periods, weekly, bi-weekly, semi-monthly, and monthly, this employee would earn $600, $1,200, $1,300, or $2,600, respectively, for each paycheck during the year.

A worker earning commission does not actually get paid based on how long it takes to do the job, but by how productive the worker is (oftentimes based on the amount of items the worker sells). Sometimes, commission is a flat fee per item sold, other times, it is a percentage of sales. For example, if an employee earned 7% commission on sales and sold $1,250 worth of merchandise on a given day, then pay would be calculated $1,250 × 7% = $1,250 × 0.07 = $87.50. Some jobs combine an hourly rate and commission—the employee earns a certain amount of money for every hour they are at the job, but then also earns commission on top of that wage to determine the total money earned.

Payroll Withholdings
Upon receipt of a paycheck or notice of direct deposit, usually the amount paid to the employee (the net pay) is less than what is calculated as his or her earnings for the pay period (the gross pay). Before being issued money, an employee may have his or her income reduced by certain amounts—some voluntary, others involuntary. In order to pay for various levels of government (and the benefits they offer), income and payroll taxes are frequently withheld from earnings. Some employees pay premiums for different insurances (such as medical, life, or disability) from their pay. Sometimes, money is withheld as a long-term savings for eventual retirement of the employee. Job-related expenses can also be withheld, such as for dues or charges for employee uniforms.

Further Reading
Booth, Phillip, Robert Chadburn, Steven Haberman, Dewi James, Zaki Khorasanee, Robert Plumb, and Ben Rickayzen. *Modern Actuarial Theory and Practice.* 2nd ed. Boca Raton, FL: CRC Press, 2004.

Bragg, Steven M. *Essentials of Payroll: Management and Accounting.* Hoboken, NJ: Wiley, 2003.

Haug, Leonard. *The History of Payroll in the U.S.* San Antonio, TX: American Payroll Association, 2000.

CHAD T. LOWER

See Also: Accounting; Budgeting; Income Tax; Money.

Pearl Harbor, Attack on

Category: Government, Politics, and History.
Fields of Study: Geometry; Measurement; Number and Operations; Problem Solving.
Summary: Mathematicians were involved in both the planning of and the response to Pearl Harbor.

The attack on Pearl Harbor, a major engagement of World War II and the impetus for the United States' entry into the war, took place early Sunday morning, December 7, 1941, on the island of Oahu, Hawaii. The Japanese Navy, commanded by Admiral Isoroku Yamamoto, planned and executed the surprise attack against the U.S. naval base and nearby army air fields. As a result, the United States declared war on Japan. In his address to Congress, President Franklin D. Roosevelt famously proclaimed December 7 "a date which will live in infamy."

Both leading up to and as a result of the attack on Pearl Harbor, mathematicians in Japan and the United States mobilized for the war. For instance, after Pearl Harbor, the American Mathematical Society and the Mathematical Association of America converted their War Preparedness Committee to a War Policy Committee to increase research on "mathematical problems for military or naval science, or rearmament" and to strengthen mathematics education in order to prepare undergraduate students for military service. The attack has also been surrounded by speculation as to how the United States could have been caught off guard so easily. The naval base had been designed as nearly impenetrable to surprise attack because of the geography and geometry of the island. However, new

A photo taken from a Japanese plane during the Pearl Harbor torpedo attack on ships moored on both sides of Ford Island.

technologies made the attack possible: the aircraft carrier could bring low-flying aircraft within attack range, and the Japanese development of shallow-running torpedoes could skim the surface of the harbor's relatively shallow water. One of the largest controversies involves U.S. efforts to decode Japanese communications that may have given forewarning of the attack.

Japanese Mathematicians

Leading up to Pearl Harbor, the number of Japanese graduate students increased and several studied in Germany. Mathematicians applied lattice theory and logic to the design of circuits. In the 1930s, both the United States and Japan successfully built a cyclotron, an early particle accelerator. Mathematics was also important in electrical engineering and airplane design. With a focus on aerodynamics and science and technology policy, the Japanese Technology Board was founded in 1941. A statistical institute contributed to war production. Japanese cryptologists also created many variations of military codes that were in use prior to Pearl Harbor, such as Kaigun Ango—Sho D, later referred to as "JN-25B" by cryptanalysts in the United States. Before committing to the attack on Pearl Harbor, the Japanese Navy conducted feasibility studies that included calculations and considerations of their current military resources; the need for a longer, circuitous route outside the customary naval traffic lanes to avoid detection by both

military and civilian ships,; the probability of encountering severe winter storms and critical data obtained from spies in Hawaii, such as the patterns of military activity at Pearl Harbor. They concluded that the attack was possible, if dangerous, and they originally intended to specifically target U.S. aircraft carriers to optimize the long-term effects of the attack. Experimentation and simulated training attacks yielded a satisfactory plan only a few weeks before the event.

U.S. Mathematicians

In the United States, mathematicians conducted ballistics research at Aberdeen Proving Ground. Max Munk used the calculus of variations in airfoil design at the National Advisory Committee for Aeronautics, a precursor to the National Aeronautics and Space Administration (NASA). Technology such as radar, developed by scientists and mathematicians including Christian Doppler and Luis Alvarez, served military uses, though it was still in its infancy at the time of Pearl Harbor. Responsibility for compiling codes for military use and using cryptology to decipher codes shifted from Military Intelligence to the Army Signal Corps in 1929. William Friedman was the chief civilian cryptologist at the Signal Intelligence Service. The U.S. Army at that time realized the importance of mathematics in deciphering, and the first three civilian cryptanalysts hired by the U.S. Army were mathematics teachers.

Forewarning

Many wonder how the United States could not have known of the impending Japanese attack, which had been planned and practiced months in advance. The new radar installation on Opana Point did, in fact, detect the incoming Japanese attack planes, but they were ultimately mistaken for a group of U.S. planes that were due to arrive from the mainland that morning. A U.S. destroyer also spotted a Japanese submarine attempting to enter the harbor, which it reported, but the information was not acted upon immediately. Both would have given at least short-term warnings to the ships and personnel. However, much of the accountability is assigned to the U.S. and Japanese intelligence and counter-intelligence efforts. Correspondence declassified many years after the war suggests that the United States could at least partially understand the codes needed to monitor Japanese naval movements on the eve of Pearl Harbor. While U.S. and British cryptanalysts had successfully broken some Japanese codes, such as the MAGIC code, the United States was not able to determine from those messages that the attack was about to happen. The broken codes were the ones used primarily for diplomatic messages sent by the Japanese Foreign Office and military strategy was rarely shared with the Japanese Foreign Office. The U.S. Navy had three cryptanalysis centers devoted to breaking Japanese naval codes. Prior to the attack, American cryptanalysts had been using traffic analysis to follow Japanese naval movements. Traffic analysis is the process of looking for patterns in communications to infer if an attack is about to occur. According to the National Security Agency, the Japanese, aware that their communications were being monitored, issued "dummy traffic to mislead the eavesdroppers into thinking that some of the ships sailing through the North Pacific were still in home waters." Additionally, as Japanese forces were preparing for the attack, radio traffic was limited, greatly reducing the ability of American intelligence to determine a pattern. These efforts to stymie cryptologists were effective in keeping the impending attack a secret from the United States. Mathematicians and historians continue to analyze whether signal intelligence techniques could have revealed Japan's intentions.

Further Reading

Booß-Bavnbek, Bernhelm, and Jens Høyrup. *Mathematics and War*. Basel, Switzerland: Birkhauser, 2003.

National Security Agency. "Pearl Harbor Review." http://www.nsa.gov/about/cryptologic_heritage/center_crypt_history/pearl_harbor_review.

PearlHarbor.org. "Why Did Japan Attack Pearl Harbor?" http://www.pearlharbor.org.

Wilford, Timothy. "Decoding Pearl Harbor: USN Cryptanalysis and the Challenge of JN-25B in 1941." *The Northern Mariner* XII, no. 1 (2002).

Calli A. Holaway
Michael G. Lovorn

See Also: Aircraft Design; Artillery; Coding and Encryption; Predicting Attacks; World War II.

Pensions, IRAs, and Social Security

Category: Business, Economics, and Marketing.
Fields of Study: Algebra; Data Analysis and Probability; Measurement; Number and Operations.
Summary: The development and allocation of retirement income can involve significant mathematical analysis.

Planning for retirement is one of the most important financial responsibilities a person faces. Ideally, after working for several decades, a person will be in a financial position to sustain a desired lifestyle during retirement. In the United States, the source of retirement income can be from a combination of one or more of the following: Social Security; an employer-sponsored pension plan; individual savings, including individual retirement accounts (IRAs); or other mechanisms. The U.S. government has provided military pensions to disabled veterans and widows since the Revolutionary War. This benefit expanded after the U.S. Civil War to include nearly any veteran who had served honorably for some minimum time. Southern states also paid Confederate veterans.

By the early twentieth century, state, municipal, and city governments were paying pensions to their employees, especially firemen and policemen. Teachers were the next large group to receive benefits. Private pensions started in the late nineteenth century with the American Express Company and several railroads. When the 1926 Revenue Act exempted pension trust income from taxes, companies had a new incentive to provide employee pensions, which became commonplace by the 1930s. Social Security was designed in 1935 to extend pension benefits to those not covered by a private pension plan. In the early twenty-first century, Social Security benefits are the main source of retirement income for most retirees, though this varies greatly depending on income from earnings, assets, and private pensions.

Each of these potential sources of retirement income can involve significant mathematical and financial analysis to estimate an individual's retirement needs, determine necessary pre-retirement financial planning, and evaluate the potential uncertainty associated with personal and economic factors. Specialized mathematicians known as "actuaries" work for governments and industries to design financially sound insurance and pension programs that help meet people's retirement needs.

At the same time, as the professionals at the American Pension Corporation (a major pension administrator) assert, good actuaries "are more than just mathematicians—[they] take great pride in [their] ability to dissect and communicate the intricacies of pension administration in layman's language."

Pensions

A pension provides a stream of income during retirement. It is typically sponsored by a person's employer—either a corporation or governmental entity. The amount and timing of the retirement income stream provided by a pension are a function of several factors, such as the worker's salary, the proportion of that salary invested into the pension plan, any matching funds or contributions to the pension fund provided by the employer, the length of the worker's tenure with the employer, and the investment performance of the pension fund. There are two types of pension plans: defined-benefit (DB), and defined-contribution (DC). DB plans, which have to some extent been phased out in the private sector but are still common in the public sector, define the benefits that will be paid to the worker during retirement. Assuming the solvency of the pension plan—a significant issue in itself—a worker covered under a DB plan is guaranteed to receive the benefits defined by the plan.

Because of potential difficulties in adequately funding DB plans, many (particularly private sector) employers converted to DC plans during the last several decades of the twentieth century. With DC plans, the retirement benefits are not specified; rather, the plan defines the periodic contributions to be invested during the worker's life, and then the retiree receives an income stream based on the actual accumulated amount of the investment fund. Relative to DB plans, this means that the employer's risk of inadequate retirement benefit funding is reduced, and that some risk has been transferred to the employee, who faces an uncertain pension income stream.

Mathematics of Pensions

The mathematics associated with pensions involves both "future (or accumulated) value" and "present

value" concepts. The general idea is that a worker (or the sponsoring employer) accumulates a retirement fund by setting aside and investing periodic amounts during the working years. Then, upon retirement, this accumulated amount ideally represents sufficient funds with which to provide the retiree an adequate stream of income until death. This retirement income stream may be obtained by leaving the funds invested and withdrawing a certain amount per year, or through the purchase of an annuity, which provides the payment stream. In most cases, these two approaches are mathematically equivalent.

While somewhat straightforward conceptually, achieving an adequately funded and effective retirement plan (especially DB plans, which generally involve more sophisticated and extensive mathematical and financial considerations than DC plans) is a challenging mathematical and actuarial problem. Some of the parameters involved in a pension analysis, and for which assumptions must be made, include the following:

1. Periodic contributions to the pension investment fund—usually expressed as a percentage of worker salary during employment.
2. Size of the retirement income stream needed or desired—generally estimated as a percentage of projected salary immediately prior to retirement.
3. Rate of return on the invested retirement funds, both before and after retirement.
4. Changes in worker salary throughout employment.
5. Impact of inflation on the worker's buying power.
6. Taxation rules and regulations, both during employment and in retirement.
7. Longevity and mortality.

Along with these assumptions, actuaries use mathematics and computer modeling to determine potential answers to questions such as how much must a worker (or employer) invest every month (or year) into a retirement plan in order to successfully achieve that worker's financial goals in retirement?

IRAs and Social Security

In addition to having an employer-sponsored pension plan, a worker can supplement retiree income with personal savings. One such mechanism is one or more types of IRA. While the rules surrounding IRAs are extensive, they can have potential advantages for some people, including certain tax-advantaged properties.

Social Security is a particularly contentious issue in the twenty-first century. Some have compared Social Security to a type of scam called a "Ponzi scheme" in which a growing pool of new investors' money is used to pay the promised returns to previous investors. Despite superficial resemblances (for example, current taxpayer money is used to pay variable benefits to others), Social Security is not a savings plan or investment account, but rather a tax, which nullifies the comparison. However, there have been proposals to replace Social Security with an investment program, using a variety of calculations and probabilistic mathematical models to try to demonstrate its cost-effectiveness and the likelihood of the system's impending failure.

Another major financial issue related to social security is the potential misuse of Social Security numbers. Initially issued to track workers for taxation and benefits, these nine-digit numbers are now assigned routinely at birth and have grown over time into the role of a unique identifier for creditors, schools, employers, and others who want to assign codes to individuals. Modern identity theft, which usually involves a person using a fake or stolen social security number to obtain credit or other benefits, has been on the rise as a result of Internet growth and the widespread collection of personal data. Mathematicians have calculated that a person making up a false social security number in 2010 has about a 50% chance of matching a real number. Faking multiple numbers results in an almost-guaranteed match very quickly.

These calculations have been used to counter thieves' assertions that they did not know numbers they were using were real. Social Security numbers themselves are not random (for example, the first three digits are a numerical code for geographic location), and mathematical and computer methods have used publicly available data, like date and place of birth, to successfully predict most or all of a person's social security number. There are also concerns that the government will run out of Social Security numbers, which are not reused after a person dies. Some calculations suggest that the supply will be exhausted early in first half of the twenty-second century. Alter-

Stochastic Variables

What makes such a quantitative analysis particularly challenging is that many of these parameters are stochastic rather than deterministic—their future values are uncertain, and they can (and do) change value over time. Data analysis and probability concepts are used to account for this uncertainty. For example, inflation and investment rate of return are both stochastic variables, with considerable uncertainty regarding their values in both the short- and long-term.

Estimates of possible future values and the relative probabilities or likelihoods of those values can be made by analyzing historical data. These estimates can then be used to project future scenarios and quantify the potential impact of possible future values on the retirement funding process.

Another critical stochastic variable in retirement planning is the age at death of the retiree. The number of years that a retiree lives beyond the date of retirement is an essential factor in determining the total amount of income needed during the retirement years. Actuaries research and analyze historical mortality data for people with various identifiable attributes. From these analyses, a probability distribution of possible ages at death, with their relative likelihoods, can be developed.

native proposals include using alphanumeric or hexadecimal strings, which offer more permutations for a series of nine "digits." Others suggest including a security checksum in the number to decrease fraudulent use.

Further Reading

Anderson, Arthur. *Pension Mathematics for Actuaries.* Winsted, CT: ACTEX Publications, 2006.

Harding, Ann, and Anil Gupta. *Modeling our Future: Population Ageing, Social Security and Taxation.* Amsterdam, The Netherlands: Elsevier Science, 2007.

Muksian, Robert. *Mathematics of Interest Rates, Insurance, Social Security, and Pensions.* Upper Saddle River, NJ: Prentice Hall, 2002.

RICK GORVETT

See Also: Budgeting; Forecasting; Loans; Money; Mutual Funds.

Percussion Instruments

Category: Arts, Music, and Entertainment.
Fields of Study: Geometry; Number and Operations; Representations.
Summary: The vibrations that emanate from percussion instruments vary mathematically based on the type of instrument.

Percussion instruments are characterized by vibrations initiated by striking a tube, rod, membrane, bell, or similar object. Percussion instruments are almost certainly the oldest form of musical instrument in human history. The archeological record of percussion instruments, in particular the *bianzhong* bells of ancient China, give clues to the history of music theory. From a mathematical point of view, percussion instruments are of special interest because—unlike other types of instruments, such as string and wind instruments—the resonant overtones typically do not follow the harmonic series. In the last half of the twentieth century, a question of great interest in applied mathematics has been the famous inverse problem: can one hear the shape of a drum?

Rods and Bars

Some percussion instruments produce a distinct pitch by the vibration of a rod or bar. Examples included the tuning fork (a U-shaped metal rod suspended at its center), a music box (a metal bar suspended at one end), and the melodic percussion instruments such as the xylophone and marimba (suspended at two non-vibrating points or "nodes" along the length of metal or wooden bars). Like vibrating strings, the frequency of the bar's vibration and the pitch of the musical sound it produces are determined by its physical dimensions. In contrast to the string in which the frequency varies inversely with the length, the vibrating bar has a frequency that varies with the square of the length. The resonant overtone frequencies f_n of the vibrating bar are related to the fundamental frequency f_1 by the formula

$$f_n = \alpha \left(n + \frac{1}{2} \right)^2 f_1$$

where the constant α is determined by the shape and material of the bar. In contrast with the harmonic overtone series of vibrating strings, $f_n = n(f_1)$, these inharmonic overtones give percussion instruments their distinct metallic timbre. The overtones of vibrating bars decay at different rates, with rapid dissipation of the higher overtones responsible for the sharp, metallic attack, while the lower overtones persist longer. The bars of the marimba are often thinned at the center,

The ancient bianzhong *bell set on display at the Hubei Provincial Museum. Each bell can produce two pitches when struck.*

effectively lowering the pitch of the certain overtones, in accord with the harmonic series.

Bells

Like the vibrating bar instruments, the classic church bell possesses highly non-harmonic overtones. These are typically tuned by thinning the walls of the bell along the circumference at certain heights. A distinctive feature in the sound comes from the fact that apart from the fundamental pitch, the predominant overtone of the church bell sounds as the minor third above the prevailing tone. This feature accounts for the somber nature of the sound.

The *bianzhong* bells of ancient China were constructed in a manner that produced two pitches for each bell, depending on the location at which it was struck. In the 1970s, a set of 65 such bells were discovered during the excavation of the tomb of Marquis Yi in the Hubei Provence. The inscriptions on the bells make it clear that octave equivalence and scale theory were known in China as early as 460 B.C.E.

Membranes

Drums are perhaps the most common percussion instrument. Consisting of vibrating membranes (called the "drum heads") stretched over one or both ends of a circular cylinder, drums exhibit a unique mode of vibration, which accounts for their characteristic sound. Mathematical models of vibrating drumheads provide a fascinating application of partial differential equations. The inharmonic overtone frequencies are distributed more densely than for vibrating strings or rods. Further, each overtone is associated with a particular vibration pattern of the drum head. These regions can be characterized by the non-vibrating curves (called "nodes") that arrange themselves in concentric circles and diameters of the drum head.

An important question in the study of spectral geometry asks: "Can one hear the shape of a drum?" In other words, can mathematical techniques be used to work backwards from the overtone frequencies to determine the shape of the drumhead that caused the vibration? The answer, as it turns out, is "not always."

Further Reading

Cipra, Barry. "You Can't Always Hear the Shape of a Drum." In *What's Happening in the Mathematical*

Sciences. Vol. 1. New Haven, CT: American Mathematical Society, 1993.
Jing, M. "A Theoretical Study of the Vibration and Acoustics of Ancient Chinese Bells." *Journal of the Acoustical Society of America* 114, no. 3 (2003).
Rossing, Thomas, D. *The Science of Percussion Instruments.* Singapore: World Scientific Publications, 2000.
Sundberg, Johan. *The Science of Musical Sounds.* San Diego, CA: Academic Press, 1991.

ERIC BARTH

See Also: Geometry of Music; Harmonics; Scales; Wind Instruments.

Perimeter and Circumference

Category: History and Development of Curricular Concepts.
Fields of Study: Communication; Connections; Geometry; Measurement.
Summary: Measuring perimeter and circumference is a geometric task with a long history of methods.

Measurements of length and distance abound in daily life, from the height of a child to the distance from home to the store. Perimeter and circumference are types of length measurements. The perimeter of a geometric entity is the path that surrounds its area. The word derives from its Greek roots *peri* (meaning "around") and from *meter* (meaning "measure"). In stricter mathematical sense, perimeter is defined as the length of the curve constituting the boundary of a two-dimensional, planar closed surface.

For example, the perimeter of a square whose side measures length a is $4a$. Perimeter is important for applications such as landscaping projects, construction, and building fences. Circumference is defined as the perimeter of a circle. The circumference of a circle of radius (r) is $2\pi r$. The circumference of a circle has played a very important role throughout history in the approximation of the mathematical constant π, which was defined as the ratio of the circumference (C) of the

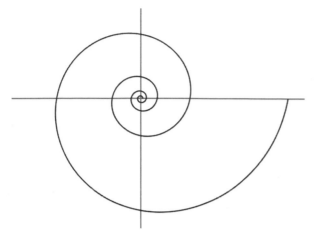

The curve that forms the shape of a nautilus shell is known as a logarithmic spiral *or* equiangular spiral.

circle to its diameter (*d*). Perhaps the most common reference to circumference that most people encounter regularly is the circumference of one's waist—the size of their waist.

The waist circumference is used as a measurement for some clothing and is also associated with type II diabetes, dyslipidemia, hypertension, and other cardiovascular diseases. Students investigate perimeter beginning in primary school, and middle grade students explore circumference. Students formulate the length of general curves, referred to as the "arc length" or "rectification," as integrals in calculus courses.

History
There is a long history of computations involving the perimeter or circumference of figures. One way was to measure length was with ropes. For instance, statements about rope measurements and the Pythagorean theorem can be found in Katyayana's *Sulbasutra*. Another way was to compare the length of two figures. A Babylonian clay tablet was discovered in 1936 and was noted as relating the hexagon perimeter to 0;57,36 (in base 60), or 24/25 times the circumference of a circumscribed circle. Mathematicians like Archimedes of Syracuse estimated the circumference or a value for π by using the perimeters of inscribed and circumscribed polygons with many sides. For example, Archimedes was known to have used 96-sided polygons. Mahavira estimated the circumference of an ellipse. In ancient times, the semiperimeter, or half

the perimeter, was useful in computing many geometrical properties of polygons such as altitude, exradius, and inradius of a triangle. The semiperimeter also appears in Heron of Alexandria's formula for the area of a triangle. The semiperimeter of a rectangle is the sum of the length plus the height and is noted as appearing on Babylonian clay tablets. Brahmagupta used the semiperimeter of a quadrilaterial in the computation of its area.

The circle is a special geometric figure, for it is the curve, given a fixed perimeter, which encompasses the maximum surface area. This is known as the isoperimetric problem. Proclus commented that, "a misconception is held by geographers who infer the size of a city from the length of its walls." The Babylonians may have worked on related problems in their investigations of solutions to quadratic equations generated by the setting of the semiperimeter and area to constants. The isoperimetric problem was partially solved by the Greek mathematician Zenodorus.

Pappus of Alexandria compared the areas of figures with a fixed perimeter. In the tenth century, Abu Jafar al-Khazin proved that an equilateral triangle has greater area than isosceles or scalene triangles of the same fixed perimeter. Many mathematicians worked on the isoperimetric problem using a variety of techniques including methods from geometry, analysis, vectors, and calculus. In 1842, a German mathematician named Jakob Steiner used geometric arguments to present five proofs of the theorem. However, Steiner had assumed that a solution was possible, which was a subtle flaw to otherwise creative arguments. Karl Weierstrass proved the existence of such solutions in 1879. Other mathematicians proved the results in a variety of other ways.

Historical Applications and Computations
One application of circumference of a circle is the computation of the Earth's circumference. Eratosthenes of Cyrene, in 240 B.C.E., computed the Earth's circumference using trigonometry and the angle of elevation of the sun at noon in Alexandria and Syene. He made an assumption that the Earth and the sun were perfect spheres and that the sun was so far away that its rays hitting the Earth could be considered parallel. By measuring the shadows thrown by sticks on the summer solstice, Eratosthenes derived a formula to measure the circumference of the Earth and deter-

mined it to be 252,000 stadia. Teachers in mathematics classrooms share Eratosthenes's calculation in order to highlight his ingenuity and showcase the power of setting up proportions and applying the congruence of alternate interior angles of parallel lines. There is debate about the value of a stadia, but historians estimate that Eratosthenes was correct within a 2% to 15% margin of error. The Indian mathematician Aryabhata made revolutionary contributions toward the understanding of astronomy at the turn of the fifth century. His calculations on π, the circumference of Earth, and the length of the solar day were remarkably close approximations.

The middle of the seventeenth century marked a fruitful time in the history of calculating the length of general curves. For instance, the curve that forms the shape of a nautilus shell is called the "logarithmic spiral" or "equiangular spiral." Evangelista Torricelli described its length using geometric methods. Christopher Wren published the rectification of the cycloid curve. Hendrik van Heuraet and Pierre de Fermat independently explored ideas that would eventually lead to the integral formula of arc length.

In the twentieth century, methods from fractals, popularized by Benoit Mandelbrot, have proven useful in modeling objects like a coastline. One example that is regularly examined in mathematics classrooms is the Koch snowflake, named for Helge von Koch, an example of a curve that bounds a region with finite area yet has infinite perimeter.

Further Reading

Blasjo, Viktor. "The Isoperimetric Problem." *The American Mathematical Monthly* 112, no. 6 (2005).

Briggs, William. "Lessons From the Greeks and Computers." *Mathematics Magazine* 55, no. 1 (1982).

Dunham, William. "Heron's Formula for Triangular Area." In *Journey through Genius: The Great Theorems of Mathematics*. Hoboken, NJ: Wiley, 1990.

Steinhaus, H. *Mathematical Snapshots*. 3rd ed. New York: Dover, 1999.

Wells, D. *The Penguin Dictionary of Curious and Interesting Geometry*. London: Penguin, 1991.

ASHWIN MUDIGONDA

See Also: Curves; Mapping Coastlines; Measurements, Length; Pi; Polygons.

Permutations and Combinations

Category: History and Development of Curricular Concepts.
Fields of Study: Communication; Connections; Data Analysis and Probability; Number and Operations.

Summary: For centuries, mathematicians have posed and studied problems that involve various arrangements or groupings of sets of objects, which are known as permutations and combinations.

In a very broad sense, combinatorics is about counting. The mathematical discipline of combinatorics addresses the enumeration, permutation, and combination of sets of objects, as well as their relations and properties. Combinatorial problems can be found in many areas of pure and applied mathematics, including algebra, topology, geometry, probability, graph theory, optimization, computer science, and statistical physics. One of the earliest problems in combinatorics is found in the work of Greek biographer Plutarch, who described mathematician Xenocrates of Chalcedon's work on calculating how many syllables could be produced by taking combinations of the letters of the alphabet. This occurred between 400 and 300 B.C.E. Millennia later, mathematician William Gowers won the 1998 Fields Medal, widely regarded as the most prestigious prize in mathematics, for his "contributions to functional analysis, making extensive use of methods from combination theory." In twenty-first-century school curricula, primary school children study number combinations to facilitate learning basic operations like addition, subtraction, multiplication, and division. High school students often study permutations and combinations as counting techniques. Permutations and combinations were fundamental for cracking the World War II Enigma code and continue to remain vital in cryptography, among other fields.

Definitions

In mathematical fields like algebraic group theory, combinatorics, or probability, the term "permutation" has several meanings that are all essentially related to the idea of rearranging, ordering, or permuting some kind of mathematical object. When paired with combinations, particularly in primary and secondary curricula, a permutation is usually thought of as an ordered arrangement of some set or subset of objects. For example, for the set of objects A, B, and C, there are six permutations of the set: {A, B, C}, {A, C, B}, {B, A, C}, {B, C, A}, {C, A, B}, and {C, B, A}. A combination is then a subset of objects selected from a larger set, where order does not matter. For example, for the set A, B, and C, one combination of two objects is {A, B}. In some applications, the objects in a combination are thought of as being chosen sequentially. However, since order does not matter, the selection {B, A} would represent the same combination as {A, B}. All possible two-object combinations are {A, B}, {B, C}, {A, C}. Mathematicians Blaise Pascal and Gottfried Leibniz used the specific term "combinations" beginning in the seventeenth century, while Jacob Bernoulli is often credited with introducing the term "permutations" a short while later. Some alternatively trace it to Thomas Strode in the seventeenth century.

History and Early Applications

The real-world motivation for many early problems involving what are now called "combinations" and "permutations" was religion. For example, Jaina, Christian, and Jewish scholars were interested in letter permutations, which some believed had spiritual power. In the ninth century, the Jaina mathematician Mahavira discussed rules for using permutations and combinations. In the tenth century, Rabbi Abraham ben Meir ibn Ezra used combinations to study the conjunction of planets. Another motivator was games of chance, which also drove probability theory. Archaeological evidence suggests that gambling has been around since the dawn of humankind, and many games rely on players achieving special combinations of symbols or objects like knucklebones, sticks, or polyhedral dice. Surviving writings show that Egyptian, Greek, Hindu, Islamic, and perhaps Chinese scholars and mathematicians studied permutations and combinations.

The Egyptian game "Hounds and Jackals" used a set of "throw" sticks that resulted in combinations of outcomes that determined how far a player might move. In the sixth century B.C.E., Hindus discussed combinations of six tastes: sweet, acid, saline, pungent, bitter, and astringent. Some consider the Chinese divination text *I-Ching* to be part of the literature on combinations and permutations since it discussed arrangements sets of trigram and hexagram symbols. Versions date to at least 400–300 B.C.E. In the sixth century, Roman philosopher Anicius Manlius Severinus Boëthius presented a rule for finding the possible combinations of objects taken two at a time from some set.

In the tenth and eleventh centuries, mathematicians like Acharya Hemachandra explored the how many combinations of short and long syllables were possible in a line of text with a fixed length, and Bhaskara's treatise *Bhaskaracharyai* contained an entire chapter devoted to combinations, among other chapters on topics like arithmetic, geometry, and progressions. Both al-Marrakushi ibn Al-Banna and Kamal al-Din Abu'l Hasan Muhammad Al-Farisi explored the relationship between polygonal numbers, the binomial theorem, and combinations. Al-Farisi used what historians consider a form of induction to show the relationship between triangular numbers (numbers that can be represented by an equilateral triangular grid of points such as, 1, 3, 6, 10), and the combinations of subsets of objects drawn from a larger set. Mi'yar al-'aqul ibn Sina (Avicenna) developed a system of combinations of "simple" machines to classify complex mechanisms.

The original concept of simple machines is attributed to mathematician Archimedes of Syracuse. A group might be machines containing rollers and levers, chosen from a larger set of possibilities that included windlasses, pulleys, rollers, levers, and other components. Starting in the Renaissance, the most commonly recognized set of six simple machines was the lever, inclined plane, wheel and axle, screw, wedge, and pulley. Students continue to discuss more complex machines as combinations of simple machines.

In Europe, beginning around the twelfth century and up through the nineteenth century, many mathematicians such as Levi ben Gerson, Bernoulli, Leibniz, Pascal, Pierre Fermat, Abraham de Moivre, George Boole, and John Venn worked on the development of combinations and permutations, frequently in the context of probability theory. For example, Johann Buteo (or Jean Borell) discussed the possible throws of four dice as well as locks with movable combination cylinders in his sixteenth-century work *Logistica*.

Bernoulli's *Ars Conjectandi* collected knowledge of permutations and combinations through the seventeenth century and was a popular combinatorics book in the eighteenth century. However, standard notation for permutations and combinations was still emerging.

Factorials

A mathematical function called a *factorial* is used to compute the number of possible permutations and combinations. Let $n!$ equal

$$n \times (n-1) \times (n-2) \times (n-3) \times \ldots \times 3 \times 2 \times 1.$$

For example, $5! = 5 \times 4 \times 3 \times 2 \times 1 = 120$. Further, $0!$ is defined to be 1. Bernoulli had proved many factorial results, like the fact that $n!$ gives the number of permutations of n objects. The use of the exclamation point to indicate a factorial, which was more convenient for printers of the day than some older notations, has been attributed to mathematician Christian Kramp. He worked in the late eighteenth and early nineteenth centuries. The general rule for finding permutations and combinations is sometimes attributed to Bernoulli and sometimes to sixteenth and seventeenth century mathematician Pierre Hérigone, who is also famed for introducing a variety of mathematical and logical notations. However, mathematicians used their own methods for indicating permutations and combinations well into the nineteenth century. For example, Thomas Harriot's seventeenth-century work *Ars Analyticae Praxis* contained unique symbolism for displaying the combinatorial process of finding binomial products.

In the notation common in the twentieth and twenty-first centuries, the number of permutations is stated as $n\mathrm{P}r$ where n is the total number of objects in a set and r is the number of objects selected from n and permuted,

$$n\mathrm{P}r = \frac{n!}{(n-r)!}.$$

The number of combinations is $n\mathrm{C}r$, which is read as "n choose r,"

$$n\mathrm{C}r = \binom{n}{r} = \frac{n!}{r!(n-r)!}.$$

The partial origins of this approach may perhaps be traced to nineteenth-century amateur mathematician Jean Argand, who used (m, n) to represent combinations of n objects chosen from a set of m objects.

Modern Developments

In the early twentieth century, mathematicians and others continued to develop theories and applications of combinatorial concepts. For example, statistician Ronald Fisher applied combinations to the design of factorial experiments, while artist Maurits Cornelius (M.C.) Escher developed his own system for categorizing combinations of shape, color, and symmetrical properties, which can be found in his 1941 notebook later referred to as a paper, *Regular Division of the Plane with Asymmetric Congruent Polygons*. Historians discuss that the sketchbooks of a typical artist contain preliminary versions of final works. Escher's book, on the other hand, appeared to form a theoretical mathematical basis for his tiling work. These combinatorial categories also influenced the field of crystallography.

Circular permutations are also common. One could think of lining up six people in a straight line to take their picture versus seating them at a round table. There are $n!$ permutations of the people lined up. However, once all six people are seated, even if they were all asked to move over one seat, they would all still be seated in the same overall order. There are therefore $(n-1)!$ ways of putting objects in a circle. Another possibility is that all items in the set are not unique, like the letters in "Mississippi," which reduces the number of unique permutations and combinations versus a set of the same length with unique components.

Permutation Groups

In a field like modern algebra, permutations can be viewed as maps that relate a set to itself. The set of permutations is then collected into an algebraic structure called a "group." One example is the various possible transformations of a Rubik's Cube puzzle, named for Erno Rubik. There are 43,252,003,274,489,856,000 permutations in the group for a 3-by-3-by-3 Rubik's Cube. Mathematicians often use software like the Groups, Algorithms, Programming (GAP) system to model and understand the transformations. Theories about permutation groups have been traced by historians to at least as far back as Joseph Lagrange's 1770 work *Réflexions sur la résolution algébrique des équations*, in which he discussed the permutations of the roots of equations and considered those roots as abstract structures. Paolo Ruffini used what would now be called *group theory* in his work, including permutation groups, and proved many fundamental theorems. In the nineteenth century, Augustin-Louis Cauchy generalized some of Ruffini's results. He studied permutation groups and proved what is now known as Cauchy's theorem. High school mathematics teacher Peter Sylow wrote his book *Théorèmes sur les groupes de substitutions* in the latter half of the nineteenth century, and it contained what are now known as the three Sylow theorems, which he proved for permutation groups. Arthur Cayley wrote about the connections between his work on permutations and Cauchy's, extended the notion of permutation groups into the broader idea of algebraic groups, and ultimately proposed that matrices and quaternions were types of groups. Some of his work served as one foundation for physicist Werner Heisenberg's development of quantum mechanics.

In the early twentieth century, George Pólya used permutation groups and other methods to enumerate isomers (compounds that have the same molecular components but different structural arrangements, or permutations) in organic chemistry. He also influenced Escher's studies of combinations. The George Pólya Prize is given every two years by the Society for Industrial and Applied Mathematics. One criterion for winning is "a notable application of combinatorial theory." Mathematicians continue to explore permutations and combination concepts in algebra and many other areas of mathematics.

Further Reading

David, F. N. *Games, Gods & Gambling: A History of Probability and Statistical Ideas.* New York: Dover Publications, 1998.

Davis, Tom. "Permutation Groups." http://www .geometer.org/mathcircles/perm.pdf.

Higgins, Peter. *Number Story: From Counting to Cryptography.* New York: Copernicus, 2008.

CARMEN M. LATTERELL

See Also: Binomial Theorem; Data Analysis and Probability in Society; Dice Games; Lotteries; Probability; Transformations.

Perry, William J.

Category: Government, Politics, and History.
Fields of Study: Connections.

Summary: Influential secretary of defense William J. Perry earned a Ph.D. in mathematics.

William J. Perry (1927–) is an American businessman, mathematician, engineer, and former secretary of defense under President Bill Clinton. William Perry received many honors and recognition for his work. In 1997, he was awarded the Presidential Medal of Freedom, and he has been decorated twice each with the Department of Defense Service Medal and the Defense Intelligence Agency's Outstanding Civilian Service Medal. He has also received many awards from foreign governments. William J. Perry's academic degrees, all of which are in pure mathematics, may seem an unlikely preparation for a successful businessman and secretary of defense. The logical mindset and the steps of analytic problem solving he learned as a student of mathematics helped Perry make rational and objective decisions about complicated situations for which only partial information was available. This connection is not an unusual; people who have been trained in mathematical reasoning before going on to careers in nonmathematical fields often cite the utility of mathematical thinking as a way to approach difficult and complex problems.

Early Life and Education

Perry was born in Vandergrift, Pennsylvania. After graduating from high school in 1945, Perry enlisted in the U.S. Army and served in Japan before attending Stanford University. Perry was always interested in both mathematics and English, but he finally settled on mathematics as a major because, "I simply had more flexibility by going into mathematics." He attributed his interest in advanced mathematics to George Polya, his advisor at Stanford, saying: "He just pushed me and gave me interesting problems to work on. And he exposed me to parts of mathematics that I had never seen before. And he was just a warm human being." He later earned a Ph.D. in mathematics from Pennsylvania State University, where his research was in the field of partial differential equations. While working on his doctorate, Perry loved teaching mathematics, and he imagined that he might become a mathematics professor. He took a part-time job as an applied mathematician at an electronics company in order to support his family and decided to concentrate on the applied side of mathematics.

Career

Perry enjoyed a successful career as an engineer and businessman. He spent 10 years as director of the Electronic Defense Laboratories of Sylvania/GTE, followed by 13 years as founding president of ESL Inc. In 1977, he became President Jimmy Carter's undersecretary of defense for research and engineering. In this position, he played an important role in developing stealth aircraft technology. In 1981, he returned to industry as the managing director of an investment bank that focused on high-technology companies. In 1993, William Perry was appointed as deputy secretary of defense under then-secretary of defense Les Aspin. The following year, he was promoted to secretary of defense, a position he would hold until 1997. He stated: "Quite clearly, knowing how to solve a differential equation is not a useful tool for me. I've never been asked to solve one since becoming the Secretary of Defense. But, the discipline of thinking, systematically approaching problems, of rigorous thinking is a useful—I would say even an indispensable tool—for a job of this sort."

"Preventive defense" was the watchword of Perry's strategy as secretary of defense: prevent threats before they happen, deter threats that are realized, and respond with decisive military force to threats that cannot be deterred. He noted that: "Analytical thinking is a good framework, a good foundation for which to approach problems." This strategy manifested itself as threat reduction programs, including the START II treaty (for which Perry advocated strongly), active opposition to nuclear proliferation, and expansion of the North Atlantic Treaty Organization. He worked hard to maintain an effective and modern military in spite of defense budget shortfalls. One of his priorities was to establish relationships with members of the military at all levels. Unlike many other secretaries of defense, William Perry was an active participant in foreign policy, traveling often to foreign countries as part of his response to the many global challenges during his tenure as secretary of defense, which included the Bosnian War, conflict in Somalia, the aftermath of the first Gulf War, North Korean nuclear aspirations, and the crisis in Haiti.

Further Reading

Albers, Donald. "The Mathematician Who Became Secretary of Defense." *Math Horizons* 4 (September 1996).

Department of Defense. "SecDef Histories—William J. Perry" http://www.defense.gov/specials/secdef _histories/bios/perry.htm.

Perry, William. "It's Your Ship: Lessons in Leadership." Speech given April 2007 at Stanford University. http://ecorner.stanford.edu/authorMaterialInfo .html?mid=1677.

MICHAEL "CAP" KHOURY

See Also: Careers; Mathematics, Applied; Strategy and Tactics.

Personal Computers

Category: Communication and Computers.
Field of Study: Algebra; Communication; Data Analysis and Probability; Number and Operations; Representations.
Summary: Advances in computing have made mathematical processing power so inexpensive that it has become more practical to do many tasks on the computer.

A computer is a device that manipulates raw data into potentially useful information. Computers may be analog or electronic. Analog computers use mechanical elements to perform functions. For example, Stonehenge in England is believed by some to be an analog computer. It allegedly uses the stones along with the positions of the sun and moon to predict celestial events like the solstices and eclipses. Electronic computers use electrical components like transistors for computations.

Many consider the first personal computer to be Sphere 1, created by Michael Wise in the mid-1970s. The Apple II was introduced in 1977, and Apple Inc. offered the Macintosh, which had the first mass-marketed graphical user interface, by 1984. IBM debuted its personal computer in 1981. "Macs" and PCs quickly became common in businesses and schools for a variety of purposes. Processing speed, size, memory capacity, and other functional components have become faster, smaller, lighter, and cheaper over time, and personal computers have evolved into a multitude of forms designed to be customizable to each user's needs.

At the beginning of the twenty-first century, desktops, laptops, netbooks, tablet PCs, palm-sized smartphones, handheld programmable calculators, digital book readers, and devices like Apple's iPad offer access to computing, the Internet, and other functions.

Mathematical History of Computers

Modern computing can be traced to nineteenth century mathematician Charles Babbage's analytical engine. Boolean algebra, devised by mathematician George Boole later in the same century, provided a logical basis for digital electronics. Lambda calculus, developed by mathematician Alonso Church in the early twentieth century, also laid the foundations for computer science, while the Turing machine, a theoretical representation of computing developed by mathematician Alan Turing, essentially modeled computers before they could be built. In the 1940s, mathematicians Norbert Wiener and Claude Shannon researched information control theory, further advancing the design of digital circuits. The Electrical Numerical Integrator and Calculator (ENIAC) was the first general purpose electronic computer. It was created shortly after World War II by physicist-engineer John Mauchly and engineer J. Presper Eckert. They also developed the Binary Automatic Computer (BINAC), the first dual-processor computer, which stored information on magnetic tape rather than punch cards, and the first commercial computer, Universal Automatic Computer (UNIVAC). Mathematician John Von Neumann made important modifications to ENIAC, including serial operations to facilitate mathematical calculations. Scientists William Bradford Shockley, John Bardeen, and Walter Brattain won the 1956 Nobel Prize in Physics for transistor and semiconductor research, which influenced the development of most subsequent electronic devices, including personal computers. During the latter half of the twentieth century, countless mathematicians, computer scientists, engineers, and others advanced the science and technology of personal computers, and research has continued into the twenty-first century. For example, Microsoft co-founder Bill Gates published a paper on sorting pancakes, which has extensions in the area of computer algorithms. Personal computers have facilitated mathematics teaching and research in many areas such as simulation, visualization, and random number generation, though the use of calculators and software like Maple for teaching mathematics generated controversy.

Devices, Memory, and Processor Speeds

The typical personal computer has devices for the input and output of information and a means of retaining programs and data in memory. It also has the means of interacting with programs, data, memory, and devices attached to the computer's central processing unit (CPU). Input devices have historically included a keyboard and a mouse, while newer systems frequently use

Many consider the Sphere to be first personal computer. Sphere had difficulty providing the product and shut down after two years.

touch technology, either in the form of a special pad or directly on the screen. Other devices include scanners, digital cameras, and digital recorders. Memory storage devices are classified as "primary memory" or "secondary" devices. The primary memory is comprised of the chips on the board inside the case of the computer. Primary memory comes in two types: read only memory (ROM) and random access memory (RAM). ROM contains the rudimentary part of the operating system, which controls the interaction of the computer components. RAM holds the programs and data while the computer is in use. The most popular types of secondary memory used for desktop computers include magnetic disk drives, optical CD and DVD drives, and USB flash memory.

The speed of the computer operation is an important factor. Computers use a set clock cycle to send the voltage pulses throughout the computer from one component to another. Faster processing enables computers to run larger, more complex programs. The disadvantage is that heat builds up around the processor, caused by electrical resistance. ENIAC was 1000 times faster than the electromechanical computers that preceded it because it relied on vacuum tubes rather than physical switches. Turing made predictions regarding computer speeds in the 1950s, while Moore's law, named for Intel co-founder Gordon Moore, quantified the doubling rate for transistors per square inch on integrated circuits. The number doubled every year from 1958 into the 1960s, according to Moore's data. The rate slowed through the end of the twentieth century to roughly a doubling every 18 months. Some scientists predict more slowdowns because of the heat problem. Others, like mathematician Vernor Vinge, have asserted that exponential technology growth will produce a singularity, or essentially instantaneous progress. Processing speed, memory capacity, pixels in digital images, and other computer capabilities have been limited by this effect. There has also been a disparity in the growth rates of processor speed and memory capacity, known as *memory latency*, which has been addressed in part by mathematical programming techniques, like caching and dynamic optimization.

Carbon nanotubes and magnetic tunnels might be used to produce memory chips that retain data even when a computer is powered down. At the start of the twenty-first century, this approach was being developed with extensive mathematical modeling and physical testing. Other proposed solutions involved biological, optical, or quantum technology. Much of the physics needed for quantum computers exists only in theory, but mathematicians like Peter Shor are already working on the mathematics of quantum programming, which involves ideas like Fourier transforms, periodic sequences, prime numbers, and factorization. Fourier transforms are named for mathematician Jean Fourier.

The Digital Divide

The digital divide is the technology gap between groups that have differential access to personal computers and related technology. The gap is measured both in social metrics, such as soft skills required to participate in online communities, and infrastructure metrics, such as ownership of digital devices. Mathematical methods are used to quantify the digital divide. Comparisons may be made using probability distributions and Lorenz curves, developed by economist Max Lorenz, and measures of dispersion such as the Gini coefficient, developed by statistician Corrado Gini. Researchers have found digital divides among different countries, and within countries, among people of different ages, between genders, and among socioeconomic strata.

The global digital divide quantifies the digital divides among countries and is typically given as the differences among the average numbers of computers per 100 citizens. In the early twenty-first century, this metric varied widely. Several concerted private and government efforts, such as One Laptop Per Child, were directed at reducing the global digital divide by providing computers to poor countries. The breakthroughs connected to these efforts, such as mesh Internet access architecture, benefited all users. The Digital Opportunity Index (DOI) is computed by the United Nations based on 11 metrics of information and communication technologies, such as proportion of households with access to the Internet. It has been found to be positively associated with a country's wealth.

Further Reading

Lauckner, Kurt, and Zenia Bahorski. *The Computer Continuum.* 5th ed. Boston: Pearson, 2009.

Lemke, Donald, and Tod Smith. *Steve Jobs, Steve Wozniak, and the Personal Computer.* Mankato, MN: Capstone Press, 2010.

Wozniak, Steve, and Gina Smith. *iWoz: Computer Geek to Cult Icon: How I Invented the Personal Computer, Co-Founded Apple, and Had Fun Doing It.* New York: W. W. Norton, 2007.

Zenia C. Bahorski
Maria Droujkova

See Also: Cerf, Vinton; Internet; Lovelace, Ada; Servers; Software, Mathematics.

Pi

Category: History and Development of Curricular Concepts.
Fields of Study: Communication; Connections; Measurement; Geometry.
Summary: The ratio of a circle's circumference to it's diameter, π, is one of the most important constants and the first irrational number encountered by most students.

By definition, pi (π) is the ratio of a circle's circumference to the diameter. This definition holds for any circle, with the value of π being the constant value $3.14159265358979\ldots$. This decimal neither terminates nor repeats, making π irrational. Mathematicians and non-mathematicians alike are intrigued by the many appearances of π in diverse situations. Capturing this apparent mysticism in the 1800s, the mathematician Augustus de Morgan wrote, "This mysterious $3.14159\ldots$ which comes in at every door and window, and down every chimney."

Values Used for Pi

Since the beginning of written mathematics, people have tried to calculate π's value. Around 2000 B.C.E., the Babylonians and Egyptians assigned values equal to 3 1/8 (3.125) and $4\left(8/9\right)^2$ (3.1605). In 1100 B.C.E., the Chinese used $\pi = 3$, a value which also appears in the Bible (I Kings 5:23). In 300 B.C.E., Archimedes of Syracuse produced the first "accurate" value, using inscribed and circumscribed 96-sided polygons to produce the approximation $3\ 10/71 < \pi < 3\ 1/7$ (or $3.140845\ldots < \pi < 3.142857\ldots$). Since that time,

multiple methods and formulas have been created to determine more exact values of π. Today, powerful computers use similar formulas to calculate values of π to extreme precision, with the current value exceeding 2.7 trillion digits (the record as of January 2010). Two examples of these formulas involving infinite series are

$$\frac{\pi}{2} = \frac{2 \times 2 \times 4 \times 4 \times 6 \times 6 \times 8 \cdots}{1 \times 3 \times 3 \times 5 \times 5 \times 7 \times 7 \cdots}$$

or $\quad \dfrac{\pi}{4} = \dfrac{1}{1} - \dfrac{1}{3} + \dfrac{1}{5} - \dfrac{1}{7} + \dfrac{1}{9} - \cdots.$

Students in the twenty-first century learn about π in elementary school, and exposure to π continues in later courses in mathematics and physics. Since spherical coordinates are used in many applications, π is found in physical formulas such as Einstein's field equations, the Heisenberg uncertainty principle, and Coulomb's law for electric force, which are named after Albert Einstein, Werner Heisenberg, and Charles-Augustin de Coulomb, respectively. Mathematicians and computer scientists describe π as a great stress test for computers because of the seemingly random aspects of its digits.

Algorithms to compute the digits of π are regarded as more important than the digits themselves. Mathematicians continue to investigate other unsolved problems related to π, including attempts to determine how random the digits are.

Applications

The number π has played important roles in multiple situations. In 1767, Johann Lambert proved that π was irrational (it could not be written as the ratio of two integers). Then, in 1882, Ferdinand von Lindemann proved that π was transcendental (it could not be constructed using geometric tools and was not a root of a non-constant polynomial equation with rational coefficients). These two discoveries provided the key to proving the impossibilities of the Greeks' three problems of antiquity—squaring a circle, trisecting an angle, and duplicating a square.

Considered by many to be a ubiquitous number, π shows up in odd situations. First, in 1777, the naturalist Georges Buffon approximated the value of π experimentally by tossing a needle (length L) on a ruled surface (parallel lines spaced at distance D). If the tossed needle touches a line S times on N tosses, then

$$\pi \approx \frac{2SL}{DN}.$$

Second, the probability that two random integers are relatively prime (they have no common divisor) is

$$\frac{6}{\pi^2}.$$

Anyone can try these experiments, either by dropping needles or taking ratios of random integers; many are surprised that both produce good approximations for π. However, complex mathematics is needed to explain "why."

In 1743, Swiss mathematician Leonhard Euler published the formula $e^{ix} = \cos(x) + i\sin(x)$, linking exponentials, trigonometric functions, and complex numbers. Substituting $x = \pi$, the result becomes the most beautiful formula in mathematics: $e^{i\pi} + 1 = 0$.

Popular Culture

The fascination with decimal expressions of π has led to competitive memorization contests. The Guinness World Records officially recognized Lu Chao as the most recent record holder in the early twenty-first century, but others have claimed more digits. Some people use piems (mnemonic poems); for example, "How I need a drink, alcoholic of course, after the heavy lectures involving quantum mechanics." In this piem, replace each word with its number of letters, producing $\pi \approx 3.14159265358979$. Hideaki Tomoyori, who held the world record of 40,000 digits memorized from 1987–1995, used a pictorial mnemonic system and explained, "I want to go on with the challenge of memorizing π, for just the same reason that people climb high mountains. I think it's a wonderful thing to challenge the limits of what we can do. . . . the more one memorizes of it, the closer one comes to the real value of the circle—closer to perfection." Researchers compared his cognitive abilities with a control group and concluded that he was not superior; they attributed his achievement to extensive practice.

The number π also is connected to some odd events. In 1897, the Indiana State Legislature almost passed a mathematically incorrect bill relating to π and squar-

ing the circle. By its definition, the value of π changes if the circle shifts out of the Euclidean world. That is, in taxicab geometry, or metric geometry on a rectangular lattice structure, the value of π is 4.

The number π is an amazing number, both in its interesting properties and the obsessive attention given it by both mathematicians and non-mathematicians. How else could one explain why on March 14 at 1:59, many people shout, "Happy Pi Day!"

Further Reading

Adrian, Y. E. O. *The Pleasures of Pi, e and Other Interesting Numbers*. Singapore: World Scientific Publishing, 2006.

Beckmann, Petr. *A History of π (Pi)*. New York: Barnes & Noble, 1971.

Berggren, Lennart, Jon Borwein, and Peter Borwein. *Pi: A Source Book*. New York: Springer-Verlag, 1997.

Blatner, David. *The Joy of π*. New York: Walker & Co., 1997.

Takahashi, Masanobu, et al. "One% Ability and Ninety-Nine% Perspiration: A Study of a Japanese Memorist." *Journal of Experimental Psychology. Learning, Memory, and Cognition* 32, no. 5 (2006).

JERRY JOHNSON

See Also: Archimedes; Numbers, Rational and Irrational; Sequences and Series; Universal Constants.

Planetary Orbits

Category: Space, Time, and Distance.
Fields of Study: Algebra; Geometry.
Summary: It took mathematicians thousands of years to accurately describe planetary motion.

For millennia, the shape of the paths in which the planets orbited was dominated by metaphysical concerns and assumed, almost without question, to be circular. It was not until the seventeenth century that science discovered the actual shape of planetary orbits, the ellipse.

Early Conceptions

In ancient Greek astronomy, it was assumed that the Earth was the center of the universe, and all of the known planets (including the sun and the moon) as well as the stars revolved around it. Furthermore, at least from the time of Pythagoras (c. 569–475 B.C.E.), these orbits were assumed to be circular. This assumption was a metaphysical one.

The Pythagoreans believed in the perfection of mathematics and held the view that the circle was perfect because of its symmetry and continuity. Therefore, the universe must surely be constructed to reflect this perfection by requiring the planets to revolve around the Earth in perfect circular motion. That influential philosophers such as Plato and Aristotle accepted the perfection of circular motion contributed to the fact that the idea went almost unchallenged for nearly 2000 years.

With the increasing ability to make accurate observations of the movements of the heavens and mathematical calculations to predict those movements, the simple assumption of perfect circular motion became more problematic. The predictions of the planetary positions did not match the actual observed locations. Eudoxus (408–355 B.C.E.) addressed this discrepancy by devising a complicated system of nested spheres in which each planet moved, maintaining circular motion of each sphere while more accurately predicting the location of the planets.

For many centuries, one man's work dominated European thinking on planetary motions. The Greek mathematician and astronomer Ptolemy (85–165 C.E.) compiled all that was known about the movements of heavenly bodies into one work that came to be known as *The Almagest*. This book employed an array of very complex geometric and trigonometric theories to describe the movement of the planets, with the Earth remaining at the center. In order for the observations to be as close as possible to the calculations, Ptolemy used epicycles (small circles revolving upon bigger circles as they revolve around the Earth) and moved the Earth away from the center of revolution of the planets.

The new center of revolution was an imaginary point some distance away from the Earth. Ptolemy's influence on Western astronomy was partially because of its general agreement with Christian doctrine. As the center of God's creation, the Earth must rest at the center of the cosmos. Furthermore, a perfect Creator would use the perfect circle to put His creation in motion.

Challenges

The most serious challenge to Ptolemaic cosmology came from the Polish church official, Nicolaus Copernicus (1473–1543), whose revolutionary work *De Revolutionibus* placed the sun, not the Earth, at the center of the universe, relegating the Earth to mere planethood. Copernicus, however, remained adamant in his belief that the planets orbited the sun in a composite of perfect circular motions. The doctrine of perfect circular motion in the heavens was finally challenged by the German astronomer Johannes Kepler (1571–1630). Kepler, after many years of tedious and painstaking calculations involving the orbit of Mars, finally determined that Mars actually orbited the sun in an elliptical orbit, not a circular one. This revolutionary idea was based in part on another discovery by Kepler that the speed of the planets varied as they orbited the sun. Later, the great British mathematician and scientist, Isaac Newton (1643–1727), used his universal law of gravitation and laws of motion to provide a mathematical explanation for Kepler's claim of elliptical orbits, finally putting an end to the ancient doctrine of circular motion in the heavens.

Mathematics continues to play an important role in modeling planetary orbits. For example, Mercury's orbit is more accurately represented with hyperbolic geometry than with Euclidian geometry. Further, the orbit of Mercury allows researchers to see the impact of the sun's gravitational field on the curvature of space.

Further Reading

Danielson, Dennis Richard. *The Book of the Cosmos: Imagining the Universe From Heraclitus to Hawking.* Cambridge, MA: Perseus Publications, 2000.

Gingerich, O. *The Eye of Heaven: Ptolemy, Copernicus, Kepler.* New York: American Institute of Physics, 1993.

Heath, Thomas L. *Greek Astronomy.* New York: E. P. Dutton, 1932.

Kopache, Gerald. "Planetary Motion: Also Featuring Some Stars, Some Comets and the Moon." Ceshore Publishing Company, 2004.

Montenbruck, Oliver, and Gill Eberhard. *Satellite Orbits: Models, Methods and Applications.* Berlin: Springer, 2000.

Pannekoek, Anton. *A History of Astronomy.* New York: Dover Publications, 1989.

Sagan, Carl. *Cosmos.* New York: Random House, 1980.

Todd Timmons

See Also: Astronomy; Conic Sections; Geometry of the Universe; Greek Mathematics.

Plate Tectonics

Category: Weather, Nature, and Environment.
Fields of Study: Data Analysis and Probability.
Summary: Tectonic plate movement is measured and analyzed using mathematics.

The ideas of plate tectonics and continental drift have been theorized by many scientists over the years. For example, in the early twentieth century, Alfred Wegener publicly presented theories regarding the existence of a supercontinent called "Pangea" that eventually formed all the known continents. He supported centrifugal force as an explanation for drift. A few years later, Arthur Holmes supported thermal convection as an explanation. At the time, there was insufficient mathematical and scientific evidence to support these theories and they were largely dismissed, in part because seeing into the depths of the oceans and into the Earth itself is often a more difficult venture than seeing galaxies at the far reaches of the universe. By the latter half of the twentieth century, discoveries such as mid-Atlantic underwater volcanic chains and the mapping and mathematical analysis of seismic activity suggested the existence of large, mobile plates in the Earth's crust.

In the twenty-first century, scientists and mathematicians are still developing new and innovative ways to collect data, model, visualize, and simulate the Earth's inner structure. For example, geophysicist Robert van der Hilst and mathematician Maarten Van de Hoop have used a mathematical technique known as "micro-local analysis," as well as statistical methods, such as confidence intervals, to explore the geom-etry of the layers near the boundary of the Earth's core and mantle. This technique extends existing methods for analyzing noisy seismic data. It produces not only an image, but also an estimate of the probability that a true layer has been discovered. Ongoing collaboration between mathematicians and geophysical scientists is crucial to address the massively scaled problems that arise in geoscience, such as continental drift. This is true not only for data collection in the field, but also for computer simulation, which is increasingly an avenue of exploration and cross-validation for theories and data. These simulations often require combining many scales of data, both macro and micro, as well as observations collected over different periods of time. Further, much of the data is noisy, incomplete, or difficult to directly measure. Mathematics is also involved in the increasingly sophisticated tools that allow scientists to visit the depths of the oceans and begin to look at some previously impenetrable layers of the Earth.

The Spreading Sea Floor

As an officer in the U.S. Navy, Harry Hess's curiosity led him to measure the ocean floor using sounding gear and magnetometers during World War II. Once the war ended, Hess developed the theory of sea floor spreading to explain his data. He proposed that magma oozed up

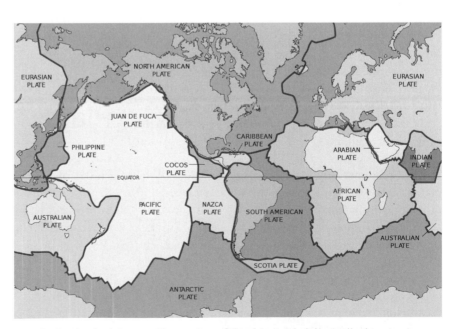

A U.S. Geological Survey illustration of Earth's rigid slabs (called tectonic plates) that are moving relative to one another.

between the plates along the ridges in the ocean floor, pushing them apart and causing the plates to move.

Strips of rock parallel to the ridges provide evidence for sea floor spreading. Strips closest to the ridge have the same polarity as the Earth (magnetic north pointing to the north pole); however, the strips moving out away from the ridge on opposite sides mirror each other and alternate between current polarity and reversed polarity as the Earth's magnetic field reversed over time. These alternating strips suggest that new rock is created along the ridges over geologic time.

Continents Adrift

Until 1912, scientists assumed that the continents were fixed in place. In that year, Alfred Wenger suggested that the continents were adrift, originally part of one large landmass. Wegner cited evidence such as matching geological formations and fossils from South America and Africa. It was not until the late 1960s that discoveries were made and measuring techniques improved to the extent that the theory of plate tectonics emerged and became widely accepted. Scientists now recognize that the continents are attached to plates and move with them rather than moving independently. Scientists also now know that the plates that make up Earth's crust and the continents attached to them are moving several centimeters per year on average as they collide, move apart, and brush up against each other.

Plate Movement

Muawia Barazangi and James Dorman (1969) charted the locations of all earthquakes occurring from 1961 to 1967 and found that most occurred in a narrow band of seismic activity. This band of high earthquake and volcanic activity, commonly called the "Pacific Ring of Fire," defines many plate boundaries around the Pacific Ocean.

Most plate movement occurs along the edges of the plates. Scientists can measure the velocity (speed and direction) of plate movement and determine how that relates to earthquake and volcanic activity. For historical information, scientists turn to ocean floor magnetic striping data and geological dating of rock formations.

Measurement techniques have improved greatly since Hess's measurements. The most common technique for measuring plate movement in the early twenty-first century is the Global Positioning System (GPS). As satellites continuously transmit radio signals to Earth, each GPS ground site simultaneously receives signals from at least four satellites. By recording the exact time and location of each satellite when its signal was received, it is possible to determine the precise position of the GPS ground site on Earth (longitude, latitude, and elevation). Regularly measuring distances between specific points allows scientists to determine if there has been active movement between plates on a scale of millimeters. Using time-series graphs and plotting vectors, it is possible to learn how the plates move.

While scientists know that most earthquakes and volcanoes occur along plate boundaries, they still cannot predict exactly when and where they will occur. By monitoring plate movement, scientists hope to learn more about the events building up to earthquakes and volcanic eruptions.

Further Reading

Barazangi, Muawia, and James Dorman. "World Seismicity Maps Compiled from ESSA, Coast and Geodetic Survey, Epicenter Data, 1961–1967." *Bulletin of the Seismological Society of Am*erica 59 (1969).

Preskes, Naomi. *Plate Tectonics: An Insider's History of the Modern Theory of the Earth*. Boulder, CO: Westview Press, 2003.

CHRISTINE KLEIN

See Also: Earthquakes; Geothermal Energy; GPS; Tides and Waves; Volcanoes.

Plays

Category: Arts, Music, and Entertainment.
Fields of Study: Communication; Geometry; Representations.
Summary: Numerous plays explore mathematical concepts and mathematicians.

The genre of "mathematical theater" is a relatively recent phenomenon. A smattering of earlier examples of mathematics appeared on stage, but the turning point was Tom Stoppard's 1993 play *Arcadia*, which opened the door to an entirely new realm of collaborative possibilities between theater and the mathematical sciences. Following on the heels of *Arcadia* was the award-winning *Copenhagen* (1998), a play by Michael Frayn about the fraught relationship between physicists Neils Bohr and Werner Heisenberg. If there were any lingering doubts as to whether mathematics was a relatable theme for theater audiences, David Auburn's Pulitzer Prize–winning play *Proof* (2000) laid them firmly to rest. The ensuing years have produced successful dramas, comedies, and biographical scripts that are marked not just by the inclusion of mathematical references, but also by the wholesale incorporation of mathematics into the content and structure of the play. Some even turn a critical lens back on traditional mathematics education and related gender issues.

Stoppard and Science

Bertold Brecht's *The Life of Galileo* (1939) gives a cursory acknowledgment of the protagonist's training as a mathematician. *The Physicists* (1962), by Friedrich Durenmatt, features Isaac Newton as a character—or rather, it features a spy who is posing as a patient in a mental institution, pretending to believe he is Newton. Terry Johnson's play *Insignificance* (1982) contains a scene where Marilyn Monroe explains special relativity to Albert Einstein.

But the best place to look for a forerunner for the substantial and explicit role of mathematics in *Arcadia* is in Tom Stoppard's earlier writing. His first major success was *Rosencrantz and Guildenstern are Dead* (1966), a dark comedy, which opens with a scene of the two Shakespearean characters trying to rectify the laws of probability with the fact that they have just witnessed nearly 100 occurrences of heads in as many flips of a coin. Zeno's paradoxes appear in *Jumpers* (1972), and there is a cameo appearance of Leonhard Euler's famous Bridges of Königsburg problem in *Hapgood* (1988), a play that also contains significant discussions of quantum mechanics. *Hapgood* comes closest to *Arcadia* in its attempt to fully integrate mathematics and science into the mechanics of the play, but this was confusing for some audiences and the reviews for *Hapgood* tended to be rather harsh. *Arcadia*, in contrast, was greeted as something of a marvel and an instant classic when the play opened in London in 1993.

The opening scene of *Arcadia* is set in 1809, where 13-year-old Thomasina Coverly is growing frustrated with her tutor, who has asked her to find a proof for Fermat's Last Theorem. Thomasina has more romantic issues on her mind ("Septimus, what is carnal embrace?" is the first line of the play), and her restlessness—and her genius—eventually lead her to discover the core principals of fractal geometry and chaos theory 150 years before their time. *Arcadia* also contains a second set of characters living in the present day in the same house, and among them is a mathematician whose expertise in dynamical systems allows him to decipher Thomansina's notebooks for the other characters—and the audience. In a clever homage to Fermat, Thomasina writes in one of her notebooks that "I, Thomasina Coverly, have found a truly wonderful method whereby all the forms of nature must give up their numerical secrets and draw themselves through number alone. This margin being too mean for my purpose, the reader must look elsewhere for the New Geometry of Irregular forms discovered by Thomasina Coverly."

A recurring theme in *Arcadia* is the juxtaposition of reasoned, classical thinking with untamed, romantic expression. With respect to the mathematics in the play, the Euclidean geometry of circles and spheres is contrasted with the fractal geometry of leaves and clouds. In a related way, the determinism inherent in Newton's Laws of Motion is challenged by the unpredictability of chaotic systems and ultimately by the Second Law of Thermodynamics. These scientific ideas provide a compelling metaphorical backdrop for the interpersonal tensions that drive the emotional arc of the script. The result is a play where the science and the storytelling work in a mutually enriching collaboration.

Copenhagen

Whereas *Arcadia* is a hybrid of mathematics and science, Frayn's *Copenhagen* is very much a "physics play," but its influence is too significant to ignore. The play is inspired by a real historical event. Werner Heisenberg had been put in charge of the Nazi nuclear program, and in 1941, he paid a visit to his mentor Neils Bohr, whose hometown of Copenhagen was under German occupation. The visit ended abruptly, and the deep friendship between these two pioneers of atomic physics ended with no clear resolution ever agreed upon as to what exactly was discussed. Frayn's play explores this question by recreating the experiment of Heisenberg's visit multiple times and, in the spirit of quantum mechanics, each

The Most Well-Known Mathematics Play: *Proof*

Even if it had not been turned into a popular film, *Proof* would still likely be the most well-known mathematics play and most frequently performed. It should be pointed out, however, that unlike *Arcadia* and *Copenhagen*, there is virtually no technical material written into the script. The central relationship in the play is between a father and daughter. The father is a brilliant mathematician who, the audience learns, has become debilitated by serious mental illness. His daughter Katherine has given up on her own education to care for her father, and upon his death, Katherine is plagued by the question of whether this was the right decision, as well as whether she has inherited her father's mental instability. A major plot twist comes when Katherine discloses the existence of a mathematical proof hidden in her father's desk, and a central issue is to determine its rightful author. The audience is never told what the theorem actually is, but is made to understand that it is a monumental result on the order of the Riemann Hypothesis.

A debate among theater critics is whether the mathematics in *Proof* is crucial to the workings of the play, or whether it is intellectual window dressing that could be replaced by some other creative art form; for example, the father might be a composer and put a symphony score in the desk drawer. Although the discussions of explicit mathematics in *Proof* are confined to a few imaginary number jokes and some witty banter about primes, there are several compelling conversations about the aesthetic beauty of mathematics and the discipline is sympathetically portrayed. Because the main questions in the play deal with degrees of certainty, there is an argument that the rigorous standard for what constitutes a mathematical proof provides a valuable point of contrast for the various investigations by the characters in the play.

run of the experiment results in a different outcome. Along the way, the fundamental ideas behind Bohr's Theory of Complementarity and Heisenberg's Uncertainty Principle are given enough explication for the audience to apply these ideas to the process of human introspection as well as to the play itself.

Hardy, Ramanujan, Turing, and Beyond

The most high-profile play about mathematics since *Proof* is *A Disappearing Number*, created and produced by a London-based company called Complicite under the leadership of Simon McBurney. *A Disappearing Number* won the 2007 Olivier Award for Best New Play, among many others, and eventually it toured internationally. The starting point for *A Disappearing Number* is G. H. Hardy's famous essay, *A Mathematician's Apology*. Hardy appears as a character as does the Indian genius Srinivasa Ramanujan. The celebrated collaboration between Hardy and Ramanujan is also the subject matter for a less well-known play called *Partition* (2003), written by Ira Hauptman, and in a less direct way it served as inspiration for *The Five Hysterical Girls Theorem* (2000) written by Rinne Groff. Whereas *Partition* is a fanciful account of a real historical friendship, *The Five Hysterical Girls Theorem* is a purely fictitious comedy about an international mathematics conference that features a protagonist loosely based on Hungarian mathematician Paul Erdös.

Biography and historical fiction are the dominant forms for most new mathematical theater. Isaac Newton is the central subject of *Leap* (2004), by Lauren Gunderson as well as *Calculus* (2003) by Carl Djerassi. *Seventeenth Night* (2004), by Doxiadis Apostolos, tells the story of the final days of logician Kurt Gödel's life in a way that is meant to illustrate the actual content of Gödel's revolutionary Incompleteness Theorems. Georg Cantor's bouts with mental illness are the subject of *Count* (2009), by John Martin and Timothy Craig, and Cantor also appears alongside his philosophical nemesis Leopold Kronecker in a scene in the experimental play *Infinities* (2002), written by John Barrow. *Infinities* actually consists of five scenes or scenarios—one features the Hilbert Hotel introduced by mathematician David Hilbert—each of which explores some paradoxical aspect of infinity.

The drama, and ultimate tragedy, of Alan Turing's life is the subject of at least four plays. The most well-known of these is *Breaking the Code* (1986) by Hugh Whitmore, which is available as an episode of *Masterpiece Theater*. The most ambitious play about Turing in terms of engaging the essence of his mathematical work is probably *Lovesong of the Electric Bear* (2003) by British playwright Snoo Wilson, which received a string of productions in the United States.

Plays By and About Women

Lauren Gunderson, who has been writing plays since she was 16 years old and is known for her interpretations of feminism, science, and history, has spoken widely on the rich intersection of science and theater. She cites *Arcadia* as a good example of the idea that "Science, like any theoretical idea, should lead to a deeper kind of play—a more layered, woven play where the science permeates the form of the play as well as the content." She also encourages playwrights to explore these themes, noting that the fundamental questions of mathematics and science do not exist in some inaccessible other world, but rather are deep and universal. One of her most well-known plays is *Emilie: Le Marquise Du Chatelet Defends Her Life Tonight*, which is about eighteenth-century woman mathematician Gabrielle Émilie Le Tonnelier de Breteuil, Marquise du Châtelet, whose many achievements include a translation and commentary on Isaac Newton's *Principia*. In 2010, Gunderson was the first Playwright in Residence at The Kavli Institute for Theoretical Physics.

Emilie du Chatelet was known for passionately pursuing mathematics in a time when many women were barely literate. Kathryn Wallet's *Victoria Martin: Math Team Queen* examines the modern-day tug of war between popularity and mathematics talent that girls often face as they move into middle school and high school. This theme is also critically explored in Gioia De Cari's autobiographical play *Truth Values: One Girl's Romp Through M.I.T.'s Male Math Maze*. The author uses her personal experiences, such as being asked to serve cookies at a seminar, for comic effect. However, the play is a serious exploration of traditional mathematics in higher education and the role of women in science and mathematics.

Further Reading

Harvey, Don, and Ben Sammler. *Technical Design Solutions for Theatre.* Burlington, MA: Focal Press, 2002.

Manaresi, Mirella. *Mathematics and Culture in Europe: Mathematics in Art, Technology, Cinema, and Theatre.* Berlin: Springer, 2007.

Shepherd-Barr, Kirsten. *Science on Stage.* Princeton, NJ: Princeton University Press, 2006.

STEPHEN ABBOTT

See Also: Literature; Movies, Mathematics in; Musical Theater.

Poetry

Category: Arts, Music, and Entertainment.
Fields of Study: Geometry; Measurement; Representations.
Summary: Rhyme schemes and meter in poetry can be mathematically analyzed and some new forms of poetry are based on mathematical priniciples.

A popular sentiment is that mathematics and poetry lie on opposite ends of some spectrum. However, both are the works of pure intellect and they share many similarities. Whether considering rhyme, rhythm, or visual layout, effective poetry is rich with patterns that may be analyzed with a mathematical eye. At the same time, succinct mathematics has often been compared to poetry. In the modern era, the connections have become explicit, as mathematics has been co-opted by poets to create new poems, while poetry has been analyzed (and occasionally written) by mathematicians.

Meter and Rhyme

Poetic meter is a formalized version of rhythm. When considering rhythm in spoken language, one can focus on syllable stresses, pitch, tone, or *morae*. Mora (plural *morae*) is a term used by linguists to denote an individual unit of sound; a long syllable (such as "math") consists of two morae, while a short syllable consists of a single mora. A poetic cadence of length n is a pattern of long and short syllables whose total number of morae is n. Cadences play an especially important role in Indian and Japanese poetry, as well as in modern free verse.

Traditional English meter, however, is usually based on stressed syllables (denoted —) versus unstressed syllables (denoted ˘). The most well-known English meters are iambic pentameter and dactylic hexameter, used extensively by William Shakespeare and Henry Wadsworth Longfellow, respectively. In each of these meters, the first word denotes the metrical foot, and the second word denotes the number of feet per line. A metrical foot is a particular pattern of stressed and unstressed syllables. It usually consists of two, three, or four syllables. For example, an iamb consists of an unstressed syllable followed by a stressed syllable. So a line of iambic pentameter is $2 \times 5 = 10$ syllables in length, and the pattern is ˘—˘—˘—˘—˘—. A dactyl consists of stressed syllable followed by two unstressed syllables. A line of dactylic hexameter is $3 \times 6 = 18$ syllables, and the pattern is —˘˘—˘˘—˘˘—˘˘—˘˘—˘˘. Simple counting shows that there are four possible disyllabic feet (pyrrhus is ˘˘, iamb is ˘—, trochee is —˘, and spondee is ——), eight possible trisyllabic feet, and 16 possible tetrasyllabic feet.

There are further formal devices used by poets, often with the aim of producing euphony, which is beautiful sound combinations: assonance (the same sound repeating within a line), alliteration (multiple words beginning with the same consonant), or specific rhyme schemes. Two examples of rhyme schemes are *ababcdcdefefgg* for a Shakespearean sonnet and *abbaabbacdecde* for an Italian sonnet. The initial lines of Shakespeare's "Sonnet 30":

When to the sessions of sweet silent thought
I summon up remembrance of things past,
I sigh the lack of many a thing I sought,
And with old woes new wail my dear time's waste:

illustrate alliteration and an "abab" rhyme in iambic pentameter—though "past/waste" is only a near rhyme.

Classical Poetic Traditions and Forms

History is rich with individuals such as Omar Khayyám, who excelled in poetry and mathematics separately without drawing a strong connection between the two. However, in at least one culture, the two disciplines were intimately interwoven. In the Indian Vedic civilization, poetic chants and hymns were utilized to pass down a vast body of knowledge. A portion of this knowledge was mathematical, including theorems in arithmetic and geometry. The method of transmission was mathematical: a single text would be recited in up to 11 different ways. Each way emphasized a different poetic approach, such as applying devices of euphony, pausing every other word, or repeating groups of words forward, backward, and in even more complicated permutations. This method is reminiscent of the error-correcting codes employed in twenty-first century CD audio discs. Just as a scratched CD will often still play seamlessly, the redundancy of the Vedic poetic chants allowed for an uncorrupted oral transmission year after year.

A poetic form offers the writer a set of constraints to which the work has to conform. There are many such prescribed forms, some very strict, and others quite open. Perhaps the best-known forms are the sonnet, ode, and haiku. The traditional Japanese haiku, for instance, comprises three lines of 5, 7, and 5 morae, respectively. In English, the syllable is used as counter instead of the mora.

A sestina is a 39-line poem consisting of six 6-line stanzas followed by a 3-line envoy. The six words ending the lines in the first stanza must end the lines in each of the subsequent stanzas, but in a fixed new order. The permutation of the words may be denoted

$$\sigma = \begin{pmatrix} 1 & 2 & 3 & 4 & 5 & 6 \\ 2 & 4 & 6 & 5 & 3 & 1 \end{pmatrix}.$$

This notation indicates that the word ending the first line must end the second line of the next stanza, the word ending the second line must next end the fourth line, and so forth. This permutation is then repeated from one stanza to the next. Mathematicians Anton Geraschenko and Richard Dore have investigated a generalized notion of a sestina to an (n-line-per-stanza) n-tina where n can be any whole number. They prove that if the n-tina is to be interesting—in the sense that the pattern does not repeat before the poem ends—then $2n + 1$ must be a prime number.

Modern Directions

In the modern era, poetry is more often read on a page than spoken aloud, and the two-dimensional geometry of the text is visible. For example, a poem in traditional meter naturally takes on a ragged-on-the-right rectangular shape. The diamond shape of a diamante poem, introduced in 1969 by Iris Tiedt, naturally results from the prescribed construction of its seven lines: one noun, two adjectives, three gerunds, four nouns, three gerunds, two adjectives, and one noun. When poetry purposefully forms a recognizable shape it is called "shape poetry," "griphi," "carmen figuratum," or "concrete poetry." The idea of a shape poem is nothing new: around 300 B.C.E. the Greek poet Simias of Rhodes wrote *Pteryges, Oon,* and *Pelekys* (*Wings, Egg,* and *Hatchet,* respectively) poems whose shape mirrored their subject. Recently, shape poetry has flourished: Lewis Carroll gave a mouse's tail; Guillaume Apollinaire, the Eiffel Tower; e e cummings, a snowflake; John Hollander, a swan with reflection; and Mary Ellen Solt, a forsythia bush. In the 1990s, Eduardo Kac moved poetry into the third dimension with his holopoetry: poetry that floats above a surface as a hologram and takes different meanings when viewed from different angles.

The group Ouvroir de Littérature Potentielle (Workshop of Potential Literature), or Oulipo for short, originated in 1960 with 10 writers, mathematicians, and philosophers. The group has the twin goals of elucidating old and creating new rigid forms for potential literature. A prototypical example of their oeuvre may be seen in Raymond Queneau's *Cent Mille Milliards de poèmes* (*One Hundred Thousand Billion Poems*). This work appears at first glance to consist of 10 sonnets. However, it also includes the instruction that the reader should consider all poems that may be formed by choosing a first line from among the 10 given, then a second line, and so forth. At each stage, the reader has 10 lines from which to choose, and there are 14 lines, so this work encompasses $10^{14} = 100,000,000,000,000$ complete sonnets.

Many forms of poetry have emerged that are very consciously mathematical. The "pioem" is a poem whose words are of length determined by the digits of π in order: 3, 1, 4, 1, 5, 9, ... The number of words in a pioem is not predetermined; it may be as long or short as the author desires. The "Fib" is a poetry form that, like the haiku, prescribes the number of syllables to appear in each line. This prescription is based upon the Fibonacci sequence 1, 1, 2, 3, 5, 8, 13, . . . in which each number is the sum of the previous two numbers. Interestingly, the Fibonacci sequence not only gives a form for poetry, but also arises in the mathematical study of poetic cadence. If C_n denotes the number of

poetic cadences of length n, Indian polymath Acarya Hemacandra showed

$$C_n = C_{n-1} + C_{n-2}.$$

This equation, known as a *recurrence relation*, generates the Fibonacci sequence. Hemacandra's observance was about 50 years prior to Leonardo of Pisa's 1202 treatise *Liber Abaci*, from which the Fibonacci sequence derives its name.

Further Reading

Birken, Marcia, and Anne Coon. *Discovering Patterns in Mathematics and Poetry.* New York: Rodopi, 2008.

Filliozat, Pierra-Sylvain. "Ancient Sanskrit Mathematics: An Oral Tradition and a Written Literature." In *History of Science, History of Text.* Edited by Karine Chemla. Boston: Kluwer Academic, 2004.

Fussel, Paul. *Poetic Meter & Poetic Form.* New York: Random House, 1965.

Robson, Ernest, and Jet Wimp, eds. *Against Infinity: An Anthology of Contemporary Mathematical Poetry.* Parker Ford, PA: Primary Press, 1979.

Wolosky, Shira. *The Art of Poetry: How to Read a Poem.* New York: Oxford University Press, 2001.

CALEB EMMONS

See Also: Literature; Permutations and Combinations; Stylometry; Vedic Mathematics.

Polygons

Category: History and Development of Curricular Concepts.
Fields of Study: Communication; Connections; Geometry.
Summary: Polygons have properties making them important in engineering, architecture, and elsewhere.

Shapes and figures define how people view the world. Polygons are special figures whose properties and relationships are prevalent in nature and are used extensively by architects, engineers, scientists, landscapers, and artists. Specifically, polygons are traditionally planar (two-dimensional) figures that are closed and comprised of line segments that do not cross. These line segments are called "edges" or "sides," and the points where the edges meet are called "vertices." Planar polygons are very important in engineering, computer graphics, and analysis because they are rigid, they work well with functions, and they are easy to transform. Other types of polygons are also useful, such as spherical, hyperbolic, complex, or near polygons.

Properties of Polygons

Polygons are named by the number of their sides. Typically, polygons with more than 10 sides are called *n*-gons.

Calculating angle sums, areas, and perimeters of polygons is important in architecture, landscaping, and interior design. Understanding properties of triangles and parallelograms facilitates these kinds of calculations. For instance, the sum of the measures of the interior angles of a polygon can be determined by realizing that a polygon with *n* sides can be divided into *n* − 2 triangles, and that the sum of the measures of the interior angles of any triangle is 180 degrees. Using these ideas, a carpenter could easily determine the angles at which, for example, the sides of a hexagonal window frame should meet. Furthermore, the ability to create polygons from triangles and the ability to rearrange or

The Rich-Twinn Octagon House in Akron, New York, built in 1849.

duplicate some polygons to form parallelograms allow the derivation of area formulas. Michael Serra describes in his 2008 book, *Discovering Geometry: An Investigative Approach*, how the area of a parallelogram can be derived from a rectangle, and the area of a triangle can be derived from a parallelogram.

Real World Examples

Polygons are prevalent in the world. Even traffic signs come in the shapes of triangles, rectangles, squares, kites, and octagons. The properties of polygons make them useful in many areas including architecture, structural engineering, nature, and art.

Polygons are sometimes used in architecture for their structural benefits. Trusses formed from triangles provide support for bridges and roofs because, unlike other polygons, triangles do not tend to deform when force is exerted on a vertex. Fences are often formed into polygons because they can be built by linking together straight segments of material that are of equal size and shape. The buildings that comprise the Pentagon building in Washington, D.C., are arranged in a pentagonal shape because, according to Stephen Vogel, walking distances between buildings are less than in a rectangle, straight sides are easier to build, and the symmetrical shape is appealing. In the 1850s, Orsen Fowler popularized octagonal-shaped houses because octagons have larger areas than rectangles with the same perimeter. Thus, octagonal houses provided maximal living space while keeping heating, cooling, and building costs similar to that of the smaller rectangular house with the same outer wall space.

Properties of quadrilaterals and triangles facilitate the creation of squares and right angles. For example, using the properties of a square's diagonals, an approximate baseball diamond could be constructed by cutting diagonals of equal length from string or rope. To form the square, the diagonals would be positioned to bisect (halve) each other at right angles. The ends of each string would then mark the square's four corners. The same format could be used to create a rectangular play area, except the diagonals would not be perpendicular. According to Sidney Kolpas, although unaware of the Pythagorean theorem, ancient Egyptians used right triangles to reconstruct property boundaries after the annual flooding of the Nile River. To create a 90 degree angle, Egyptians would create a 3-4-5 right triangle by tying 13 equally spaced knots in a rope, placing stakes at knots 4 and 8, then drawing the ends of the rope at knots 1 and 13 to meet.

Polygons are prevalent in nature. Mineral crystals often have faces that are triangular, square, or hexagonal. The cross section of the Starfruit is shaped like a pentagonal star. Katrena Wells describes practical applications of hexagons, such as the often hexagonal shape of snowflakes and the hexagonal markings on many turtles' backs.

Tessellations of polygons are arrangements of polygons on a plane with no gaps or overlaps. These are also seen frequently in nature. Marvin Harrell and Linda Fosnaugh discuss many examples, including the facts that bees use a hexagonal tessellation for their honeycomb, some plant cell structures form hexagonal tessellations, and cooling lava may have formed the tessellating hexagonal columns of basalt rock at the Giant's Causeway in Ireland. Interestingly, a giraffe's skin is covered with a tessellation of various approximate polygons.

When creating sketches of objects or animals, artists often use polygons as the basis of their work by breaking the figure down into polygons and circles, then smoothing and filling in the details of the drawing after the rough polygonal sketch is created. Michael Serra explains how artist M.C. Escher used tessellations of triangles, squares, and hexagons as a framework, then rotated or translated various drawings along the sides of each polygon in the tessellation to create marvelous patterns of reptiles, birds, and fish. Islamic artists covered their buildings with ornate tessellations of polygons. A prime example is the Alhambra Palace in Grenada, Spain.

Investigating polygons as they exist in the world is one method of introducing geometry and instilling a value of geometry to people of all ages. Examining polygons with hands-on learning activities and real-world examples provides students with opportunities to investigate the characteristics and properties among polynomial shapes and helps them grasp an understanding of geometry at a higher level.

Development of Polygons

Planar polygons have been important since ancient times. Up until the seventeenth century, polygons that inscribed and circumscribed a circle were used by Archimedes and many others to estimate values of π. In 1796, at the age of 19, Carl Friedrich Gauss constructed a 17-sided polygon using a compass and straight edge. A year earlier, he had described the area of a polygon, which is often referred to as the "Surveyor's formula," although this concept also is attributed to A. L. F. Meister in 1769. The concept of a tiling or tessellation also requires polygons, and these have a long history of represention in art, weaving, architecture, and mathematics. Johannes Kepler studied the coverings of a plane with regular polygons, and in 1891, crystallographer E. S. "Yevgraf" Fedorov proved that there are 17 different types of symmetries that can be used to tile the plane. Planar polygons also star as main characters in Edwin Abbott's 1884 novel *Flatland* and the subsequent twenty-first-century movies. In the early twenty-first century, young children investigate the mathematical properties of planar polygons in primary school.

Other types of polygons are also interesting and useful. Non-convex polygons like a star polygon, where line segments connecting pairs of points no longer have to remain inside the polygon, were studied systematically by Thomas Bredwardine in the fourteenth century. Generalized polygons in the twentieth century include complex polygons investigated by Geoffrey Shephard and H. S. M "Donald" Coxeter; Moufang polygons, named after Ruth Moufang; and near polygons. In 1797, Norwegian surveyor Caspar Wessel explored planar and spherical polygons in his theoretical investigation of geodesy. M. C. Escher represented hyperbolic polygons in his tessellated artwork. Some twenty-first-century college geometry texts contain spherical and hyperbolic polygons.

Further Reading

Bass, Laurie E., Basia R. Hall, Art Johnson, and Dorothy F. Wood. *Geometry: Tools for a Changing World*. Upper Saddle River, NJ: Prentice Hall, 1998.

Botsch, Mario, Leif Kobbelt, Mark Pauly, Pierre Alliez, and Bruno Levy. *Polygon Mesh Processing*. Natick, MA: A K Peters, 2010.

Cohen, Marina. *Polygons*. New York: Crabtree Publishing, 2010.

Fowler, Orson. *The Octagon House: A Home for All*. New York: Dover Publications, 1973.

Guttmann, A. J. *Polygons, Polyominoes and Polycubes*. Berlin: Springer, 2009.

Harrell, Marvin E., and Linda S. Fosnaugh. "Allium to Zircon: Mathematics." *Mathematics Teaching in the Middle School* 2, no. 6 (1997).

Icon Group International. *Polygons: Webster's Timeline History, 260 B.C.–2007*. San Diego, CA: ICON Group International, 2009.

Kolpas, Sidney J. *The Pythagorean Theorem: Eight Classic Proofs*. Palo Alto, CA: Dale Seymour, 1992.

Serra, Michael. *Discovering Geometry an Investigative Approach*. Emeryville, CA: Key Curriculum, 2008.

van Maldeghem, Hendrik. *Generalized Polygons*. Basel, Switzerland: Birkhäuser, 1998.

Vogel, Stephen. "How the Pentagon Got Its Shape." May 2007. http://www.washingtonpost.com/wp-dyn/con tent/article/2007/05/23/AR2007052301296_4.html.

Wells, Katrena. "Hexagons in Nature—6-Sided Shapes Made Fun: Teach Practical Application of Natural Beauty of the Hexagon." http://primaryschool .suite101.com/article.cfm/hexagons-in-nature --6-sided-shapes-made-fun.

LINDA REICHWEIN ZIENTEK
BETH CORY

See Also: Archimedes; Bees; Pi; Polyhedra; Ruler and Compass Constructions.

Polyhedra

Category: History and Development of Curricular Concepts.
Field of Study: Communication; Connections; Geometry.
Summary: Regular solid shapes play important roles in nature and geometry.

People frequently encounter objects in polyhedral shapes, such as buildings that have cubic or prismatic shapes and geodesic domes or dice that are shaped like polyhedra. This prevalence is partly because of their aesthetic appeal and partly because of their practical properties. Polyhedra also appear in nature; many crystals have the shapes of regular solids, particularly of tetrahedron, cube, and octahedron, and virus capsids can be icosahedral. Furthermore, carbon atoms can form a type of molecule known as "fullerenes," which are in the form of a triangulated truncated icosahedron. A polyhedron is a solid in space with polygonal faces that are joined along their edges. If the faces consist of regular polygons, then it is called a "regular polyhedron." A polyhedron is convex if the line segment joining any two points lies on or inside it. Regular convex polyhedra are particularly important for their aesthetic value, symmetry, and simplicity. There are only five of them: the tetrahedron, cube or hexahedron, octahedron, dodecahedron, and icosahedron. Beginning in primary school, students investigate and classify geometric shapes, including polyhedra. In middle school and high school, students explore area and volume measurements as well as transformations and cross-sections.

History

Some of the earliest known polyhedra are the Egyptian pyramids. The five regular solids appear as decorations on Scottish Neolithic carved stone balls, which date to 2000 B.C.E. There are also examples of cuboctahedra worn by east-African women around the ankle and a variety of polyhedral earrings in medieval Europe. The Greeks are thought to have first studied the mathematical properties of regular solids, particularly the Platonic solids, named for Plato. The last book of Euclid of Alexandria's *Elements* is devoted to the study of the properties of these solids, including detailed descriptions of their construction. The book is based on the work of Theaetetus of Athens. There is some evidence that Hippasus of Metapontum may have been the first to describe the dodecahedron. Hypsicles of Alexandria inscribed regular polyhedra in a sphere in his treatise. The Platonic solids also represented physical aspects: Earth was associated with the cube, air with the octahedron, water with the icosahedron, fire with the tetrahedron, and the dodecahedron with the universe. Plato noted: "So their combinations with themselves and with each other give rise to endless complexities, which anyone who is to give a likely account of reality must survey."

The Kepler–Poinsot polyhedra are named for the 1619 work of Johannes Kepler and the 1809 work of Louis Poinsot. They constructed four regular "stellated" polyhedra. These new solids were obtained by extending the faces. In the twentieth century, Donald Coxeter classified and studied the stellation process and described many stellated polyhedra.

Figure 1. Five Platonic solids.

Properties

One common classroom investigation that relates to polyhedra is the Euler characteristic χ, named for Leonhard Euler. It is an equation that combines the number of vertices (V), edges (E), and faces (F) of a polyhedron as $\gamma = V - E + F$. All convex polyhedra have the same Euler characteristic: 2. René Descartes discovered the polyhedral formula in 1635, and Euler discovered it in 1752. In the nineteenth century, Ludwig Schläfli generalized the formula to polytopes and Henri Poincaré proved the result.

The shape of a polyhedron lends itself to a very convenient symbolic or combinatorial description, called the "Schläfli symbol" of the polyhedron. Let $\{n, p\}$ represent a regular polygon with n-gon faces, p of them meeting at each vertex. For example $\{4, 3\}$ would represent a cube because three squares meet at each vertex This symbolic representation is particularly useful if one would like to express various quantities like the dihedral angle, angular deficiency, radii of inscribed and circumscribed spheres, and surface area. For instance, the surface area of a Platonic solid $\{n, p\}$ can be expressed by

$$S = nF\left(\frac{a^2}{2}\right)\cot\left(\frac{\pi}{n}\right)$$

where F is the number of faces and a is the side length.

Mathematically, polyhedra are very appealing for their fine properties such as duality, symmetry, and versatile constructability. The dual of a polyhedron is constructed by taking the vertices of the dual to be the centers of the faces of the original figure by interchanging faces and vertices. For instance, the dodecahedron and the icosahedron are duals. Many polyhedra are highly symmetrical, and in the nineteenth century, Felix Klein investigated them. The groups of symmetries are algebraic structures consisting of reflections and rotations. One can also generate new polyhedra from old by truncating the vertices of polyhedra, a process known and studied since antiquity. Some of the truncated polyhedra are also known as the "Archimedean solids," named for Archimedes of Alexandria, whose faces consist of two or more types of regular polygons.

There are 13 Archimedean solids, and there are 53 other semiregular, non-convex polyhedra, which are non-Archimedean. The collection of all Platonic, Kepler–Poinsot, Archimedean, and semiregular, non-convex polyhedra together with prisms form the family of polyhedra called "uniform polyhedra."

Non-Euclidean polyhedra took on a prominent role in some theories of a spherical dodecahedral universe at the beginning of the twenty-first century. There are also non-Euclidean polyhedra with no flat equivalents. For instance, a spherical hosohedron with Schläfli symbol $\{2, n\}$ is shaped like a segmented orange or beach ball with lune faces. The name "hosohedron" is attributed to Coxeter.

There have been many artistic and physical models of polyhedra in mathematics classrooms. With the advent of perspective, polyhedra were easier to draw and mathematicians and artists designed and collected polyhedral models. Albrecht Dürer introduced polyhedral nets in his 1525 book. Students continue to use nets to build models. In 1966, Magnus Wenninger published a work on polyhedral models for the classroom through the National Council of Teachers of Mathematics. Wenninger noted that the popularity of the book reflected the continued interest in polyhedra. In the twenty-first century, origami polyhedra have also become important in mathematics and computer science classrooms.

Further Reading

Artmann, Benno. "Symmetry Through the Ages: Highlights From the History of Regular Polyhedra." In *Eves' Circles*. Edited by Joby Anthony. Washington, DC: Mathematical Association of America, 1994.

Cromwell, Peter. *Polyhedra*. New York: Cambridge University Press, 1997.

Demaine, Erik, and Joseph O'Rourke. *Geometric Folding Algorithms: Linkages, Origami, Polyhedra.* New York: Cambridge University Press, 2007.

Gabriel, Francois. *Beyond the Cube: The Architecture of Space Frames and Polyhedra.* Hoboken, NJ: Wiley, 1997.

Dogan Comez
Sarah J. Greenwald
Jill E. Thomley

See Also: Crystallography; Greek Mathematics; Polygons; Symmetry; Transformations.

Polynomials

Category: History and Development of Curricular Concepts.
Fields of Study: Algebra; Communication; Connections.
Summary: Polynomial functions have long been studied by mathematicians and have interesting and important applications.

Polynomials have a broad array of theoretical and real-world applications and are widely used by mathematicians, scientists, and engineers to mathematically model data and explore many mathematical and scientific concepts. Technologies that transmit electronic signals, ranging from deep space probes communicating with Earth, to home DVD players, commonly use polynomial error-correcting codes, like the Reed–Solomon codes, named for mathematicians Irving Reed and Gustave Solomon. Cryptographic algorithms that help ensure secure data transmission also rely on polynomials to represent and manipulate data. Calculators may use approximations called "Taylor polynomials," named for mathematician Brook Taylor, for functions like square roots. Civil engineers model and estimate properties, such as volume for lakes and other irregular natural features, with polynomials. Orthogonal polynomials provide the foundation for many multivariate statistical procedures. In twenty-first-century classrooms, polynomials are typically part of advanced middle school or high school curriculums, though linear functions and comparisons of linearity versus nonlinearity are common in middle school, and some of the basic concepts of functions are introduced in the elementary grades.

Early in their mathematical studies, students learn that the graph of the squaring function is a parabola, and that the plot of $y = p(x) = x^2$ is shown in Figure 1, which is the first natural function to consider beyond ones that generate straight lines.

There is an entire family of functions like the squaring function, the cubing function, the fourth power function, and more. If indexed, one could call

$$\text{the squaring function } p_2(x) = x^2,$$

$$\text{the cubing function } p_3(x) = x^3,$$

$$\text{the fourth power function } p_4(x) = x^4,$$

and, in general, the nth power function $p_n(x) = x^n$.

The family of power functions also includes the zero power function $p_0(x) = 1$ and the first $p_1(x) = x^1$. These power functions are the building blocks of "polynomial functions," functions that are made from taking sums and constant multiples of power functions. As such, these functions are especially simple because

Figure 1.

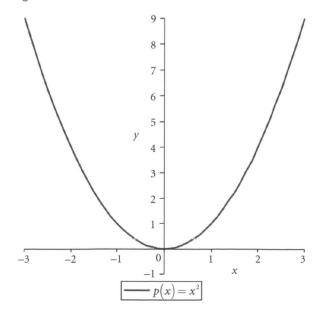

$p(x) = x^2$

Figure 2.

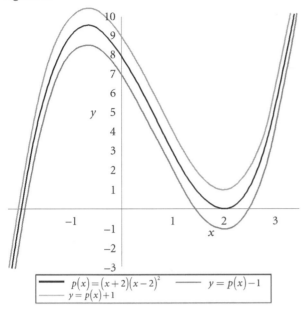

$p(x) = (x+2)(x-2)^2$ $y = p(x) - 1$
$y = p(x) + 1$

their formulas only involve addition and multiplication. The first understanding of these power functions is generally credited to Abu Bekr ibn Muhammad ibn al-Husayn Al-Karaji, who lived c. 1000 C.E. in what is now Iraq. In particular, he made advances in the use of variables and humankind's ability to think of arithmetic operations on "placeholders," instead of simply on individual numbers.

Finding the Zeros
Consider this example, $p(x) = x^3 - 2x^2 - 4x + 8$: this function is obtained by taking the cubing function, subtracting twice the squaring function, subtracting 4 times the first power function, and finally adding 8. Regardless of the power functions chosen and the constants multiply by, a polynomial is built. That is, polynomials are functions that have the form

$$p(x) = a_n x^n + \cdots + a_2 x^2 + a_1 x + a_0$$

where $a_0, a_1, \ldots, a_n$ are real numbers. Provided that a_n is not zero, it is stated that p is a degree n polynomial; the degree represents the highest power of x that is present. Much of the modern notational perspective on these functions is due to the work of René Descartes, who in the early 1600s did important work that popular-

ized not only the notation above using subscripts and superscripts but also offered a visual perspective on polynomial functions through their graphs.

Going back to the first polynomial example, $p(x) = x^3 - 2x^2 - 4x + 8$, one can rewrite this sum of multiples of power functions in the formula as a product of even simpler functions. Specifically, it is possible to show that

$$p(x) = x^3 - 2x^2 - 4x + 8 = (x+2)(x-2)(x-2).$$

One can easily observe that $p(-2) = 0$ and $p(2) = 0$. Mathematicians call -2 and 2 the *zeros* or *roots* of $p(x)$; since the $(x-2)$ factor, which leads to the zero 2, appears twice, mathematicians say that "2 is a double root" or "2 is a zero of multiplicity two." The graph of the polynomial in Figure 2 is also enlightening as it shows that the zeros of the function lie where the function crosses or touches the horizontal axis:

If one shifts the graph of the degree 3 polynomial $p(x)$ (in black) slightly up, the new graph (top line in light gray) will have just one real zero, while if one shifts the graph slightly down, the new function (bottom line in medium gray) will have three distinct real zeros. This illustration demonstrates an important fact about degree 3 polynomials: every degree 3 polynomial has 1, 2, or 3 distinct real zeros. Indeed, the Fundamental Theorem of Algebra, which was proved in its earliest form in 1799 by the great mathematician Carl Friedrich Gauss, states that every polynomial of degree n has at most n distinct real zeros.

If one is willing to permit zeros to be complex numbers and count zeros by their multiplicity, a much stronger version of the Fundamental Theorem of Algebra (which was also known to Gauss) can be proved: every polynomial of degree n has exactly n zeros, provided one counts them according to their multiplicity and allows zeros to be complex. The Fundamental Theorem of Algebra asserts only that n roots of a polynomial function of degree n exist; it does not tell what those roots are.

Quadratic, Cubic, and Quartic Formulas
The search for the zeros of polynomial functions attracted many great minds. The quadratic formula, which calculates the zeros of any degree 2 polynomial, was understood in certain forms by Babylonian mathematicians as early as 2000 B.C.E. The quadratic

formula asserts that in order for $ax^2 + bx + c = 0$, it must be the case that

$$x = \frac{-b \pm \sqrt{b^2 - 4ac}}{2a} \ .$$

For cubic equations and their roots—finding where a polynomial of degree 3 is zero—it took another 3500 years for mathematicians to fully understand the situation. Following contributions from ancient Greeks, Indians, and Babylonians, as well as Persians in the eleventh and twelfth centuries, a group of Italian mathematicians in the 1500s (Scipione del Ferro, Niccolo Tartaglia, and Gerolamo Cardano) proved that there is a cubic formula. In other words, based on the coefficients of a degree 3 polynomial, there is a very complicated formula involving cube roots that calculates the locations of the polynomial's zeros.

Mathematicians were able to take these discoveries a step further. Near the mid-1500s, Ludovico Ferrari found a way to solve quartic equations. This quartic formula is incredibly complicated and represents a major feat in the understanding of polynomial functions. Interestingly, these general formulas cease to exist beyond polynomials of degree 4. In 1824, Neils Abel and Paolo Ruffini published a theorem, based on the work of Evariste Galois, proving that there was no general formula for the roots of a degree 5 polynomial or higher. This latter work on polynomials ended up founding an entire new branch of mathematics called *modern algebra*. Sometimes in mathematics, the quest to solve one problem leads to a whole host of other interesting problems or even a new collection of coherent ideas.

Applications

Polynomial functions demonstrate all sorts of interesting patterns and properties and have long been studied because they are interesting in their own right. But even more than this, polynomials play important roles in other areas of mathematics and in applications. For example, polynomial functions spawned the subject of modern algebra, and key ideas in modern algebra are used in the field of public key cryptography—the science of keeping important information private in such essential settings as Internet commerce.

A more direct application of polynomial functions comes in the design of fonts that appear on computer screens. So-called Bezier curves, named for mathema-

tician Pierre Bezier, are degree 3 polynomial functions that can be easily spliced together to form elegant shapes. For instance, at right is the letter S in the Palatino font.

Each piece of the S—the portion of the curve between consecutive squares that represent points on the curve—consists of a degree 3 parametric polynomial. There is deep and elegant mathematics behind why Bezier curves work so well and why they are particularly suited to computer graphics. This is just one example of how substantial ideas and applications in mathematics often emerge from simple beginnings.

Polynomial functions are the simplest of all functions, can be used to approximate more complicated functions that are not polynomials, and often emerge in important applications. They are indeed some of the key building blocks of mathematics.

Further Reading

Barbeau, Edward. *Polynomials*. New York: Springer, 1989.

Kalman, Dan. *Polynomia and Related Realms*. Washington, DC: Mathematical Association of America, 2009.

Kushilevitz, Eyal. *Some Applications of Polynomials for the Design of Cryptographic Protocols*. Berlin: Springer-Verlag, 2002.

Strogatz, Steven. "Power Tools." *New York Times*. http://opinionator.blogs.nytimes.com/2010/03/28/power-tools/.

Matt Boelkins

See Also: Coding and Encryption; Exponentials and Logarithms; Functions.

Popular Music

Category: Arts, Music, and Entertainment.
Fields of Study: Algebra; Measurement; Number and Operations; Representations.

Summary: Popular music can be analyzed and enhanced by mathematical techniques and to some degree the popularity of music can be predicted mathematically.

The interaction between mathematics and popular music goes far beyond the popularity of numbers in song titles, like Tennessee Ernie Ford's "16 Tons" or 2gether's "U + Me = Us (Calculus)." Mathematics is fundamental to musical theory and composition. The twentieth-century subgenres math rock and mathcore are perhaps the most explicitly mathematical compositions, but there are also songs about mathematics concepts. These are usually intended to be humorous or educational, such as "That's Mathematics" by mathematician and musician Thomas Lehrer. Mathematics is also increasingly important to recording and analyzing popular music, including its potential effects on learning. Experimental electronic artist Jamal Moss, founder of record label Mathematics, notes: "Mathematics is the body of sound knowledge centered on such concepts as quantity, music, structure, space, and change—and also the academic discipline that studies them."

Popular Artists

Mathematics in popular music reflects society's often polarized opinions on mathematics. For example, Jimmy Buffet's song "Math Suks" expressed the singer's feelings about the difficulty of mathematical concepts like fractions, algebra, and geometry. Other singers and groups embrace mathematics, like the Texas indie rock band named "I Love Math." Mathematics is often found in album cover art. British band Coldplay's 2005 *X & Y* album featured a cover with colored blocks that spell out "X and Y" in the binary code developed in 1870 by Emile Baudot for use with telegraph systems. Coldplay's lead guitarist Jonny Buckland studied astronomy and mathematics at University College London. Some artists have been criticized for incorrectly using mathematics. Pink Floyd's very popular 1973 album *Dark Side of the Moon* features cover art showing a prism and spectrum. It is correct in depicting some facts, like violet light refracting the most and red the least, but some other aspects are not accurate, such as the relative dispersion of the different colors. Mariah Carey's 2009 album $E = MC^2$, borrowed from Albert Einstein's well-known theory of relativity.

Mathematical Subgenres of Popular Music

Avant-garde composer Iannis Xenakis and post-rock subgenres math rock and mathcore are prominent examples of popular music that relies heavily on mathematics. Xenakis was one of the most significant avant-garde composers of the twentieth century and a grandfather of modern electronic music. His work incorporated mathematical models, such as probability theory, stochastic processes, group theory, set theory, game theory, and Markov chains. He developed algorithms to produce computer-generated music using probability theory and stochastic functions in the 1960s. In his 1966 cello solo "Nomos Alpha," he divided the 24 sections of the piece into two layers. The first layer, consisting of every section not divisible by four, is determined by the 24 orientation-preserving elements of the octahedral group, while the second layer is a more traditional structure. The work has been compared to a musical kaleidoscope, and its structure likened to a fractal.

In the 1990s, post-rock like Slint's *Spiderland* became a dominant genre in experimental rock. Critic Simon Reynolds coined the term "math rock" to describe music that "uses rock instrumentation for non-rock purposes, using guitars as facilitators of timbre and textures rather than riffs and power chords." Math rock bands began to explore the use of dramatically alternating dynamic shifts and unusual time signatures and dissonance, and songs tend to avoid the verse-chorus-verse structure of pop songs. Mathcore developed largely independently of math rock, growing out of hardcore punk and extreme metal, with a huge debt to hardcore pioneers Black Flag.

Mathematics Songs

As of 2010, the Web site M A S S I V E: Math And Science Song Information, Viewable Everywhere is part of the National Science Digital Library and contains over 2,800 mathematical and scientific songs. Popular YouTube songs include mathematical raps and parodies, like "I Will Derive." Hard 'n Phirm's song "Π" rose in popularity because of the 2005 music video by award winning director Keith Schofield. Some songs help students learn mathematics concepts, like multiplication. Other songs showcase the mathematicians who love to sing. The Klein Four Group is a Northwestern University a cappella group who sing about undergraduate and graduate level mathematics. They are most known for their song "Finite Simple Group (of Order Two)."

Self-proclaimed "mathemusician" Lawrence Lesser writes educational songs in order to increase mathematics awareness. Educators often incorporate mathematics songs into their classrooms to enhance student learning of specific concepts and many students use music of various kinds to help them focus while they study mathematical concepts, but these effects are not yet definitively supported or refuted. One study that investigated using jingles to teach statistics concepts found that students who sung several jingles versus reading aloud definitions for the same concepts performed better as a group on a follow-up test. On the other hand, a study that compared classical, popular, and no music to enhance learning found that the students in the three groups performed no differently on a mathematics placement test. This matched findings regarding the effect of music on other academic areas.

Audio Processing

While music production techniques have always allowed a certain amount of alteration and error correction by adjusting the relative levels and balance of the recorded elements, twenty-first century software capabilities have progressed to the point where lower-quality vocals can be processed to professional-sounding quality.

The software package most associated with this is Auto-Tune, released in 1997, and developed by Exxon engineer Harold "Dr. Andy" Hildebrand, who applied seismic data interpretation methods to the analysis and modification of musical pitch. Auto-Tune is an enhancement of existing phase vocoder technology, which uses short-time Fourier transforms, named after mathematician Jean Fourier, to convert time domain representations of sound into time-frequency representations that can be modified before being converted back. Extreme changes can leave tell-tale artifacts in recordings, in the form of a warble like a degenerating audiocassette tape. Audio processing has become standard in many pop albums and on television shows, such as *Glee*. Some well-established singers regularly use Auto-Tune for both albums and in live performances. Other musicians have refused to do so out of fear that it will change the sound enough to make them unrecognizable.

Predicting Popular Song Success

In 2010, Platinum Blue and Music Intelligence Solutions specialize in mathematically predicting hit songs, while services like iTunes and Music IP create suggested playlists or make recommendations. Platinum Blue CEO Mike McCready explained that he and others dis-

covered mathematical patterns in hit songs while trying to build an automated recommendation platform. The algorithm his company uses is based on roughly 30 song traits that are quantified mathematically, such as melody, harmony, beat, tempo, and rhythm. These traits are analyzed for patterns, resulting in groups of songs that are ranked according to probability of success. Hit songs tend to have identifiable similarities, but falling into a particular category is not a guarantee of success. For example, lyrics are an influential song component that are not reliably quantifiable, and aggressive marketing can have an effect not captured by the algorithm. McCready noted: "We figured out that having these optimal mathematical patterns seemed to be a necessary, but not sufficient, condition for having a hit song."

Further Reading

Crowther, Greg, and Wendy Silk. "M A S S I V E: Math And Science Song Information, Viewable Everywhere." *National Science Digital Library*. http://www.science groove.org/MASSIVE/.

Lesser, Lawrence. "Sum of Songs: Making Mathematics Less Monotone!" *Mathematics Teacher* 93, no. 5 (2000).

VanVoorhis, Carmen. "Stat Jingles: To Sing or Not to Sing." *Teaching of Psychology* 29, no. 3 (2002).

Waldman, Harry. "Tom Lehrer: Mathematician and Musician." *Math Horizons* 4 (April 1997).

Xenakis, Iannis. *Formalized Music: Thought and Mathematics in Composition*. 2nd ed. Hillsdale, NY: Pendragon Press, 2001.

BILL KTE'PI

See Also: Composing; Geometry of Music; Harmonics; Pythagorean and Fibonacci Tuning; Scales.

Predator–Prey Models

Category: Weather, Nature, and Environment.
Fields of Study: Algebra; Data Analysis and Probability; Number and Operations.
Summary: The interaction between the population sizes of a predator species and a prey species can be modeled using systems of equations.

Predator–prey models are systems of mathematical equations that are used to predict the populations of interacting species, one of which—the prey—is the primary food source for the other—the predator. One famous example that has been extensively studied is the relationship between the wolves and moose on Isle Royale in Lake Superior.

The Isle Royale populations are well suited for modeling the predator–prey relationship because there is little food for the wolves other than the moose and there are no other predators for the moose. In addition, the geographic isolation limits other factors that would complicate the mathematics in the equations, such as hunting or migration. This predator–prey interaction has been carefully studied since the 1950s and continues to be investigated into the twenty-first century.

Modeling Predator–Prey Populations

Most predator–prey models are composed of two equations, the first representing the change in the prey population, and the second the change in the predator population. Each equation has the following form: birth function minus death function.

If $X(t)$ represents the quantity of prey at time t, and $Y(t)$ represents the quantity of predators at time t, then the instantaneous rate of change in prey is

$$\frac{dX}{dt} = f_1 - f_2$$

and the instantaneous rate of change in predators is

$$\frac{dY}{dt} = f_3 - f_4$$

where f_1 is the mathematical term that describes the births in the prey population, f_2 describes the deaths in the prey population, f_3 describes the births in the predator population, and f_4 describes the deaths in the predator population.

There have been many predator–prey models proposed since the beginning of the twentieth century. The most famous and the earliest known is the Lotka–Volterra system, named for the two scientists who developed the same mathematical model independently, American Alfred Lotka (1880–1949) publishing the equations in 1925 and Italian Vito Volterra (1860–1940) publishing them in 1926. Lotka had degrees in physics

and chemistry, and he believed that one could apply physical principles to biological systems. His work on predator–prey interactions is just part of extensive work he published in 1925 in the text titled *Elements of Physical Biology*. Lotka used a chemical reaction analogy to justify the terms in the model.

In the absence of predators, the prey should increase at a rate proportional to the current quantity of prey, X. In other words, more moose around to mate without being hunted means more calves would be born. Likewise, in the absence of prey, the predators should die off at a rate proportional to the current predator population, Y. In other words, with many wolves and no moose for food, more wolves would starve.

Lotka used a chemical reaction analogy to explain prey deaths and predator births: when a reaction occurs by mixing chemicals, the rate of the reaction is proportional to the product of the quantities of the reactants. Lotka argued that prey should decrease and predators should increase at rates proportional to the product of the quantity of prey and predators, XY. In other words, the moose deaths should be closely related to the rate of interaction of wolves and moose, and the wolf births should be as well because wolves need the moose for food to be healthy and have pups. The equations can be written as

$$\frac{dX}{dt} = aX - bXY \text{ and } \frac{dY}{dt} = cXY - dY$$

for non-negative proportionality constants a, b, c, and d.

Volterra arrived at the same model using different reasoning. Volterra was a physicist whose daughter and son-in-law were biologists. While looking for a mathematical explanation for a problem his son-in-law was working on, Volterra became very interested in interactions of species and spent the rest of his professional life looking for a mathematical theory of evolution.

The Lotka–Volterra predator–prey model can be solved without a computer and yields a graph that makes sense. The population of the predator oscillates as does that of the prey, with the predator population trailing slightly behind. Too many prey results in more predators, who swamp the prey causing a decrease in prey. As the prey become scarce, the predators also start to die out, and the cycle begins again (see Figure 1).

Figure 1. Predator–prey interaction.

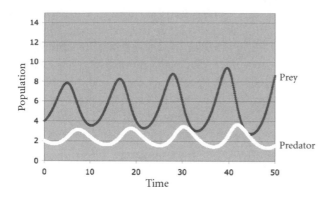

While this result has reasonable qualitative behavior, many scientists have objected to the equations in this form. Some of the concerns about the model have included the following:

- If there are no predators, the prey population would grow arbitrarily large
- A reduction in the number of prey should cause more predator deaths rather than fewer predator births
- For a fixed number of predators, the number of prey eaten is proportional to the number of prey present, implying that predators are always hungry and eat the same proportion of the prey no matter how large the number of prey gets
- The food for the prey plays a role in the births and deaths of the prey, and should be included in the model
- No spatial considerations are incorporated in the model, so factors such as migration or seeking safety in herds are ignored
- These equations do not take into account gestation periods and seasonal changes in birth rates
- The constants a, b, c, and d are difficult to estimate for a given situation without a large amount of data collected from field observations

Much work has been done since the 1930s to modify the equations to address these concerns and to apply the equations to data from specific situations, such as the moose and wolves of Isle Royale. In the twenty-first century, scientists use sophisticated computer models to model predator–prey interactions using increasingly intricate equations to incorporate more realistic assumptions in the mathematics.

Further Reading

Kingsland, Sharon E. *Modeling Nature, Episodes in the History of Population Ecology*. Chicago: University of Chicago Press, 1995.

Lotka, Alfred J. *Elements of Physical Biology*. Baltimore, MD: Williams and Wilkins Publishers, 1925.

Volterra, Vito. "Variations and Fluctuations of the Number of Individuals in Animal Species Living Together." In *Animal Ecology*. Edited by R. Chapman. New York: McGraw-Hill, 1926.

Vucetich, John A. "The Wolves and Moose of Isle Royale." http://www.isleroyalewolf.org/wolfhome/home.html.

Holly Hirst

See Also: Animals; Fertility; Social Networks.

Predicting Attacks

Category: Government, Politics, and History.
Fields of Study: Algebra; Data Analysis and Probability.
Summary: Predictive mathematical models can be used to attempt to foresee and counter various types of attacks.

An increasing area of interest in mathematics is the use of algorithms and computer models to predict attacks—military attacks, terrorist attacks, and even attacks on Web servers. As with meteorology, a model is a probabilistic statement; the future cannot be predicted with absolute certainty but probable causes, patterns, and outcomes can be quantified and mathematically modeled to extrapolate the likelihood of new events. Humankind has been trying to predict attacks ever since one group first fought another using some combination of observation and subjective judgment. However, formal prediction of attacks using mathematical methods appears to have originated only within the last two centuries and has escalated with advances in technology and data gathering.

Mathematician Lewis Richardson made contributions to many areas within and outside mathematics, such as numerical weather prediction, in the first half of the twentieth century. The Richardson iteration is one method for solving systems of linear equations, while the Richardson effect refers to the apparently infinite limit of coastline lengths as the unit of measure decreases, a precursor to the modern study of fractals. Richardson spent many years analyzing data on wars from the early nineteenth century onward, using mathematical methods such as probability theory and differential equations, often quantifying psychological variables, such as mood. He identified several patterns in war and identified some variables likely to prevent conflict. He is often credited with first introducing the notion of power laws to relate conflict size, frequency, and death toll. At the start of the twenty-first century, models had grown in complexity. In 2009, a University of Maryland team developed a model that uses 150 variables and data accumulated from the activity of 100 insurgent groups in the Middle East in order to model their reactions to Israeli activities. Other models have been developed to attempt to predict violence and attacks in Iraq and continue to be refined. Statistical methods like data mining and power law functions are prevalent in modern predictive modeling.

Data Mining

Data mining is the process of extracting patterns from large to enormous bodies of data. Isaac Asimov's *Foundation* stories, the first of which was published in 1942, depicted a future where "psychohistory" was the study of the future using the body of history as data from which to extrapolate the future. Modern data mining is quite similar to Asimov's predictions and may be accomplished by many mathematical methods. For example, many use artificial neural networks, which are computational models that mimic neuron behavior. Genetic algorithms, credited to scientist John Holland, are search heuristics inspired by the processes of gene recombination and evolution. Decision trees may be used to determine conditional probabilities. In the 1980s, support vector machines (SVMs) were developed to analyze data to find patterns for statistical classifica-

tion. All of these developments greatly advanced the state and potential of machine learning and facilitated rapid processing of increasingly larger and frequently interlinked databases from sources such as credit card companies, telecommunications businesses, and government intelligence agencies. Within the U.S. government, the Department of Defense began using data mining in the late 1990s in its Able Danger program, which gathered counterterrorism data, including data about the Al Qaeda terrorist group. Some asserted that the program uncovered the names of four of the alleged September 11, 2001, hijackers a year before the attacks. In February 2002, the U.S. Office of Science and Technology Policy convened a panel of government and industry leaders to discuss data mining as a counterterrorism tool. While it is now widely used, some criticize it because the sparsity of some information and the relative infrequency of terrorist attacks make identifying statistically significant patterns, which are critical to finding the anomalies that signal an attack, prone to unacceptable levels of error.

Cyber Security

Mathematicians, computer scientists, and others are continually working on new methods to predict and counter attacks on Web servers, e-mail, and digital records of all kinds. The Internet is filled with malicious activity, from phishing and identity theft to distributed denial of service attacks. Electronic attacks are facilitated by the same computer technology that is used to predict attacks. The traditional guard has been to block a source of malice after the attack, by e-mail as spam or blocking an IP address after harmful activity originates from it. These methods are commonly known as *blacklists* and are now widely compiled and shared. However, they are by definition reactive measures to attacks. Just as e-mail spam filters have become preemptive, marking mail as "spam" automatically based on a number of factors, IP-blocking can also be conducted preemptively.

The method of predictive blacklisting uses shared attack logs as the basis for a predictive system, like the customer recommendation systems employed by Amazon or Netflix. Computer scientists Fabio Soldo, Anh Le, and Athina Markopoulou developed what is known as an "implicit recommendation system"—implicit because ratings are inferred rather than given directly by the subjects of the model. Their multilevel prediction model uses mathematical methods, such as time series analysis and neighborhood models, adjusted specifically for attack forecasting. Inputs to the model include factors such as attacker-victim history and interactions between pairs or groups of attackers and victims. Similar models—using different types of data—can be built to predict terrorist attacks and the behavior of enemy forces, and such models are included in the standard order of battle intelligence reports used by the U.S. Army.

The data needed to predict attacks are not restricted to private databases. Information is widely available from the Internet or the scrolling news banners of 24-hour news networks. Neil Johnson used a variety of sources to investigate insurgent wars, employing some of the same mathematical techniques as Richardson in his analyses and modeling. After gathering and analyzing data for almost 60,000 insurgent attacks occurring in multiple conflicts around the world, he and his collaborators discovered similarities between the frequency and intensity of attacks in all conflicts. Further, they found that the statistical distribution for insurgency attacks was significantly different from the distribution of attacks in traditional war. The model quantifies connection between insurgency, global terrorism, and ecology, and counters the common theory of rigid hierarchies and networks in insurgencies. Johnson notes:

Despite the many different discussions of various wars, different historical features, tribes, geography and cause, we find that the way humans fight modern (present and probably future) wars is the same, just like traffic patterns in Tokyo, London, and Miami are pretty much the same.

Further Reading

Jakobsson, Markus, and Zulfikar Ramzan. *Crimeware: Understanding New Attacks and Defenses.* Boston: Addison-Wesley, 2008.

Memon, Nasrullah, Jonathan Farley, David Hicks, and Torben Rosenorn. *Mathematical Methods in Counterterrorism.* New York: Springer, 2009.

BILL KTE'PI

See Also: Intelligence and Counterintelligence; Predicting Preferences; Spam Filters; Vietnam War.

Predicting Divorce

Category: Friendship, Romance, and Religion.
Fields of Study: Algebra; Communication: Data Analysis and Probability.
Summary: Statistical data analysis and mathematical models can be used to predict the likelihood of divorce.

There is a common misconception that one out of every two marriages ends in divorce. The 50% number comes from dividing the number of divorces in a given year (about 1.3 million) by the number of marriages in that same year (about 2.6 million). The mistake is failing to realize that, in any given year, the people getting divorced are probably not the same as those getting married, because the average length of a marriage before a divorce is about eight years (the overall length of marriage, on average, is about 24 years). Hence, those getting married in any given year have an eight-year lag in their projections for divorce. This lag means that the numerator and denominator of the above ratio are not comparable. Instead, experts suggest that about two out of every five marriages end in divorce (or about 40%).

Because of the propensity for some to remain married, for some to divorce more than once, and for some to never marry, only about one out of every five people are predicted to experience a divorce in their lifetime. However, these figures mask the distribution of divorce rates by category—40% of all first marriages end in divorce, 60% of second marriages end in divorce, and 73% of all third marriages end in divorce. There are also some differences by age group, with divorce rates highest for those in their early 20s and declines steadily in subsequent age groups.

There are two main ways to predict divorce: empirical (or statistical) methods that take advantage of data gathered on married and divorced couples; and mathematical models that try to make a priori predictions of future divorce using features of existing marriages or theoretical assumptions based on extensive work in the area.

Empirical Methodology

Empirical work suggests that indicators predicting divorce can be separated into two groups: factors present before marriage and factors that occur within the marriage. Some of the more common risk factors brought into a marriage include parental history of divorce (children of divorced parents are more likely to divorce), educational attainment (those with lower levels of education are more likely to divorce), and age (those who marry younger are more likely to get divorced). The risk factors that arise within the marriage include communication styles (couples with poor or destructive communication have a greater chance of divorce), finances (couples with financial problems, including a large disparity in spending habits, disposable income, and wealth goals, are at a greater risk for divorce), infidelity, commitment to the marriage (a lack of commitment or a dissimilarity in the amount of commitment often leads to divorce), and dramatic change in life events.

Mathematical Models

Mathematical models seek to discover features of current relationships that will put a couple at risk for future divorce. Professor John Gottman argues that the way couples communicate can often predict divorce. His research, which is based on analyzing hundreds of videotaped conversations between married couples, claims a 94% accuracy rate. The work also monitors pulse rates and other physiological data that, when combined with the observations, leads to what he calls the "bitterness rating." The rating is based on six signs. The first sign posits that when a conversation starts with accusations, criticisms, or negativity, the discussion is likely to end badly. However, he argues that the opposite is also true. The second sign encompasses four patterns of negative interaction that can be deleterious to a marriage: criticism, contempt, defensiveness, and stonewalling. The third sign is "flooding," in which negativity of one partner overwhelms the positive feelings of the spouse until there is virtually nothing left but discontent. The fourth sign recognizes that physiological changes, such as increases in adrenaline and blood pressure, often lead to feelings of entrapment and serve to poison an otherwise benign conversation. The fifth sign identifies the fact that some marital discord is unchanged by the repeated attempt by one partner to repair the damage done to the relationship. Finally, the sixth sign involves one or both people rewriting the history of their relationship to be largely negative. Once people reach the sixth sign, Gottman argues, divorce is likely.

Further Reading

Booth, Alan, and John N. Edwards. "Age at Marriage and Marital Instability." *Journal of Marriage and the Family* 47 (1985).

Gottman, John, and Nan Silver. *The Seven Principles for Making Marriage Work*. London: Orion, 2004.

Martin, Teresa Castro, and Larry L. Bumpass. "Recent Trends in Marital Disruption." *Demography* 26 (1989).

South, Scott, and Glenna Spitze. "Determinants of Divorce Over the Marital Life Course." *American Sociological Review* 51 (1986).

Wolfinger, Nicholas H. "Trends in the Intergenerational Transmission of Divorce." *Demography* 36 (1999).

CASEY BORCH

See Also: Mathematics, Applied; Measurement in Society; Psychological Testing.

Predicting Preferences

Category: Business, Economics, and Marketing.
Fields of Study: Data Analysis and Probability; Geometry; Measurement.
Summary: Psychology of choice and predictive models of preferences are exciting areas of mathematics blending social science, economics, and commerce.

Mathematically, preference is an ordering of alternative possibilities. It can refer to conscious choices based on ideas and beliefs, positive emotional responses or liking, or biologically mandated behaviors. Preferences are usually determined statistically: for individuals, based on multiple instances of decisions over time; and for groups, based on aggregated data of members. In 2009, the Netlix Prize contest awarded a team called BellKor's Pragmatic Chaos $1 million for their preference-predicting algorithm.

Theoretical and Behavioral Economics

Among all sciences that deal with predicting preferences, such as social psychology and education theory, the most developed mathematical apparatus can be found in economics. As any branch of mathematics, theories of economic preferences start with axiomatic assumptions. These abstract axioms do not always apply to all real situations. Economic theories that take into account psychological factors, such as cognitive limitations and emotions, are developed within an interdisciplinary area called "behavioral economics."

Most abstract theories of preference prediction assume most parts of the so-called total order, which is a group of mathematical axioms and properties from set theory. Let A, B, and C be different choices. Total order assumes that either $A \leq B$ or $B \leq A$. In real life, this assumption is a statistical statement at best: today a person can prefer apples, but might prefer bananas tomorrow. The property of transitivity says that if $A \leq B$ and $B \leq C$, then $A \leq C$. This property works in some situations; for example, if one prefers $20 over $10, and $100 over $20, it is likely the person will prefer $100 over $10. However, in complex situations with multiple choices, such as elections, transitivity fails to describe real human behavior. Experiments show that, given a choice between one pair of candidates at a time, people may prefer Beth over Alice, Carol over Beth, and Alice over Carol. One axiom of total order, called "antisymmetry," that almost never makes sense in preference theories is that if $D \leq E$ and $E \leq D$, then $E = D$. For example, when group data shows that people think diesels are worse or the same than electric cars, and electric cars are worse or the same as diesels, it does not mean that diesel cars and electric cars are the same entity. It means that people prefer them about the same. Economic theories call this situation "indifference" and use a separate symbol for it: $E \sim D$.

Another assumption frequently made in economic preference predictions comes from topology and is called "continuity." It is the assumption that if A is preferred over B, then an option that is very similar (close) to A will also be preferred over at option that is very similar to B. Many complex phenomena, including preferences, are discontinuous. They exhibit various "tipping points," near which minute differences cause radical changes in preferences. These non-continuous phenomena are studied using models from calculus or chaos theory, a branch of differential equations. One frequent example of noncontinuous preference is price near powers of 10: many people choose to buy an object that costs $999 over a similar object that costs $1,001 even though the difference in prices is minuscule compared to the total. Behavioral economics explains this by cognitive limitations: people see 1001 as thousands

Psychology of Choices

Statistical analysis of real situations, such as elections, as well as results of experiments and questionnaires, allow scientists to aggregate increasingly sophisticated knowledge of human mechanisms of choice and preference. For example, from the purely mathematical viewpoint, gaining an amount and avoiding loss of the same amount are equivalent. However, most people regret loss more strongly than they regret missed opportunity—a fact extensively used in advertisements of savings and discounts.

Preferences are very strongly influenced by power over the situation. Most people accept much higher risks for given gains if they enter the situation of their free will, compared to risks of mandated behaviors. This phenomenon comes up, for example, when mandatory immunizations are proposed—the fact that people would not have a choice makes very small risks unacceptable.

and 999 as hundreds, which is technically correct but makes less of a difference in this case than intuition leads one to believe.

Paradoxical Preferences
A paradox is a false or contradictory statement that logically follows from a set of true statements. Preference prediction leads to several types of paradoxes.

A very frequent type is the situation when an initial model describes the reality well, but its mathematical corollaries do not. Another type, a true logical paradox, occurs when mathematical corollaries contradict one another.

For example, the expected value is the sum of products of probabilities and payoffs. Suppose a fair coin is flipped in a hypothetical game and the player is paid $10 if the coin lands on heads and $20 if it lands on tails. The expected value of winning is $15 because $0.5(10) + 0.5(20) = 15$. When the same game is played many times, it is rational to prefer options with higher expected values. Under this assumption, it is better to play the game where the player is paid nothing for heads and $40 for tails than the first game, because the expected value of winning is higher: $0.5(0) + 0.5(40) = 20$. However, in real life, risk aversion will make many people choose the first game.

To resolve this and other related paradoxes, many preference models account for risk aversion as a separate variable. A utility function is the measure of relative satisfaction of a range of choices. An assumption that people will only want to maximize utility is not realistic, because it does not account for risk aversion. Because marginal choices usually come with higher risks, the utility function that accounts for risk aversion will look like a hump, being concave.

Bounded rationality principle is commonly used to explain paradoxical preferences by taking into account limited information, time, and cognitive abilities of people. Models based on bounded rationality include human limitations, such as computational capacity, and are based on computer science, statistics, and psychology.

Information Theory and Aesthetic Preferences

Information theory is a mathematical science that studies storing, compressing, and processing of data. In the 1990s, its branch called "algorithmic information theory," which deals with the complexity of algorithms, was applied to explain some aspects of the human sense of beauty and of aesthetic preferences. According to this theory, objects that have shorter algorithmic descriptions in terms of observer's knowledge will seem more beautiful, compared to objects with longer algorithmic descriptions. For example, it is easier to remember an object with mirror symmetry because only half of the information is original—symmetry provides information compressibility. Therefore, symmetric objects, as well as objects with patterns or fractal self-similarity, are seen as more beautiful.

Algorithmic information theory also models preferences by interest, which are separate from preferences based on beauty. Within these models, interest can be compared to the first derivative of beauty, showing the observer's perception of change in understanding. People prefer an experience on the basis of interest when it involves better compressibility or predictability of information than before. For example, noticing a new pattern (and therefore better organizing an image) is preferred because it is interesting.

Preferences, Desires, and Motivation

Many preferences and choices are based on needs, wants, and desires, which are explained in theories of motivation. Researching motivation is challenging because of individual differences among people, as well as language ambiguity. There are disagreements among researchers even over relatively straightforward terminology, such as intrinsic and extrinsic motivation. Many motivation theories include taxonomies of needs and desires. For example, in Maslow's hierarchy, named after Abraham Maslow, unsatisfied physiological needs, such as hunger or thirst, have higher priority than unsatisfied self-esteem needs, such as recognition. Some theories identify long lists of motivators, such as curiosity, tranquility, order, and independence. Other theories only define a few broad classes of needs.

Each category of need can be considered a variable. Graphs of values of these variables versus levels of motivation often demonstrate the characteristic "mirrored C" shape called a "backward bending curve." For example, as activities provide more order, they first become more motivating (and preferred), but beyond a certain point, more order becomes less motivating. This curve is famously described in the baseball manager Lawrence "Yogi" Berra's joke about a restaurant: "Nobody goes there anymore. It's too crowded." People usually prefer restaurants that are not too empty or too full.

Preferences and Demographics

A number of statistical studies find significant differences in preferences of different demographics within populations, such as males and females, socioeconomic classes, ages, and political affiliations. Because statistical packages make many types of mathematical and statistical analyses of databases very easy, there are many results that demonstrate significant differences in preferences among different demographics. However, determining meanings of these differences is a significantly more difficult research problem. Demographic differences in preferences can also vary from culture to culture. In some cultures, for example, more females than males prefer bright colors in clothes, and in other cultures, it is reversed.

Further Reading

Anthony, Martin, and Norman Biggs. *Mathematics for Economics and Finance: Methods and Modeling.* New York: Cambridge University Press, 1996.

Berry, M. A. J., and G. Linoff. *Data Mining Techniques For Marketing, Sales and Customer Support.* Hoboken, NJ: Wiley, 1997.

Netflix Prize. http://www.netflixprize.com.

MARIA DROUJKOVA

See Also: Data Mining; Expected Values; Mathematical Modeling, Predicting Attacks; Predicting Divorce; Raghavan, Prabhakar.

Pregnancy

Category: Medicine and Health.
Fields of Study: Algebra; Data Analysis and Probability.
Summary: Various mathematical models help describe issues related to conception, diseases associated with pregnancy, and population dynamics.

Much of the conclusions drawn in medicine, in particular in obstetrics and gynecology, are often based on heuristics, limited observations, and sometimes even biased data. Mathematicians and statisticians have recently attempted to develop general theoretical models that can be adapted to specific situations in order to facilitate the understanding of various aspects of human pregnancy. Specifically, more recent studies have been conducted regarding conception time, dis-

ease prediction related to pregnancy, and the effect of pregnancy on population growth.

Modeling the Most Efficient Time to Conceive

One of the most fundamental and important research topics in the study of human pregnancy is the so-called time-to-pregnancy (TTP). TTP can be defined scientifically as the number of menstrual cycles it takes a couple engaging in regular sexual intercourse with no contraception usage to conceive a child. Fittingly, statisticians attempt to generate as much data as possible from various couples regarding their personal TTP experiences. The data are collected in a way that is as unbiased as possible—it is intended to accurately represent couples in the general population attempting to conceive a child. From the data, both qualitative and quantitative statistical methods are implemented in order to ascertain the most efficient method to achieve conception.

For example, some social trends increase the age at which a woman attempts to become pregnant. When this situation arises, women are often concerned about achieving conception before the onset of infertility, which proceeds menopause. In fact, couples that are unsuccessful in conceiving within one year are clinically classified as *infertile*. When this condition occurs, medical doctors often recommend that the couple engage in assisted reproductive therapy (ART). However, ART can be very expensive and often increases the risk of adverse outcomes for the offspring, including various birth defects. Therefore, statistical models have been developed that pose an alternative to ART. These models are developed using Bayesian decision theory, named for Thomas Bayes, and search for optimal approaches for a couple to time intercourse in order to achieve conception naturally, without the potentially disadvantageous ART. These models quantitatively incorporate various biological aspects, including menstrual cycles and basal body temperature, as well as the monitoring of electrolytes—among other phenomena—in order to be as efficient as possible.

Predicting Diseases Associated With Pregnancy

Medical evidence supports the notion that women often repeat reproductive outcomes. In particular, women with a history of bearing children with adverse outcomes often have up to a two-fold increase in sub-

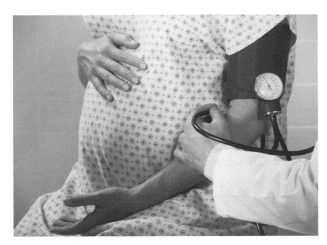

Mathematical epidemiology is used to predict conditions like pre-eclampsia during pregnancy.

sequent risk. Therefore, researchers in the mathematical and statistical sciences realized the necessity for statistical analyses that address this issue. In fact, statistical research has been conducted in order to promote a consistent strategy that assesses the risks each woman may face in a subsequent pregnancy. The goal is for these types of models to become increasingly more accurate, as they incorporate statistical data regarding the recent reproductive history of the woman, among other biological factors, which were not fully taken into account in previous studies.

Mathematical epidemiology (the study of the incidence, distribution, and control of diseases in a population) attempts to better comprehend, diagnose, and predict various diseases incorporated with pregnancy, and this field is ever-expanding. By designing and implementing various statistical approaches and mathematical models to better predict realistic outcomes, mathematicians and statisticians have studied congenital defects and growth restrictions, as well as preterm delivery, pre-eclampsia, and eclampsia.

For example, pre-eclampsia is a pregnancy condition in which high blood pressure and high levels of protein in urine develop toward the end of the second trimester or in the third trimester of pregnancy. The symptoms of this condition may include excessive weight gain, swelling, headaches, and vision loss. In some cases this condition can be fatal to the expectant mother or the child. The exact causes of pre-eclampsia are unknown at the beginning of the twenty-first cen-

tury, and the only cure for the disease is the delivery of the child. Therefore, it is apparent that determining which women are prone to develop pre-eclampsia is an exceedingly important area of research.

Empirical evidence indicates that a woman's heart rate is a deterministic factor in the prediction of pre-eclampsia. In recent times, statisticians have therefore developed a novel and non-invasive approach to detect abnormalities in pre-eclamptic women that distinguishes from women with non-pre-eclamptic pregnancies. This approach is accomplished by comparing the dynamical complexity of the heart rates of women that are pre-eclamptic with those that are non-pre-eclamptic. The analysis revealed that the heart rate of pre-eclamptic women demonstrated a more regular dynamic behavior than those women that were not pre-eclamptic, which substantiates the empirical notion that diseased states may be associated with regular heart rate patterns.

Population Dynamics

Mathematicians have long developed models to analyze population dynamics. One contemporary model also incorporates how pregnant women directly influence such dynamics. This model consists of an equation that describes the evolution of the entire population and an equation that analyzes the evolution of pregnant women. These equations are coupled—they are studied simultaneously. Moreover, this particular system of equations can be analyzed as a linear model (not sensitive to initial data), with or without diffusion (permitting members of the population to travel large distances), or as a nonlinear model (sensitive to initial data) without diffusion. The asymptotic behavior of the solutions to this system (the long-term behavior of the population) was also addressed.

Further Reading

Fragnelli, Ginni, et al. "Qualitative Properties of a Population Dynamics System Describing Pregnancy." *Mathematical Models and Methods in Applied Sciences* 4 (2005).

Germaine B., et al. "Analysis of Repeated Pregnancy Outcomes." *Statistical Methods in Medical Research* 15 (2006).

Salazar, Carlos, et al. "Non-Linear Analysis of Maternal Heart Rate Patterns and Pre-Eclamptic Pregnancies." *Journal of Theoretical Medicine* 5 (2003).

Savitz, David A., et al. "Methodologic Issues in the Design and Analysis of Epidemiologic Studies of Pregnancy Outcome." *Statistical Methods in Medical Research* 15 (2006).

Scarpa, Bruno, and David B. Dunson. "Beyesian Methods for Searching for Optimal Rules for Timing Intercourse to Achieve Pregnancy." *Statistics in Medicine* 26 (2007).

Scheikle, Thomas H., and Niels Keiding. "Design and Analysis of Time-to-Pregnancy." *Statistical Methods in Medical Research* 15 (2006).

Daniel J. Galiffa

See Also: Disease Survival Rates; Drug Dosing; Fertility; Genetics; Mathematical Modeling; Ultrasound; Viruses.

Prehistory

Category: Government, Politics and History.
Fields of Study: Measurement; Number and Operations.
Summary: Historians believe that even the earliest people used mathematics.

Many books on the history of mathematics begin with the ancient Egyptians and Babylonians, but those civilizations did not begin until about 5000 years ago. Although historians do not know many details, human life had been progressing for several millennia prior to that time. Even archeology offers little detail on the earliest mathematics, so most knowledge comes from speculation. However, from what is known about human beings in general, and especially about prehistoric life, even the earliest people must have known and used some mathematics.

The use of "mathematics" probably even precedes the development of modern human beings. Studies of animal behavior have shown that animals, and especially birds, seem to possess limited number sense, recognizing the difference between groups of two and three and even larger sets. Bees can recognize and even communicate information about the location of orchards and fields for pollination, displaying a sense of space

that could be called "geometry." Even more spectacular are the long migratory trips of herd animals, flocks of birds, and groups of butterflies, often traveling thousands of miles to return to the same fields every year. These examples certainly do not represent a sophisticated concept of mathematics and are instinctual, but they show a mathematical organization in the brain.

Language, Counting, and Quantities

The earliest humans (wherever the line is drawn between pre-human and human) continued the mathematical thinking shown in animals. As their brains developed, their mathematics also grew stronger and more sophisticated. This progression continued as early grunts become proto-languages, for a key part of mathematics is not only having the concepts in one's head, but also representing and communicating the concepts to others. Hence, language was a key ingredient in prehistoric mathematics (as it remains today).

A concept of counting must have come early, as people began to distinguish quantity. Even if they did not have linguistic terms for numbers beyond three or four, they would at least be able to make rough comparisons of large quantities and much larger quantities—consider that even modern humans often need notations, pictures, or concrete examples to handle specific large quantities, but certainly can tell the difference between a dozen and a hundred and a million. Many aspects of life require at least limited counting—to make sure all one's goats (or children) are present, to share items fairly in a group or to calculate the size of a load to be carried, and many other applications.

It is only a small jump of abstraction to begin to record quantities with tally marks. It is likely that people first collected stones or other small objects to represent quantities and later began to "write" them as tallies. Tally marks have been found in many parts of the world scratched on cave walls or carved onto wooden sticks and were also likely written in sand or clay, which shifted to destroy the writing. Probably the most famous prehistoric mathematical object is the Ishango bone, found in south-central Africa, and thought to be at least 15,000 years old. The bone has several sets of tallies scratched onto it—some have pointed out that they are mostly prime numbers, but that is probably a coincidence. Using tallies quickly leads to a problem: a long line of marks is hard to deal with, even if one had some limited counting words. Probably, many people around the world recognized that some structure helped handle large quantities of tally marks—especially collecting them into groups of the same size. Not only does this make counting more efficient but it also leads to the concept of multiplication. In nearly all modern languages—most derived from ancient or even prehistoric languages—the higher counting words use a system of groups and groups of groups, now called "place-value," but they reach back to the prehistoric convenience of putting tally marks together.

Measurement and Geometry

Closely tied to counting was the use of comparative relationships—especially large and small, tall (or long) and short, and even old and young. These may have come when exact counts were difficult, but the comparisons were obvious and usually visual. A tall stack of blocks would easily be seen to have more items than a short stack; a long line of tally marks (grouped or ungrouped) was a greater quantity than a short line. As actual counting developed and numbers were applied to comparisons, the beginnings of measurement occurred—measurement is really just comparisons of quantities where one side of the comparison is a defined unit. To make comparisons easier, certain items of specific size or quantity became units, and as people reached farther to wider audiences, units became at least roughly standardized. Often, body parts were used both for counting tabulations and as "standard" units. For example, the distance from the elbow to the fingertips was approximately the same for most adults, so in the Middle East, this length became the "cubit."

Geometry also has deep roots in the human story. Circles must have been recognized in the shape of the sun and full moon and the apparent edge of the horizon. Efficiency caused people to arrange objects to fit together well in patterns—often circular but sometimes rectangular. The first tools used sharp angles, heavy weights, and tall, thin cylinders. The beginnings of farms led to more organized geometrical arrangements in the shapes of fields and structures. Often, the "invention" of the wheel is considered one of the big milestones of the start of civilization, and this represents a practical understanding of the geometry of circles. As objects became more sophisticated—woven mats, farming tools, larger structures, and even bridges—many more geometrical relationships and properties were discovered. These might be considered the beginnings of engineering—using mathematical properties in practical applications.

Pure Mathematics

Archeologists have also noted some prehistoric mathematics that may have been closer to pure mathematics. Cave paintings, carved sculptures, and textile patterns show contemporary mathematical objects such as circles, triangles, parallel lines, quadrangles, symmetric patterns, and the crosshatch. However, no one has yet deciphered what the geometric signs meant to prehistoric peoples. Some symbols appeared repeatedly in various parts of the world. They may have served practical or religious values, but they also were art—perhaps art for its own sake, for beauty. Certain numbers may have had mystical meanings that were seemingly less useful for day-to-day activity but important for esthetics and spirituality.

The overlap between this pure mathematics and the practical needs of early farmers was the use of mathematics in astronomy and calendars. Could the gods show the times for planting and harvesting? Could humans discern the plans of these gods and use them in practice? Most of the spectacular prehistoric structures, from Stonehenge in England to the huge geometrical patterns of Nazca in Peru, have been linked to measures of the sun's movement and the seasons. Mathematics led prehistoric peoples in solving their daily problems and to thinking of the universe and infinity. Mathematics still serves modern humans in the same ways.

Further Reading

Boyer, Carl. *A History of Mathematics*. Hoboken, NJ: Wiley, 1991.

Burton, David M. *The History of Mathematics: An Introduction*. New York: McGraw-Hill, 2007.

Eves, Howard. *Introduction to the History of Mathematics*. New York: Saunders College Publishing, 1990.

Ifrah, Georges. *The Universal History of Numbers*. Hoboken, NJ: Wiley, 1994.

Von Petzinger, Genevieve. "Geometric Signs in Rock Art & Cave Paintings." http://www.bradshawfoundation.com/geometric_signs/geometric_signs.php.

LAWRENCE H. SHIRLEY

See Also: Animals; Basketry; Bees; Calendars; Number and Operations in Society

Probability

Category: History and Development of Curricular Concepts.

Fields of Study: Communication; Connections; Data Analysis and Probability.

Summary: Humans have implicitly understood concepts of probability and randomness since antiquity, but these concepts have been more formally studied since the seventeenth century.

Throughout history, humans have used many methods to try to predict the future. Some believed that the future was already laid out for them by a divine power or fate, while others seem to have believed that the future was uncertain. There are still debates on the extent to which people were able to speculate on the future prior to the development of statistics in the seventeenth and eighteenth centuries. Some assert that such speculations were impossible, yet other historical evidence suggests that at least some people must have been able to perceive the world in terms of risks or chances, even if it was not in quite the same way as later mathematicians and statisticians. The Greek philosopher Aristotle proposed that events could be divided into three groups: deterministic or certain events, chance or probable events, and unknowable events. The idea of "randomness" is often used to indicate completely unknowable events that cannot be predicted. In mathematics, the long-term outcomes of random systems are, in fact, "knowable" or describable using various rules of probability. Probability distributions, expressed as tables, graphs, or functions, show the relationship between all possible outcomes of some experiment or process, like rolling a die, and the chance that those outcomes will happen. For example, lotteries state the chances of winning various prizes, and people seeking medical treatment might be told the odds of success. Random processes and probability can run counter to human intuition and the way in which human brains perceive and organize information, which is perhaps another reason that quantifying ideas of probability is still an ongoing endeavor. Students are often introduced to probability concepts in the earliest elementary grades, such as basic binary classifications of outcomes as "likely" or "unlikely" and the notion of probabilities as experimental frequencies. More formal axioms of probability may be introduced in the middle grades. Probability theory and probability-based mathematical statistics are typically studied in college, though they may be included in advanced high school classes. Some elements of probability theory and applications are also taught in other academic disciplines, like business, genetics, and quantum mechanics.

Early History

Archaeological evidence, such as astragalus bones found at ancient sites, suggests that games of chance have been around for several millennia or longer. Egyptian tomb paintings show astragali being used for games like Hounds and Jackals, much like the way twenty-first-century game players use dice. The ideas of randomness that underlie probability were often closely tied to philosophy and religion. Many ancient cultures embraced the notion of a deterministic fate. The Greek pantheon was among those that included deities associated with determinism, literally known as the Fates. The popular goddess Fortuna in the Roman pantheon suggests a recognition of the role of chance in the world. Jainism is an Indian religion with ancient roots, whose organized form appears to have originated sometime between about the ninth and sixth centuries B.C.E. The Jainist logic system known as *syadvada* includes concepts related to probability; its sanskrit root word *syat* translates variously as "may be" or "is possible." Probability is also a component of the body of Talmudic scholarship; for example, the notion of casting lots, used in some temple functions. Babylonians had a type of insurance to protect against the risk of loss for sea voyages, called "bottomry," as did the Romans and Venetians.

Origins of Study in the Seventeenth Century

Given the near omnipresence of probability in the ancient world, it seems reasonable to think that there were some efforts to estimate or calculate probabilities, at least on a case-specific basis; for example, those who issued maritime insurance would have assigned some type of monetary values for cost and payoff. There is relatively little evidence of broad mathematical research on probability before about the fifteenth century, though some analyses for specific cases survive. For example, a Latin poem by an unknown author called "De Ventula" describes all the ways that three dice can fall. Mathematician and friar Luca Paccioli wrote *Summa de arithmetica, geometria, proportioni e proportionalita* in 1494, which contains some discussion of probability. A few other works address dice rolls and related ideas. Historians tend to agree that the systematic mathematical study of probability as it is now known originated in the seventeenth century. At the time, considerable tensions still existed between the philosophies of religion, science, determinism, and randomness. Determinists asserted that the universe was the perfect work of a divine creator, ruled by mathematical functions waiting to be discovered, and that any apparent randomness was because of faults in human perception. Many emerging scientific theories, like the heliocentric model of the universe advocated by mathematician and astronomer Nicolaus Copernicus, challenged this view by explicitly exploring and quantifying variation and deviations in observations. Astronomy and other sciences, along with the rise of combinatorial algebra and calculus,

would ultimately prove to be very influential in the development of probability theory. Changes in business practices also challenged notions of risk, requiring new methods by which likelihood and payoffs could be determined. Harkening back to ancient human activities, however, the most popular story for the origin of probability theory concerns gambling questions posed to mathematician Blaise Pascal by Antoine Gombaud, Chevalier de Méré.

In 1654, the Chevalier de Méré presented two problems. One concerned a game where a pair of six-sided dice was thrown 24 times, betting that at least one pair of sixes would occur. Méré's attempts at calculation contradicted the conventional wisdom of the time and purportedly led him to lose as great a deal of money. The second problem, now called the Problem of Points or Problem of Stakes, concerned fair division for a pot of money for a prematurely terminated game between equally skilled players where the winner of a completed game would normally take the whole pot. Spurred by de Méré's queries, Pascal and Pierre Fermat exchanged a series of letters in which they formulated the fundamental principles of general probability theory.

At the time of its development, Pascal and Fermat's burgeoning theory was commonly referred to as "the doctrine of chances." Inspired by their work, mathematician and astronomer Christian Huygens published *De Ratiociniis in Ludo Aleae* in 1657, which discussed probability issues for gambling problems. Jakob (also known as James) Bernoulli explored probability theory beyond gambling into areas like demography, insurance, and meteorology and he composed an extensive commentary on Huygen's book. One of his most significant contributions was the Law of Large Numbers for the binomial distribution, which stated that observed relative frequencies of events become more stable, approaching the true value, as the number of observations increases. Prior definitions based on gambling games tended to assume that all outcomes were equally likely, which was generally true for games with inherent symmetry like throwing dice. This extension allowed for empirical inference of unequal chances for many real-world applications. Bernoulli also wrote *Ars Conjectandi*. Influenced by this work, mathematician Abraham de Moivre derived approximations to the binomial probability distribution, including what many consider to be the first occurrence of the normal probability distribution, and his *The Doctrine of*

The cube design of dice allows for each of their sides to have an equal probability of being rolled.

Chances was the primary probability textbook for many years.

Objective and Subjective Approaches

Historically and philosophically, many people have asserted that to be objective, science must be based on empirical observations rather than subjective opinion. Estimating probabilities through direct observations is usually called the "frequentist approach." The method of inverse or inductive probability, which allows for subjective input into the estimation of probabilities, is traced back to the posthumously published work of eighteenth-century minister and mathematician Thomas Bayes. Conditional probabilities had already been explored by de Moivre, providing the basis for what is known as "Bayes theorem" (or "Bayes rule"). In Bayes's inductive framework, there is some probability that a binary event occurs. A frequentist would make no assumptions about the probability and carry out experiments to attempt to determine the true probability value. Using Bayes's approach, some probability value can be arbitrarily chosen, and then experiments conducted to ascertain the likelihood that the value is in fact the correct one. In later interpretations and applications of the method, the initial value might be chosen according to experience or subjective criteria. His work also produced the Beta probability distribution. Bayes's writings contained no data or examples, though they were extended upon and presented by minister Richard Price. At the time, they were relatively less influential than frequentist works, though Bayesian methods have generated much discussion and saw a great resurgence in the latter twentieth century.

Applications

Like Bernoulli, Pierre de Laplace extended probability to many scientific and practical problems, and his probability work led to research in other mathematical areas such as difference equations, generating functions, characteristic functions, asymptotic expansions of integrals, and what are called "Laplace transforms." Some call his 1812 book, *Théorie Analytique des Probabilités*, the single most influential work in the history of probability. The Central Limit Theorem, named for George Pólya's 1920 work and sometimes called the DeMoivre–Laplace theorem, was critical to the development of statistical methods and partly validated the common practice at the time (still used in the twenty-

first century) of calculating averages or arithmetic means of observations to estimate location parameters. Error estimates were usually assumed to follow some symmetric probability distribution, such as rectangular, quadratic, or double exponential. While they had many useful properties, they were mathematically problematic when it came to deriving the sampling distributions of means for parameter estimation. Laplace's work, which he proved for both direct and inverse paradigms, rectified the problem for large-sample cases and formed the foundation for large sample theory.

Normal Distribution

The normal distribution is among the most central concepts in probability theory and statistics. Many other probability distributions may be approximated by the normal because they converge to the normal as the number of trials or sample sizes approach infinity. Some of these include the binomial and Poisson distributions, the latter named for mathematician Simeon Poisson. The Central Limit Theorem depends on this principle. Mathematician Karl Friedrich Gauss is often credited with "inventing" the normal (or Gaussian) distribution, though others had researched it and Gauss's own notes refer to "the elegant theorem first discovered by Laplace." He can fairly be credited with the derivation of the parameterization of the distribution, which relied in part on inverse probability. Mathematician Robert Adrain, who was apparently unaware of Gauss's work, discussed the validity of the normal distribution for describing measurement errors in 1808. His work was inspired by a real-world surveying problem. However, Gauss tends to be credited over Adrain, perhaps because of his many publications and the overall breadth of his mathematical contributions.

The fact that Laplace and Gauss worked on both direct and inverse probability was unusual from some perspectives, given the philosophical divide between frequentist and Bayesian practitioners even at the start of the twenty-first century. Later, both would gravitate toward frequentist approaches for minimum variance estimation, which is seen by some as a criticism of inverse probability. Other mathematicians, such as Poisson and Antoine Cournot, criticized inverse methods, while Robert Ellis and John Venn proposed defining probability as the limit of the relative frequency in an indefinite series of independent trials—essentially, the frequentist approach. The maximum likelihood estimation method proposed by Ronald Fisher in the early twentieth century was interpreted by some as melding aspects of frequentist and inverse methods, though he adamantly denied the notion, saying, "The theory of inverse probability is founded upon an error, and must be wholly rejected." This may explain the essential absence of inverse or Bayesian probability concepts in the body of early statistical inferential methods, which were heavily influenced by Fisher.

Mathematician and anthropometry pioneer Adolphe Quetelet brought the concept of the normal distribution of error terms into the analysis of social data in the early nineteenth century, while others like Francis Galton advanced the development of the normal distribution in biological and social science applications in the latter half of the same century. Many mathematicians, statisticians, scientists, and others have contributed to the development of probability theories, far too many to exhaustively list, though recognized probability distributions are named for many of them, such as Augustin Cauchy, Ludwig von Mises, Waloddi Weibul, and John Wishart. Pafnuty Chebyshev, considered by many to be a founder of Russian mathematics, proved the important principle of convergence in probability, also called the Weak Law of Large Numbers. Andrei Markov's work on stochastic processes and Markov chains would lead to a broad range of probabilistic modeling techniques and assist with the resurgence of Bayesian methods in the twentieth century.

Some historians have suggested that one difficulty in developing a comprehensive mathematical theory of probability, despite such a long history and so many broad contributions, was difficulty agreeing upon one definition of probability. For example, noted economist John Keynes asserted that probabilities were a subjective value or "degree of rational belief" between complete truth and falsity. In the first half of the twentieth century, mathematician Andrey Kolmogorov outlined the axiomatic approach that formed the basis for much of subsequent mathematical theory and development. Later, Cox's theorem, named for physicist Richard Cox, would assert that any measure of belief is isomorphic to a probability measure under certain assumptions. It is used as a justification for subjectivist interpretations of probability theory, such as Bayesian methods. There are variations or extensions on probability with many applications. Shannon entropy, named for mathematician and information theorist Claude Shannon

and drawn in part from thermodynamics, is used in the lossless compression of data. Martingale stochastic (random) processes, introduced by mathematicians such as Paul Lévy, recall the kinds of betting problems that challenged de Méré and inspired the development of probability theory. Chaos theories, investigated by mathematicians including Kolmogorov and Henri Poincaré, sometimes offer alternative explanations for seemingly probabilistic phenomena. Fuzzy logic, derived from mathematician and computer scientist Lotfali Zadeh's fuzzy sets, has been referred to as "probability in disguise" by Zadeh himself. He has proposed that theories of probability in the age of computers should move away from the binary logic of "true" and "false" toward more flexible, perceptual degrees of certainty that more closely match human thinking.

Further Reading

Devlin, Keith. *The Unfinished Game: Pascal, Fermat, and the Seventeenth-Century Letter That Made the World Modern*. New York: Basic Books, 2008.

Gigerenzer, Gerd. *The Empire of Chance: How Probability Changed Science and Everyday Life*. Cambridge, England: Cambridge University Press, 1990.

Hacking, Ian. *The Emergence of Probability: A Philosophical Study of Early Ideas About Probability, Induction and Statistical Inference*. Cambridge, England: Cambridge University Press, 2006.

———. *An Introduction to Probability and Inductive Logic*. Cambridge, England: Cambridge University Press, 2001.

Hald, Anders. *A History of Probability and Statistics and Their Applications Before 1750*. Hoboken, NJ: Wiley-Interscience, 2003.

Sarah J. Greewald
Jill E. Thomley

See Also: Closed-Box Collecting; Data Analysis and Probability in Society; Game Theory; Normal Distribution; Randomness; Sample Surveys.

Probability in Society

See *Data Analysis and Probability in Society*

Problem Solving in Society

Category: School and Society.
Fields of Study: Connections; Problem Solving.
Summary: Mathematics is used to find and solve problems, often spurring new mathematical investigations.

Problem solving is fundamental not only to the learning and application of mathematics as a student, but to all walks of life. Many people consider mathematics and problem solving synonymous. However, there are many mathematicians who do not solve problems or who do more than solve problems. Some work to build new theories or advance the language of mathematics. Others unify or explain previous results, sometimes from many fields of mathematics. Yet others consider the very nature and philosophy of mathematics as a discipline. In twenty-first-century society, mathematics teaching at all levels seeks to develop students' abilities to effectively address a wide variety of mathematics problems, including proving theorems; reducing new problems to previously solved problems; formulating and solving both real-life and abstract word problems; finding and creating patterns; interpreting figures, graphs, and data; developing geometric constructions; and doing appropriate computations or simulations, often with computers or calculators.

Problem solving is also an instructional approach in which students actively learn fundamental concepts through their contextualization within problems rather than from a passive lecture. What fundamentally connects these activities, beyond the mathematics techniques and skills necessary to solve them, is the framework of "how to think." Students must have the necessary tools and techniques at their disposal through a solid education in the fundamentals. They must also be able to either analyze the characteristics and requirements of a problem in order to decide which tools to apply, or know that they do not have the appropriate tool at their disposal. Further, students must practice with these mathematical tools in order to become skilled and flexible problem solvers, in the same way that athletes or craftsmen practice their trades. As Hungarian mathematician George Pólya expressed, "If you wish to become a problem solver, you have to solve problems." This idea extends to the notion that problem solving is by its nature cyclic and dynamic. In many cases, the solution to a problem results in one or more new problems or opens the path to solving an older problem for which a solution has previously proven elusive. Sometimes, mathematics problems have real and immediate applications, and many new mathematical disciplines, like operations research or statistical quality control, have developed from these sorts of problems. In contrast, there are many issues in theoretical mathematics that do not appear to have any immediate benefit to society. In some cases, people question the need to explore such abstract problems when there are more immediate needs. Often, these abstract problems turn out to have very concrete applications decades or even centuries after their initial introduction. Even if that is not the case, theoretical problem solving adds to the growing body of mathematics knowledge and, just as importantly, shows people yet another way to think about the world.

History

The mathematics body of knowledge is not static; it has been evolving with humans. As soon as humans organized themselves into communities attached to the land, benefits rapidly emerged. Certainly, an advantage was an increase in agricultural and livestock productivity. As a result, part of the harvest and the cattle was accumulated for worse times. Accumulation demanded certain mechanisms to identify the ownership and use of the land (the process of land surveying) and to record who contributed to what was collected (the system of counting). The success of such social structure allowed skilled individuals to take advantage of their abilities to exchange the resultant products for food surplus (the beginnings of commerce). The development of commerce demanded a new tool to register the commercial operations in order to recognize who was implicated and the amount involved. This tool was based in a new kind of language (mathematics) able to do operations such as additions, subtractions, iterative sums, and partitions that natural languages were unable to support. As with any language, it consisted of two elements: notation to represent ideas (numbers) and syntax to manipulate these ideas (calculation).

After the accumulation of goods came the capability to organize collective efforts. It was possible to build massive public works. Warehouses, markets, fortresses, temples, aqueducts, and even pyramids were constructed in urban centers and their surroundings. Construction presented a new problem related to the manipulation and combination of forms. Early exercises were based on rules used for land surveying; for instance, to calculate areas and volumes. Additional difficulties arose when public works increased their complexity; hence, the application of forms and their interactions to develop better habitats gave rise to the development of architecture as an independent discipline. The Greeks separated land surveying from the study of spatial relations and forms; as a result, geometry was born. This discipline was used to solve abstract mathematical problems. For instance, Pythagoras recognized the relation between the sides of a right triangle as $a^2 + b^2 = c^2$ (the Pythagorean Theorem), and Archimedes studied the relation between the circle's circumference and its diameter. The latter is known as *pi* (π), an irrational number with the value of 3.1415926535897932384626433832795502 . . . .

The Problem of Representation and the Dynamics of Change

With the accelerated increase of richness and variety in social interactions, intractable problems of representation appeared. Operations were required to record social experiences from an ever growing dynamism. This endeavor made limitations in the notation systems available at that moment evident. Hindis and, afterwards, Arabs and Muslims developed the positional decimal system still in use in the twenty-first century. The decimal system allows the representation of arithmetic operations without the need to use an abacus. Changes in quantities demanded introducing a general notation for variable and constant amounts, which were linked by operators to form different sentences, called "equations." The study of these relations is known as "algebra."

The capability to represent abstract ideas and their relations allowed mathematicians at the beginning of the sixteenth century to discuss problems related to the dynamics of change. In fact, the field of astronomy proposed new challenges to mathematics. Between 1507 and 1532, Copernicus presented a series of works where he substituted the traditional viewpoint, which located the Earth at the center of the universe (the geocentric view), with another where the sun was at the focus (the heliocentric view). This view helped to explain inconsistencies in the stellar movement, such as the retrograde displacement of planets. Around 1605, Johannes

Kepler empirically discovered the elliptic orbit of planets around the sun. He also noticed that the line that joins each planet with the sun (called the "radius vector") sweeps the same area in the same period of time. Galileo focused his telescope to Jupiter, and, in 1610, he posited that the lights surrounding that planet were, in fact, satellites. To demonstrate all of this in mathematical terms demanded the study of change in relation to time, something impossible to solve at that moment. Isaac Newton and Gottfried Leibnitz simultaneously developed a useful procedure known as "calculus." When it is used to represent the change of a certain quantity in relation to another in terms of infinitesimal moments, it is called "differential calculus." Interestingly enough, this procedure can be reversed to reckon space sections bounded by different functions. The general procedure consists on dividing them into additive infinitesimal blocks—a process named "integral calculus." Both procedures operate in an inverse manner through the fundamental theorem of calculus.

The Problem of Estimation

In the seventeenth century, additional problems appeared when the practical world confronted an impossible question. How can one characterize something that is not stable enough to be counted? For instance, in order to establish public policies, politicians need to know what resources are at their disposal—the demographic and economic capabilities, which can be determined in a census. The main problem with exhaustive counting of populations is that they change. There are births and deaths. In order to solve this issue, one method is to select a fraction (called a "sample") of the object of study (called the "population"), to identify the sample characteristics and to generalize them to the population. Advantages for this sampling procedure are lower costs and faster data collection than following a comprehensive census. But there is an important difficulty: how to guarantee that the characteristics of the sample are the same as those of the entire population. One needs to estimate the sampling error because of selecting a sample that does not represent the population and to define a confidence interval by identifying the reliability of the estimate. The part of mathematics interested in this kind of problems is known as "statistics."

Statistics helps to solve many technical problems. Statisticians may need to (1) estimate the size of a population, as Laplace did in 1786 for France, by using a sample; (2) describe a population in terms of different numerical relations, such as its expected value (called "average"), its most frequent value (called "mode"), the limits of the data series (called "range"), the value that separates the higher half of the data series from the lower half (called "median"), and the data dispersion (called "standard deviation"); (3) test a hypothesis as J. H. Jagger did in 1873 at the Beaux-Arts Casino at Monte Carlo, when he collected results from a roulette wheel to prove that it was fraudulent; (4) estimate if a process needs products of a certain quality or it requires to be fixed, as in statistical quality control; (5) identify if changes in a process result in a positive outcome (called "correlation"), such as the Hawthorne study done in a working line to correlate the increase in illumination with workers' productivity; (5) predict and forecast future outcomes by means of recognizing patterns of behavior, what is known as "regression"; (6) extrapolate future data through the analysis of previous results; (7) reconstruct incomplete series data by means of that which is known and available, through interpolation; or (8) model the behavior of an entity in order to transform data into valuable information (called "data mining").

The Problem of Decision Making

The Industrial Revolution introduced a massive change in the social order. Early stages of the period witnessed the substitution of agricultural workers with machines by the thousands. It represented an increase in the productivity for many industries and services, mainly textiles and transportation, to levels never before seen. It surpassed the previous cumulative capacity of mankind. It also implied a surplus of energy with the use of internal combustion engines and electrical power generation. However, finding the equilibrium in this new social order was not an easy endeavor. Two world wars witnessed this planetary enterprise, and the postwar era during two different visions of the best way to organize the global society developed into a mortal conflict: capitalism versus communism.

At the beginning, the Industrial Revolution promised benefits with no end, although it made the medieval work system based on guilds inoperative. Groups of artisans loyal to a closed system of hierarchical progression were substituted by interchangeable clusters of men and machines located at industrial centers with short-term economic success as its main performance criterion. These were operationalized in terms of effec-

The Problem of Distributions

Although knowing certain characteristics from the population allows one to make more informed decisions, it does not solve particular cases. For instance, if 80% of people in a community prefer vanilla flavor rather than chocolate ice cream, will Ms. X like it? If the identification of general preferences does not ensure that individual expectations will be fulfilled, how can one propose the best offer to an individual in particular? How can one quantify the chance of an event happening? The study of the individual behavior from a collective characterization is known as *probability*. It is important to note that probability has to do with descriptions from populations and not from individuals.

Probability studies began with Blaise Pascal and Pierre de Fermat (1654), when the former was approached by a gamester, the Chevalier de Méré, to solve a game problem—how to divide the stakes between two players who want to leave the table before finishing their game. It was not until the nineteenth century, again in the field of astronomy, that the potential use of probability was recognized. In 1801, Giuseppe Piazzi discovered the first asteroid, Ceres, but he had so few observations that he was unable to determine its orbit. A mathematician, Carl Friederich Gauss, analyzed the data available and, in order to correct the observational errors, he supposed that they would follow a normal distribution. This distribution is one of the most well-known among probability users. Probability has been used for hypothesis testing according to different probability distributions, statistical mechanics, probabilistic processes, the random movement of particles suspended in a fluid, and options' valuation.

tiveness and efficiency, and optimization was the prime improving activity. Methods based on empiricism and not on tradition acquired a new value. For instance, in 1840, Charles Babbage realized a study about mail classification and transportation; the result was the institution of the Uniform Penny Post; a taxation procedure by which a letter not exceeding half an ounce in weight could be sent from any part of the United Kingdom to any other part for one penny. In 1911, Taylor proposed a series of managerial principles that were the foundations of what is currently known as "management science" or "operational research." This Science of the Better consists in the application of advanced analytical methods to help make better decisions.

Operational Research took shape just prior to World War II. At the beginning, exercises were focused on solving problems of fighter direction and control in the British air defense system. The new radar system acted as an early warning system that was able to identify German aircraft before they would bomb air bases, ports, industrial areas, and cities. Success demanded, later during the war, to extend these exercises to the Atlantic Ocean. Massive ship losses because of the attacks of U-boats (German submarines) put Allied supplies to Europe and North Africa at risk. Accordingly, different analyses were conducted to increase the U-boat sinking rate. Different criteria were mathematically explored and solutions were implemented, including (1) identifying which kind of aircraft was the best suited to chase German submarines; (2) reckoning the time at which depth charges should explode, and (3) defining the size of merchant fleets that minimizes Allied losses when crossing the Atlantic.

From the success of analyzing the performance of military operations, this field of mathematics was extended to other industrial and social activities. Many different problems have been studied and alleviated by this approach, including (1) community development, in order to organize collectives, support strategies that deal with social dissatisfactions, help groups in rural communities and developing countries, and create the social conditions for effective public policies; (2) criminal justice, to maintain a safe society by optimizing the use of resources allocation that enforce the law and reduce spaces for organized crime and to assess policy impact; (3) education, to evaluate teaching quality, students learning experiences, and assessment procedures; (4) efficiency and productivity analysis; (5) healthcare services; (6) logistics and supply chains; (7) quality control; (8) security and defense; (9) scheduling; (10) strategic management; and (11) transport.

The Problem of Prediction in a Complex World

The acquisition, distribution, and use of knowledge are key factors for the development of individuals and society, an idea that has shifted social structures to more complex levels of organization. The introduction of concepts such as "entrepreneurship" (a wild spirit who causes creative destruction by innovation and disruption) or "leadership" (a process of social influence and emotional contagion) are the result of recognizing that people's actions affect many others in non-evident ways. Economy, ecology, management, and politics require new approaches as these phenomena develop with intensities never before expected. The limitation of resources demands humans to use them responsibly and to make decisions for a better future. The main difficulty consists of predicting the future from the present. How can a person predict future concequences of actions to recognize good actions from bad ones?

Advising people on how to act is an age-old business. For a long time, the unique sources at disposal were divinely inspired or supported by powerful collectives. However, since the 1800s, the emphasis shifted toward scientific study of the environment regarding which actions take place. Prediction was focused on learning from the past and expecting the future to behave similarly, what is known as "time-series procedures." These can be useful where individual decisions have little impact on the overall behavior; for example, the results of the lottery or the weather conditions for the next few days.

Accordingly, different patterns can be found in the data (such as horizontal, seasonal, cyclic, or trend), but no explanations for the phenomenon under study have been developed. Explanatory models require assuming a relationship between what one wants to forecast (called the "dependent variable") and something one knows or controls (called the "independent variable"). Through a regression analysis one may minimize differences between observations and the points from an expected trend, linear or not, which can be adjusted to indicate certain seasonality. For more complex phenomena, one may introduce additional independent variables in order to conduct multiple regression analysis. In certain conditions, this approach enhances information for a better decision-making process but assumes the non-evolutionary viewpoint that the best model for the future is the one which better fits historical data. This approach also reduces the size of phenomena under scrutiny because modeling a real complex phenomenon such as the world's climate goes easily beyond twenty-first-century computers' capabilities and human understanding.

Complexity is related to many things such as size, difficulty, variety, order, or disorder. However, it has nothing to do with complication. Anything complicated can be solved, usually by introducing more resources to crack current problems. Conversely, complexity is associated with the impossibility of guaranteeing future behaviors based on current ones. The mathematical treatment of complexity introduced a discipline known as "chaos theory." It is a collection of mathematical, numerical, and geometrical techniques that allow mathematicians to deal with non-linear problems that do not have explicit general solutions. It is based in the use of differential equations to analyze dynamic behaviors extremely sensitive to initial conditions. In this context, predicting the future has to do with recognizing stable equilibrium points (called "fixed point attractors"), those that appear when dynamic systems stop. An attractor indicates the natural tendency of a system to behave in a certain way in the long-term future, if nothing else disturbs it. Common physical examples of this kind of behavior are pendulums and springs. Attractors are used for decision making in different fields, such as finance, where investors try to identify stock market tendencies. Some major applications related to its origins are weather prediction, solar weather prediction models, and predicting fisheries dynamics.

The increase of computing power allows mathematicians to run mathematical models based in little pieces of code that represent specific behaviors (called "intelligent agents"). Agent-based models can be used to study complex behaviors to simulate individual behaviors, such as people's movements inside stadiums or automobiles avoiding traffic jams. Other studies related to self-organized and self-organizing behaviors can also be conducted as they can represent phenomena from economy and financial markets; opinion dynamics; emergency of social rules and institutions; creation or disappearance of companies; and technology innovation, adoption, and diffusion.

To recognize stability areas and patterns in complex behaviors resulting from a multiplicity of agents interacting is then at the basis of the next social challenge, and procedures to deal with this are at the edge of twenty-first-century capabilities. The study of elements and their interactions have developed new viewpoints to observe reality. To visualize problems as a myriad of elements richly interconnected with unseen behaviors and consequences has introduced notions such as "systems" and "networks" in discourse. In 1950, Ludwig von Bertalanffy, a biologist, recognized similar fundamental conceptions in different disciplines of science, irrespective of the object of study. He tried to represent those rules through a language to describe such entities, which he named the "General System Theory."

A year before, Werner introduced the notion of communicative control in machines and living beings by looking at the effects of feedback on future behaviors. He named it "cybernetics." Based on this, in 1956, Ashby provided a single vocabulary and a single set of concepts suitable for representing the most diverse types of systems. Since then, different researchers have developed alternative methodologies to describe phenomena not in terms of problems and solutions, but in terms of satisfaction and alleviation. This has been used to deal with non-technical problems—those considered impossible to solve only through analytical tools, as they include humans' interactions. In this context, relations between individuals are diagramed and studied in terms of bunches of nodes interconnected by links. From this viewpoint the image of a "network" emerges. This notion has been developed, for instance to measure the "distance" between two persons from different places and contexts and reckoned that the average number of intermediate people between them is 5.5, hence the phrase "six degrees of separation." Network analysis is important as it can be used to model and study phenomena such as the Internet and its vulnerability to hackers, viruses and their uncontrollable expansion, or technology innovation and its diffusion. Future developments on this area are expected.

Further Reading

Koomey, Jonathan, and John Holdren. *Turning Numbers into Knowledge: Mastering the Art of Problem Solving.* 2nd ed. Oakland, CA: Analytics Press, 2008.

Körner, T. W. *Naive Decision Making: Mathematics Applied to the Social World*. Cambridge, England: Cambridge University Press, 2008.

Polya, George. *How to Solve It: A New Aspect of Mathematical Method*. Princeton, NJ: Princeton University Press, 2004.

Schoenfeld, Alan. *How We Think: A Theory of Goal-Oriented Decision Making and Its Educational Applications*. New York: Routledge, 2010.

Tao, Terence. *Solving Mathematical Problems*. 2nd ed. Oxford, England: Oxford University Press, 2006.

Wilson, James, et al. "Research on Mathematical Problem Solving." In *Research Ideas for the Classroom*. Edited by P. S. Wilson. New York: Macmillan, 1993.

Zaccaro, Edward. *Becoming a Problem Solving Genius: A Handbook of Math Strategies*. Bellevue, IA: Hickory Grove Press, 2006.

Zeitz, Paul. *The Art and Craft of Problem Solving*. 2nd ed. Hoboken, NJ: Wiley, 2006.

Eliseo Vilalta-Perdomo

See Also: Mathematics, Applied; Mathematics, Defined; Mathematics, Elegant; Reasoning and Proof in Society.

Producers

See *Writers, Producers, and Actors*

Professional Associations

Category: Mathematics Culture and Identity.
Fields of Study: Communication; Connections.
Summary: Professional mathematical associations help mathematicians advocate, share ideas, and organize.

Organizations are a fundamental component of society, in part because of the human need to connect around similar interests. Professional associations form in response to individual and societal needs and concerns, and in turn impact society. Mathematics students, teachers, and researchers may join professional mathematics organizations to feel like a part of the larger mathematics community and make a difference beyond their school or university. There are international associations with worldwide memberships, like the International Mathematical Union, as well as associations that are organized by geographical region. National and regional associations in countries around the world address many of the same issues as mathematics associations in the United States. These issues include teaching, research, service, and the mathematics profession.

Mathematical associations may advocate for the mathematical sciences, engage in public policy discussions, and promote collaboration among specialized subgroups. They may provide professional development to mathematicians and engage in public mathematics outreach. Professional associations organize regional, national, or international conferences; fund professional development and outreach; publish a diverse array of books and journals on mathematical topics; and facilitate peer review and curricular changes. Philosophers and mathematicians like Paul Ernest and Reuben Hersh have written about the social and ethical responsibility of mathematicians, and mathematicians may work toward the greater good within the structure of professional organizations. As officers and committee members, mathematicians also run these associations.

Mathematical Organizations
The American Statistical Association (ASA) was formed in Boston, in 1839, by members with diverse interests. ASA's Web site states the following:

> Present at the organizing meeting were William Cogswell, teacher, fund-raiser for the ministry, and genealogist; Richard Fletcher, lawyer and U.S. Congressman; John Dix Fisher, physician and pioneer in medical reform; Oliver Peabody, lawyer, clergyman, poet, and editor; and Lemuel Shattuck, statistician, genealogist, publisher, and author of perhaps the most significant single document in the history of public health to that date.

From the beginning, the ASA had close ties with the government on statistical issues like those surrounding the census. ASA is international and comprised of professionals from industry, government, and academia in fields ranging from pharmaceuticals, health policy, agriculture, business, education, to technology. It promotes statistical knowledge through meetings, publications, membership services, education, accreditation, and advocacy.

In the United States, two well-known mathematics organizations are the American Mathematical Society (AMS) and the Mathematical Association of America (MAA). Both publish research journals, host professional conferences, and engage in student, community, and public policy outreach, although they have different focuses. The AMS originated as the New York Mathematical Society in 1888, and, in 1894, it became a national organization that concentrated on research. Teacher Benjamin Finkel created the *American Mathematical Monthly* in 1894, stating the following:

> Most of our existing journals deal almost exclusively with subjects beyond the reach of the average student or teacher of mathematics or at least with subjects with which they are familiar, and little, if any, space, is devoted to the solution of problems.

In 1915, when managing editor H. E. Slaught unsuccessfully tried to bring the *Monthly* to the AMS, the society instead recommended that there should be a different organization devoted to the journal. The AMS continues to focus primarily on mathematics research and scholarship, while the MAA promotes communication, teaching, learning, and research in mathematics and its applications, especially at the collegiate level. The mission of the MAA incorporates five core interests of education, research, professional development, public policy, and public appreciation. The MAA sponsors the highly regarded William Lowell Putnam Competition for undergraduate students and the American Math Olympiad mathematics competitions. The AMS and MAA join at the Joint Mathematics Meetings each January.

The National Council of Teachers of Mathematics (NCTM) was created in 1920, in part, to counter the efforts of social efficiency experts who believed that school curricula should emphasize fostering job-related skills and knowledge. Its membership includes mathematics teachers, mathematics teacher educators, and mathematics education researchers. It is perhaps most well-known for publishing one of the earliest sets of K–12 mathematics standards. NCTM's stated objectives are to develop effective curriculum and instruction, ensure equity in mathematics education, shape public policy, produce high quality mathematics education research, and provide professional development opportunities for mathematics educators. NCTM publishes works like the *Principles and Standards for School Mathematics* and *Curriculum Focal Points*, in addition to journals such as the *Mathematics Teacher, Teaching Children Mathematics, Mathematics Teaching in the Middle School*, and the *Journal for Research in Mathematics Education*. State, regional, and local affiliates also work to carry out NCTM's mission through annual conferences and other professional development opportunities. Similarly, trainers of mathematics teachers assemble in the Association of Mathematics Teacher Educators, and supervisors assemble in the National Council of Supervisors of Mathematics.

The Society for Industrial and Applied Mathematics (SIAM) originated in the early 1950s to represent mathematicians working in industry. Their numbers had grown as a result of the importance of mathematics in military research during World War II and the evolution of computers. SIAM seeks to advance applied mathematics, promote practical research, and encourage the exchange of applied mathematical ideas. Annual meetings, subject-specific workshops and conferences, and discipline-specific activity groups allow members to develop new applied mathematical ideas and techniques.

Organizations designed to promote minorities in mathematics include what is now known as the National Association of Mathematicians (NAM), which started as an informal group at the Annual Meeting of the American Mathematical Society in 1969. Lee Lorch recalled:

> In 1960, when A. Shabazz and S.C. Saxena, both on the faculty of Atlanta University (now Clark-Atlanta), and their graduate student W.E. Brodie were subjected yet again to Jim Crow treatment at the spring meeting of the Southeastern Section of MAA. . . . This, it should be noted, was several years after AMS and MAA commitments to the contrary. They had not been warned in advance that such discourtesy would be in store. The three left in protest. And so in 1969 the National Association of Mathematicians (NAM) came into being to address the needs of the Black mathematical community. This

was a turbulent period. A group of more or less left-oriented mathematicians established the Mathematicians Action Group (MAG) that same year. We were motivated largely by concern over the Vietnam war, the militarization of mathematics, the lack of democracy in the AMS, the existence of racism and sexism, and related social issues as they impinged on mathematicians and vice versa.

NAM focuses on education, career development, research, student development, and databases. NAM also publishes a newsletter and organizes a lecture series. The Benjamin Banneker Association was founded in 1986 to concentrate on the mathematics education of African Americans. There are also many associations that focus on science, like the Society for the Advancement of Chicanos and Native Americans in Science.

Organizations like Association for Women in Mathematics, European Women in Mathematics, and Korean Women in the Mathematical Sciences were created to support and promote female students, teachers, and researchers via social events, sponsored talks or conferences, workshops, and contests. Many attribute the beginning of the Association for Women in Mathematics to events in Boston and Atlantic City. In the late 1960s, Alice Shafer and Linda Rothschild organized a mathematics women's group in the Boston area. At a 1971 conference in Atlantic City, Joanne Darken suggested that women already at the Mathematics Action Group remain to form a caucus. As noted by president Lenore Blum:

What I remember hearing about Mary Gray and the Atlantic City Meetings, indeed what perked my curiosity, was an entirely different event, one that was also to alter dramatically the character of the mathematics community. In those years the AMS was governed by what could only be called an 'old boys network,' closed to all but those in the inner circle. Mary challenged that by sitting in on the Council meeting in Atlantic City. When she was told she had to leave . . . she responded she could find no rules in the by-laws restricting attendance at Council meetings. She was then told it was by 'gentlemen's agreement.' Naturally Mary replied 'Well, obviously I'm no gentleman.' After that time, Council meetings were open to observers and the process of democratization of the Society had begun.

Mary Gray placed an official announcement about the organization in the Notices of the American Mathematical Association and created its first newsletter in 1971.

Other notable mathematical organizations include the American Mathematical Association of Two Year Colleges (AMATYC), which was founded in 1974. AMATYC organizes conferences and workshops and publishes books and proceedings related to mathematics education in the first two years of college.

Mathematicians create other professional organizations under the umbrella of a wide variety of interests and themes. They assemble in national and international subject-specific societies that focus on areas such as linear algebra, mathematical physics, or mathematics and art, including the International Linear Algebra Society and the Association for Symbolic Logic, or through special interest groups at the Mathematical Association of America. Mathematical organizations that are related to religion or sexual orientation include the Association of Christians in the Mathematical Sciences and the Association of Lesbian, Gay, Bisexual and Transgendered Mathematicians. National and international mathematics honor societies include Kappa Mu Epsilon and Pi Mu Epsilon. Mathematicians interested in the advancement of science policy participate in advocacy groups such as the Triangle Coalition for Science and Technology Education and the American Association for the Advancement of Science.

Further Reading

American Mathematical Society. "Math on the Web: Societies, Associations and Organizations." http://www.ams.org/mathweb/mi-sao.html.

American Statistical Association. "History of the ASA: What do Florence Nightingale, Alexander Graham Bell, Herman Hollerith, Andrew Carnegie, and Martin Van Buren Have in Common?" http://www.amstat.org/about/history.cfm.

Archibald, Raymond. *A Semicentennial History of the American Mathematical Society, 1888–1938.* Providence, RI: American Mathematical Society, 1938. http://www.ams.org/publications/online-books/hmreprint-index.

Ball, John. "The IMU and You." *Notices of the American Mathematical Society* 52, no. 10 (2005).

Blum, Lenore. "A Brief History of the Association for Women in Mathematics: The Presidents' Perspectives"

Notices of the American Mathematical Society 38, no 7 (September 1991). http://www.awm-math .org/articles/notices/199107/blum/.

Kalman, Dan. "The Mathematics Tribe." *Math Horizons* 3 (September 1995).

Lorch, Lee. "The Painful Path Toward Inclusiveness." In *A Century of Mathematical Meetings*. Edited by Bettye Anne Case. Providence, RI: American Mathematical Society, 1996. http://www-users.math.umd.edu/~rlj/Lorch.html.

O'Connor, John, and Edmund Robertson. "MacTutor History of Mathematics Archive: Professional Societies." http://www-history.mcs.st-and.ac.uk/Societies/.

Straley, Tina. "A Brief History of the MAA." http://www .maa.org/aboutmaa/maahistory.html.

National Council of Teachers of Mathematics. "About NCTM." http://www.nctm.org/about/default .aspx?id=166.

CHRISTOPHER J. STAPEL

See Also: Clubs and Honor Societies; Curricula, International; Ethics; Government and State Legislation; North America.

Proof

Category: History and Development of Curricular Concepts.
Fields of Study: Communication; Connections; Problem Solving; Reasoning and Proof.

Summary: The product of deductive reasoning, the nature of proof has long been fundamental.

Deductive proofs have been an essential part of mathematics for over 2000 years, and some equate proof ability with competence in mathematics. Mathematician David Henderson defines an effective proof as a convincing communication that answers "why." Thus, a proof may connect ideas within a mathematical system or illuminate both the "how" and the "why" underlying the conjecture. Since a proof depends on the accepted standards of the audience or society through some type of peer review, there is also a long history of concerns about the nature of proof. For example, Galileo Galilei and Christoph Clavius debated the legitimacy of pictures, and Leopold Kronecker criticized the use of nonconstructivist methods. In the twentieth and twenty-first centuries, philosophical concerns about proofs continued as mathematicians considered the role of computers or empirical aspects and the implications of Kurt Gödel's groundbreaking work on consistency. While reasoning and proof have long been a part of mathematics curricula, the concepts took on an increased importance in the United States during the era of Sputnik and the space race, when many different types of proofs were emphasized. In the early twenty-first century, proofs remain fundamental in education, beginning in primary school. In pure mathematics, new research depends on proofs. The notion of proof has been clarified by mathematicians in the field of logic, who explore the foundations of proof.

Brief History
Early civilizations developed sophisticated notions of mathematical argumentation, as documented by evidence such as cuneiform tablets, papyri, and mathematical texts from ancient Babylon, China, and Egypt. The idea of a formal deductive proof arose as a distinct part of ancient Greek mathematics. Greek mathematicians studied and generalized mathematical ideas, using proofs to justify their claims. Mathematics historians theorize that the prevalence of debate in Greek society provided a conducive environment for the development of axiomatic argumentation. Euclid's *Elements* became "the" model for using a small set of axioms to deduce a large system of theorems and knowledge, now known as "Euclidean geometry."

Logic
Logical systems become the foundational structures necessary to create a proof. First, mathematicians use logic tools to argue that one mathematical statement follows as a logical consequence from other mathematical statements, and then they use logic tools to establish a formal proof by building a chain of consequent statements from initial assumptions. These logic tools include connectives (negation, conjunction, disjunction, conditional implications, and equivalence), quantifiers, truth statements, tautologies, and inferential structures (such as *modus ponens* and *reductio ad absurdum*).

Direct Proofs

Mathematical proofs can be done in diverse ways, all reflecting different inferential structures. Starting with an initial conjecture (such as $H{\rightarrow}C$) involving a hypothesis H and a conclusion C, Direct Proofs build logical chains of compound statements, using conditional implications as the links. They are usually written in the traditional two-column format using high school geometry. The following illustrates the use of a Direct Proof, though it is not entirely rigorous.

Conjecture: If a and b are prime numbers greater than 2, then their sum $a + b$ is composite.
  Proof:

Statement	Justification
1. a and b are prime numbers > 2	1. Given
2. a and b are odd numbers	2. The only even prime is 2
3. Let $a = 2m + 1$ and $b = 2n + 1$ for $m, n > 1$	3. Definition of an odd number
4. Sum $a + b = (2m + 1) + (2n + 1)$	4. Substitution
5. Sum $a + b = 2(m + n + 1)$	5. Properties of arithmetic operations
6. Sum $a + b$ is even	6. Definition of an even number
7. Sum $a + b > 2$	7. $m + n + 1 > 1$
8. Sum $a + b$ is composite	8. The only even prime is 2

A Proof by Contrapositive is very similar to a Direct Proof, with the difference being the format of the conjecture itself. While a Direct Proof proves the conjecture $H \rightarrow C$, a Proof by Contrapositive uses a Direct Proof to prove the contrapositive, ${\sim}C \rightarrow {\sim}H$. In the previous example, a Proof by Contrapositive would prove the statement "If the sum $a + b$ is prime, then a and b are not both prime numbers greater than 2."

Indirect Proofs

In contrast to Direct Proofs, an Indirect Proof assumes the negation of the conclusion ${\sim}C$ to be true and then uses a Direct Proof to prove the truth of the negation ${\sim}S$ for some true statement S. By the Law of Logic Contradiction (S and ${\sim}S$ cannot both be true), which implies that the original conclusion C must be true. The following illustrates the use of an Indirect Proof.

Conjecture: The $\sqrt{2}$ is an irrational number.
  Proof:

Statement	Justification
1. Assume $\sqrt{2}$ is a rational number	1. Negation of conclusion
2. $\sqrt{2} = a/b$ where $\gcd(a, b) = 1$ and a, b positive integers	2. Definition of the rationals and greatest common divisor
3. $2 = a^2/b^2$	3. Squaring both sides
4. $2b^2 = a^2$	4. Multiplying both sides by b^2
5. a^2 is even	5. Definition of an even number
6. a is even	6. Squares of odd integers are odd
7. $a = 2m$ for $m > 1$	7. Definition of an even number
8. $2b^2 = (2m)^2 = 4m^2$	8. Substitution
9. $b^2 = 2m^2$	9. Multiplying both sides by 1/2
10. b^2 is even	10. Definition of an even number
11. b is even	11. Squares of odd integers are odd
12. $\gcd(a, b) \geq 2$	12. Definition of gcd
13. Original assumption is false	13. Contradicting assumption $\gcd(a, b) = 1$

Using the idea of infinite descent, this Indirect Proof is considered to be one of the most "beautiful" proofs by the mathematical community. Though not as elegant, it would be possible to prove this same conjecture using a Direct Proof. Also, it is important to note that this Indirect Proof uses some outside knowledge from number theory (such as, squares of odd integers are odd), which would have to be proved prior to its use as justification within the Indirect Proof.

Deduction and induction constantly play important roles in the proof process. For example, in the previous Direct Proof, a considerable number of examples could be systematically examined: $3 + 5 = 8$, $3 + 7 = 10$, $5 + 11 = 16$, $23 + 47 = 70$, and so on. These examples provide inductive evidence that the conjecture is true, nothing more. That is, the cumulative contribution of the examples is only increased confidence that the conjecture is true and that a formal deductive proof is needed. And, a Proof by Exhaustion of All Cases is not possible because the number of pairs of primes to consider is infinite. Nonetheless, a Proof by Induction is often possible in situations involving an infinite number of examples, as illustrated by the following proof.

Conjecture: $1 + 2 + 3 + 4 + \cdots + n = \dfrac{n(n+1)}{2}$.

Proof:
Case $n = 1$: Substituting, $1 = \dfrac{1(1+1)}{2} = 1$.

Assume case for k is true, need to show case for $(k + 1)$ is true: Given the assumption

$$1 + 2 + 3 + \cdots + k = \frac{k(k+1)}{2}. \text{ Then,}$$

$$1 + 2 + 3 + \cdots + k + (k+1) = \frac{k(k+1)}{2} + (k+1)$$

$$= (k+1)\left(\frac{k}{2} + 1\right)$$

$$= (k+1)\left(\frac{k}{2} + \frac{2}{2}\right)$$

$$= \frac{(k+1)(k+2)}{2}$$

$$= \frac{(k+1)\left[(k+1) + 1\right]}{2}.$$

For some conjectures, a visual "Behold!" Proof is possible. A common example is this proof of the Pythagorean Theorem.

Conjecture: In any right triangle, the square of the hypotenuse c (side opposite the right angle) equals the sum of the squares of the other two sides a and b: $c^2 = a^2 + b^2$.

Proof: Behold!

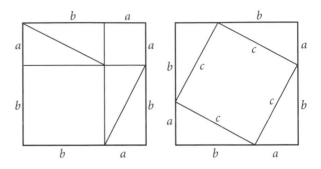

As in most proofs, effort is needed to understand a "Behold!" Proof. In this example, focus on the common areas and rearrangement of the four triangles. The first large square involves two smaller squares (areas a^2 and b^2) and four triangles, while the second large square involves one small square (area c^2) and the same four triangles. Noting that this proof has been traced back to early Chinese mathematics, it is important to add that more than 360 different proofs of the Pythagorean Theorem are known.

Despite their connection to truth, proofs can create mathematical fallacies. Examples include the use of Mathematical Induction to prove that, "All horses are of the same color," the misleading dependence on a geometrical diagram to prove that all triangles are isosceles or even proofs that disguise computational errors, such as the following:

Conjecture: $1 = 2$
Proof:

Statement	Justification
1. Let $n = m > 0$	1. Assumption
2. $n^2 = mn$	2. Multiplying both sides by n
3. $n^2 - m^2 = mn - m^2$	3. Subtracting m^2 from both sides
4. $(n+m)(n-m) = m(n-m)$	4. Factoring both sides
5. $(n+m) = m$	5. Dividing both sides by $(n-m)$
6. $(m+m) = m$	6. Substitution as $n = m$

| 7. $2m = m$ | 7. Simplification |
| 8. $2 = 1$ | 8. Dividing both sides by m |

In supporting this obviously wrong conclusion, this proof relies on the reader's literal acceptance of each statement and its justification. That is, the proof seems "true" unless the reader notices that statement five involves division by zero, which is not possible.

When constructing proofs of mathematical conjectures within a system, mathematicians are concerned with many issues related to the logical structure. Is the system "consistent," in that no proven theorem contradicts another? Is the system "valid," in that no mathematical fallacies or false inferences will be created? Is the system based on underlying axioms or initial assumptions that are reasonable? And, is the system "complete," in that every conjecture can be proven either true or false? In the 1930s, logician Kurt Gödel shocked the mathematical world when he proved that a "powerful" mathematical system cannot be both complete and consistent at the same time. For some mathematicians, Gödel's theorems weakened the foundational structure of mathematics, while others felt that it strengthened it. Mathematicians also debate about the role of computers in proofs. In the seventeenth century, Gottfried Leibniz predicted an automatic counting machine that would vastly improve reasoning. In the twentieth century, Herbert Gelernter wrote a program to prove theorems from Euclid's *Elements*, but critics noted the dependence on programmer-supplied rules. Some mathematicians do not accept proofs such as the first proof of the Four Color Theorem in 1977, which depended on an analysis of many cases by a computer.

Formal proof is a special technique within the realm of mathematics, which is why the public views mathematics as the prime model for establishing truth via argumentation. The idea of proof is invoked in other fields, but with a more limited meaning. For example, in courtrooms, the element of truth is replaced with the phrase "beyond reasonable doubt" given the available evidence. In the sciences, proof is desired but cannot be established by experimental data; at best, the data can support the creation of hypotheses and theories, which will be either further verified or discounted by new experiments and new data.

Further Reading

Cupillari, Antonella. *The Nuts and Bolts of Proofs*. Belmont, CA: Wadsworth Publishing, 1989.

Laczkovich, Miklós. *Conjecture and Proof*. Washington, DC: Mathematical Association of America, 2001.

MacKenzie, Donald. *Mechanizing Proof: Computing, Risk, and Trust*. Cambridge, MA: MIT Press, 2004.

Nelson, Roger. *Proofs Without Words: Exercises in Visual Thinking*. Washington, DC: Mathematical Association of America, 1997.

Polster, Burkard. *Q.E.D.: Beauty in Mathematical Proof*. New York: Walker & Company, 2004.

Solow, Daniel. *How to Read and Do Proofs: An Introduction to Mathematical Thought Processes*. Hoboken, NJ: Wiley, 1982.

Velleman, Daniel. *How to Prove It: A Structured Approach*. Cambridge, England: Cambridge University Press, 1994.

JERRY JOHNSON
SARAH J. GREEWALD
JILL E. THOMLEY

See Also: Mathematics, Elegant; Reasoning and Proof in Society.

Proof in Society

See *Reasoning and Proof in Society*

Psychological Testing

Category: Medicine and Health.
Fields of Study: Algebra; Data Analysis and Probability; Representations.
Summary: Though they often require a subjective element, psychological tests make every effort to generate useful quantitative data.

Testing is used for many different purposes within psychology—among them to evaluate intelligence, diagnose psychiatric illness, and identify aptitudes and interests. Although the results of testing are rarely used as the sole criterion to make a diagnosis or other decision about an individual, they are often used in conjunction with information gained from other sources, such as interviews and observations of behavior. There are many types of psychological tests, but most share the goal of expressing an essentially unobservable quality such as intelligence or anxiety in terms of numbers. The numbers themselves are not meant to be taken literally—no one seriously believes that a person's intelligence is equivalent to their IQ score, for instance. Instead the numbers are useful tools that help evaluate a person's situation; for instance, how does the intellectual development of one particular child relate to that of other children of his age? Of course, the results of psychological testing should be evaluated with the social context of the individual in mind and with full respect for human diversity.

Most psychological tests try to translate unobservable qualities such as intelligence or anxiety in terms of numbers.

Psychometrics

Psychometrics is a field of study that applies mathematical and statistical principles to devise new psychological tests and evaluate the properties of current tests. Psychologist Anne Anastasi was often known as the "test guru" for her pioneering work in psychometrics. In her 1954 book *Psychological Testing*, she discussed the ways in which trait development is influenced by education and heredity as well as how differences in training, culture, and language affect measurement. The two most common approaches to psychometrics in the twenty-first century are classical test theory and item response theory (IRT).

Classical test theory is the older approach and the calculations required can be performed with a pencil and paper, although twenty-first-century computer software is often used. Classical test theory assumes that all measurements are imperfect and thus contain error: the goal is to evaluate the amount of error in a measurement and develop ways to minimize it. Any observed measurement (for instance, a child's score on an intelligence test) is made up of two components: true score and error. This may be written as an equation: $X = T + E$, where X is the observed score, T is the true score (the score representing the child's true intelligence), and E is the error component (resulting from imperfect testing). Classical test theory assumes that that error is random and thus will sometimes be positive (resulting in a higher observed score than true score) and sometimes negative (resulting in a lower observed score than true score) so that over an infinite number of testing occasions, the mean of the observed scores will equal the true score. Although normally a test is administered only once to a given individual, this is a useful model that facilitates evaluation of the reliability and validity of different tests.

Item response theory (IRT) is a different approach to psychological testing and assumes that observed performance on any given test item can be explained by a latent (unobservable) trait or ability so that individuals may be evaluated in terms of the amount of that trait they contain, and items may be evaluated in terms of the amount of the trait required to answer them positively. For an item on an intelligence test (intelligence being the latent trait), persons with higher intelligence should be more likely to answer the question correctly. The same principle applies to IRT-based tests evaluating other psychological characteristics; for instance, if an item in a psychological screening test is meant to diagnose depression, a person with more depressive symptoms should be more likely to answer it positively. IRT is a mathematically complex method of analysis that depends on the use of specialized computer software and has become a popular means to evaluate psychological tests as computers have become more affordable. Although the mathematical models of IRT differ from that of classical test theory, the goals are the same: to devise tests that measure characteristics of individuals with a minimum of error.

Reliability and Validity

The term "reliability" refers to the consistency of a test score: if a test is reliable it will yield consistent results over time and without regard to the irrelevant conditions such as the person administering the test. Internal consistency is considered an aspect of reliability: it means that all the items in a test measure the same thing. Temporal reliability is also called "test-retest reliability" because it is typically evaluated by having groups of individuals take the same test on several occasions and seeing how their scores compare Some differences are expected because of the random nature of the error component, but there should be a strong relationship between the observed scores of individuals on multiple occasions.

The term "inter-rater reliability" refers to the consistency of a test or scale regardless of who administers it. For instance, psychiatric conditions are often evaluated by having an observer rate an individual's behavior using a scale, and the results for different observers evaluating the same individual at the same time should be similar. For instance, three psychologists using a scale to evaluate the same child for hyperactivity should reach similar conclusions. Both types of reliability are typically evaluated by correlating test results on different occasions (temporal) or the scores returned by different raters (inter-rater).

Internal consistency can be measured in several ways. The split-half method involves having a group of individuals take a test, then splitting the items into two groups (for instance, odd numbered items in one group and even in the other) and calculating the correlation between the total scores of the two groups.

Cronbach's alpha (or coefficient alpha) is a refinement of the split-half method: it is the mean of all possible split-half coefficients. The measure was developed and named "alpha" by Lee Chronbach, an educational psychologist and measure theorist who began his career as a high school mathematics and chemistry teacher.

The term "validity" refers to whether a test measures what it claims to be measuring. Three types of validity are typically discussed: content, predictive, and construct. Content validity refers to whether the test includes a reasonable sample of the subject or quality it is intended to measure (for instance, mathematical aptitude or quality of life) and is usually established by having a panel of experts evaluate the test in relation to its purpose. Predictive validity means that test scores correlate highly with measures of similar outcomes in the future; for instance, a test of mechanical aptitude should correlate with a new hire's success working as an auto repairman. Construct validity refers to a pattern of correlations predicted by the theory behind the quantity being measured: the scores on a test should correlate highly with scores on other tests that measure similar qualities and less highly with those that measure different qualities.

Further Reading

Embretson, Susan E., and Steven P. Reise. *Item Response Theory for Psychologists*. Mahwah, NJ: Erlbaum, 2000.

Furr, R. Michael, and Verne R. Bacharach. *Psychometrics: An Introduction*. Thousand Oaks, CA: Sage, 2007.

Gopaul McNicol, Sharon-Ann, and Eleanor Armour-Thomas. *Assessment and Culture: Psychological Tests with Minority Populations*. Burlington, MA: Elsevier, 2001.

Kline, Paul. *The Handbook of Psychological Testing*. 2nd ed. New York: Routledge, 2000.

Wood, James M., Howard N. Garb, and M. Teresa Neszworski. "Psychometrics: Better Measurement Makes Better Clinicians." In *The Great Ideas of Clinical Science: 17 Principles That Every Mental Health Professional Should Understand*. Edited by Scott O. Lilienfeld and William T. O'Donohue. New York: Routledge, 2007.

Sarah Boslaugh

See Also: Diagnostic Testing; Educational Testing; Intelligence Quotients.

Pulleys

Category: Architecture and Engineering.
Fields of Study: Algebra; Geometry.
Summary: Pulleys provide mechanical advantage and help people do work.

A pulley is a simple machine consisting of a cylinder, called a "drum," "wheel," or "sheave," rotating on an axle, and a rope, chain, or belt running over the cylinder without sliding. Pulley drums often have grooves and ribs that prevent their ropes from sliding over the edge. People use pulleys in three ways: to change directions of forces, to change magnitude of forces, and to transmit power. Pulleys are used in building and construction, ship rigging, and within belt-driven mechanisms.

Mathematicians have investigated many aspects of pulleys. There is evidence that Archimedes of Syracuse used a compound pulley to move a ship and studied the related theories. He famously expressed: "Give me a place to stand and I will move the Earth." While his mechanical inventions brought him recognition among his contemporaries, he seems to have preferred pure mathematics. Guidobaldo del Monte reduced systems of pulleys to levers. Guillaume de l'Hôpital investigated the equilibrium of a pulley system, and mathematicians continue to explore his pulley problem using algebra, geometry, trigonometry, and calculus. A mechanical tide-predicting machine, which incorporated pulleys, is attributed to William Thomson, who later became Lord Kelvin.

Changing Directions of Forces

In an example of this use of pulleys, construction workers often attach pulleys to roofs of buildings. A builder standing on the ground can pull down on one end of the pulley's rope and a weight on the other end will move up as the drum rotates.

The vectors of input and output forces always go along the two ends of the pulley's rope. This means that a pulley can change the direction of a force within the plane that is perpendicular to the pulley's axle but not sideways from that plane. The builder can also stand inside the building, pulling the rope through a window, or on the roof pulling horizontally, as long as the triangle formed by the worker, the weight, and the pulley's drum is perpendicular to the pulley's axle.

Changing Magnitudes of Forces

When a pulley is used to change the magnitude of a force, its axle is attached to the weight, and the pulley moves up together with the weight. For example, a sailor can attach one end of a line to a yardarm, string it around a pulley's drum attached to a weight, and pull the other end up, standing on the yardarm. The sailor will only have to apply the force equal to one-half of the weight.

Does the other half of the force disappear, breaking the conservation of energy law and the work-energy theorem? No, it is distributed to the other, attached end of the rope. Moreover, the sailor will use half the force, but pull enough line to cover twice the distance the weight is lifted. The total work, which is equal to the product of the force and the distance, will be the same as in the fixed pulley case:

$$W = F \times d = \frac{1}{2} F \times 2d.$$

Changing Directions and Magnitudes of Forces: Blocks and Tackles

Because it is much easier to work for longer than to increase one's force, movable pulleys are widely used. A block and tackle is a pulley system where the rope zigzags through movable and fixed pulleys. Depending on the way the tackle is rigged, it can provide a force advantage with the factor of two, as in the example above, or 3, 4, 5 and so on. At first sight, it would seem that a block and tackle can reduce the force required to lift weights by any factor. However, friction interferes increasingly with more pulleys used.

Marine cadets memorize rigging of common block and tackle systems, and the names of tackles corresponding to force advantage factors: factor 2: "gun"; factor 3: "luff"; factor 4: "double"; factor 5: "gyn."

Drums for tackles may have multiple grooves to reduce rope friction. When tackles are combined, for example, a double tackle upon a luff tackle, their force advantage factors multiply, in this case, creating the force advantage of $3 \times 4 = 12$.

Transmitting Power

A belt or a chain going in a loop over two or more pulley drums makes all of them rotate when one is rotated. For example, a bicyclist rotates the special pulley drum, called a "crank," to which pedals are attached. The rotation of this crank is transmitted to the rotation of the rear wheel's crank, which makes the bicycle move. Using drums of different diameters, such as cranks on a sports bicycle drivetrain, can produce a force advantage.

Until the mid-twentieth century, factories typically used belts distributing power to individual machines from one central rotating drum, connected to a large steam, turbine, or animal-powered capstan engine. This power transmission system is called "line shaft." Because most industries have switched to compact electric motors, one is currently more likely to meet this type of a pulley in a museum or a history book. A human-powered capstan is also a popular science or historical fiction trope, used to demonstrate oppression, for example, in *Conan the Barbarian* and *Captain Blood*.

Further Reading

Boute, Raymond. "Simple Geometric Solutions to De l'Hospital's Pulley Problem." *College Mathematics Journal* 30, no. 4 (1999).

Hahn, Alexander. *Basic Calculus: From Archimedes to Newton to its Role in Science*. Emeryville, CA: Key College Publishing, 1998.

Rau, Dana. *Levers and Pulleys: Super Cool Science Experiments*. Ann Arbor, MI: Cherry Lake Publishing, 2009.

MARIA DROUJKOVA

See Also: Archimedes; Bicycles; Elevators.

Puzzles

Category: Games, Sport, and Recreation.
Fields of Study: Algebra; Geometry; Number and Operations.
Summary: Because problem solving is a core activity of mathematics, it lends itself well to puzzles.

A puzzle is a question, problem, or contrivance designed to challenge and expand the mind and perhaps test ingenuity. Puzzles have been found in virtually all cultures and all historic periods, even in mythology. According to legend, the Sphinx prevented anyone from entering Thebes who failed to find the correct answer to the question: What is it that has four feet in the morning, two at noon, and three at twilight?

Mathematicians have long created puzzles and explored their solutions for research and applications. They have also created puzzles for purely recreational purposes. Teachers in many subjects within and outside mathematics use puzzles in the classroom.

There are a number of ways in which words and arrangements of letters or objects are used to create puzzles. Some problems in the Rhind Mathematical Papyrus (1650 B.C.E.) are seen as puzzles. One example is a rhyme that also appears in Leonardo Pisano Fibonacci's 1202 work *Liber Abaci* and is still popular today. Here is a modern version:

> As I was going to St. Ives,
> I met a man with seven wives.
> Each wife had seven sacks,
> Each sack had seven cats,
> Each cat had seven kits.
> Kits, cats, sacks, wives,
> How many were going to St. Ives?

One may only assume that the narrator was going to St. Ives, not necessarily the other travellers. Mathematically, logic, branching diagrams, multiplication, and addition can be used to determine the final solution.

Traditional in several cultures, namely in Africa, is the Crossing Problem. The following is a version from Alcuin of York (735–804):

> A man wishes to ferry a wolf, a goat, and a cabbage across a river in a boat that can carry only the man and one of the others at a time. He cannot leave the goat alone with the wolf nor leave the goat alone with the cabbage on either bank. How will he safely manage to carry all of them across the river?

To solve this problem, one must recognize that the man may carry an item back and forth across the river as many times as needed and ultimately find appropriate combinations and sequencing. Dynamic versions of this game appear online, adding visual and tactile components to the solving process. Extensions of this problem include adding more items to the list, increasing the size of the boat to carry more items, and adding an island in the middle of the river where objects may be placed. Mathematicians such as Luca Pacioli, Niccolo Tartaglia, Claude-Gaspar Bachet, and Edouard Lucas investigated this problem. A well-known medieval task consisted of arranging men in a circle so that when every k-th man is removed, the remainder shall be a certain specified man. Several authors commented on this, from Girolano Cardano in the sixteenth century to Donald Coxeter in the twentieth century.

Word Puzzles

Anagrams have a long and mysterious history, being seen as source of ludic pleasure but are also believed by some to possess mystic powers. Inside a word or phrase, another one is hiding that one can get by permuting the letters in a different order. For instance, the letters in the word "schoolmaster" may be rearranged to form the related phrase "the classroom."

Lewis Carroll (1832–1898) invented a forerunner of the crossword: the "doublet." There are two words presented to the solver, who is required to change one word to the other by replacing only one letter at a time, forming a legitimate word with each transformation. One of his examples is to change "HEAD" into "TAIL," which can be done via the following sequence: "HEAL," "TEAL," "TELL," "TALL," and "TAIL."

Visual Puzzles

Visual puzzles are also popular, such as optical illusions, which have long been investigated by mathematicians. Some of these address mathematical questions in disciplines like geometry and visualization, including figures that appear to be impossible.

Figure 1. Are the two dark lines parallel?

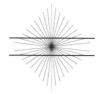

Figure 2. An illustration of an "impossible" object.

Samuel Loyd (1841–1911) is referred to by some as "America's greatest puzzlist." He reputedly created thousands of puzzles. Some of his inventions were very original, like the Get Off the Earth puzzle. There are 13 men in the figure on the left. Rotating the puzzle, as shown in the figure on the right, produces a drawing that has 12 men. What happened to the 13th man?

Figure 3. The Get Off The Earth puzzle.

Arithmetic Puzzles

Numerical relations and arithmetical principles are often found in puzzles. "Magic squares," which are square arrays of consecutive numbers with constant sum in columns, rows, and diagonals, illustrate this clearly. One of the oldest, the Chinese *lo-shu*, dates back thousands of years. Leonhard Euler's (1707–1783) work on Latin Squares, which are arrays of symbols with no repetitions in rows or columns, is one of the foundations of *Sudoku* puzzles, which appeared in a U.S. magazine in the 1970s but became famous first in Japan and then in the world. Tartaglia (1500–1557) presented the following numerical problem: A dying man leaves 17 horses to be divided among his three sons in the proportion $1/2 : 1/3 : 1/9$. Can the brothers carry out their father's will? Since 17 is not a multiple of 2, 3, or 9, there is no solution that would give all of the sons a whole number of horses.

Some authors shared problems, even if they lived in different centuries. Fibonacci (1170–1250), Tartaglia, and Bachet (1581–1638) all investigated the question:

> If you have a balance, what is the least number of weights necessary to weigh any integer number of pounds from 1 to 40? (Assume you can put weights in either side of the balance.)

"Cryptarithms," created for training the calculating mind in 1913, were very popular in the twentieth century. In a cryptarithm, one is asked to find the digits erased from a valid calculation. Later, prolific English puzzle inventor, Henry Dudeney (1857–1930), substituted letters for the unknown numbers to create another layer of meaning. In his first example of an "alphametic" is the equation: SEND + MORE = MONEY, where each letter represents a different digit, and the addition is correct.

Rearrangement Puzzles

Some dissection and rearrangement puzzles are based on mathematical principles. Archimedes of Syracuse (287–212 B.C.E.) may have created a 14-piece puzzle, the "Stomachion," as part of his research. It resembles a version of a "Tangram," a Chinese puzzle that became very popular in the nineteenth century in the West and is often used in mathematics classrooms in the twenty-first century to investigate dissections and concepts like the Pythagorean Theorem, named for Pythagoras of Samos. The Fibonacci sequence relation

$$\left(F_n\right)^2 = F_{n-1}\,F_{n+1} + \left(-1\right)^{n-1}$$

with $n = 6$ can be used to create a dissection puzzle. Larger values of n generate similar, more impressive puzzles, where the difference of area between a large square and a large rectangle is always included. Some dissection puzzles may lead to optical illusions when the pieces do not fit exactly together, leading to two figures composed of the same pieces that have different areas.

Figure 4. An 8-by-8 square and 5-by-13 rectangle made with the same pieces?

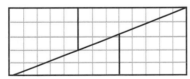

Topological Puzzles

Ring and string puzzles as well as knotted puzzles are examples of topological puzzles, where no discontinuous deformations like cutting the string are allowed. In his *De Viribus Quantitatis* (c. 1500), cited as the oldest book in recreational mathematics, Luca Pacioli (1445–1517) describes the Chinese Rings, a topological puzzle still popular in the twenty-first century.

Figure 5: A modern version of the Chinese Rings puzzle

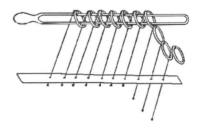

Euler's name is linked to several puzzles. He solved the Bridges of Konigsberg Problem, and this work of his is usually seen as the starting point of topology and graph theory.

Figure 6: The Bridges of Konigsberg Problem: is it possible to cross all the bridges only once?

The concept of the Eulerian graph is rooted in Euler's resolution of the Bridges of Koenigsberg problem.

Movement Puzzles

Numerous puzzles involve patterned movement within some type of framework, and solutions sometimes involve mathematical techniques like numbering, recursion, group theory, and determinants. Peg Solitaire traces its origins from seventeenth-century France. It is a game where a board has all its holes occupied with pegs except for the central one. The objective is, making valid moves (small jump capture), to empty the entire board but for a solitary peg in the central hole (see Figure 7).

The Towers of Hanoi is a puzzle invented in 1883 by N. Claus, a pseudonym of the mathematician Edouard Lucas (1842–1891). A pile of discs of decreasing radius lays on one of three poles. Moving one disc at a time, without letting a bigger disc rest on a smaller one, the solver is asked to change the pile from one pole to another (see Figure 8).

Figure 7. Peg Solitaire: starting and target position.

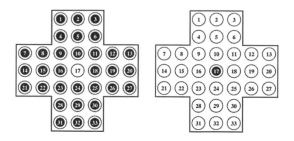

Figure 8. Towers of Hanoi: starting and target positions.

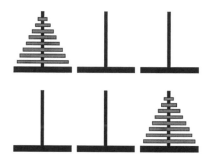

The recursive character of the solution to this puzzle makes it somewhat similar to the Chinese Rings.

Other Puzzles

The chessboard is a rich source of puzzles that attracted many mathematicians. In the Knight Tour problem, a knight must visit all the squares of the board just once. Euler is one mathematician who published a solution. Mathematician Johann Carl Friedrich Gauss (1777–1855) was attracted by the 8-Queen Problem, in which eight queens must be placed on a chessboard so they cannot capture any other queen. Some mathematicians have used determinants to solve this problem.

Figure 9. The 8-Queen Problem: one solution.

The nineteenth century produced a popular puzzle named "15." It consists of a sliding device, a 4-by-4 array with the numbers one through 15 and an empty cell. The puzzle was scrambled and the solver was required to transform the scrambled order back to the natural order with the empty cell in the last position. Sam Loyd offered $1,000 to whoever could reorder a scrambled 14 and 15 in an otherwise solved puzzle. The prize was never claimed. The impossibility of this challenge can be understood when phrased in the language of group theory.

Figure 10: The "impossible" task.

1	2	3	4
5	6	7	8
9	10	11	12
13	15	14	

Another very mathematical puzzle that captivated the world was Rubik's Cube, created by Hungarian architect Erno Rubik in the 1970s that became the best selling puzzle in history. A 3-by-3-by-3 cube, with differently colored faces, moves by slices, getting scrambled with just a few moves. To find the way back to the starting position is an incredible challenge. This toy puzzle is used to illustrate many group theory concepts. On the other hand, knowledge of group theory facilitates the understanding of the puzzle itself.

Since ancient times, descriptions of "mazes" that must be traversed in a particular pattern of moves have abounded in legend and literature. The Minotaur–Theseus tale is one such example. Stone and hedge labyrinths may still be found in places like Europe and many puzzle books contain paper mazes. Some mazes can be understood using what is known as "level sequences."

The "jigsaw puzzle" was invented in England in the mid-1870s as a pedagogical device. Children were asked to rebuild maps. In the twentieth and twenty-first centuries, jigsaw puzzles expanded to include three-dimensional jigsaw puzzles, including spherical three-dimensional puzzles, and two-dimensional jigsaw puzzles that are all one color that have all the pieces cut to the same shape. This last style of puzzle is related to tiling. Another mathematical question is how to optimally and efficiently design and cut out puzzle pieces according to certain specifications.

Puzzle designer Scott Kim is considered by some to be a master of symmetry. He has diverse interests in many fields, including mathematics, computer science, puzzles, and education. When discussing these interests, he emphasizes the ties between them rather than their differences. One of his creations is an ambigram to honor of the great Martin Gardner (1914–2010), who invented many puzzles and is known for his recreational mathematics works. An ambigram is a figure that appears the same when rotated 180 degrees or viewed upsidedown.

Further Reading

Danesi, Marcel. *The Puzzle Instinct: The Meaning of Puzzles in Human Life.* Bloomington: Indiana University Press, 2002.

Dedopulos, Tim. *The Greatest Puzzles Ever Solved.* London: Carlton Books, 2009.

Olivastro, Dominic. *Ancient Puzzles: Classic Brainteasers and Other Timeless Mathematical Games of the Last 10 Centuries.* New York: Bantam Books, 1993.

Petkovic, Miodrag. *Famous Puzzles of Great Mathematicians.* Providence, RI: American Mathematical Society, 2009.

Sam Loyd's Puzzles. http://www.samuelloyd.com/gallery.html.

Scott Kim Puzzlemaster. "Inversions Gallery." http://www.scottkim.com/inversions/gallery/gardner.html.

Slocum, Jerry, and Jack Botermans. *New Book of Puzzles: 101 Classic and Modern Puzzles to Make and Solve.* New York: W.H. Freeman, 1992.

———. *The Tangram Book: The Story of the Chinese Puzzle With Over 2,000 Puzzles to Solve.* New York: Sterling Pub. 2003.

Slocum, Jerry, and Dic Sonneveld. *The 15 Puzzle: How It Drove the World Crazy; The Puzzle That Started the Craze of 1880; How America's Greatest Puzzle Designer, Sam Loyd, Fooled Everyone for 115 Years.* Beverly Hills, CA: Slocum Puzzle Foundation, 2006.

Spencer, Gwen. "A Conversation with Scott Kim." *Math Horizons* 12 (November 2004).

Jorge Nuno Silva

See Also: Acrostics, Word Squares, and Crosswords; Board Games; Coding and Encryption; Dice Games; Mathematical Puzzles; Optical Illusions; Sudoku.

Puzzles, Mathematical

See *Mathematical Puzzles*

Pythagorean and Fibonacci Tuning

Category: Arts, Music, and Entertainment.
Fields of Study: Algebra; Measurement; Representations.
Summary: The relationship between mathematics and music led to several tuning systems.

A musical scale is a sequence of ordered notes used to construct music compositions. Scales can be classified according to their starting point, the intervals between their notes, or the number of notes they contain. Instruments may be tuned according to many possible systems. There are close mathematical connections between musical scales, tuning systems, and number theory, as well as dynamical systems. Mathematics also plays a critical role in designing playable and efficient keyboards for instruments that will be tuned to something other than the standard eight-note Western scale.

Most Western music uses an eight-note "octave" scale (do, re, mi, fa, sol, la, ti, do), where the two "do" notes have the same tone but different pitches. The piano keyboard is set up in the C major key, where the white keys starting with C correspond to the eight notes in the octave.

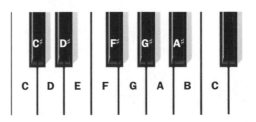

Figure 1.

F♯	G	G♯	A	A♯	B	C	C♯	D	D♯	E	F	F♯	G	G♯	A	A♯	B	C
11	3	14	6	17	9	1	12	4	15	7	18	10	2	13	5	16	8	19

There are also tones between some of the notes on the scale, represented on the piano by the black keys. Counting from C to B, there are 12 equal semitones in the chromatic scale of Western music.

To tune an instrument with strings, the lengths of the strings are adjusted to produce the correct pitch. Pythagoras of Samos (570–495 B.C.E.) is credited with realizing two things that allowed him to calculate the string lengths for the 12 semitones of the chromatic scale:

1. A string that is half as long produces the tone that is one octave higher. A string that is twice as long produces a tone that is one octave lower.
2. A string that is two-thirds as long produces a tone that is up five notes (called a *fifth*, or *do-sol* interval), seven semitones higher in the 12-tone chromatic scale.

Pythagoras saw that seven and 12 share no common factors and that he could use this fact to generate the lengths of all 12 strings in the chromatic scale.

1. Start with a string that sounds like a C note.
2. Cut a string that is two-thirds of the C string to give G.
3. Cut a string that is twice as long as G, yielding the same tone down an octave.
4. Cut a string two-thirds of this new lower G to give D.
5. Cut a string two-thirds as long as D to give A.
6. Cut a string twice as long as A, yielding A down an octave.

7. Cut a string two-thirds of the lower A to give E.
8. Cut a string two-thirds of E to give B.
9. Cut a string twice as long as B, yielding B down an octave.

Continue in this pattern, shortening a string to two-thirds to produce new higher notes and doubling the string when needed to avoid going past the top of the octave. After 19 steps, all of the strings of the C to C octave are determined, as well as a few extra notes below C (see Figure 1).

Called the "circle of fifths," this method of tuning by shortening the string to move up seven semitones (and back 12 when needed) would not work if the two numbers involved shared a common factor, such as four and 12. Not all of the semitones would be "hit" in that case.

Equal Tuning

Pythagoras was a little off when he assumed that a string two-thirds as long would produce the seventh semitone. In actuality, using irrational numbers (something Pythagoras did not believe in), the lengths of string needed to produce all of the semitones can be found more precisely. Starting with a string of length two, one can factor two into 12 equal parts or "twelfth roots." This method of tuning, used in the twenty-first century for most music, is called "equal tuning" (see Figure 2). The values of these irrational numbers to three decimal places show that the fifth note (or seventh semitone) string, G, is actually slightly more than two-thirds of the C string: two-thirds of a string of

Figure 2.

C	C♯	D	D♯	E	F	F♯	G	G♯	A	A♯	B	C
$\left(\sqrt[12]{2}\right)^{12}$	$\left(\sqrt[12]{2}\right)^{11}$	$\left(\sqrt[12]{2}\right)^{10}$	$\left(\sqrt[12]{2}\right)^{9}$	$\left(\sqrt[12]{2}\right)^{8}$	$\left(\sqrt[12]{2}\right)^{7}$	$\left(\sqrt[12]{2}\right)^{6}$	$\left(\sqrt[12]{2}\right)^{5}$	$\left(\sqrt[12]{2}\right)^{4}$	$\left(\sqrt[12]{2}\right)^{3}$	$\left(\sqrt[12]{2}\right)^{2}$	$\sqrt[12]{2}$	1
2	1.888	1.782	1.682	1.587	1.498	1.414	1.335	1.260	1.189	1.122	1.059	1

length 2 would yield a G string of length 1.333 rather than the equal tuning length of approximately 1.335. This little bit of difference is magnified when the circle of fifths technique is used to tune the strings, yielding notes that sound flat.

Other Tuning Systems

Between Pythagoras's time and the twenty-first century, a number of other tuning strategies were developed as music and mathematics knowledge grew. Popular in the medieval age, for example, was "just" tuning, which differs from both Pythagorean and equal tuning. To use equal tuning in the twenty-first century, one does not have to physically measure strings precisely; equipment can be used to measure the fundamental frequency (related to the pitch) of the sound wave generated by the string in order to tighten the string to the correct length.

There is also a method of tuning based on the Fibonacci series of Leonardo Pisano Fibonacci, which has been analyzed by English mathematician Sir James Jeans. The numbers in the musical Fibonacci series (2, 5, 7, 12, 19, . . .) can be generated by increasingly long series of musical fourths and fifths from the octave scale. An interval of two tones that are a fifth apart, such as F and C, have a frequency ratio of three-halves. The next fifth is a G, which is musically very close to the original F, but an octave higher, so the two-tone scale is left as F and C. Extending the fifths to a five-tone scale gives F, C, G, D, and A. This would be followed by E, which is again almost the initial F. A slight modification made by slightly raising all the tones (after the initial F) would create a five-note equal tuning scale. Increasingly larger scales can be made by continuing this pattern.

Further Reading

Ashton, Anthony. *Harmonograph: A Visual Guide to the Mathematics of Music*. New York: Walker & Co., 2003.

Hall, Rachel W., and Kresimir Josic. "The Mathematics of Musical Instruments." *American Mathematical Monthly* 108, no. 4 (2001).

Jeans, James. *Science and Music*. New York: Dover Publications, 1968.

HOLLY HIRST

See Also: Geometry of Music; Harmonics; Popular Music; Scales.

Pythagorean School

Category: Friendship, Romance, and Religion.
Fields of Study: Algebra; Communication; Geometry; Number and Operations.
Summary: Religious devotees of mathematics, the Pythagoreans could not accept irrational numbers but made lasting contributions.

The Pythagorean School is the name given to a number of mathematicians and followers of Pythagoras. Pythagoras founded the school in the sixth century B.C.E. in what is now southern Italy. It appears to have been a religious sect built around the proposition that reality was revealed through numbers. It was one of the earliest philosophical schools, and at the time there were no rigid boundaries between philosophy, religion, and politics.

The school had aspects of all three and was a major political force in some Greek cities. To some extent, it was thought of as a secret society. Initiates are said to have taken a vow of silence. This fact and many others about the school are difficult to verify because of a lack of sources from this time. Most of what is known about the Pythagoreans comes in fragments from later philosophers like Plato or Aristotle. Much of the detail about Pythagoras's life is revealed from even later sources, in the works of Diogenes Laertius, Iamblichus, and Porphyry, who wrote many centuries after his death. As a result, much information about the school is ancient hearsay that embellishes what was already a peculiar belief system. The Pythagorean habit of attributing discoveries to Pythagoras, as well as the silence, also makes it hard to distinguish the discoveries of the man from his school. Nevertheless, the influence of his school and mathematical philosophy can still be felt in the twenty-first century concept of "the liberal arts."

Pythagoras and the Foundation of the School

Pythagoras lived from around 580 to 500 B.C.E., but the exact dates are uncertain. He was the son of a leading citizen of Samos (an island in the Aegean), and it is possible that his political significance led Pythagoras to leave the city during the rule of Polykrates the Tyrant. He does not seem to have become prominent until around 530 B.C.E. in the city of Croton, on the southern shore of Italy. The ancient authors account for his life before then by a journey gathering the wisdom of other cultures, such as the Egyptians and Babylonians.

The wisdom he gained is said to have given him many powers. For example, he claimed to recall his previous incarnations, such as his life as the Trojan hero Euphorbus. He is also said to have appeared talking to friends at Metapontium in southern Italy and Tauromenium, on Sicily, on the same day, despite this being impossible with the transport of the day. The same chapter of Porphyry's *Life of Pythagoras* also recounts that a river spoke very clearly to say "Hail Pythagoras!" as he crossed it. Some ancient authors, such as the philosopher Heraclitus, were unconvinced. These and similar tales show not only that he was seen as a divine figure in the ancient world but also that the ancient sources are not wholly reliable. Some modern historians go so far as to discount any mathematical achievements being the work of the actual Pythagoras. Instead, they argue that the achievements of Pythagoreans were attributed to Pythagoras to add luster to his memory.

The school was extraordinarily egalitarian for its era. It admitted both men and women at a time when women were not considered citizens and were usually treated in the same manner as children. The school spread as a society throughout southern Italy and seems to have become a potent political force. Eventually, the power of the school was challenged by the non-Pythagoreans, and violence ensued. Polybius, writing in the second century B.C.E., described the chaos as being a maelstrom of murder, sedition, and "every kind of disturbance." There are several conflicting stories of the death of Pythagoras, but the oddest is that it occurred because he refused to cross a field of beans when an angry mob was chasing him. This behavior was eccentric even by the standards of ancient Greece and only makes sense in light of the Pythagorean beliefs taught at the school.

Pythagorean Beliefs

Pythagoreans believed that numbers were a fundamental property of the universe and that the cosmos operated in harmonies that could be represented as ratios of whole numbers. The purpose of life was to achieve harmony with the universe through a process of purification to counter the corrupting influence of the body. One of the features of this purification was that Pythagoreans were vegetarian—a strong political statement. At this time, one of the duties of a citizen was to participate in civic religious events. Avoiding such events or refusing to perform them properly could draw the ire of the gods. Almost all festivals required the sacrifice of an animal, usually an ox or a goat. The fat and bones would be offered to the gods on the altar and meat would be part of a communal meal. A vegetarian was therefore separating himself from the community.

As for the material that made the cosmos, Pythagoras thought it was governed by numbers. He is said to have come to this conclusion after discovering that musical harmonies can be represented as ratios of whole numbers. The connection between two such different practices such as music and mathematics led Pythagoras to believe that there must be something cosmically significant about numbers. These ratios are embedded in the tetractys symbol—a triangle of 10 dots in four rows, one dot at the top, then two dots, then three, and finally four. The ratios of the motions of the planets were also assumed to be harmonious, and it is said that Pythagoras claimed to be able to hear "the music of the spheres," the harmonies generated by these motions. Numbers that could not be represented by ratios of whole numbers were therefore a serious problem in Pythagorean cosmology.

The Pythagorean Legacy

It is hard to be sure that the theorem that bears his name was actually a Pythagorean concept. While 3-4-5 triangles were used before Pythagoras's time, he may have been the first to prove the Pythagorean theorem, or this might be a later proof attributed to the inspiration of the school. However, there is reason to consider the interest in irrational numbers to be a Pythagorean innovation. Quite how this was discovered is uncertain. The Pythagorean theorem can be used to prove that $\sqrt{2}$ is irrational, but irrationality can also be found in the "pentalpha," a five-pointed star more commonly known as "pentagram," adopted as a symbol by the Pythagoreans. The discovery of irrational numbers is sometimes credited to Hippasus of Metapontum. Usually in these tales, Hippasus meets a grisly end at the hands of Pythagoras who resents the existence of irrational numbers. While this might be a fantastical tale, it is believed that the Pythagoreans were sworn to secrecy concerning the existence of irrational numbers because it was a significant threat to their belief system.

A celebrated legacy of the Pythagorean school is that its approach to applying mathematics to the natural world led to the establishment of the quadrivium: arithmetic, astronomy, geometry, and music that, with grammar, logic, and rhetoric, formed the "liberal arts" that were the foundation of medieval university courses. While the philosophy of liberal arts has changed in modern times, mathematics remains an important feature, as it can be found in many areas in higher education.

Further Reading

Burkert, W. *Lore and Science in Ancient Pythagoreanism.* Translated by Hans Carl Verlag. Cambridge, MA: Harvard University Press, 1972.

Kahn, Charles H. *Pythagoras and the Pythagoreans.* Indianapolis, IN: Hackett Publishing, 2001.

Riedweg, Christoph. *Pythagoras: His Life, Teaching and Influence.* Translated by Steven Rendall. Ithaca, NY: Cornell University Press, 2005.

ALUN SALT

See Also: Greek Mathematics; Harmonics; Numbers, Rational and Irrational; Pythagorean and Fibonacci Tuning; Pythagorean Theorem.

Pythagorean Theorem

Category: History and Development of Curricular Concepts.
Fields of Study: Communication; Connections; Geometry.
Summary: The Pythagorean theorem is a fundamental theorem of mathematics and has numerous applications in number theory and geometry.

The Pythagorean theorem stands as one of the great theorems of mathematics. Ancient peoples appear to have used the Pythagorean theorem to calculate the duration of lunar eclipses or to create right angles in their pyramids or buildings. Archeological evidence suggests that the truth of the result was known in Babylon more than 1000 years before Pythagoras, approximately 1900–1600 B.C.E. Mathematicians and historians continue to debate the early history of the theorem and whether it was dis-

Figure 1.

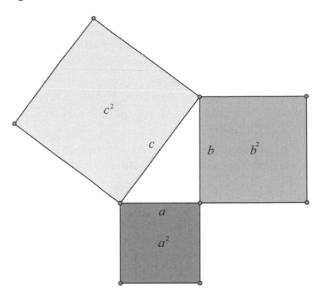

covered independently in such places as Mesopotamia, India, China, and Greece. For instance, some theorize that Pythagoras may have learned the theorem during a visit to India, which in turn may have been influenced by Mesopotamia. The theorem is the culminating proposition of the first book of Euclid's *Elements*. While Euclid (c. 350 B.C.E.) did not mention Pythagoras, later writers such as Cicero and Plutarch referred to it as his discovery. As phrased in the twenty-first century, the theorem states the following:

In any right triangle, the square of the hypotenuse c is equal to the sum of the squares of the legs a and b. That is, $a^2 + b^2 = c^2$.

The theorem has inspired countless generations, and it is useful in a wide variety of contexts and applications, such as in chemistry cell-packing and music.

In Pythagoras's day, humankind had not yet invented algebra. As such, this theorem was not viewed with algebraic perspective but rather in a distinctly geometric way. Visually, as shown in Figure 1, on the right triangle with legs a and b and hypotenuse c, the sum of the areas of the darker gray squares is equal to the area of the lightest gray square.

Proofs

Among the many remarkable features of the Pythagorean theorem, one of the most prominent is that the result admits so many different proofs, including one by former U.S. President James Garfield in 1876. Some of the shortest representations of the Pythagorean theorem are geometric figures called "dissections." For example, Indian mathematician Bhaskara's dissection figure was accompanied by the word "Behold." The Chinese also presented dissection figures that are now called "Pythagorean," and some theorize that these may have led to the development of tangram puzzles. Complete Pythagorean proofs based on dissection figures often combine algebra and geometry.

Given a right triangle with legs of length a and b, construct a square of side length $a + b$. Then, along each side, mark a point that lies a units along the side. If consecutive pairs of these points are connected with line segments, four identical (congruent) copies of the original triangle have been constructed inside the large square (see Figure 2).

In addition, these four line segments have generated a quadrilateral (a four-sided polygon) in the interior of the large square. This quadrilateral's sides each have length c, which is the hypotenuse of the given right triangle. Further, a straightforward argument involving angle measurements in the triangles shows that each of the four angles in the interior quadrilateral measures 90 degrees. Hence, the inside quadrilateral is in fact a square.

Consider the area of Figure 2 in two different ways. First, the area A of the entire outside square, which has sides of length $a + b$, must therefore be $A = (a + b)^2$.

Figure 2.

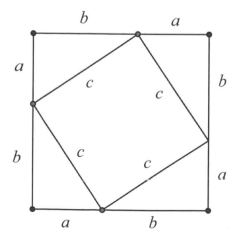

At the same time, one can view the area of the outside square as having been subdivided into five parts. Four of those pieces are congruent right triangles whose area is each $ab/2$. The fifth part is the interior square, whose area is c^2. Thus, the area A of the outer square also satisfies the relationship that

$$A = \frac{4ab}{2} + c^2.$$

Equating the two different expressions for A, one finds

$$(a + b)^2 = \frac{4ab}{2} + c^2.$$

Expanding the left side and simplifying the right, it follows that $a^2 + 2ab + b^2 = 2ab + c^2$.

Finally, subtracting $2ab$ from both sides, the conclusion of the Pythagorean Theorem follows: $a^2 + b^2 = c^2$.

Applications

Furthermore, the Pythagorean theorem is rightly viewed as one of the most central results in Euclidean geometry. Its statement is equivalent to Euclid's parallel postulate, and therefore is directly tied to the truth of a large number of other key results.

In addition to the geometric ideas the Pythagorean theorem evokes, it generates key new ideas and questions about numbers. For instance, if one takes the legs of a right triangle to each have length 1, then it follows that the hypotenuse c is a number such that $c^2 = 2$. There is no rational number (that is, no ratio of whole numbers) whose square is 2. This situation forced Greek mathematicians to reconsider their original conviction that all numbers were "commensurable": that any possible number must be able to be expressed as the ratio of whole numbers. Remarkably, it took mathematicians another 2000 years to put the so-called real numbers, the set of numbers on which calculus is based, on solid footing.

Another Pythagorean idea that has generated a remarkable amount of mathematics is the notion of a "Pythagorean Triple," which is an ordered triple of whole numbers like (3, 4, 5) that represents a solution to the Pythagorean theorem, since $3^2 + 4^2 = 5^2$. A Babylonian clay tablet, named the "Plimpton 322 Tablet," contains many Pythagorean triples. Some suggest that these were a set of teaching exercises, though historians and mathematicians continue to debate their role. Euclid is credited with the development of a formula that will generate a Pythagorean triple, given any two natural numbers. Indeed, there are even infinitely many "primitive" Pythagorean triples, triples in which a, b, and c share no common divisor. Algebraic extensions include investigating solutions to Pythagorean-like equations with other powers, such as $a^3 + b^3 = c^3$. Remarkably, no three positive numbers satisfy such equations; Pierre de Fermat, a French lawyer in the seventeenth century, wrote this note (as translated by historians) in the margins of Diophantus of Alexandria's *Arithmetica*:

> I have discovered a truly marvelous proof that it is impossible to separate a cube into two cubes, or a fourth power into two fourth powers, or in general, any power higher than the second into two like powers. This margin is too narrow to contain it.

No one ever discovered Fermat's proof, yet Fermat's Last Theorem stimulated the development of algebraic number theory in the nineteenth century, and many results in mathematics were shown to be true if Fermat's Last Theorem was true. Andrew Wiles finally proved it to be true near the end of the twentieth century.

There are many other extensions of the Pythagorean theorem. Pappus of Alexandria generalized the theorem to parallelograms. In the 1939 film *The Wizard of Oz*, the Scarecrow recites a version using square roots instead of squares. The Scarecrow's theorem is false in planar geometry, but it can hold in spherical geometry. However, the Pythagorean theorem does not hold on a perfectly round planet. In this case, $a^2 + b^2 > c^2$. Writers for the animated television show *Futurama* named this the Greenwaldian theorem, after mathematician Sarah Greenwald. In the twenty-first century, physicists and mathematicians investigate whether the Pythagorean theorem holds in our universe.

The Pythagorean theorem is also a fundamental idea in several other areas of mathematics and applications. Essentially all of plane trigonometry rests on the Pythagorean Theorem as its starting point, and the modern notion of "orthogonality" in linear algebra is an extension and generalization of the work of Pythagoras. Both trigonometry and orthogonality lead to a wide range of interesting and important applications, including the theory of wavelets and Fourier analysis, mathematics that enables prominent image compression algorithms to help the Internet function.

Its own inherent beauty, the multitude of possible proofs, the rich mathematical ideas it spawns, and the applications that follow all contribute to making the Pythagorean theorem one of the genuine masterpieces in all of mathematics.

Further Reading

MacTutor History of Mathematics Archive. "Pythagoras's Theorem in Babylonian Mathematics." http://www-history.mcs.st-andrews.ac.uk/HistTopics/Babylonian_Pythagoras.html.

Maor, Eli. *The Pythagorean Theorem: A 4000-Year History*. Princeton, NJ: Princeton University Press, 2007.

Posamentie, Alfred. *The Pythagorean Theorem: The Story of Its Power and Beauty*. Amherst, NY: Prometheus Books, 2010.

Matt Boelkins

See Also: Geometry and Geometry Education; Geometry of the Universe; Mathematics, Elegant; Parallel Postulate; Pythagorean and Fibonacci Tuning; Pythagorean School; Wiles, Andrew.

Quality Control

Category: Business, Economics, and Marketing.
Fields of Study: Data Analysis and Probability; Measurement.

Summary: Industrial productions and processes can be mathematically studied to help ensure quality.

Statistical quality control, or more broadly, quality assurance, seeks to improve and stabilize the production and delivery of goods and services. A central concern of quality control is the testing and reporting of measurements of quality—typically as part of a monitoring process—to ensure that the quality of the item being studied meets certain standards.

Quality standards are determined by those who produce the goods or services. Some standards are specification limits imposed by engineering or design concerns that define conformance to a standard. For example, in making airplane engines, a certain part may need to have a diameter between 12 and 14 millimeters or it will not fit into a housing. However, for many processes, there are no specification limits and quality standards may be defined internally from data on past behavior of a process that is judged to be "in control" or "stable." For example, in examining the safety of a large production line, it may be that in each week of the last five years, the average number of person hours lost to accidents has been 1.3. There is no specification limit for this quantity, but control limits can be based on this historical average.

In order to analyze a process for statistical quality control effectively, a process must first be declared to be "in control." To be in statistical control, the vast majority of the products or services must be of sufficient quality for the producers to be satisfied. Moreover, the process must be stable (the mean and variance of the quality measurements must be roughly constant). If a process is in control, then statistical analysis can provide meaningful control limits to the process for monitoring. Graphical methods play a significant role in statistical quality control.

History

Some measure of quality control was in evidence during the building of the Great Pyramids of Egypt. Archeologists have long been impressed not only with the complexity of the construction process, but also by its precision. In the Middle Ages, medieval guilds were formed, in part, to ensure some level of quality of goods and services. The use of statistical methods in quality control—also called "statistical process control" or (SPC)—is more recent, with most of the development in the twentieth century. Graphical methods for quality control were introduced in a series of memos and papers in the 1920s by Walter E. Shewhart of Bell

Telephone Laboratories. The charts he developed and promoted are known today as "Shewhart control charts." H. F. Dodge and H. G. Romig, also of Bell Laboratories, applied statistical theory to sampling inspection, defining rules for the acceptance of many products. Joseph M. Juran, whose focus was more on quality management, rather than SPC, was another early quality pioneer at Bell Laboratories and later Western Electric.

W. Edwards Deming applied SPC to manufacturing during World War II and was instrumental in introducing these methods to Japanese industry after the war ended. He and Juran are generally credited with helping Japanese manufacturing shed the negative image that "made in Japan" had in the 1950s and transforming the country into a source of high quality goods consumed all over the world. In the early twenty-first century, quality control issues continue to appear in the media as concerns proliferate over the quality of goods produced in China.

Common-Cause and Special-Cause Variation

Shewhart and Deming defined two types of variation that occur in all manufacturing and service processes in their 1939 book *Statistical Methods from the Viewpoint of Quality Control*. A certain amount of variation is a part of all processes and can be tolerated even when the goal is to produce goods and services of high quality. This variation is called "common-cause variation," and it comprises all the natural variation in the process. The second variation, called "special-cause variation," is unusual and is not part of the natural variation. Special-cause variation needs to be detected as soon as possible. Quality control charts are designed to detect special-cause variation and distinguish it from common-cause variation.

Quality Control Charts

A quality control chart plots a summary of the quality measurements from each item (or a sample) in sequence against the sample number (or time). A center line is drawn at the mean, or at the desired center of this statistic. Upper and lower control limits are drawn indicating thresholds above or below which will signal an "out of control" measurement. Sometimes, various warning lines are drawn as well, and a variety of rules for deciding if the measurement is really out of control are available. The simplest chart, called an "individual" (or "runs") "chart," plots a single measurement for each item. The control limits are based on the Normal probability model, which implies that for a process in control, only 0.27% of the observations will lie more than three standard deviations (σ) from the center. Therefore, if the process stays in control, a false alarm will occur only once in about 1/0.0027 or once every 370.4 observations. The central idea of a control chart is that a special cause will cause the mean to shift (or the standard deviation to increase), and so the measurement will fall outside the 3σ limits with higher probability. If the shift is great enough, the time to detection will be very short. However, if the special cause results in a subtle shift, it may take many observations before such a signal is detected. Various other types of charts are available that have generally better performance in terms of both false alarm rates and failure to detect shifts.

Total Quality Management and Philosophy

The ideas of Deming, Juran, Shewhart, and others have inspired numerous other people and quality movements. One such movement is total quality management (TQM) also known as "total quality" and "continuous quality improvement." As the name implies, this approach to quality involves more than the monitoring of manufacturing or service processes. It includes all parts of the organization and, specifically, the role of management to help ensure that in providing goods or services, that "all things are done right the first time." Implementing these ideas throughout a large organization gave rise to an abundance of books, experts, and quality "gurus" in the latter part of the twentieth century. One approach to total quality focuses on reducing variation (decreasing σ). If the common-cause variation can be reduced enough, while the process is in control, essentially no measurements will fall outside the 3σ limits. This notion is the essential idea behind the 6σ approach, first popularized by the Motorola company and later the General Electric Company in the 1980s. By the late 1990s, a majority of the Fortune 500 companies were using some form of the 6σ approach.

Further Reading

Deming, W. Edwards. "Walter A. Shewhart, 1891–1967." *American Statistician*, 21 (1967).

——. *Out of the Crisis*. Cambridge, MA: MIT Press, 2000.

Juran, Joseph M. *Quality Control Handbook*. New York: McGraw-Hill, 1999.

———. *Management of Quality Control.* New York: Joseph M. Juran, 1967

Snee, Ronald D., and Roger W. Hoerl. *Leading Six Sigma: A Step-by-Step Guide Based on Experience With GE and Other Six Sigma Companies.* Upper Saddle River, NJ: FT Press, 2002.

RICHARD DE VEAUX

See Also: Deming, W. Edwards; Normal Distribution; Scheduling; Water Quality.

Quilting

Category: Arts, Music, and Entertainment.
Fields of Study: Geometry; Measurement; Representations.
Summary: Quilting can incorporate and help teach mathematical concepts, such as symmetry and tessellations.

Quilting is a needlework technique in which two layers of fabric are sewn together, usually with an inner layer of padding (called "batting") between them. Often, one or both outer layers are formed by sewing together (or "piecing") smaller pieces of fabric. Sometimes, designs are appliquéd (sewn onto a larger piece of fabric) or embroidered on the quilt. The quilting itself (the stitches holding the layers together) is often also decorative. Many traditional quilt designs display mathematical concepts, such as symmetry and tessellations, that generalize into the abstract mathematics of group theory and tiling theory. In diverse parts of the world, people create quilts not only to warm the body at night, but also to use as clothing, furnishings, or to share family or cultural history. A carving of an ancient Egyptian Pharaoh figure containing what may be a quilt and a quilted carpet found in the mountains of Mongolia dates to approximately the first century. Directions can be found to quilt coded designs that may have been used on the Underground Railroad.

Quilt Designs

Some traditional quilts are "crazy quilts" in which scraps of fabric are sewn together in no particular pattern. Others are formed of similar or identical square "blocks," each of which may be pieced together. Often, quilt patterns involve careful measurement (using common fractions) in the cutting and sewing of the pieces.

Quilt designs are often symmetrical—the entire design can be folded in half along a line such that one half falls directly onto the other half. Each half is a reflection of the other along that line, which is called a "line of symmetry." These lines may be vertical, horizontal, or diagonal. Some quilt blocks, such as the traditional Amish Star, are symmetric along many lines. Quilts and quilt blocks may also have rotational symmetry—the design can be rotated around a point through less than a full rotation in a way that leaves the overall design unchanged. Quilts in the Hawaiian Islands are known for their distinctive radial symmetry.

Mathematics generalizes this everyday concept of symmetry. A mathematical object (not necessarily a geometric shape) is symmetric with respect to a particular mathematical operation if the operation, applied to the object, preserves some property of the object. A mathematical group consists of a set of operations that preserve a given property of a given object. Group theory is central to abstract algebra and has many applications.

Fabric quilts, construction paper versions, or computerized models of quilt designs have been used to introduce students as early as elementary school to geometric concepts, such as symmetry and transformations. They help children develop, at a basic level, fundamental algebraic properties, such as inverse, identity, and equivalence. Students also make quilts to explore many other concepts, such as the Pythagorean theorem, polar coordinates, group theory, the Fibonacci sequence, and Pascal's triangle, named after mathematician Blaise Pascal.

Tessellations

A tessellation (or tiling) is an infinitely repeating pattern composed of polygons covering a plane without any openings or overlaps. Many quilt designs are formed from tessellations. A regular tessellation uses one polygon with equal sides and equal angles, such as equilateral triangles, squares, or regular hexagons. For example, the traditional Grandmother's Flower Garden and Honeycomb quilt designs use tessellations of regular hexagons. Many modern watercolor quilts use tessellations of one-inch squares.

A semi-regular tessellation uses a combination of squares, triangles, and hexagons that are arranged

Many traditional quilt designs display mathematical concepts, such as symmetry and tessellations, that generalize into the abstract mathematics of group theory and tiling theory.

identically around each vertex. Demi-regular tessellations, with two vertices in each repetition, form more complicated quilt patterns. Many quilt blocks, such as Log Cabin variations, consist of non-regular tessellations.

Mathematicians have generalized tiling theory to higher dimensional Euclidean spaces and to non-Euclidean geometries. These generalizations reveal links to group theory and to classical problems in number theory. Much of the art of M.C. Escher is based on non-Euclidean tessellations.

Other Designs

Contemporary quilters like mathematician Irena Swanson have also incorporated other mathematical concepts in their designs, such as infinite geometric series and fractals, as well as portraits of mathematicians. Mathematician Gwen Fischer created quaternionic quilts to visually showcase the algebraic structure of the group. For example, the lack of reflection symmetry across the main diagonal highlights the lack of commutativity of the group elements.

Further Reading

Fisher, Gwen. "Quaternions Quilt." *FOCUS* 25, no. 1 (2005).

Meel, David, and Deborah Youse. "No-Sew Mathematical Quilts: Needling Students to Explore Higher Mathematics." *Visual Mathematics* 10, no. 2 (2008).

Paznokas, Lynda. "Teaching Mathematics Through Cultural Quilting." *Teaching Children Mathematics* 9 (2003).

Rosa, Milton, and Daniel Orey. "Symmetrical Freedom Quilts: The Ethnomathematics of Ways of Communication, Liberation, and Art." *Revista Latino Americana de Etnomatemática* 2, no. 2 (2009).

Venters, Diana, and Elain Ellison. *Mathematical Quilts—No Sewing Required.* Emeryville, CA: Key Curriculum Press, 1999.

Bonnie Ellen Blustein

See Also: Escher, M.C.; Symmetry; Textiles; Transformations.

R

Racquet Games

Category: Games, Sport, and Recreation.
Fields of Study: Algebra; Data Analysis and Probability; Geometry.
Summary: The equipment, game play, and scoring of racquet sports can be analyzed using mathematical concepts, such as vector operations and probability.

Racquet games include sports such as tennis, badminton, squash, and table tennis, as well as other less popular games like real tennis, racquets, and racquetball. Mathematics has many roles to play in these games—from equipment testing and court marking to training and analysis of play.

For example, the scoring system in tennis is not a simple counting or linear progression. Mathematicians model a ball's spin in multiple axes, along with trajectories and deflections, as functions of other variables. Markov chains and vector operations can be used to analyze the progression of games and both probability and statistical methods are used to describe performance, seed players for competition, and predict outcomes of matches.

Racquets

Racquet weight distribution, shape, and string material are important factors in the resultant power, accu-racy, and comfort of a racquet. Increasing power, for example, can lead to a decrease in accuracy and it is important to balance these properties. Computer-aided design is the natural choice for this process because of its fast and powerful recalculation abilities.

Projectiles

Racquet sport projectiles such as balls and shuttlecocks are subject to strict regulations and must adhere to these for as long as possible at the highest levels of play. For example, the World Squash Federation allows balls that are 40 millimeters in diameter and each must be tested at 23 degrees Celsius (73 degrees Fahrenheit) and 45 degrees Celsius (113 degrees Fahrenheit), room temperature and play temperature, respectively. There are several dot grades according to level of rebound but an average squash ball rebounds at around 30% (dropped from a height of 3.2 feet, it should reach 12 inches on the bounce). A tennis ball rebounds at around 50%, although changes in ambient air pressure (because of altitude) can affect this figure. Table tennis balls rebound at 85%.

A popular way to gauge the overall performance of these projectiles is to measure their maximum speed. Tennis balls seem to hold the record for the being the fastest, and indeed Andy Roddick can propel a tennis ball very fast (152 miles per hour). However, the fastest badminton stroke left the racquet at over 186 miles per hour. This figure seems counterintuitive because a shuttlecock slows down much more quickly than a tennis ball.

Training

One of the most important roles for mathematics in racquet sports is in training. Sports science researchers study muscle and joint strain and develop nutritional guidelines that allow the player to remain comfortable and energetic during play. Of the racquet sports, squash is regarded as the most intense—players burn roughly 50 percent more calories per hour than badminton or tennis. However, tennis games can run several hours, whereas badminton and squash games are typically decided in under an hour. The total number of calories burned is the product of the calories per hour and the number of hours.

Scoring

In all of the major racquet sports (and many others), a feature of the scoring system may mean that the player who wins more individual points or rallies can still lose the match. Consider the scores of the 1972 British Open final decided by the best of five games, each played to nine points: 0–9, 9–7, 10–8, 6–9, and 9–7. The loser (Geoff Hunt) scored 40 points and won two games; the winner (Jonah Barrington) scored 34 points, won three games and the title.

The same quirk appears in any scoring system where victory is decided by the most wins over a specific number of games. In tennis, this feature exists on two levels. It is possible to win more points and more games but still lose the match. For example, if a match ends 6–4, 0–6, 6–4, 0–6, 6–4, the winner wins 18 games, the loser wins 24 games. The maximum difference in points or rallies in this case is 60 (72–132) in favor of the loser.

Further Reading

Gallian, Joseph. *Mathematics and Sports*. Washington, DC: Mathematical Association of America, 2010.

Havil, Julian. *Nonplussed! Mathematical Proof of Implausible Ideas*. Princeton, NJ: Princeton University Press, 2007.

Lees, A., D. Cabello, and G. Torres, eds. *Science and Racket Sports IV*. New York: Routledge, 2009.

Lees, A., J. F. Kahn, and I. W. Maynard, eds. *Science and Racket Sports III*. New York: Routledge, 2004.

Sadovskii, L. E., and A. L. Sadovskii. *Mathematics and Sports*. Providence, RI: American Mathematical Society, 2003.

EOIN O'CONNELL

See Also: Hitting a Home Run; Hockey; Probability; Rankings; Tournaments.

Radar

See *Doppler Radar*

Radiation

Category: Weather, Nature, and Environment.
Fields of Study: Algebra; Data Analysis and Probability; Measurement; Number and Operations.
Summary: Radiation research has a heavy mathematical component, especially in modeling distribution of or shielding from radiation.

Radiation is the transmission of energy via waves or particles, such as energetic electrons, photons, or nuclear particles. These waves or particles, called "quanta," travel radially in all directions from the source, leading to the name "radiation." Radiation exists everywhere, from both natural sources, like the sun, and many manmade sources, like radio stations and particle accelerators. The various types of radiation that exist may be harmful or beneficial to people, depending on source and application. Ionizing radiation contains enough energy per quantum to detach electrons from atoms, like X-rays or the radiation emitted by particle accelerators. High energy particles are created constantly by all luminous objects in the universe. Most of these particles never reach the surface of Earth. They may be deflected by magnetic fields or interact with atmospheric particles. Common types of nonionizing radiation include visible light, radio waves, and microwaves.

Many mathematicians have contributed to radiation research, like Wilhelm Wien, who derived a distribution law of radiation and won a Nobel Prize for his work on heat radiation. Physicist Max Planck used some of Wein's mathematics as the basis for quantum theory. Paul Ehrenfest contributed to quantum statistics, in part by applying Plank's quantum theory to rotating bodies. Subrahmanyan Chandrasekhar won the Royal Society Copley Medal for his work in mathematical astronomy, including the theory of radiation. Victor Twersky was widely regarded as an expert on radiation scattering. His work has been used in diverse applications, such as studying the effect of atmospheric dust on light propagation. Mathematicians continue to work on radiation problems, including applications such as detecting radiation or shielding satellites from the harmful effects of cosmic radiation, as well as creating mathematical methods for formulating and investigating radiation problems, such as Monte Carlo simulations.

Properties

Properties of radiation waves can be used to determine their potential effects on people and objects or their usefulness for applications. Wavelength is the length of one cycle of the wave, or the distance from one peak to the next. Frequency is the number of cycles of the wave that travel past a fixed point along its path per unit time. All electromagnetic waves travel in a vacuum at a speed of about 3×10^8 meters per second. A fundamental relationship between wavelength and frequency is that wave speed is the product of wavelength and frequency, which means that greater wavelengths correspond to lower frequencies. The energy of electromagnetic photons is the product of wave frequency and Planck's constant, so higher frequencies produce greater photon energies. Among the common types of EMR radiation, radio waves have the longest wavelengths, resulting in low frequencies and low energies. Higher frequency ultraviolet radiation has the most energy and is the most harmful component of the cosmic radiation that penetrates Earth's atmosphere. X-rays, discovered by physicist Wilhelm Röntgen, occur naturally when solar wind is trapped by Earth's magnetic field in the Van Allen belts, named for physicist James Van Allen.

Black holes are also sources of X-rays in the universe. While photons have no mass, some forms of radiation are particles with positive mass produced in the atomic decay of radioactive materials. For example, beta radiation is composed of high-energy electrons, which are dangerous because they can penetrate skin to the layer where new cells are produced. Mathematician Jesse Wilkins's work on mathematical models to compute the penetration and absorption of electromagnetic gamma rays has been used in the design of nuclear radiation shields.

Further Reading

Dupree, Stephen, and Stanley Fraley. *A Monte Carlo Primer: A Practical Approach to Radiation Transport.* New York: Springer, 2001.

Knoll, Glenn. *Radiation Detection and Measurement.* Hoboken, NJ: Wiley, 2010.

Electromagnetic Radiation

Electromagnetic radiation (EMR) includes both ionizing and nonionizing forms of radiation. EMR waves result from the coupling of an electric field and a magnetic field. The fields are perpendicular to one other and to the direction of energy propagation. Electromagnetic radiation behaves like both a wave—with properties including reflection, refraction, diffraction, and interference—and a particle, because its energy occurs in discrete packets or quanta. Maxwell's equations, named for physicist and mathematician James Maxwell, are cited as the most elegant way to express the fundamentals of electromagnetism. The set of four equations, which have integral and differential forms include: Gauss's laws for electricity and magnetism, named for mathematician Carl Freidrich Gauss; Faraday's law of induction, named for physicist and chemist Michael Faraday; and Ampere's law with Maxwell correction, named for physicist and mathematician Andre-Marie Ampere. Many have derived theories and applications from these building blocks, such as mathematician Josef Stefan, who showed that total radiation from a blackbody is proportional to the fourth power of its absolute temperature.

U.S. Environmental Protection Agency. "Radiation
Protection." http://www.epa.gov/radiation/
programs.html.

Sarah J. Greenwald
Jill E. Thomley

See Also: EEG/EKG; Elementary Particles; Energy;
Light; Medical Imaging; Microwave Ovens.

Radio

Category: Communication and Computers.
Fields of Study: Algebra; Measurement;
Representations.
Summary: Radio waves have numerous applications
and are described, analyzed, encoded, and "jammed"
using mathematics.

Radio is a means of sending information by transmitting signals using radio waves, which are a type of electromagnetic radiation with frequencies in the spectrum
of approximately 3 kilohertz (kHz) or 1000 cycles per
second, to 300 gigahertz (GHz), or 1 billion cycles per
second. These units are named for German experimental physicist Heinrich Hertz. Radio waves are used not
only to carry radio and television signals but are also
used in many other common technologies including
wireless computer networks, wildlife tracking systems,
cordless and cellular phones, baby monitors, and garage
door openers. One interesting way that mathematics connects to radio is through mathematically based
radio shows, like *Math Medley*, which was hosted by
Patricia Kenschaft. Mathematicians have also spoken on
programs like National Public Radio's *Science Friday*.

Radio waves are sinusoidal, meaning that they are
characterized by a smooth, repetitive oscillation whose
function at time t can be described algebraically as

$$y(t) = (A)\sin(\omega t + \phi)$$

where A is the wave's amplitude (peak deviation), ω
is the wave's angular frequency (described in radians
per second), and ϕ is the wave's phase (where the wave
cycle is at time $t = 0$).

Brief History and Unique Properties
In 1864, the British physicist James Clerk Maxwell predicted the existence of radio waves as part of his theory
of electromagnetism. Hertz confirmed Maxwell's theory between 1886 and 1888 and is generally credited
with being the first person to send and receive radio
waves. Several individuals played an important role in
developing a practical system of radio transmission
including the Serbian-American engineer Nikola Tesla,
who demonstrated wireless radio communication in
1893; the British physicist Oliver Lodge, who demonstrated the transmission of Morse Code using radio
waves in 1894; and the Italian physicist Guglielmo
Marconi, who in 1896 was granted the first patent for
a radio. Radio communications between ships and
coastal stations were in use by 1897, and the first radio
time signal (used to synchronize clocks) was transmitted from a U.S. Naval Observatory clock in 1904.

Radio waves may be broadcast over long distances because of the Heaviside Layer (also called the
"Kennelly–Heaviside layer"), a conducting layer in the
ionosphere predicted independently in 1902 by the
British mathematician and physicist Oliver Heaviside
and the British physicist Arthur Edwin Kennelly. The
existence of the Heaviside Layer was established in
1924 by the British physicist Edward Appleton, who
also determined that the height of this reflective layer
was about 100 kilometers (62 miles) above the Earth's
surface. The Heaviside Layer allows radio signals to
follow the curvature of the Earth (rather than disappearing into space) because they are reflected by the
Heaviside layer and thus "bounce back" to Earth.

Applications
Radio astronomy, which led to the discovery of
objects such as pulsars and quasars, dates from the
1931 discovery by American physicist Karl Guthe
Jansky of radio waves emitted from the Milky Way
galaxy. American astronomer Grote Reber created
the first radio frequency sky map in 1941, and in
the 1950s, the British astronomers Martin Ryle and
Antony Hewish produced two notable catalogues of
celestial radio sources.

Historically, most radio broadcasts used one of two
techniques for sending their signals: amplitude modulation (AM) or frequency modulation (FM). AM is the
older technology (the first AM broadcast took place in
1906) and it was the dominant radio technology for

most of the twentieth century. AM encodes information by modifying the amplitude of the transmitted signal. The technology for FM broadcasting, which encodes information by varying the frequency of the transmitted signal, was developed in the 1930s and became common by the late 1970s. The information in these analog signals is inherently part of the signal itself—the information influences the wave's shape, and thus information loss can occur with any disruption of the signal. One example is the audible static that occurs when a radio receiver begins to travel beyond the range of a radio transmitter. In the twenty-first century, digital modulation has been increasingly used to minimize this problem. Digital modulation transfers digitized information using a broad spectrum of radio frequencies—far more than the AM or FM systems. Further, each signal is sent many times, reducing the chance of interference and signal loss because separate bits from many streams may be pieced together. Further, since the radio waveforms are not altered by the information, multiple signals may be carried at the same time in the form of one composite signal that is decoded by the receiver, a technique called "multiplexing." Satellite radio systems take advantage of multiplexing and the wider angle of coverage to offer many hundreds of specialized channels across broad geographic areas. Television is also transitioning from analog to digital signals.

Radio transmissions are used for communication during wartime, but because a radio signal may be picked up by anyone with a receiver, various coding methods have been developed. One famous example is the code talkers used by the American Army during World War I and World War II. This program capitalized on the fact that Native-American languages such as Navajo and Choctaw were almost unknown outside those tribes and also developed a simple code for terms like "tank" and "submarine," which allowed them to code and encode messages rapidly and with little risk of comprehension by the enemy. Also in World War II, the German Army used mechanical circuits to encrypt information. Although supposedly unbreakable because of the large number of combinations possible, the British mathematician William Tutte was able to deduce the pattern of the encoding machines after British intelligence intercepted two long coded messages, each of which was transmitted twice (the second time with corrected punctuation).

Interference

Radio waves can be blocked by weather formations, geographic features, and many other natural phenomena. Further, if several stations are broadcasting on a similar frequency, they may interfere with each other. Use of an antenna tuned to a particular frequency (so it will pick up the signal at the frequency more strongly than signals at other frequencies) and aimed at the source of the signal can improve reception. Radio signals can be deliberately jammed by broadcasting noise on the same frequency as the signal. For example, the Soviet Union regularly jammed broadcasts by Radio Free Europe and Voice of America.

To minimize unintentional interference, different parts of the radio spectrum are reserved for different uses and broadcast stations are assigned specific frequencies for their use. In the United States, AM radio uses frequencies from 535 to 1700 kHz, and FM radio uses frequencies between 88 megahertz (mHz) and 108 mHz. A radio station that identifies itself as "90.7 FM" is broadcasting at the frequency of 90.7

The Radio Astronomy Explorer was a radio telecope placed in a moon orbit in 1973 to obtain radio measurements of the planets.

mHz, or 90,700,000 cycles per second (technically, 90.7 mHz is the station's mean frequency). Other parts of the spectrum are reserved for other uses. For instance, 30–30.56 mHz is allocated for military air-to-ground and air-to-air communications systems for tactical and training operations and for land mobile radio communication in support of wildlife telemetry and natural resource management.

Further Reading

Regal, Brian. *Radio: The Life Story of a Technology*. Westport, CT: Greenwood Press, 2005.

Richards, John. *Radio Wave Propagation: An Introduction for the Non-Specialist*. New York: Springer, 2008.

Weightman, Gavin. *Signor Marconi's Magic Box: The Most Remarkable Invention of the 19th Century and the Amateur Inventor Whose Genius Sparked a Revolution*. Cambridge, MA: Da Capo Press, 2003.

Sarah Boslaugh

See Also: Satellites; Television, Mathematics in; Televisions; Tides and Waves; Wireless Communication.

Raghavan, Prabhakar

Category: Friendship, Romance, and Religion.
Fields of Study: Communications; Connections.
Summary: Prabhakar Raghavan has made important contributions to Internet and Web analysis, as well as online social networking, through his work at Yahoo! Research Labs.

Prabhakar Raghavan is the head of Yahoo! Research Labs, where he pursues research in text and Web mining and algorithm design, in addition to overseeing the lab's work. He has received honors like being elected as a member of the National Academy of Engineering and as a fellow of both the Association for Computing Machinery and the Institute of Electrical and Electronic Engineers. He is listed as a Consulting Professor of Computer Science at Stanford University and a member of the editorial board of *Internet Mathematics*, a journal on the mathematics of managing huge databases like the Internet. Beginning in 2007, Raghavan served as a member of the board of trustees for the Mathematical Sciences Research Institute. Raghavan attended the Indian Institute of Technology in Madras, where he earned his Bachelor's of Technology in Electrical Engineering in 1981, before coming to the United States to complete his education with a Master's of Science in Electrical and Computer Engineering from the University of California at Santa Barbara and a Ph.D. in Computer Science from University of California at Berkeley. While at Berkeley, Raghavan won the 1986 Machtey Award, given by the annual IEEE Symposium on Foundations of Computer Science, for his paper "Probabilistic Construction of Deterministic Algorithms: Approximating Packing Integer Programs."

Career

After graduate school, he worked for IBM's T. J. Watson Research Center and Almaden Research Center before becoming vice president and chief technology officer at Verity, Inc., an intellectual capital management software developer. Verity had first made its name with a text retrieval system called Topic that allowed users to search for the information they were looking for based on conceptual keywords, rather than being limited to searching for words actually in the text—much like Yahoo!'s later hierarchical organization of Web sites by topic. In 2005, Raghavan was hired to head the newly established Yahoo! Research Labs, the same year that Verity was bought out by rival Autonomy Corporation.

As head of Yahoo!'s labs, Raghavan has spoken of the need to determine the science and mathematics underlying online communities and social networks, saying: "Is it better to pay a celebrity $10,000 to tweet about your product, or find 10,000 non-celebrities to tout you? The nascent research suggests your money is better spent on the crowd—but the key is finding the people who are slightly more influential than most." Mathematicians, computer scientists, and social scientists work to understand the motivations and responses of online users. "We have this huge mountain of data, and it raises fascinating questions about how we can use that to better the experience for our users," says Raghavan, who refers to researchers in this area as "Internet social scientists," who combine mathematical analysis of large databases and algorithmic understanding with techniques from the social sciences and economics, including sociology and psychology. He notes that while his computer science education was heavily grounded in mathematics and engineering, he also believes that these other disciplines should become a fundamental component of a computer science education. The science of optimizing monetization of Internet services is better understood; although still in development, it is a significant interest of Raghavan's as he seeks to monetize social networks. An eclectic group of computer scientists and social scientists came together and figured out how to take computing from a "glass house" to where a billion people could use it. In the twenty-first century, the ways people interact with computers are becoming mundane. How people will interact with each other to create rich social experiences is the crux of this new and ever-expanding science.

Further Reading

Raghavan, Prabhakar. "IBM Research: How Social Collaboration Makes Chatter Lucrative." http://domino.research.ibm.com/comm/research.nsf/pages/d.compsci.prabhakar.raghavan.html.

Raghavan, Prabhakar, C. D. Manning, and H. Schutze. *Introduction to Information Retrieval.* New York: Cambridge University Press, 2008.

Singel, Ryan. "Yahoo Wants to Blind the Competition With Science/Wired.com." http://www.wired.com/epicenter/2010/08/yahoo-science/#ixzz13xIcLWU2.

BILL KTE'PI

See Also: Internet; Predicting Preferences; Search Engines; Social Networks.

Randomness

Category: History and Development of Curricular Concepts.
Fields of Study: Communication; Connections; Data Analysis and Communication.
Summary: While a seemingly simple idea, the concept of randomness has been studied by mathematicians for thousands of years and has many modern applications.

The philosophical concept of determinism supposes that all events that occur in the world can be traced back to a specific precipitating cause and denies the possibility that chance may influence predestined causal paths. Mathematical determinism similarly states that, given initial conditions and a mathematical function or system, there is only one possible outcome no matter how many times the calculation is performed.

Historical Studies of Randomness and Certainty

Many ancient cultures embraced the idea of fate. For example, the Greek pantheon included goddesses known as Fates. At the same time, the existence of ancient gambling games and deities like the Roman goddess Fortuna suggest that these people understood the notion of randomness or chance on some level. Around 300 B.C.E., Aristotle proposed dividing events into three different categories: certain events, which were deterministic; probable events, which were because of chance; and unknowable events.

In the 1600s, the work of mathematicians such as Blaise Pascal and Pierre de Fermat laid some foundations for modern probability theory, which quantifies chance. Abraham de Moivre published *The Doctrine of Chances* in 1718. Around the same time, Daniel Bernoulli investigated randomness in his *Exposition of a New Theory on the Measurement of Risk*. Nonetheless, determinism continued to maintain a prominent place in mathematics and science. Researchers often assumed that seemingly observed randomness in their data was because of measuring error or a lack of complete understanding of the phenomena being observed.

The emergence of fields like statistics and quantum mechanics in the nineteenth century helped drive new work on randomness. Mathematician Émile Borel wrote more than 50 papers on the calculus of probability between 1905 and 1950, emphasizing the diverse ways in which randomness could be applied in the natural and social sciences as well as in mathematics. Applied probabilistic modeling grew very quickly after World War II.

Randomness in Society

In twenty-first-century colloquial speech, the word "random" is often used to mean events that cannot be predicted, similar to Aristotle's unknowable classification. However, probability theory can model the long-term behavior of random or stochastic systems using probability distribution functions, which are

Casinos monitor roulette wheel performance and rebalance and realign them to keep results random.

essentially sets of possible outcomes having mathematically definable probabilities of occurring. They describe the overall relative frequencies of events or ranges of events, though the specific sequence of individual events cannot be completely determined. Stochastic behavior is observed in many natural systems, such as atmospheric radiation, consumer behavior, the variation of characteristics in biological systems, and the stock market. It is also connected to mathematical concepts like logarithms and the digits of π. Elementary school children discuss some of the basics of randomness when studying data collection methods, like surveys and experiments. Formal mathematical explorations typically begin in high school and continue through college.

Society depends on the use of randomness or the assumption that randomness is involved in a given process. Examples include operating gambling games and lotteries; encrypting coded satellite transmissions; securing credit card data for e-transactions; allocating drugs in experimental trials; sampling people in surveys; establishing insurance rates; creating key patterns for locks; and modeling complex natural phenomena such as weather and the motion of subatomic particles.

Generating Randomness

Generating random numbers, however, is very different from observing random behavior. For example, in 1995, graduate students Ian Goldberg with David Wagner discovered a serious flaw in the system used to generate temporary random security keys in the Netscape Navigator Web browser. Almost every civilization in recorded history has used mechanical systems, such as dice, for generating random numbers and randomness has close ties with gaming and game theory. Physical methods are not generally practical for quickly generating the large sequences of random numbers needed for Monte Carlo simulation and other computational techniques. Flaws in shuffling and physical characteristics, like a worn-down corner on a die, or deliberate human intervention, can also introduce bias. In fact, some people have proven their ability to flip a coin in a predetermined pattern. Motivated by the mathematical unreliability of these physical systems, mathematicians and scientists sought other reliable sources of randomness. Leonard Tippet used census data, believed to be random, to create a table of 40,000 random digits in 1927. Ronald Fisher used the digits of logarithms to generate additional random tables in 1938. In 1955, RAND Corporation published *A Million Random Digits with 100,000 Normal Deviates*, which were generated by an electronic roulette wheel. Random digit tables are still routinely used by researchers who need to perform limited tasks like randomizing subjects to treatment groups in experimental designs as well as in many statistics classes.

The development of computers in the middle of the twentieth century allowed mathematicians, such as John von Neumann, and computer scientists to generate "pseudorandom" numbers. The name comes from the fact that the digits are produced by some type of deterministic mathematical algorithm that will eventually repeat in a cycle, though relatively shorter runs will display characteristics similar to truly random numbers. Using very large numbers, or trigonometric or logarithmic functions, tends to create longer non-repeating sequences. Linear feedback shift registers are frequently used for applications such as signal broadcast and stream cyphers. Linear congruential generators produce numbers that are more likely to be serially correlated, but they are useful in applications like video games, where true randomness is not as critical and many random streams are needed at the same time. Hardware random number generators, built as an alternative to algorithm-driven software generators, are based on input from naturally occurring phenomena like radioactive decay or atmospheric white noise and produce what their creators believe to be truly random numbers.

Randomness Tests

Mathematicians and computer scientists are perpetually working on methods to improve pseudorandom number algorithms and to determine whether observed data are truly random. Randomness can be counterintuitive. For example, the sequences 6, 6, 6, 6, 6, 6 and 2, 6, 1, 5, 5, 4 produced by fair rolls of a six-sided die are equally likely to occur, but most people would say that the first sequence does not "look" random. Irregularity and the absence of obvious patterns are useful ideas, but they are difficult to measure. Distinctions between local and global regularity must also be made, which include the ideas of finite sets and infinite sets. Irenée-Jules Bienaymé proposed a simple test for randomness of observations on a continuously varying quantity in the nineteenth century. Florence Nightingale David published a power function for randomness tests shortly after World War II. Another technique from information theory measures randomness for a given sequence by calculating the shortest Turing machine program that could produce the sequence. The National Institute of Standards and Technology recommends many such tests, including binary matrix rank, discrete Fourier transform, linear complexity, and cumulative or overlapping sums. As of 2010, the digits of π had passed all commonly used randomness tests.

Classical probability theory is not the only way to think about randomness. Claude Shannon's development of information theory in the 1940s resulted in the entropy view of randomness, which is now widely used in many scientific fields. By the latter half of the twentieth century, fuzzy logic and chaos theory also emerged. Fuzzy logic was initially derived from Lot-fali Zadeh's work on fuzzy sets and non-binary truth values, while chaos theory dates back to Henri Poincaré's explorations of the three body problem. Bayesian statistics, based on the eighteenth-century work of Thomas Bayes, challenges the frequentist approach by allowing randomness to be conceptualized and quantified as a partial belief, which shares characteristics with fuzzy logic. Spam filtering is one application that relies on Bayesian notions of randomness.

Further Reading

Bennett, Deborah. *Randomness*. Cambridge, MA: Harvard University Press, 1998.

Mlodinow, Leonard. *The Drunkard's Walk: How Randomness Rules Our Lives*. New York: Pantheon Books, 2008.

RANDOM.ORG. http://www.random.org.

Sarah J. Greenwald
Jill E. Thomley

See Also: Coding and Encryption; Probability; Sample Surveys.

Rankings

Category: Games, Sport, and Recreation.
Fields of Study: Number and Operations; Measurement.
Summary: Ranking is a widely used to create ordered lists of people or objects, and there are many ways to assign and analyze ranks.

Throughout human history, people have been ordering objects into hierarchies based on criteria such as measurements or qualitative properties. In the twenty-first century, people rank many objects, such as quarterbacks, political candidates, and restaurants. Every spring, high school seniors eagerly wait to see who will be the valedictorian, or top-ranked student, of their high school class. However, there is not usually a single unique ranking for a set of objects, since ranks depend on the criteria selected and the specific method in which they are combined. *US News and World Report* aggregates multiple quantitative and qualitative indicators in its annual ranking of colleges. Mathematicians use a variety of techniques to study ranking, such as algebra, geometry, graph theory, game theory, operations research, and numerical methods. An entire subset of statistical techniques based on ranks, called *nonparametric* or *distribution-free tests*, are used to transform and analyze data that do not conform to the assumptions or parametric tests.

These techniques are often used in the social sciences. There are also debates about whether ranks are true numbers, given that the spacing between ranks need not be equal in the manner of most common measurement scales. For example, the difference between one inch and two inches is the same as between two inches and three inches. The difference between first and second place, however, is not necessarily quantitatively or qualitatively the same as the difference between second and third place.

Sports

Athletic competitions are one very visible use of rankings. During the ancient Olympic Games, athletes would compete in events, such as running, boxing, and the pentathlon, to determine which athletes were better than others. Ultimately, they would be ranked by their performance in these events. Even during the modern Olympics, though the events are more numerous and athletes generally compete in only a few events, the result is a ranking of the best athletes, with prizes being awarded to the top three finishers. There are rankings for other sports as well. For example, the Associated Press ranks the top 25 NCAA football teams by polling sportswriters across the nation. Each writer creates a personal, subjective list of the top 25 teams from all eligible teams (more than 25). The individual rankings are then combined to produce the national ranking by giving a team 25 points for a first place vote, 24 points for a second place vote, and so on down to one point for a 25th place vote. Teams are also regularly ranked by their number of wins or other game-related metrics, as are individual players.

Tests

Rankings also occur on standardized tests. Rather than give each individual a unique rank, tests such as the SAT separate the scores into percentages and then rank test takers according to the percentage they fall into. Percentile ranks can also be seen in other places,

Tiebreakers

Some ranking strategies result in ties between one or more individuals. Sometimes there is a tiebreaker, and other times there is not. The ranking of items occurring after the tie can vary depending on the type of ranking used. The most common is called *standard competition ranking*, where a gap is left in the numbering after the tie takes place corresponding to the number of elements in the tie. For example, if there were six items and a three-way tie for second occurred, the ranking would be given as "1, 2, 2, 2, 5, 6" with third and fourth place omitted. Some methods, especially those used in statistical analysis, assign an average rank. In a three-way tie for second place out of six objects, the assigned rankings would be "1, 3, 3, 3, 5, 6," since the average of 2, 3, and 4 is 3.

such as height and weight charts for children. Whereas many rankings place an emphasis on small numbers (it is better to be ranked first or second than twenty-fifth), percentiles are considered in the opposite manner—a larger value percentile ranking is a better rank. Percentiles indicate what percentage of the test-taking group performed the same or worse than a test-taker in that percentile. For example, being in the 57th percentile would indicate that 57 percent of the test takers scored the same or worse. When considering rakings, it is important to determine how the ranking is arranged to properly interpret the data.

Other Mathematical Connections

The word "rank" carries many specific definitions in various fields of mathematics. For example, the rank of a matrix is the number of linearly independent rows or columns. In graph theory, the rank of a graph is the number of vertices minus the number of connected components. Other definitions of rank can be found in set theory and Lie algebra (named for mathematician Sophus Lie). In chess, a game studied by many mathematicians, a rank is a row on the chessboard.

Further Reading

Gupta, Shanti, and S. Panchapakesan. *Multiple Decision Procedures: Theory and Methodology of Selecting and Ranking Populations.* Philadelphia: Society for Industrial Mathematics, 2002.

Marden, John I. *Analyzing and Modeling Rank Data.* New York: Chapman & Hall, 1995.

Winston, Wayne. *Mathletics: How Gamblers, Managers, and Sports Enthusiasts Use Mathematics in Baseball, Basketball, and Football.* Princeton, NJ: Princeton University Press, 2009.

CHAD T. LOWER

See Also: Growth Charts; Infinity; Measurement in Society; Nielsen Ratings; Statistics Education; Voting Methods.

Rational Numbers

See *Numbers, Rational and Irrational*

Reasoning and Proof in Society

Category: School and Society.
Fields of Study: Connections; Reasoning and Proof.
Summary: Many aspects of society have inherited from mathematics the desire for a method of proof that is demonstrable and irrefutable.

Reasoning and proof are fundamental components of human existence. Children begin applying reasoning as soon as they can make connections between actions and consequences. They then go on to explore more formal methods of reasoning and proof throughout their educational careers, not just in mathematics. Although people often associate mathematics solely with deductive proofs, many other types of reasoning are important to mathematics, including inductive logic, evidence-based reasoning, and computer-assisted arguments. Furthermore, the concept of truth being produced by reasoning and proof also pervades other fields, including philosophy, the natural and social sciences, and political and legal discourse.

Origins of Mathematical Proof

What proves a statement? Generally, it is believed that statements are proved by deducing the statement as a logical consequence of something already believed to be true. One might think that proofs are necessary only when what is being proved is not apparent. The Greeks, however, did not limit proving to non-obvious statements; they gave a logical structure to all of geometry, assuming as its basis the smallest possible number of "already believed" statements. They also employed a method called "proof by contradiction" in which a truth is not demonstrated directly, but rather by showing that its opposite cannot be maintained.

Why did Greek culture give geometry this kind of logical structure, and why did the Greeks think that doing so was significant? The question is important because the causes that produced mathematical proof still exist in the twenty-first century, where they continue to operate and promote the use of proof.

First, proofs give a way to reconcile discordant opinions. Greek mathematics was heir to two earlier traditions, Egyptian and Babylonian mathematics, whose results did not always agree. For instance, in studying circles, the Babylonians approximated π first as 3, and later as 3.125. Egyptian computations give a value for π of about 3.16. The Greeks wanted to know π's true value. One way to avoid having multiple answers to the same question is to make no assumptions other than those with which nobody could disagree, like "all right angles are equal," and then deduce other facts solely from those un-doubtable assumptions. What is amazing is how many results this approach produced.

Second, proofs are a natural outcome of the search for basic principles. The pioneering Greek philosophers of nature of the fifth and sixth centuries B.C.E. sought simple explanatory principles that could make sense out of the entire universe. Thales, for instance, said that "everything is water," and Anaximenes claimed that "everything is air." The Pythagoreans asserted that "all is number," while Democritus said that "everything is made of atoms." As in nature, so in mathematics, the Greeks wanted to develop explanations based on simple first principles, on the so-called elements.

Third, the logic of proofs can arise from the process of discovery. One effective way to solve a problem is to reduce it to a simpler problem whose solution is already known. For instance, Hippocrates of Chios in the fifth century B.C.E. reduced finding the area of some lunes (areas bounded by two circular arcs) to finding the area of triangles. In reducing complicated problems to simpler problems, and then reducing these to yet simpler problems, the Greek mathematicians were creating sets of logically linked ideas. If such a set of linked ideas is run in reverse order, a proof structure emerges—simple statements on which rest more complex statements on which rest yet more complex statements. The simplest statements at the beginning are called the "elements"; the intermediate ones are the fruitful results that are now called "lemmas"; and these in turn demonstrate the final and most important results.

Fourth, logical reasoning played essential roles in classical Greek society. In the sixth and fifth centuries B.C.E., Greece was largely made up of small city-states run by their citizens. Discourse between disputing parties, from the law courts to the public assemblies, required and helped advance logical skills. A good way, then and now, to persuade people is to understand their premises, and then construct one's own argument by reasoning from their premises. A good way to disprove someone's views is to find some logical consequence of those views that appears absurd. These techniques are beautifully illustrated in Greek legal proceedings and political discourse, as well as in the dialogues of Plato.

Finally, Greek mathematics developed hand in hand with philosophy. Greek philosophers began by trying to logically refute their predecessors. Zeno, for instance, presented his paradoxical arguments not to prove that motion is impossible but to challenge others' intuition and common-sense assumptions. That Plato wrote in dialogue form both illustrates and demonstrates that Greek philosophy was as much about the method of logical argument as it was about conclusions. Aristotle wanted every science to start, like geometry, with explicitly stated elementary first principles, and then to logically deduce the key truths of the subject. Greek philosophy issued marching orders to mathematicians, and men like Euclid followed these orders.

Philosophy returned the favor. Plato made mathematics the center of the education of the rulers of his ideal Republic and mathematics has remained at the heart of Western education. Plato championed mathematics because it exemplified how, by reasoning alone, one could transcend individual experience. Such transcendence is most striking in the case of proof by contradiction. The argument form, "If you accept A, then you must also accept B, but B contradicts C," was part and parcel of the educated Greek's weapons of refutation. But proof by contradiction is not merely destructive, it also allows people to rigorously test conjectures that cannot be tested directly and, if they are true, to demonstrate them.

For example, Euclid defined parallel lines as lines in the same plane that never meet. But it can never be shown directly that two lines can never meet. However, it can be assumed that the two lines do, in fact, meet and then prove that this assumption leads to a contradiction. This process made Euclid's theory of parallels possible.

As another example, consider the Greek proof that $\sqrt{2}$ cannot be rational (it cannot be the ratio of two whole numbers). Because the Pythagorean theorem holds for isosceles right triangles, $\sqrt{2}$ must exist.

But no picture of an isosceles right triangle can allow one to distinguish a side of rational length from one of irrational length.

Nor can one hope to prove the irrationality of $\sqrt{2}$ by squaring every single one of the infinitely many rational numbers to see if its square equals 2. However, if one assumes that there is a rational number whose square is two, logic then leads to a contradiction, so it is proved that $\sqrt{2}$ cannot be rational.

By such means the Greeks proved not only that $\sqrt{2}$ was irrational but also that a whole new set of mathematical objects existed: "irrational numbers."

Proof in general, and proof by contradiction in particular, transformed the nature of mathematics. Logic lets people reason about concepts that are beyond experience and intuition—about ideas that cannot be observed. Mathematics had become the study of objects transcending material reality, objects visible only to the eye of the intellect. There could be truths about such objects and such truths could be proved. These developments had profound consequences far beyond mathematics.

Beyond Mathematics

The ideal of logical proof in mathematics took on a life of its own. Since mathematicians apparently had achieved truth by means of proof, practitioners of other areas of Western thought wanted to do the same. So in theology, politics, philosophy, and science people tried to imitate the mathematicians' method.

In 1637, Rene Descartes wrote in his *Discourse on Method*, "Those long chains of reasoning . . . which enabled geometers to reach the most difficult demonstrations, made me wonder whether all things knowable to men might not fall into a similar logical sequence." If so, he continued, there cannot be any propositions that cannot be eventually discovered and proven.

Building on Descartes's ideas, Baruch Spinoza in 1675 wrote a book called *Ethics Demonstrated in Geometrical Order*. Like Euclid, Spinoza first explicitly defined his terms, including "God" and "eternity." He then stated axioms about existence and causality. On the basis of his list of definitions and axioms, Spinoza logically demonstrated his philosophical conclusions, including the existence of God.

Isaac Newton wrote his great *Principia* in 1687. This work includes Newton's laws of motion and theory of gravity. He did not structure the *Principia* like a modern physics book; he gave it the same definition-axiom-theorem structure that Euclid had given the *Elements*. Newton expressly called his famous three laws "Axioms, or Laws of Motion." From these axioms, Newton logically deduced the laws of the universe, including universal gravitation, just as Euclid had deduced his own theorems.

The American Declaration of Independence of 1776 also pays homage to the ideal of Euclidean proof. The principal author, Thomas Jefferson, was well versed in the mathematics of his time. Jefferson began with axioms, saying, "We hold these truths to be self-evident," including the axioms "that all men are created equal" and that, if a government does not preserve human rights, "it is the right of the people to alter or abolish it, and set up new government." The declaration then says that it will "prove" that King George III's government had not protected human rights. Once Jefferson proved this, the Declaration of Independence concludes: "We therefore . . . publish and declare that these United Colonies are and of right ought to be free and independent states." Indeed, Jefferson could have ended his argument, as had Spinoza and Newton, with the geometer's "QED."

Jefferson's argument exemplifies the characteristic program of Enlightenment philosophy—using reason to reach conclusions on which everyone will agree. This program is epitomized in the words of Voltaire in his *Philosophical Dictionary*: "There is but one morality, as there is but one geometry."

Abstraction, Symbolism, and their Power

Logical proof in mathematics and the use of mathematical models of reasoning in the larger intellectual world were not limited to geometry. In mathematics in the seventeenth and eighteenth centuries, proof methods moved beyond the geometric to include the algebraic. This shift began when François Viète, in 1591, first introduced general symbolic notation in algebra, an idea with incredible power.

School children learn that for every pair of distinct numbers, not only does $9 + 7 = 16$, so does $7 + 9$. Viète's general symbolic notation allows one to write down the infinite number of such facts all at once: $B + C = C + B$.

A century later, Isaac Newton summed up the power and generality of Viète's idea by calling algebra "universal arithmetic." Newton meant that one could prove algebraic truths from the universal validity of the symbolic manipulations that obey the laws of ordinary arithmetic. For instance, consider the quadratic equation $2x^2 - 11x + 15 = 0$. Simply stating, "3 and 2 1/2 are the solutions" gives no information about how those answers were obtained. But every quadratic equation has the general form of $ax^2 + bx + c = 0$. Solving that general equation by the algebraic technique of completing the square gives the well-known quadratic formula for the general solution:

$$x = \frac{-b \pm \sqrt{b^2 - 4ac}}{2a}.$$

This general solution contains the record of every operation performed in getting it. The original example had $a = 2$, $b = -11$, $c = 15$. As such, it is known exactly how the answers, 3 and 2 1/2, are obtained from the coefficients in the equation. More important, this process proves that these and only these must be the answers.

In the seventeenth century, Gottfried Wilhelm Leibniz was so inspired by the power of algebraic notation to simultaneously make and prove mathematical discoveries that he invented an analogous notation for his new differential calculus. Furthermore, he envisioned an even more general symbolic language that would, once perfected, find the indisputable truth in all areas of human thought. Once such a language existed, Leibniz said, if two people were to disagree, one could say to the other, "let us calculate, sir!" and the disagreement would be resolved. This idea made Leibniz the prophet of modern symbolic logic.

By the eighteenth century, many mathematicians thought discovery and proof should be based on abstract symbolic reasoning. Imitating mathematics, scientists introduced analogous notations in other fields. For instance, Antoine Lavoisier and Claude-Louis Berthollet developed a new chemical notation that they called "chemical algebra," which is used when balancing a chemical equation.

These ideas, both within and beyond mathematics, led the Marquis de Condorcet to write in 1793 that algebra contains within it the principles of a universal instrument, applicable to all combinations of ideas. Such an instrument, he said, would eventually make the progress of every subject embraced by human intelligence as sure as the progress of mathematics.

In the nineteenth century, George Boole produced the first modern system of symbolic logic and used it to analyze a wide variety of complicated arguments. His system, developed further, underlies the logic used by digital computers in the twenty-first century, including applications embodying Condorcet's dream, from automated theorem-proving to translators, grammar checkers, and search engines.

Non-Euclidean Geometry:
The Triumph of Euclidean Logic

Unthinkable as it may have been to Enlightenment philosophers like Voltaire, there are alternatives to Euclid's geometry. But non-Euclidean geometry was not invented by imaginative artists or by critics of mathematics speculating about alternative realities. Like irrational numbers, non-Euclidean geometry was discovered by mathematicians. Its discovery provides another example of human reason and logic trumping intuition and experience and it—like Euclid's geometry—has had a profound effect on other areas of thought.

Non Euclidean geometry grew out of attempts to prove Euclid's parallel postulate:

If a straight line falling on two straight lines makes the interior angles on the same side less than two right angles, then the two straight lines, if produced indefinitely, meet on that side where the angles are less than two right angles.

Such attempts were made because the postulate seemed considerably less self-evident than his other postulates. From antiquity onward, mathematicians felt that it ought to be a theorem rather than an assumption, and many eminent mathematicians tried to prove it from the other postulates. Some attempted to prove it indirectly; assuming it to be false, they deduced what appeared to be absurd consequences from that assumption. For instance, that parallel lines are not everywhere equidistant, and that there is more than one line parallel to a given line through a point in the same plane. These results contradict our deep intuitive sense of symmetry.

But in the nineteenth century, three mathematicians independently realized that these conclusions were not absurd at all, but were perfectly valid theorems in an alternative geometry. Nicolai Ivanovich Lobachevsky, by analogy with imaginary numbers, called his new subject "imaginary geometry." Janos Bolyai more theologically called it "a new world created out of nothing." But Carl Friedrich Gauss, acknowledging the logical move that made it possible, called the new subject "non-Euclidean geometry."

The historical commitment of mathematicians to the autonomy of logic and to logical proof enabled them to overcome their scientific, psychological, and philosophical commitments to Euclidean symmetry to create this new subject. Logical argument once again let mathematicians find and demonstrate the properties of something neither visual nor tangible—something counter-intuitive. Non-Euclidean geometry is the ultimate triumph of the Euclidean method of proof. But there are wider implications.

From this discovery, nineteenth-century philosophers concluded that the essence of mathematics (as opposed to the natural sciences) is its freedom to choose any consistent set of axioms that meets the mathematician's sense of what is important, beautiful, and fruitful—just as long as the logic is right. There could even be real-world applications of systems that contradict all past mathematical orthodoxies. In physics, for instance, the type of non-Euclidean geometry studied by Bernhard Riemann in the 1850s turned out to be exactly what Albert Einstein needed for his general theory of relativity; the new mathematics can

Proof and the Law

Though the goal of legal arguments is persuasion as well as proof, legal arguments require evidence, and thus require discerning what follows logically from such evidence. Some legal thinkers have carried this view quite far.

For instance, Christopher Langdell, pioneer of the case method in legal education and Dean of Harvard Law School in the 1870s, saw law as a science. By analogy with geometry, law, according to Langdell, is governed by a consistent set of general principles. The correct legal rules should be logically deduced from those general principles and then applied to logically produce the correct legal ruling in line with the facts of a particular case.

Most Anglo-American legal theorists do not follow Langdell's "classical orthodoxy," agreeing instead with Oliver Wendell Holmes that the life of the law has not been logic but experience. Yet Holmes, too, employed logical argument within every case he discussed. For instance, he used a proof by contradiction to argue that freedom of speech is not absolute when he famously said that the most stringent protection of free speech would not protect a man in falsely shouting "fire!" in a theater and causing a panic.

Finally, the adversary system of Anglo-American law not only allows but also requires, that in order for a case to prevail in court, the winning argument must not only support that case but also explicitly answer the arguments on the other side, with these counter-arguments presented as strongly as possible. Thus, logical proof pervades all legal argument.

explain gravitation, describe the curvature of space, and account for black holes.

Knowing that alternative systems of mathematical thought are logically possible has also had philosophical and social implications. José Ortega y Gasset, for instance, contrasted the view of the old geometry (interpreted as saying that nations may perish but principles will be kept) to the new perspective, which he interpreted as saying that people must look for such principles as will preserve nations, because that is what principles are for.

Proof and the Citizen

Citizens of democracies need to be able to evaluate arguments presented to them, whether by friends, adversaries, politicians, or advertisers. In the words of Jacques Barzun, "The ability to feel the force of an argument apart from the substance it deals with is the strongest possible weapon against prejudice."

Citizens also need to be free to work out the logical implications of the principles they treasure. In the words of Winston Smith, a character in George Orwell's novel 1984, "Freedom is the freedom to say that two plus two make four. If that is granted, all else follows." This kind of "proving" has driven the progress of the idea of universal human rights. For instance, building on the Declaration of Independence, Elizabeth Cady Stanton, a pioneer in fighting for women's rights in America, wrote in the Seneca Falls Declaration of 1848, "We hold these truths to be self-evident; that all men *and women* are created equal." Similarly, Martin Luther King, Jr., in his "I Have a Dream" speech, spoke of "the promise that *all* men, yes, black men as well as white men, would be guaranteed the unalienable rights of life, liberty, and the pursuit of happiness."

Now, just as in ancient Greece, the ability to reason and prove and the liberty of expressing and acting upon the results of proofs are essential to a free and democratic society. The historical function of proof in mathematics has not been just to prove theorems but also to exemplify and teach logical argument in areas such as philosophy, law, politics, religion, and every area of modern life.

Further Reading

Berman, Harold J. *Talks on American Law; A Series of Broadcasts to Foreign Audiences by Members of the*

Harvard Law School Faculty. 2nd ed. New York: Vintage, 1971.

Erduran, Sibel, et al. *Argumentation in Science Education.* Berlin: Springer, 2007.

Gold, Bonnie, and Roger Simons. "Proof and Other Dilemmas: Mathematics and Philosophy." Washington, DC: Mathematical Association of America, 2008.

Grabiner, Judith V. "The Centrality of Mathematics in the History of Western Thought," *Mathematics Magazine* 61 (1988).

Levi, Mark. *The Mathematical Mechanic: Using Physical Reasoning to Solve Problems.* Princeton, NJ: Princeton University Press, 2009.

Rips, Lance. *The Psychology of Proof: Deductive Reasoning in Human Thinking.* Cambridge, MA: MIT Press, 1994.

JUDITH V. GRABINER

See Also: Geometry and Geometry Education; Mathematical Certainty; Parallel Postulate; Proof; Strategy and Tactics.

Recycling

Category: Weather, Nature, and Environment.
Fields of Study: Algebra; Data Analysis and Probability; Number and Operations.
Summary: Efficient recycling requires the use of sophisticated mathematical models to maximize product use and reuse and minimize energy consumption.

Recycling is the extraction of usable materials out of used objects. Materials that are often recycled at the start of the twenty-first century include metal, paper, glass, and plastic. One important mathematical problem of recycling is the comparison of environmental and monetary costs of recycling and virgin production. Mathematicians are also involved in developing new methods for recycling and modeling both economic and environmental impacts. The notion of "algorithm recycling" applies to resources used in some mathematical investigations. For example, statistical boot-strap recycling reuses samples to minimize demands on computational resources. Some mathematicians, scientists, educators, and others use recycling for education, recreation, and art. Mario Marin has designed polyhedral outdoor play spaces and kinetic sculptures from recycled and remaindered materials and has published many creative ways to recycle household objects, like plastic bottles, into interesting polyhedral structures. With regard to learning, some have even suggested a concept called "neuronal recycling," which refers to adaption of neuronal circuits for new uses.

Proportion-Based Regulations and Labeling

To motivate recycling, companies and governments set rules that demand the recycling of a certain proportion of materials and the use of a certain proportion of recycled material in production. Because recycling is the third desirable option in the waste management hierarchy, after reduction of waste and reusing of objects and materials, setting high recycling quotas is never a goal in its own right. However, recycling is often preferable to disposal. Governments sometimes directly mandate minimum recycled content in certain classes of manufactured goods. Labeling laws, which require companies to display the percent of recycled content in goods and packages, may also promote recycling if consumers support it, or hinder recycling if consumers do not find recycled goods in this particular industry appealing. Companies advertise their recycling efforts—typically by disclosing the percent of recycled material—to present ecofriendly images to their customers.

A common scheme to promote the recycling of packaging is to include a refundable fee in the price of the product. Once the customer returns the packaging to the store, the fee is refunded.

Measuring Efficiency

Because recycling is a complex process, there are ecological and economical costs involved in it. For recycling to make sense, the benefits have to outweigh the costs. Computing costs and benefits is a complex problem. Costs are incurred at all stages of recycling: collecting materials, sorting them, and re-making them. Benefits include the reduction of landfill costs, reduction of pollution, and revenues from the use of recycled materials. In the cases of nonrenewable natural resources, recycling is the only option to keep using these resources in the future.

Metal Recycling

Because of the relative difficulty and high cost of mining and smelting of metals, and the ease of collecting and recycling, metals are the most recycled materials in the world. For example, recycling aluminum takes only 5% of the energy that it would take to make it from the raw materials. About three-quarters of steel and a third of aluminum is recycled in the United States as of 2010. Some applications of science and mathematics metal recycling involves the separation of impurities, such as paint.

Paper Recycling

One category of paper recycling, post-consumer paper, is familiar to most people because paper is ubiquitous in modern society. "Mill broke" is scraps that pulp mills accumulate from making paper, which they can also recycle. Preconsumer paper is scraps collected and recycled in paper mills. Unlike metal recycling, where the cost-benefit ratio is low, paper recycling is more complicated and controversial. For example, burning paper for energy may be more environmentally sound than recycling it and harvesting and replanting forests may be cheaper than recycling.

Estimates for energy saving are 40% to 65% for recycled paper, compared to creating new paper. However, pulp mills frequently produce energy by burning roots, bark, and other byproducts, whereas recycling plants have to be close enough to collection (usually urban) areas to minimize transport cost and frequently

Atlas Recycled *by Tom Tsuchiya is a sculpture made of used atlases and maps that also serves as a recycling receptacle for bottles and cans—one of the works from the EcoSculpt 2010 exhibition in Cinncinnati, Ohio.*

depend on fossil fuels for energy. Thus, the environmental costs of conserving the same amount of energy is different, as one process uses renewable resources and the other uses nonrenewable resources. Water and air pollution benefits of paper recycling are more pronounced than energy benefits because of highly toxic bleaching used in making new paper.

Plastic Recycling

Recycling of plastics involves a scientific challenge not found in recycling of other materials. Because of the ways polymer chains are formed in plastics, different plastics do not blend well. Removing dyes, glue, paper stickers, and other impurities is also difficult. Plastics are coded with the Resin Identification Codes, numbers 1–7, inside the triangular recycle symbol.

There are several processes for recycling plastic. The most straightforward is melting similar plastics together, with some steps to remove impurities. Heat compression mixes all types of plastics in high-heat, high-pressure drums. Thermal depolymerization is currently an experimental procedure that "reverses" the process of making plastic and turns it into a substance similar to crude oil. Another experimental procedure, called "monomer recycling," reverses plastic-making halfway, turning polymers into the mix of monomer chemicals that formed them.

The short-term cost-benefit analysis may not support plastic recycling because of the high energy and labor requirements of the known processes. However, crude oil (the raw material of plastic) is a nonrenewable resource, which makes plastic recycling attractive in the long term.

Glass Recycling

The main benefits of glass recycling are saving landfill space and saving energy on producing new glass. However, because glass is sturdy and easy to clean, glass container reuse is vastly preferable to recycling. Through changing their infrastructures, along with using clear bottle standards and monetary incentives, some countries can reuse more than 95% of their glass bottles.

Crushed glass can be added to concrete. This process can be considered reuse rather than recycling because the glass is serving a different purpose. Measurements of glass-infused concrete include its insulation properties and strength properties, both of which are improved by the addition of glass. Also, concrete with glass is more aesthetically pleasing and can be used for countertops and other highly visible places.

Mathematical Modeling

Mathematical models are widely used in logistics—controlling the efficient flow and storage of goods, services, and information from the point of origin to the point of consumption. Reverse logistics is the extension of this principle that addresses concepts such as returns, source reduction, recycling, and reuse. Mathematicians have researched models for logistics that address these reversals of flows. For example, Italian researchers created a staged mathematical model of the options for recycling a broad range of appliances, electronic equipment, and other household items commonly thrown away. The model suggested that recycling can offer what is known as *economies of scale* to businesses, which are increasingly being held liable for end-of-life product disposal.

Others have used techniques such as dynamic quantitative models to simulate recycling systems and flows to better understand the driving variables and relationships among the activities and participants. These models can aid planners in making decisions about recycling policies and procedures. Nutrient recycling for trees, which has implications for issues such as global warming, has been modeled using linear and quadratic functions, along with data-based numerical simulations. However, some scientists argue that mathematical models must be contextually evaluated and used with caution for decision making and legislation. Models based on limited data may generate what appear to be useful results, but extrapolation or subsequent modeling can create bias and propagation of errors.

Further Reading

Environmental Protection Agency. "Wastes—Resource Conservation—Reduce, Reuse, Recycle." http://www.epa.gov/osw/conserve/rrr/recycle.htm

La Mantia, Francesco. *Handbook of Plastics Recycling.* Shropshire, England: Smithers Rapra Technology, 2002.

Mancini, Candice. *Garbage and Recycling (Global Viewpoints).* Farmington Hills, MI: Greenhaven Press, 2010.

Newton, Michael, and Charles Geyer. "Bootstrap Recycling: A Monte Carlo Alternative to the Nested Bootstrap." *Journal of the American Statistical Association* 89, no. 427 (1994).

Schlesinger, Mark. *Aluminum Recycling*. Oxfordshire, England: Taylor & Francis Group, 2007.

MARIA DROUJKOVA

See Also: Advertising; Carbon Footprint; Deforestation; Fuel Consumption; Green Design; Green Mathematics; Synchrony and Spontaneous Order.

Relativity

Category: Space, Time, and Distance.
Fields of Study: Algebra; Geometry; Measurement; Representations.
Summary: Albert Einstein's theory of relativity is one of the most well-known theories in physics and helps describe the nature of the universe.

Albert Einstein's theory of relativity forms one of the two pillars of modern physics, the other being quantum mechanics. It consists of two parts: the special theory of relativity from 1905, and the general theory of relativity from 1915, which both rely on significant mathematics.

The special theory of relativity describes how space and time are perceived by observers in different inertial systems. Einstein derived this theory from a single physical principle of relativity. It was discovered in 1632 by Galileo Galilei that the laws of mechanics are the same in all inertial systems—a discovery, known as "Galileo's principle of relativity," that constituted a radical break with the prevailing Aristotelian physics. Einstein's principle of relativity generalized this concept to all laws of nature, including Maxwell's laws of electromagnetism, which govern the propagation of light. It thus follows from Einstein's principle of relativity that the speed of light is the same in all inertial systems, a central result in the theory of relativity. Prior to Einstein, it was believed that light propagates through a luminiferous aether in the same way as sound propagated through air, but all attempts to measure the speed of the Earth relative to this aether, such as the Michelson–Morley experiment in 1887, failed. Special relativity explained the negative results of these experiments and made the aether hypothesis superfluous.

The general theory of relativity unifies special relativity with Isaac Newton's law of universal gravity. Its basis is Einstein's equivalence principle, according to which an accelerated system of reference (such as a so-called Einstein elevator) is indistinguishable from a system at rest in a gravitational field. Mathematically, Einstein's field equations describe how the presence of mass, energy, and momentum gives rise to a curvature of space and time. Although this idea has little significance in weak gravitational fields, such as that of the Earth, general relativity is essential in the study of the universe as a whole. For example, Karl Schwarzschild in 1915 found an exact solution to Einstein's equations that explains the existence of black holes.

The many surprising consequences of the theory of relativity have been described in numerous popularizations, most notably by George Gamow. Einstein's theory must not be confused with the various relativist positions in philosophy, such as aesthetic, moral, cultural, or cognitive relativism.

Special Relativity

The Lorentz transformation forms the basis of the special theory of relativity. It is a set of equations describing how to translate suitably chosen coordinates of space and time between two inertial systems (S) and (S') moving with the speed (v) relative to one another:

$$x' = \gamma(x - vt) \text{ and } t' = \gamma\left(t - \frac{vx}{c^2}\right)$$

where c denotes the speed of light of 299,792,458 meters per second, and the dimensionless number

$$\gamma = \frac{1}{\sqrt{1 - \frac{v^2}{c^2}}}$$

is the so-called Lorentz factor. In 1908, Hermann Minkowski gave a mathematical description of the Lorentz transformation as a rotation of the coordinate axes in four-dimensional space-time.

When v is much smaller than c, the Lorentz factor is close to 1, and the Lorentz transformation reduces to the classical Galilean transformation. When v approaches c, however, the Lorentz transformation has a number of consequences that radically contradict classical physics as well as common sense. For example, clocks in motion are slowed down (called "relativistic time dilation"), objects in motion are contracted in the direction of movement (called "relativistic length contraction"), and clocks in motion that are seen as synchronized by an observer moving with the clocks are seen as nonsynchronized by an observer at rest (called "relativity of simultaneity").

It is another consequence of special relativity that no material objects—or signals of any kind—can travel faster than light. This "speed limit" exists because anything traveling faster than light relative to one observer would appear to be traveling backwards in time relative to another observer, thus leading to paradoxes regarding cause and effect. There is a quantum-mechanical phenomenon, the so-called Einstein–Podolsky–Rosen paradox, that seems to contradict this principle. According to quantum mechanics, the wave function of two entangled particles is affected by a measurement of the state of one of the particles, causing an instantaneous change to the state of the other, even if the two particles are located in different galaxies. But this phenomenon, which has since been verified experimentally, does not really contradict relativity since it cannot be used to transmit information from one galaxy to the other.

Special relativity dictates that mass and energy are connected by the equation $E = mc^2$, undoubtedly the most famous formula in all of physics. Any particle with mass m has a rest energy given by this equation. If the same particle is accelerated to the speed v, its energy is multiplied by the Lorentz factor γ, and its kinetic energy is found as the difference between total energy and rest energy, expressed algebraically as

$$E_{kin} = \gamma mc^2 - mc^2 \approx \frac{1}{2}mv^2.$$

The approximation, valid for v much smaller than c, equals the expression for kinetic energy in classical mechanics. This formula shows that it would require an infinite amount of energy to accelerate a particle with positive mass to the speed of light.

General Relativity
Einstein noted that special relativity implies that space appears to be curved, or "non-Euclidean," to observers in accelerated systems (for example, on a rotating disc) and inferred from the equivalence principle that the same must be true in gravitational fields. However, after realizing this fundamental principle in

1907, it took him eight years to find the field equations that describe the exact curvature of space-time. The idea that physical space might be curved was not new. Already in 1823, Carl Friedrich Gauss investigated this question empirically by measuring the sum of angles of a triangle formed by three mountaintops but found no curvature. Bernhard Riemann further developed the mathematics of curved space in 1854 and this work would become an essential part of Einstein's theory.

General relativity predicts that a body falling freely in a gravitational field, such as the Earth in its orbit around the sun, follows a "geodesic" in curved space-time. This geodesic is called the body's "world-line." In a curved space, geodesics are the least curved lines, in the same way as the equator is a least curved line on the surface of Earth. Although the predictions of general relativity are nearly the same as those of classical mechanics for bodies in weak gravitational fields, the interpretation of gravity is radically different: whereas classical mechanics explains the elliptical orbit of the Earth as a consequence of a gravitational force emanating from the sun, general relativity postulates that the mass of the sun gives rise to a curvature of space-time, and that the world-line of Earth is in fact a geodesic.

It is a consequence of general relativity that clocks in gravitational fields are slowed down. This effect is called "gravitational time dilation." For a clock at rest in the gravitational field of Earth, the dilation factor is

$$\sqrt{1 - \frac{2GM}{rc^2}} \approx 1 - \frac{GM}{rc^2}$$

where G is Newton's gravitational constant, M is the mass of Earth, and r is the distance between the clock and the center of Earth.

Proofs and Applications of Relativity
Einstein showed in 1915 that general relativity explains the perihelion precession of the planet Mercury. This phenomenon, which had mystified astronomers since its discovery in 1859, is that the elliptical orbit of Mercury rotates around the sun with 43 arc seconds per century.

Also in 1915, Einstein predicted that light emitted from distant stars is deflected when passing through the gravitational field of the sun. Although this effect had previously been derived from Newtonian gravity alone, Einstein showed that the angle of deflection following from general relativity is twice the angle following from classical physics. Einstein's prediction was confirmed dramatically by Arthur Eddington during the total solar eclipse of May 29, 1919.

Contrary to quantum mechanics, the technological implementations of which are ubiquitous, relativity has few practical applications. One notable exception is the global positioning system (GPS). GPS satellites revolve around the Earth twice per sidereal day at a height of about 20,000 kilometers (12,400 miles) and with a speed of about 4 kilometers (2.5 miles) per second. Because of the speed and altitude, the atomic clocks aboard the satellites are subject both to relativistic time dilation and to a reduced gravitational time dilation.

The first effect amounts to a loss of 7 microseconds per day, the second to a gain of 45 microseconds per day. In total, therefore, the atomic satellite clocks gain 38 microseconds per day relative to clocks on the ground. Failure to take these relativistic effects into account would render GPS useless since the resulting positional error would accumulate to 11 kilometers (6.8 miles) per day.

Further Reading

Einstein, Albert. *Relativity: The Special and General Theory*. New York: Henry Holt, 1920.

Feynman, Richard, Robert Leighton, and Matthew Sands. *The Feynman Lectures on Physics*. Reading, PA: Addison-Wesley, 1964.

Gamow, George. *Mr. Tompkins in Wonderland*. New York: Macmillan, 1946.

Grøn, Oyvind, and Sigbjorn Herv. *Einstein's General Theory of Relativity: With Modern Applications in Cosmology*. New York: Springer Science+Business Media, 2007.

Møller, Christian. *The Theory of Relativity*. Oxford, Egland: Oxford University Press, 1952.

Russell, Bertrand. *The ABC of Relativity*. London: Kegan Paul, Trench, Trubner, 1925.

Ungar, Abraham. *Analytic Hyperbolic Geometry and Albert Einstein's Special Theory of Relativity*. Singapore: World Scientific Publishing, 2008.

DAVID BRINK

See Also: Black Holes; Einstein, Albert; GPS; Geometry of the Universe; Gravity.

Religion, Mathematics and

See *Mathematics and Religion*

Religious Mathematicians

See *Mathematicians, Religious*

Religious Symbolism

Category: Friendship, Romance, and Religion.
Fields of Study: Communication; Geometry; Number and Operations; Representations.
Summary: Many religious symbols are mathematical in nature.

Archaeological research suggests that religion predates people's ability to read and write but that symbols were often used to express religious ideas and to convey meaning. In this context, such symbols might be pictures, geometric objects, or numbers that hold a particular meaning within a given faith. Long after the introduction of the written word, symbols still hold a powerful place in most religions. There are many highly recognizable symbol forms that are used in various ways by different faiths around the world, though they often share similar underlying structures, themes, or meanings. Symmetry is common in religious symbolism, as are certain numbers or concepts that some believe to have special significance beyond mathematical interpretations. For example, some have proposed a stylized version of the empty set symbol to represent atheism.

Stars

Stars have been used for millennia in a variety of religions. The most common is a five-pointed star, also known as a "pentagram" (*penta* means "five"). At times, the five points have represented the five senses (vision, hearing, touch, smell, and taste). Wiccans use the points to represent five elements (spirit, fire, air, water, and earth) as do Taoists (fire, earth, metal, water, and wood). Other times, it has represented the human body with the "points" of the body relating to the head, arms, and legs outstretched, as seen in the Baha'i Faith where the pentagram is its official symbol. Christians have used the pentagram to denote the wounds (five stigmata) received by Jesus Christ when he was crucified—hands, feet, and side. Judaism has used the pentagram to represent the Pentateuch (Genesis, Exodus, Leviticus, Numbers, and Deuteronomy) and later, Solomon's Seal. Muslims use a crescent moon and a pentagram to denote the religion of Islam.

Some religions have used a point-down pentagram as part of their symbolism. For example, Anton LaVey's Satanists (who have nothing to do with Satan—they do not believe Satan exists) use the upside down pentagram for their symbol and often impose a goat's head in the symbol with the upper points representing horns, the side points being the ears, and the lower point as the chin and beard area. Mormons (belonging to the Church of Jesus Christ of Latter-Day Saints) have used

the inverted symbol in some temple architecture as representing the "morning star" (Venus' path in the sky).

Another star variation is the six-sided star, sometimes referred to as a "hexagram" (*hexa* means "six") or the Star of David as a symbol of Judaism. Most often, this star is drawn as two equilateral triangles drawn on top of each other with one pointing up, the other down, and slightly offset. Hindus have a variation of the hexagram called the *Shatkona*, which show the triangles weaved together denoting the interlocking of fire and water, or male and female. The hexagram is also a symbol for Rastafarians and is usually solid black. The Raelism Movement uses a different variation of the hexagram as their official symbol; it contains a right-facing swastika embedded in the center of the star.

A seven-pointed star (called a "heptagram") is sometimes used by Jews and Christians to denote a seven-day creation. Faery Wiccans and Blue Star Wiccans also use the seven-pointed star, but the Blue Star Wiccans refer to it as a "septagram" instead.

There are a few variations of eight-pointed stars. Islam has a star referred to as *rub el hizb*, which appears as two squares superimposed with one slightly offset the other. It is used to help facilitate the recitation of the Qur'an. The same shape (without the center circle) is referred to as the Star of Lakshmi by Hindus, where it represents the eight forms or kinds of wealth. This shape is referred to as an 8/2 "octagram" (*oct* means "eight"). The "8/2" indicates that there are eight sides on the star and every second point (or vertex) is connected with a line. An 8/3 octagram would have every third vertex connected to each other. This symbol has been used by Christians to represent baptism and resurrection. Ancient Mesopotamia calls their eight-pointed star the Seal of Shamash. The center was a circle representing the sun (Shamash) with eight points emanating from the center. Most likely, the vertical and horizontal points represent the four directions of the compass while the diagonal points represent the equinoxes and solstices.

Although the Baha'i uses a pentagram for their official symbol, a nine-pointed star is more commonly associated with the religion. The star is often drawn similar to the hexagram, but with three equilateral triangles slightly offset and a single point at the top of the star, but without the inner lines. The Baha'i Faith also uses another version of the nine-pointed star with symbols of the "nine world religions" at each point.

From top left: Christian cross, Jewish Star of David, Hindu Aumkar, Islamic star and crescent, Buddhist wheel of Dharma, Shinto torii, Sikh Khanda, Bahá'í star, and Jain ahimsa symbol.

Crosses

The cross is sometimes thought of as a universal symbol for Christianity since, in the Christian faith, Jesus is believed to have been crucified on a Roman cross. However, there are many types of "Christian crosses" and many religious crosses that are not Christian at all. The original Christian cross probably resembled an "X" for the first Greek letter in the word "Christ." It is not related to the crucifixion and came much later than Jesus's death, as many early Christians opposed its use. When placed so that its arms pointed vertically and horizontally, the meaning was the four directions of the compass—where the gospel should be spread. Eventually, the Greek cross made way for the Latin cross, which resembles a lower case "t." Orthodox Christians add a small horizontal line above the arms of the cross denoting the sign hung by Pilate, and a small diagonal line below the arms of the cross denoting a footrest. Other denominations, like Methodists, show a flame behind the cross indicating the Holy Spirit. Sometimes, the cross is displayed upside down, known as a reversed cross or the cross of Saint Peter. Although the original meaning for this cross probably originated from Peter's request to be crucified upside down (so was Christian in origin), many have associated it with the occult and Satanism. Because satanists inverted the Christian pentagram, people believe they inverted the cross as well.

The ankh has a cross for a base, but an oval in place of the head of the cross. Sometimes, the ankh is referred to as an *ansata*, or handle, cross. This symbol was primarily used in Egypt as a symbol of life and fertility. Since its context was often in regards to resurrection, this symbol was used by Gnostic sects of early Christians to symbolize the resurrection of Christ. The ankh was actually used by Christians before the Latin cross. Wiccans currently use this symbol today to mean immortality and completion.

Another misunderstood religious symbol is the swastika. The swastika is a cross with its arms bent at right angles, most commonly so that the top arm is bent to the right and each remaining arm is bent in a similar clockwise direction (from the center) to give the impression of movement. When the arms are bent in the other direction, it can be called a "swastika" or it is sometimes referred to as a "sauwastika." The name is Sanskrit in origin and can be loosely translated as "good luck charm." Historical records show that the swastika is an ancient symbol (older than the ankh). Hindus use both forms of the swastika; the right facing means the evolution of the universe, whereas the left facing indicates the involution of the universe. Together, both versions are thought of as a balance of opposites. Buddhists primarily used the right facing swastika, although recently they have changed to using the left facing version, as the right facing version has become known as an anti-Semitic hate symbol since World War II. The swastika used by the Nazis was right facing but also rotated 45 degrees and appears different from the religious symbols. In Jainism, the swastika is the symbol for their seventh saint (or Jina). Jainists draw swastikas using rice to begin and end ceremonies around altars and idols. The swastika has also been used by Native Americans to represent the sun, the four directions, and the four seasons. Raelians use the swastika in a hexagram to denote that time is infinite.

Further Reading

Grünbaum, Branko, and G. C. Shephard. *Tilings and Patterns*. New York: W. H. Freeman, 1987.

Liungman, Carl G. *Dictionary of Symbols*. Santa Barbara, CA: ABC-CLIO, 1991.

———. *Symbols: Encyclopedia of Western Signs and Ideograms*. Stockholm: HME Media, 2004.

Rest, Friedrich. *Our Christian Symbols*. New York: Pilgrim Press, 1982.

CHAD T. LOWER

See Also: Houses of Worship; Incan and Mayan Mathematics; Mathematicians, Religious Mathematics and Religion; Numbers and God.

Religious Writings

Category: Friendship, Romance, and Religion.
Fields of Study: Connections; Communication; Number and Operations; Representations.
Summary: Mathematics and religious thought have been driven by the same motive: the need to better understand the nature of life and the universe.

In addition to its computational and problem-solving power, mathematics has long been joined to religious faith to form systems of mutual support. Evidence of the most productive relationships can be found in a variety of texts that call attention to mathematical concepts and knowledge as part of a religious or theological treatise. In other cases, the purported significance of mathematics to religion is a cause for antagonism and tension. Among the most persistent relationships evoked in writings that combine mathematics and religion is one that is understood to exist between their particular ways of knowing. Whether by way of analogy or more direct linkages, predominate characteristics of mathematical knowledge—its clarity, certainty, and timelessness—have often been called upon to serve theological contemplation.

Plato

Several Platonic dialogues feature extended discussions of mathematical knowledge in relation to philosophical and cosmological considerations, most notably the *Meno*, the *Timaeus*, and the *Republic*. Coaxing a geometric argument from an unsuspecting slave boy in the *Meno* serves as an epistemological lesson in humankind's ability to access certain and timeless knowledge. In the *Timaeus*, the power of mathematics as a system that provides a way of comprehending the physical world legitimizes adopting a cosmological perspective organized around identifiable characteristics, such as intelligence and goodness. The significance of mathematics to training philosopher-rulers as presented in the *Republic* is predicated on their need to reason effectively about ideal forms such as morality and justice. Although it would be incorrect to refer to them as "religious" in a strict sense, these Platonic dialogues establish a crucial link between mathematical and metaphysical contemplation frequently reflected in later theological writing.

Gregory of Rimini

Gregory of Rimini (c. 1300–1342) followed an Aristotelian mode of thinking, according to which abstract mathematical concepts exist only in the mind of mathematicians. Unlike their characterization within the Platonic tradition, mathematical entities have no existence independent of the objects that possessed them in terms of size, quantity, or other qualitative features. Even so, Gregory of Rimini's compiled *Lectures* undertake discussions of the continuum that ultimately challenge Aristotle's opinion on the impossibility of infinity

as an actual or completed notion. This work intertwines discussions of divine omniscience, the temporal and spatial characteristics of angels, and the divisibility of the continuum, placing it squarely in a scholastic tradition that incorporates mathematical considerations within commentaries that focus primarily on religious subject matter.

Nicholas Cusanus

Although the author of several texts dedicated to Classical problems, such as squaring the circle, the philosopher and theologian Nicholas Cusanus (1401–1464) explicitly elaborated on the connection between mathematics and religion in *Learned Ignorance* (c. 1440). The significance of mathematical reason to theological contemplation discussed in this text is founded upon its ability to provide reliable and infallible knowledge about objects that transcend direct human experience. For Casanus, relations that exist between all things meant that one is able to develop an appreciation of unknowable objects based on other, better-understood objects. Polygonal approximations to a circle underscore this relationship. At the same time, Cusanus was aware that obtaining knowledge in this way depended on using various symbols and symbolic relationships in consistent and correct ways. The study of mathematics employed immutable symbols that avoided interpretive ambiguity and, thus, appealed to Cusanus as an appropriate framework for working with them.

Michael Stifel

In his 1532 *Book of Arithmetic About the Antichrist, A Revelation in the Revelation*, Michael Stifel (1468–1567) used computation skills and numerological inclinations to predict the end of the world. By doing so, he contributed to the fervor of the Reformation by associating the pope with the antichrist of the Book of Revelations. Indicative of his talents as a mathematician who pursued a lifelong fascination with numbers and their meaning, Stifel's 1544 book, *Arithmetica Integra* is considered his major achievement. In it, he explores and extends Pythagorean number theory, the construction of magic squares, the theory of irrationals, and the algebra of quadratic equations.

Galileo Galilei

Galileo Galilei (1564–1642) articulated a connection between mathematics and the divine that many found problematic. Like others before him, much of his writing asserted the superiority of mathematical reasoning, acknowledging it as the most certain way to both read and describe truths pertaining to the natural world. However, his praise of mathematics went considerably further in some texts, including the 1632 book, *Dialogues Concerning the Two Chief World Systems*. Specifically, Galileo maintained that human knowledge was indistinguishable from divine knowledge regarding those areas of mathematics to which it turned its attention. Consequently, mathematical reasoning provided unmitigated and unparalleled access to God's designs. As a threat to longstanding theological hierarchies, Galileo's pronouncements on mathematics were part of the indictments brought against him by church inquisitors.

René Descartes

A mathematical approach to reasoning is evident in the prescriptions set down by René Descartes (1596–1650) in his book, *Discourse on the Method*. Compelled by both skepticism and consistent criteria, he promoted a reductive framework for investigating problems that requires breaking up the analysis into pieces. Examining and understanding the simplest of the pieces would then lead to a solution. The first principle of this analytic approach allows one to establish a simple truth by virtue of its evident nature. Using mathematics as an exemplar for all reasoning therefore demanded an assurance of certainty. Descartes addresses this requirement in his 1641 book, *Mediations*. In particular, this work contains proofs of the existence of a benevolent and non-deceiving God, by virtue of which humans are able to recognize eternal truths for themselves. Although not above philosophical criticism, Descartes's work embraces mathematical and religious concerns of the time.

George Berkeley

George Berkeley (1685–1753) adopted a significantly antagonistic perspective on mathematics and theology. Some of his early writing evidences his affinity and appreciation for mathematics. However, later commentaries published while he served as the Bishop of Cloyne criticized mathematicians. Most notable among these are the 1732 book *Alciphron, or the Minute Philosopher* and the 1734 book, *The Analyst, or a Discourse Addressed to an Infidel Mathematician*. Berke-

ley asserted that mathematicians made unjust claims to exactness. His belief that the persuasive power of its problematic reasoning undermined the precepts of revealed religion only exacerbated this concern. Associating it with dogmatism and obscurantism, Berkeley was particularly hostile to the use of fluxions and infinitesimals, respectively, in the Newtonian and Leibnizian developments of calculus. One of his overarching objections pertained to the unacceptable admission of infinity in mathematics. Consequently, he attempted to establish the rule for computing the derivative of x^n in the *Analyst* by avoiding the use of either fluxions or infinitesimals.

Charles Babbage

Exemplary of natural theology in the nineteenth century, the *Bridgewater Treatises* were intended to provide commentary on modern scientific discoveries in relation to the Creation. In all, eight manuscripts were commissioned that discussed topics such as chemistry, geology, meteorology, and physiology. Mathematics was not one of the subjects included in the original commission, and Charles Babbage (1791–1871) took its omission as an opportunity to write his *Ninth Bridgewater Treatise*. Considered the father of modern mechanical computing, Babbage dedicated much of his life to designing the difference and analytic engines. His treatise highlights this work by arguing that events appearing miraculous can be accounted for as part of a grand design. As consummate a promoter as he was a mathematician, Babbage publicly illustrated this point several times with a model of the difference engine. These demonstrations involved programming the machine to break an identifiable recursive pattern at a moment that defied explanation by his audience.

Edwin Abbott

The enduringly popular 1884 book, *Flatland: A Romance of Many Dimensions*, introduced the concept of higher dimensional space to a wide readership. As its author, Edwin Abbott (1838–1926), drew upon his strengths as an educator, an expositor, and a theologian to convey multiple messages that relate to the mathematical imagination. Among these, scholarship has focused attention on progressive theological imperatives that he developed elsewhere and subtly incorporated into *Flatland*. Specifically, Abbott was keen to promote a form of theology that would be able to respond positively to new scientific attitudes and investigations. Mathematical research provided an ideal vehicle for Abbott, as discussions of non-Euclidean geometries suggested a loss of certainty within the discipline concomitant with a loss of religious certainty. Though perhaps the best known, Abbott joined and influenced other writers who used new developments in geometry as the impetus for renewed spiritual reflection that continued into the twentieth century, including Charles Hinton, Arthur Schofield, Peter Ouspensky, and Claude Bragdon.

Other Connections

There are other ways in which religion and mathematics are connected in writing. For example, mathematician Blaise Pascal produced many specifically religious writings, including *Provincial Letters* and the *Pensées*. Literary and religious scholars continue to study not only these works but also his mathematical and scientific writings to gain greater insight into his religious beliefs. A systematic study of the contributions of people from other cultures and religions to mathematics, such as Muslims or Hindus, or the geometric discussions in rabbinical writings also interest historians and mathematicians. Finally, while there are countless historical examples of mathematicians whose religious beliefs and mathematical work are philosophically intertwined, philosopher and mathematician Bertrand Russell's 1927 lecture, and later essay, *Why I Am Not a Christian*, has been called "devastating in its use of cold logic" in critiquing religious beliefs. A book containing this and related essays was included in the New York Public Library's list of the most influential books of the twentieth century.

Further Reading

Koetsier, T., and L. Bergmans, ed. *Mathematics and the Divine: A Historical Study*. Oxford, England: Elsevier, 2005.

Swade, Doron. "'It Will Not Slice a Pineapple:' Babbage, Miracles and Machines." In *Cultural Babbage: Technology, Time and Invention*. Edited by Francis Spufford and Jenny Uglow. London: Faber, 1996.

Valente, K. G. "Transgression and Transcendence: *Flatland* as a Response to 'A New Philosophy.'" *Nineteenth-Century Contexts* 26 (2004).

K. G. VALENTE

See Also: Greek Mathematics; Infinity; Mathematical Certainty; Mathematicians, Religious; Mathematics and Religion; Numbers and God; Proof.

Renaissance

Category: Government, Politics, and History.
Fields of Study: Geometry; Representations.
Summary: The Renaissance's resurgence in humanism also benefited mathematics and engineering.

The Renaissance or Rinascimento (both words mean "rebirth") was a flourishing of philosophy, art, architecture, science, and high culture more generally beginning in fourteenth-century Europe. Renaissance thinkers thought of themselves as restoring the civilization of Greece and Rome after what they called "the Middle Ages." The Renaissance saw the rise of humanism, hermeticism, Neoplatonism, and realist art involving optical perspective; the decline of feudalism; increased circulation of ideas due to printing; the Protestant Reformation; a strong interest in classical literature and history; a strengthened interest in science and mathematics and their applications; and world exploration.

Early Renaissance (c. 1300–1450)

The Renaissance can be traced back to the thirteenth-century writings of Dante Alighieri, Francesco Petrarca, and Brunetto Latini and the paintings of Giotto di Bodone. Such work was sponsored by bankers, merchants, and industrialists who rose to great wealth and influence, displacing the Church and landed nobility as primary sponsors of high culture.

Starting in the mid-fourteenth century, humanist scholars searched libraries to recover the lost texts of classical Rome. Many edited texts went to print, increasing their accessibility at (relatively) low cost. After approximately 50 years, attention turned to recovering the Greek heritage, which—though mostly lost in the West—had continued on in Byzantium. Many Greek scholars migrated west at this time, bringing their expertise and manuscripts to Venice, in particular. The recovery and translation of Plato's works, along with several tracts in neoplatonism and hermeticism,

The mid-Renaissance Basilica di San Lorenzo has a geometric regularity and an open lightness.

century. Political philosophy, exemplified by Niccolò Machiavelli's *Prince* and *Discourses on Livy,* attempted a rational analysis of political structures contextualized by cultural difference and the practicalities of everyday life. Vernacular languages came to be used for scholarly writing, making texts more widely readable as did printing, which advanced rapidly with the establishment of fine publishing houses in the Veneto. Examples include the Aldine Press, where italic typefaces were invented and Erhard Ratdolt's press, which pioneered the printing of mathematical diagrams when producing the first edition of Euclid's *Elements* in 1482.

The mid-Renaissance was centered on the Republic of Florence, largely sponsored by a powerful banking family, the Medici. The ideals of this period are expressed in Florentine architecture, such as Filippo Brunelleschi's Church of San Lorenzo, which has a legible geometric regularity, bright and even light, openness, and a delicately balanced stillness. Ideals in painting included realism based on optical theory. Artists could occupy the leading edge of mathematical research; Piero della Francesca, for example, produced treatises on perspective theory in addition to painting with perspective techniques. Sculpture also developed a scholarly foundation through both historical study of the classical texts that had survived and hands-on dissection of fresh cadavers.

High Renaissance (c. 1500)

The High Renaissance lasted only briefly before transforming into Mannerism. It was focused on Rome, owing to the patronage of Pope Julius II. Art gained a level of dynamism best known through the works of Rafaello Sanzio (Raphael) and Michelangelo Buonarotti in Rome, and Tiziano Vecelli (Titian) and Giorgione in Venice. Leonardo da Vinci's *Last Supper,* Raphael's *School of Athens*, and Michelangelo's ceiling in the Sistine Chapel were painted during the High Renaissance.

Further north, the Renaissance adapted to local cultures and circumstances. In Germany, for example, goldsmiths crafted clocks, automata, and mathematical and astronomical instruments for their patrons. Reformation printers published a wide range of medieval texts alongside Lutheran tracts, largely shedding the refined typography of Venice in favor of speed and quantity. Gothic elements remained strong in the art and architecture of England, the Netherlands, and

fueled an interest in applying simple numerical ratios and geometric regularity in fields as diverse as art and architecture, cosmology, alchemy, and musical tuning. The intentions included occult efforts to replicate cosmic structures, invoking astral influences at the human scale. More visceral results were achieved by composers, such as Josquin des Prez, who brought polyphonic techniques to Italy from the Low Countries, laying foundations for important Italian composers (such as Giovanni Pierluigi di Palestrina) toward the end of the sixteenth century.

Renaissance (c. 1450–1500)

The Renaissance spread north from Tuscany and across the Alps during the second half of the fifteenth

Scandinavia and Renaissance influences reached those countries only after they had become Mannerist. Because of Protestantism, secular authorities replaced the Catholic Church as the primary sponsor of cultural works.

Renaissance Science and Mathematics

Renaissance scholars initially reacted against Scholastic natural philosophy by turning to Neoplatonism, taking an often mystical and magical approach to nature, often with practical goals. This shift can be seen in the intertwining of alchemy and astrology, for example, and in the wide range of applications described in Giambattista della Porta's 1558 book *Natural Magic*. The title reflects a distinction drawn between natural magic, which invoked empirical knowledge of nature to achieve results; in contrast to spiritual magic, which regulated astral influence using amulets and talismans; and demonic magic, which invoked supernatural beings.

The Church's need for calendrical reform led Nicolaus Copernicus to develop heliocentric astronomy as an improvement upon the Hellenistic methods maintained and developed throughout the Middle Ages. Astronomy was favored also in Protestant territories owing to the educational reformer Philip Melanchthon arguing that it was an ideal way to learn about divine creation.

Artillery motivated studies in ballistics, leading to stellated polygonal designs for fortresses, such as Naarden in the Netherlands and the Kronborg in Denmark. Aristotelianism, however, still provided qualitative theory for ballistics and other practical endeavors, such as hydraulic engineering.

The development of machines and engineering techniques inspired efforts to classify and theorize about them, as shown by the published "theaters of machines" by Jacques Besson and Agostino Ramelli.

The influences of exploration can be dated at least as far back as 1488, when Bartholomeo Dias found a connection between the Atlantic and the Indian Ocean that led to trade routes established beginning in 1498 with Vasco da Gama's arrival in Calicut, six years after Christopher Columbus found the West Indies. Such journeys motivated developments in navigation and shipbuilding as well as an outward-looking attitude. Trade expanded, especially in Spain, Portugal, and—as the new knowledge spread north—the Netherlands.

Descriptions and specimens brought back from foreign regions caused disputes and reforms in biological taxonomy that were eventually settled in the eighteenth century by Charles Linnaeus.

Progressive rational problem-solving, combined with the growth of theoretical method and a growing preference for naturalistic rather than occult explanations, provided many elements needed for the eventual emergence of modern empirical science.

Mathematics was boosted early by the ascendance of merchants and bankers who needed computational methods to manage money and later to solve problems in navigation and cartography. Some advanced material was assimilated from Arabic sources, such as geometric methods and high-precision trigonometric tables. Solving polynomial equations became a display of virtuosity; the quadratic had been solved in antiquity, now Girolamo Cardano and other mathematicians developed solutions for cubics and higher order problems. As algebra developed, many algebraic symbols were invented and evolved into the forms used today. Hindu-Arabic numerals replaced Roman numerals but the calculation of the products, ratios, and square roots of large numbers in astronomy and navigation was still onerous and error-prone. These operations were facilitated by conversion into addition and subtraction problems using prosthaphaeresis (based on trigonometric transforms), and later through the invention of logarithms.

Further Reading

Field, J. V. *The Invention of Infinity: Mathematics and Art in the Renaissance*. Oxford, England: Oxford University Press, 1997.

Goulding, Robert. *Defending Hypatia: Ramus, Savile, and the Renaissance Rediscovery of Mathematical History*. New York: Springer, 2010.

Hall, Marie Boas. *The Scientific Renaissance, 1450–1630*. New York: Dover Publications, 1994.

Hay, Cynthia. *Mathematics from Manuscript to Print 1300–1600*. Oxford, England: Oxford University Press, 1988.

Alistair Kwan

See Also: Algebra in Society; Castles; Exponentials and Logarithms; Marine Navigation; Middle Ages; Multiplication and Division.

Representations in Society

Category: School and Society.
Fields of Study: Connections; Representations.
Summary: Symbols, equations, and images are all used to teach mathematical concepts and to convey mathematical information in society.

Representations are at the forefront of the focus standards of the National Council of Teachers of Mathematics to improve mathematics teaching and learning. Representations allow students to see and experience mathematics from different perspectives. The role of multiple representations in promoting students' conceptual understanding of mathematics has long been emphasized by researchers. Thus, representations are among the essential parts of mathematics lessons. Further, in the twenty-first century, even people who had very little exposure to mathematics in school will encounter various mathematical representations in their daily lives. Familiarity with mathematical representations or representational literacy has become an essential skill. Many mathematical concepts are defined in terms of representations. A function may be represented by a Taylor series of infinite terms, which is named after Brook Taylor. There is also an entire branch of mathematics called "representation theory" that expresses algebraic structures using linear transformations.

Representations

Mathematics has its own native beauty and inspirational aesthetic to represent the physical world and the world of intellect. One of the strengths of mathematics is its resources to seek for new solutions and explore frameworks to answer problems related to the real world. To achieve this goal, mathematical representations in society should be explored and important ideas of modern mathematics should be communicated properly. Representations in mathematics can be described as constructs that symbolize or correspond to real-world mathematical entities, features, or connections. Gerald Goldin broadly defined representations as any configuration of characters, images, or concrete objects that can symbolize or represent something else. Representations take various forms, such as informal representations used in preschool settings or more formal representations used in mathematics classrooms or by mathematicians. For example, children represent groups of five with their hand or, even further, they develop proportional thinking as they relate five fingers to one hand and 10 fingers to two hands. More formally, mathematics students or mathematicians use mathematical equations, for example, to represent curves or relationships among financial variables.

Internal and External Representations

Representations can be both internal and external in nature and can be created by forming individual representations, such as letters, numbers, words, real-life objects, images, or mental configurations. Internal representations are mental images or cognitive constructs of individuals that relate to external representations or to experiences in the external world. James Kaput referred to internal representations as *mental structures* and defined them as instruments that are used to organize and manage the flow of an individual's experience. Internal representation systems exist within the mind of an individual and consist of constructs to assist in describing the processes of human learning and problem solving in mathematics. Internal representations of mathematical concepts can take various forms, such as individual visualization of mathematics concepts, idiosyncratic notation systems, or attitudes toward mathematics.

External representations, on the other hand, include all external entities or symbols. External representations provide a medium to communicate mathematical ideas, concepts, or constructs. Richard Lesh defined external representations as the embodiment of internal systems of thought. Lesh also referred to external representations as mathematical representations that are simplifications of external systems. Learners use external representations, such as marks on paper, sounds, or graphics on a computer screen, to organize the creation and elaboration of their own mental structures. Unlike internal representation systems, external representation systems can be easily shared with and seen by others.

Multiple Representations in Mathematics Education

In mathematics education, there has been a shift from classic to nontraditional teaching and learning practices with multiple representations, where educators use various representations to effectively present information. Multiple representations refer to different kinds of representations that present the same mathematical ideas from different perspectives or representations that present different aspects of the same mathematical concept. For example, teaching fractions concepts using multiple representations may involve presenting fractions in real-life contexts such as partitioning a pizza or a pie, allowing students to explore equivalent fractions using kinesthetic or virtual manipulatives, or providing students with pictorial representations of fraction operations in addition to formal mathematical representations. Teaching and learning with various kinds of representations provide students with hands-on and minds-on experiences and support a better understanding of mathematical concepts. Also, using multiple representations in mathematics education can help to alter the focus from a computational or procedural understanding to a more comprehensive understanding of mathematics using logical reasoning, generalization, abstraction, and formal proof. A substantial amount of research has demonstrated the effectiveness of multiple representations in enhancing students' conceptual understanding of mathematical concepts.

The notion of multiple representations in mathematics education commonly refers to external representations. However, one of the essential goals of mathematics education is to develop internal representation systems that interact well with external representation systems. James Kaput identified five interacting types of internal and external representations: (1) mental representations—internal representation—that learners construct by reflecting on their experiences; (2) computer representations that model mental representations through computer programs, which allow for arrangement and manipulation of information; (3) explanatory representations consisting of models or analogies that create the interaction between mental and computer representations; (4) mathematical representations, where one mathematical structure is represented by another mathematical structure; and (5) symbolic representations, such as formal mathematical notations.

To understand James Kaput's taxonomy of internal and external representations, consider the different types of representations related to the concept of "slope." When learning about positive slopes, a student might internally imagine a hill, which constitutes an internal (or mental) representation. This mental representation can be replicated on a computer screen. The student can create a unique model that incorporates the mental representation through a computer representation. If the model is viable, then it can be an explanatory representation for the concept of "slope." The student, then, can sketch a similar mathematical graph of the hill and can name the steepness of the hill with the mathematical notation, "slope." This graphical representation of slope can, then, provide support to represent the slope in a symbolic form as a rate of change ($y = mx + b$, where slope is represented with m and indicates the ratio of change on the y-axis to the change on the x-axis). As portrayed in this example, internal and external representations are not separate. Rather, they

Children learn groups of five with their hands or may develop proportional thinking as they relate five fingers to one hand and 10 fingers to two hands.

are intrinsically connected, and they interact continuously. Furthermore, a concept like slope is itself a type of alternative representation. In calculus, a curve is represented by the changing nature of its tangent vector, where the solution to the first derivative at a particular point is the slope of the tangent vector.

Translational Skills Among Different Modes of Representations

In addition to the importance of the effective interactions between internal and external representations in the acquisition and use of mathematical knowledge, it is essential that students develop fluency among different external representations. Richard Lesh enumerated multiple modes through which representations can be constructed: manipulatives, pictures, real-life context, verbal symbols, and written symbols. To demonstrate deep understanding of mathematics, students need to represent their mathematical ideas with different modes of representations and smoothly translate within and between those modes. For example, in algebra, students should be able to make the connection between graphical and algebraic or symbolic representations of equations. Similarly, students need to link what they learn using concrete or virtual manipulatives to both pictorial representations and abstract symbols. For instance, students who initially learn fraction operations using concrete or virtual manipulatives should be able to relate this knowledge when they later on learn fraction operations using symbolic and more abstract mathematical representations. Connecting different modes of representation simultaneously has been demonstrated to improve conceptual understanding as well as positive attitudes toward mathematics.

In mathematics education research, there is strong evidence that students can grasp the meaning of mathematical concepts by experiencing different mathematical representations and making connections and translations between these modes of representations. Using translational skills among different representational modes encourages students not to merely memorize theorems and facts but also to think analytically to reproduce and use them in real life problems or even in pure mathematical problems.

To deepen students' understandings, teachers should provide students with multiple representations of a single mathematical concept and focus on students' transition ability from one representation to another. Teachers need to be able to present one concept in multiple modes without relying on a single mode and provide students with appropriate transitions among these representations. Teachers should provide also students with ample opportunities to represent mathematical concepts in multiple ways and to connect these representations, thereby developing representational fluency. For example, asking a student to restate a problem in unique words, to draw diagrams to illustrate the concept, or to act out the problem are some ways to provide students with opportunities to translate among representations. If teachers fail to implement the transitioning among different representations, students will be less likely to see how different representations are related and will be more likely to develop misconceptions.

Multiple modes of representation can be used by teachers and students to enhance understanding of mathematics. Most research has shown that providing students with accurate representations improves student learning. However, different representational modes might have different impacts on student understanding. One mode might be more relevant or effective than another for teaching a specific concept. Or, some representational modes can be more appropriate at different developmental stages of the same concept. For example, research on teaching and learning of fractions has shown that students should be given the opportunity to develop mental representations of fractions using manipulatives before they are presented with symbolic representations. Thus, in addition to using multiple representations, choosing effective and appropriate presentations of information is crucial in teaching and learning. Representations that allow students to actively interact with the subject matter are more effective in student learning than representations that do not support students' active involvement.

Despite the research support for development of higher order thinking skills afforded by different representational forms, little is understood about how students interact with multiple representations in various learning environments. Even though each representation provides similar information, the strain that each representation puts on students' cognitive resources may differ. Not only do individual representations have different impacts on students' conceptual understanding but integrating multiple representations may also result in interaction effects among different modes presented. Therefore, integration of multiple representations becomes an important consideration in the design of instructions. Educators should employ caution as they integrate different modes into instruction, because delivering redundant information with different modes might interfere with learning.

Mathematical Thinking and Representations in the Twenty-First Century

An increasing number of daily activities in the twenty-first century require familiarity with mathematical representations and mathematical thinking. Mathematical thinking, which is a crucial tool for every member of society, includes skills such as pattern recognition, generalization, abstraction, problem solving, proof, and analytical thinking. Most companies prefer employees who are equipped with mathematical literacy or general mathematical skills. However, many students either do not necessarily understand these qualifications or do not value them enough. It is important to emphasize that all humans use mathematical thinking tools in their every day lives and workplaces, with or without noticing they are doing so.

It is not very hard to realize the extent to which mathematical representations are integrated into mundane objects and activities. Consider the number of newspaper columns that provide their readers with different kinds of mathematical representations to explain current issues. Topics in such columns include sports, economics, advertisements, and weather reports. For example, the growth of players, the statistics and ranking of teams, and teams' transfer budgets are represented in several representational modes, such as tabular data, textual information, visual representations, or graphical interpretation. Not only do sports fans need to understand the mathematical information provided readily to them but they also may need to use the mathematical information in problem solving situations, such as estimating the chances of their team's victory. More surprisingly, when a rivalry game is present, the provided data get even more complicated to analyze the chances of each team.

Even though the use of mathematical representations and information in economic and weather columns in various modes is apparent, the ones used within advertisements or political columns may be overlooked. Understanding the mathematical information included in advertisements and deciding which product to buy requires effective use of mathematical thinking tools. In most advertisements, companies present several payment options with different price ranges instead of giving just one price for a product. In particular, mortgage plans to buy houses and installment plans to buy cars require serious analyses of options to choose the best for a given budget. In political columns, on the other hand, one would not be surprised to see percentages representing the proportion of the population that supports various political parties in a country or the votes of a poll. Such information is not only presented as tabular data, visual charts, or graphs, but also as textual information, which is another mode of mathematical representation.

Representations in Problem Solving

Problem solving is one of the essential tools for mathematical thinking. A person equipped with problem solving skills does not necessarily need to have the knowledge base for the solution to each problem encountered but needs to know how to approach problems, locate and access information from different resources, and process information to solve the problem. For example, when one faces a novel problem, an approach to solving that problem can be forming an analogy between the new problem and another, previously solved problem. In other words, known information from an earlier problem can be mapped onto the novel problem. Brainstorming may be another valuable approach to gather different ideas on solution paths to unfamiliar problems. If a problem is too complex, problem solvers can try to break it down into more manageable parts (more solvable problems). One approach to problem solving is solving the problem step-by-step and taking an action at each step to get closer to the goal. Another solving approach can be conducting extensive research to analyze existing ideas and then adjusting possible solutions to the problem in hand. Finally, trial-and-error may be an approach to find a solution to an existing problem. It is emphasized in problem solving that there are many solution paths to a problem and a willingness to try multiple approaches is encouraged. Multiple approaches and strategies may be available and some of these approaches may be more efficient than the others.

Problem solving in mathematics, and in other fields as well, requires both knowledge of different representational systems and representational fluency that enables flexible use of various representational systems. For example, when solving a mathematical problem that asks how many quarters there are in 2 1/2, various strategies that involve different representations exist to approach the problem. A student may choose to translate this problem, which is represented in words, into a real-life context, such as how many quarter slices of pizza there are in 2 1/2 pizzas. Another student may opt to draw a picture that represents the given problem and solve the problem using the pictorial representation. Or, some students may represent the problem using symbolic representations and solve the problem accordingly. There may be other approaches where students start with a real-life context and then translate it to a pictorial representation, or where students come up with various relevant representations and choose the most efficient one for them. In more complex problems, different parts of the problems may require different representations. Thus, representational fluency is an essential part of problem solving.

Problem solving is such an important skill that is not only required to help students solve mathematical problems but also provides them with necessary tools to approach and solve problems in the real world. Because the real word does not have recipes to solve a problem, and problem solving requires structured, thoughtful, and careful analysis of problems (especially ill-defined problems) in various situations, people equipped with problem-solving skills are highly valued by employers.

Mathematics as a Language

Mathematics is, to some extent, a language that is universal and can be understood in any part of the world without much difficulty. The mathematics language, which consists of both symbolic and verbal languages, has evolved as the most efficient medium to communicate mathematical ideas and information. Mathematics language also includes graphical images to effectively communicate mathematical concepts and ideas. Thus, different representational modes are used in communicating mathematical ideas and concepts. For example, when a mathematics teacher writes an equation and explains the equation in spoken language to a class, both verbal and written representational forms are in play. Communication in mathematics often involves a constant representational translation between symbolic and verbal representations. Symbolic and verbal languages of mathematics help to express ideas in a

Representational Skills

The National Council of Teachers of Mathematics presents representation as an important skill needed for students and teachers in teaching and learning mathematics in *Principles and Standards for School Mathematics*. Students should lucidly and coherently be able to express mathematical ideas through various representational modes, especially in writing and speaking. Through representational skills, abstract concepts can be manipulated into concrete concepts. Developing appropriate representation manipulation skills is necessary to improve conceptual understanding. Further, using various modes of representations, such as graphics, tabular data, mental images, physical objects, mathematical symbols and notations, drawings, and textual information, provides students with organizational skills to systematize their thinking and approach a concept from multiple views, leading to a more coherent understanding. With this ability, students can represent phenomena in a way that is meaningful to them. More importantly, the capability of representing a concept in numerous modes eliminates possible communication problems.

meaningful and efficient way. The evolution of mathematics language has been in progress for thousands of years. The goal of this progress is to improve the efficiency of communication, which is central to learning and using mathematics.

Before the emergence of mathematical notations and symbols, mathematicians found it difficult to share their knowledge with the community, even with other mathematicians. Even if a mathematician were able to prove a theorem, for example, geometrically without using mathematical notations and symbols, the mathematician might not have easily written down the proof to share it with others. Difficulties in representing mathematical ideas (writing in a concise and meaningful way using various mathematical notations and symbols) forced mathematicians to seek alternative (especially short and easy) forms to present their knowledge. The need for an effective and efficient mode of communication to convey mathematics ideas resulted in the development of the symbolic mathematical language.

Although the symbolic mathematical language is universal, the verbal mathematical language differs across societies or cultures. For example, although the American and the Japanese use the same symbolic notations to convey mathematical ideas, the verbal language each of these nations uses to communicate about mathematics is different. Differences in verbal languages to communicate mathematics have implications for teaching and learning mathematics. Verbal languages that are clearer about mathematical terms or that relate better to mathematical entities or ideas can support mathematical understanding. For example, counting in the verbal Chinese language is based on the concept of base-10 system. In Chinese, the number 11 is not an arbitrary word in the verbal language. Rather, in Chinese, 11 is "ten-one," 12 is "ten-two," 21 is "two-ten-one," 22 is "two-ten-two," and so on. In other words, the Chinese verbal language clearly conveys that there is one 10 and one 1 in 11 or there are two 10s and one 1 in 21. Such a clear relation between mathematical ideas and verbal language can be an important cognitive tool that supports mathematical understanding.

Further Reading

Curtis, Charles. *Pioneers of Representation Theory: Frobenius, Burnside, Schur, and Brauer*. Providence, RI: American Mathematical Society, 2003.

Goldin, Gerald A. "Representation in School Mathematics: A Unifying Research Perspective." In *A Research Companion to Principles and Standards for School Mathematics*. Edited by J. Kilpatrick, W. G. Martin, and D. Schifter. Reston, VA: National Council of Teachers of Mathematics, 2003.

Kaput, James. "Representation Systems and Mathematics." In *Problems of Representation in the Teaching and Learning Mathematics*. Edited by C. Janvier. Oxfordshire, England: Erlbaum, 1987.

Lesh, Richard. "The Development of Representational Abilities in Middle School Mathematics." In *Development of Mental Representation: Theories and Applications*. Edited by I. E. Sigel. Oxfordshire, England: Erlbaum, 1999.

Lesh, Richard, Kathleen Cramer, Helen M. Doerr, Thomas Post, and Judith S. Zawojewski. "Using a Translational Model for Curriculum Development and Classroom Instruction." In *Beyond Constructivism: Models and Modeling Perspectives on Mathematics Problem Solving, Learning, and Teaching*. Edited by R. Lesh and H. M. Doerr. Oxfordshire, England: Erlbaum, 2003.

National Council of Teachers of Mathematics. "Illuminations: Resources for Teaching Math." http://illuminations.nctm.org.

Utah State University. "National Library of Virtual Manipulatives." http://nlvm.usu.edu/cn/nav/siteinfo.html.

SERKAN OZEL
ZEYNEP EBRAR YETKINER OZEL

See Also: Communication in Society; Connections in Society; Geometry in Society; Mathematical Modeling.

Revolutionary War, U.S.

Category: Government, Politics, and History.
Fields of Study: All.
Summary: The American Revolutionary War saw advances in mathematics cryptography and education.

The American Revolutionary War was a political and armed conflict between Great Britain and the British colonies on the North American continent between 1775 and 1783. Colonists who sought to end British rule and declare their political and economic independence supported the establishment of 13 colonial governments, each of which in turn sent representatives to Philadelphia to set up the Second Continental Congress.

This congress debated the state of political and economic ties to Britain, plied for support from other European powers, and discussed the possibilities and potential of a collective effort to make the separation official. Shortly after its inception, the Second Continental Congress formed a Continental Army and issued the Declaration of Independence. These actions announced the birth of a new nation: the United States of America. The "War of American Independence," as the American Revolutionary War is also called, saw fierce fighting in a wide variety of locations throughout the new nation and on the soil of virtually every new state. Some key battles were fought in Lexington, Concord, and Boston, Massachusetts; Saratoga and Ticonderoga, New York; Trenton, New Jersey; King's Mountain and Cowpens, South Carolina; and Yorktown, Virginia; among many other places.

The war lasted almost a decade and ended with the Treaty of Paris, which was signed at the Palace of Versailles in 1783 and recognized the sovereignty of the United States of America. There are many statistics available that relate to aspects of the war, including casualties and cost. For instance, some report that the British spent about £80 million while incurring a national debt of 250 million pounds, while the United States spent approximately $135 million, of which $37

million became the national debt. Mathematics was used in a wide variety ways, including in the design and implementation of artillery and in planning strategy and tactics. Mathematicians fought in the war, conducted surveys, and created and decoded ciphers. The mathematics educational system also changed significantly as a result of the war.

Louis-Antoine de Bougainville

Many historians agree that the Americans would have been unable to win the war without the political and military support of France and other allies. Louis-Antoine de Bougainville was a French mathematician who became the first Frenchman to sail around the world. In 1752, he wrote a calculus book, *Traité du calcul–intégral*, which brought him recognition within the mathematical community for his clear exposition and updates to differential and integral calculus. After a second edition and election to the Royal Society of

Cryptography

Early U.S. military intelligence began during the Revolutionary War. Paul Revere, William Dawes, and others used light signals to warn of invading forces before the battles at Lexington and Concord, which are generally considered to be the first military engagements of the war. James Lovell, who has been called the "father of American cryptanalysis," broke the British ciphers, which were rearrangements of letters. He used a method known as *frequency analysis*, which involves determining letters based on the frequency of symbols in the coded message.

Lovell discovered that the British often changed ciphers by shifting them instead of creating a new rearrangement and this made them easier to decode. Lovell also created his own cipher forms but these were deemed too

Paul Revere's ride used light signals to warn the public.

confusing for those wanting to send and receive messages.

This belief was even true for Benjamin Franklin, who was well versed in mathematics and enjoyed magic squares recreationally. Franklin commented, "If you can find the key & decypher it, I shall be glad, having myself try'd in vain." American diplomats began to rely increasingly on replacements of words and other techniques instead of alphabet substitutions, and spies for both sides conveyed information about supplies and troop movements using codes. For instance, U.S. spy Benedict Arnold used book ciphering, in which a word is represented by a number that corresponds to a location in a book, in his communication with British intelligence officer John Andre.

London in 1756, he turned to a career in which he participated in numerous wars, including the Revolutionary War.

His astronomical observations became important to later explorers. He stated, "geography is a science of facts: one cannot speculate from an armchair without the risk of making mistakes which are often corrected only at the expense of the sailors." During the Revolutionary War, he was a commodore who supported the U.S. side.

Simeon DeWitt

U.S. Army geographer Simeon DeWitt subscribed to *The Mathematical Correspondent*, generally regarded as the first U.S. special-interest scientific publication. DeWitt was a student at Rutgers University when British troops burned the college buildings. He continued his study of mathematics and surveying on his own and was appointed the geographer of the army by General George Washington. After the war, he became surveyor-general of New York State.

Education

Mathematics education changed dramatically in the United States during and after the war. Before the war, students usually learned mathematics from British works, although Americans like Isaac Greenwood had written arithmetic texts. Advanced mathematics included algebra, geometry, trigonometry, calculus, and surveying techniques. Many colleges were shut down during the war because students and professors served as soldiers, and buildings were used for other purposes. However, some members of the army were trained in mathematics during the war. After the war, new primary schools and colleges were established. Between 1776 and 1815, numerous mathematics texts were published in the United States. Some of these were reprints of English works, and others were compilations or new works by American writers. In 1788, American Nicholas Pike published his text, *The New and Complete System of Arithmetick: Composed for the Use of the Citizens of the United States*, which contained both arithmetic and geometry. It was popularized by patriotic recommendations. There was also a change in the education of women. Prior to the war, it was thought that mathematics beyond simple arithmetic was unnecessary for women. After the war, mathematics educational opportunities began slowly to increase, as women were educated in mathematics to help in family businesses.

Further Reading

Weber, Ralph. "James Lovell and Secret Ciphers During the American Revolution." *Cryptologia* 2, no. 1 (1978).

Tarwater, Dalton. *The Bicentennial Tribute to American Mathematics*. Washington, DC: The Mathematical Association of America, 1977.

Tolley, Kim. *The Science Education of American Girls*. New York: Routledge, 2003.

Zitarelli, David. "The Bicentennial of American Mathematics Journals." *The College Mathematics Journal* 36, no. 1 (2005).

CALLI A. HOLAWAY
MICHAEL G. LOVORN

See Also: Artillery; Coding and Encryption; Strategy and Tactics.

Ride, Sally

Category: Space, Time, and Distance.
Fields of Study: Communication; Connections.

Summary: The first American woman in space, Sally Ride was a Mission Specialist and has become a science and mathematics education advocate.

Sally Kristen Ride, the first American woman in space, was born May 26, 1951, in Los Angeles, California. She attended Stanford University, and in 1973, earned Bachelor's degrees in physics and English. By 1978, Sally had earned Master's and Doctorate degrees in physics. After answering a newspaper advertisement for space program applicants, she was selected to complete the National Aeronautics and Space Administration's (NASA) rigorous astronaut training program. Upon completion, she served as capsule communicator on early space shuttle missions.

Time in Space

On June 18, 1983, Ride became the first American woman in space, serving as a mission specialist aboard the space shuttle *Challenger* for STS-7, commanded by Captain Robert L. Crippen and piloted by Captain Frederick H. Hauck. Soon after this historic 146-hour mission, Ride was selected as a mission specialist for STS 41-G. On October 5, 1984, again aboard the space shuttle *Challenger*, she began a mission that logged an additional 197 hours in space. Ride was training for her third space flight when the space shuttle challenger accident occurred in January 1986. As a result, her mission was cancelled but she was appointed to the Presidential Commission investigating the accident. After the investigation, Ride was assigned to NASA Headquarters in Washington, D.C., where she helped found NASA's Office of Exploration. Later, she worked at the Stanford University Center for International Security and Arms Control.

Post-Astronaut Career

In 1989, Dr. Ride accepted a faculty position at the University of California, San Diego, as a professor of physics, and she was appointed director of the California Space Institute. More than a decade later, she founded Sally Ride Science, an innovative science education company dedicated to supporting girls' and boys' interests in the

Astronaut Sally Ride monitors control panels from the pilot's chair. Floating in front of her is a flight procedures notebook.

sciences, mathematics, and technology. The company designs science education projects for elementary and middle school students. Ride has also authored several science books for elementary and middle school students, including *To Space and Back* (1989), *Voyager* (2005), *The Third Planet* (2004), *The Mystery of Mars* (1999), and *Exploring Our Solar System* (2003).

In 2003, Ride was assigned to the Space Shuttle Columbia Accident Investigation Board, and has since been named to several national committees, including the President's Committee of Advisors on Science and Technology, the National Research Council's Space Studies Board, and the Review of United States Human Space Flight Plans Committee. She has also served on the boards of the Congressional Office of Technology Assessment, the Carnegie Institution of Washington, the NCAA Foundation, the Aerospace Corporation, and the California Institute of Technology.

The Sally Ride Science Academy, which was created in 2009, focuses on training teachers to increase their students' interest in science and mathematics by changing the image of scientists. As Ride told *USA Today*, the perception that a scientist "is some geeky-looking guy who looks like Einstein, wears a lab coat and pocket protector . . . [is] not an image that an 11-year-old girl or a 10-year-old boy aspires to." In particular, Ride asserts that girls have difficulty seeing themselves as scientists: "A girl doesn't look at that stereotype and say, 'That's what I want to be when I grow up.'" The Academy trains teachers on how to utilize readings that show scientists and mathematicians in real world roles, which helps students to visualize themselves as being able to take on those roles. Ride believes that society's view that girls are not good at mathematics and science is persistent and needs to be rectified. In order for girls to become interested in mathematical and scientific careers, society needs to portray those careers as "normal" for girls to pursue. Ride views herself as a role model, particularly for girls, and describes herself as "a pretty normal 10-year-old girl who grew up to be an astronaut."

In addition to having been inducted into the National Women's Hall of Fame and the Astronaut Hall of Fame, Ride has been the recipient of numerous honors and awards. She has received the NASA Space Flight Medal, the Jefferson Award for Public Service, the von Braun Award, the Lindbergh Eagle, and the NCAA's Theodore Roosevelt Award.

Further Reading

Sally Ride Science. http://www.sallyridescience .com.

Steinberg, Stephanie. "1st Woman in Space Sally Ride Launches Science Academy." *USA Today* (August 2, 2010). http://www.usatoday.com/news/education/2010-08-02-SallyRide02_ST_N.htm.

U.S. National Aeronautics and Space Administration (NASA). "Astronaut Biographical Data: Sally Ride." http://www.jsc.nasa.gov/Bios/htmlbios/ride-sk.html.

CALLI A. HOLAWAY
MICHAEL G. LOVORN

See Also: Spaceships; Weightless Flight; Women.

Risk Management

Category: Business, Economics, and Marketing.
Fields of Study: Algebra; Problem Solving.
Summary: Effectively assessing and mitigating risk can involve sophisticated mathematical analysis and modeling.

A feeling of security is essential for the welfare of all people, ancient or modern. There are many threats in the twenty-first century that can reduce the feeling of security, including financial problems, diseases, and crime. Threats feature different causes, which may be grouped into two main categories: natural (random), and intentional (malicious).

Natural causes are independent from human will (for example, natural disasters), while intentional causes relate to the action of some adversary (for example, a terrorist). Some origins of threats, such as illness or accidents, are not completely random; though an actual intentionality is missing, correlations can be found between human behavior and the unwilled events. It is clear that the intention of any intelligent being, humans in particular, is to maximize one's own benefit throughout an entire lifetime on the base of trade-offs between expenses and medium or long term returns. This goal justifies, among other risk management strategies, the common use of insurance policies and alarm systems.

Risk Assessment

In order to predict human behavior with respect to issues of risk, as well as to support the choice of protection strategies of any nature, risk assessment is employed. In order to assess the risk, a mathematical model is required. The most common and simple mathematical model for risk assessment consists of the following formula: $R = P \cdot V \cdot D$.

Risk (R) with respect to a specific threat (T) is a combination of three different factors:

- P, the expected probability of the occurrence of T (how probable is the threat?)
- V, the expected vulnerability with respect to T (how probable is it that T will cause the expected consequences?)
- D, the expected damage caused by T (if the consequences caused by the threat are endured, how damaging are the consequences?)

Note that the combination operator "$\cdot$" is not necessarily a multiplier. Depending on the criteria used for the analysis and on the type of scale (linear or logarithmic), it can play different roles (even as a sum).

Risk can be evaluated both using qualitative and quantitative approaches. Qualitative indices use reduced scales of values of intuitive meaning; for instance: low, medium, and high. The advantage is that estimations can be more straightforward (though rougher) and computations can be easier. The disadvantage is that results are usually less rigorous, and the combination of qualitative indices is questionable. Quantitative approaches, on the other hand, use and produce values of parameters using well-specified metrics. The disadvantage is the difficulty of getting input data, which—being produced by expert judgments, statistical analyses, and stochastic modeling—are always affected by more or less relevant uncertainty errors. The advantage is that quantitative approaches enable possible automatic optimizations using appropriate algorithms.

In some approaches the $P \cdot V$ factor is compacted into a single factor, which will be defined as the frequency (F) of "successful" threats, expressed algebraically as $F = P \cdot V$.

An example of qualitative risk evaluation using associative matrices is reported in Table 1 using the estimated values of F and D to obtain R.

In quantitative approaches, risk is evaluated using a more formal approach, defining rigorous metrics for the three factors P, V, and D of the risk formula; for instance as follows:

- P is measured in number of threat events per year.
- $V = P(T \text{ success} \mid T \text{ happens})$, which is the conditional probability that a threat will succeed given that it happens.
- D is measured in monetary damages.

Therefore, in this case, the "$\cdot$" operator is actually a multiplier, and the risk can be measured; for example, in dollars per year, which is a measurement of an expected periodic monetary loss. The input values of the risk formula can be obtained in several ways, including statistical approaches and stochastic process modeling.

Risk Mitigation

In order to reduce the risk, several mechanisms can be adopted. The (possibly iterative) process of assessment and mitigation is sometimes referred to as "risk management." The objective of this process is to find an optimal trade-off between the expense in protection mechanisms and the expected risk reduction.

Countermeasures can be very different, depending on the type of risk being faced. They include organizational modifications, periodic diagnostic checks, norms, insurance policies, patrols of agents and first responders, sensors and alarm systems, preventive maintenance, early warning, mechanisms for delaying the threat, emergency preparedness, and disaster management.

With reference to the risk formula, a countermeasure should be able to significantly reduce P, V, or D, or all of them at once. For example, in the case of a viral epidemic, a behavioral change (such as staying at home, using cars instead of public transportation, and frequently washing hands) can reduce P, a vaccine or a strengthening cure can reduce V, while warmth, rest, and medicines can reduce D.

Cost-Benefit Optimization

Countermeasures employed to reduce the risk feature their own cost. While the objective of organizations (such as companies, enterprises, or countries) is to maximize the so-called return on investment, the objective of human beings is to maximize their average welfare throughout their lives. Therefore, countermeasures are adopted whose cost and effectiveness is judged to be "adequate." A more formal approach consists in analytically predicting the benefits resulting from the selected countermeasures, which needs appropriate mathematical models. In quantitative approaches, the periodic Expected Benefit (EB) is defined as $EB = RR - CC$, where RR is the expected risk reduction in a specified time slot, and CC is the countermeasures cost in a specified time slot.

The RR parameter is evaluated using standard risk assessment methodologies. Depending on the countermeasures, the CC can depend on the length of the time slot. For instance, a vaccine can last a whole lifetime with no additional costs, while insurance has periodic costs; alarm systems have an initial expense for the buying and installation of devices and additional costs because of maintenance and power consumption. Furthermore, a reliable payback analysis requires considering not only the initial investment but also the financial

Table 1. Qualitative risk evaluation using associative matrices.

F \ $D \rightarrow$	Low	Medium	High
Low	Low	Low	Medium
Medium	Low	Medium	High
High	Medium	High	High

concepts of cash flow, opportunity cost, and final value of the capital invested.

Once a suitable mathematical model for computing the *EB* has been defined, it is possible to perform a set of analyses, including parameter sensitivity and automatic optimizations.

The parametric sensitivity analysis aims to evaluate the impact of data uncertainty on the computed results. To be performed, it requires that input data are modified (increased or decreased by a certain percentage) and that corresponding results are evaluated. Depending on the results of the sensitivity analysis, models can be assessed as more or less robust to certain input parameters: the more the results are affected by variations in input parameters, the less the model is suitable to be evaluated using uncertain data.

Automatic optimizations can be performed using appropriate algorithms with the aim of maximizing the *EB* with possible external constraints, like a limited budget. For linear problems, operations research provides a set of algorithms, which can be suitable for multi-variable and multi-objective optimization of a specific function. For large non-linear problems, genetic algorithms, which mimic the evolution of live beings, can be adopted. Genetic algorithms, in particular, are based on the concepts of populations of solutions, selection, crossover, and mutations. Genetic algorithms have proven useful in solving a large number of optimization problems, including the ones regarding risk minimization, which are difficult or impossible to manage using traditional approaches.

In conclusion, when security relates to personal benefit maximization, mathematical techniques are involved, which can be very complex since they fall in the area of multi-objective optimization with external constraints and contrasting requirements. Operations research has investigated similar problems, which have even attracted interest from the communities of researchers in statistics and probabilistic modeling. In particular, Bayesian networks are among the formalisms suitable for the stochastic cause–consequences modeling using a graph-based approach, which can also be extended with decision and cost nodes (in such a case, they are named "influence diagrams"). Bayesian networks are direct acyclic graphs (DAGs) in which nodes represent random variables, and arcs represent stochastic dependencies quantified by conditional probability tables (CPTs). It can be formally demon-

strated that a well-formed Bayesian network represents the joint probability density function of the problem described by the network. Several user-friendly graphical tools are available for the solution of Bayesian networks. However, solving algorithms belong to the NP-hard class, therefore, their efficiency tends to significantly worsen as the size and complexity of the network increases.

Further Reading

Hillier, Frederick S., and Gerald J. Lieberman. *Introduction To Operations Research*. New York: McGraw-Hill, 1995.

Jensen, Finn V., and Thomas D. Nielsen. *Bayesian Networks and Decision Graphs*. 2nd ed. New York: Springer Science+Business Media, 2007.

Lewis, Ted G. *Critical Infrastructure Protection in Homeland Security: Defending a Networked Nation*. Hoboken, NJ: Wiley, 2006.

Goldberg, David E. *Genetic Algorithms in Search, Optimization, and Machine Learning*. Philadelphia: Addison-Wesley Professional, 1989.

FRANCESCO FLAMMINI

See Also: Earthquakes; Floods; Insurance; Life Expectancy; Mathematics Research, Interdisciplinary.

Robots

Category: Architecture and Engineering.
Fields of Study: Algebra; Data Analysis and Probability; Geometry; Number and Operations.
Summary: Robots, their motion driven by mathematical algorithms and coordinate or polar geometries, have long been incorporated into society and popular culture.

Robots and robotic systems are increasingly commonplace in many areas of daily life, such as manufacturing, medicine, exploration, security, personal assistance, and entertainment. In general, a robot is a mechanical device that can perform independent tasks guided by some sort of programming. Sometimes, robots are intended to replace humans in tedious or hazardous tasks. In others tasks, such as some surgeries, robots may actually exceed human capabilities. For many, the word "robot" brings to mind both futuristic androids, which are robots that are designed to look human and cyborgs, which contain both mechanical and biological components. Robots used in many industrial applications, such as in medicine, bomb disposal, and repetitive jobs, rarely resemble humans. However, several humanoid robots and robots that realistically mimic the look and behavior of animals have been produced. In 2008, a Japanese play was written and produced for both robots and human actors, and robot animals have sometimes been marketed as replacements for biological pets. The word "robot" can also refer to software-like Web crawlers that run automated tasks over the Internet to gather data, though "bot" is a more common name. The field of robotics generates many interesting problems in both theoretical and applied mathematics and benefits from the contributions of mathematicians. For some, the ultimate quest in the twenty-first century and beyond is to develop materials, technology, and algorithms to create robots that meet or perhaps exceed human levels of perception, behavior, and intelligence. Nano-robots, which are ultra-small robots about the size of a nanometer, might one day be developed for tasks like hunting and destroying cancer cells.

Brief History

Playwright Karel Capek is typically credited with introducing the word "robot" from the Czech word for "laborer," in his 1920 play *R.U.R.* (Rossum's Universal Robots). Another writer who popularized robots was Isaac Asimov, who introduced the term "robotics" in his 1941 short story *Runaround*. However, robotic devices can be found much farther back in history. One early robotic device was a water clock produced by the Babylonians, which used the mathematics of volumes and rates of water flow to calculate time. Greek mathematician Hero of Alexandra described the use of weights and ropes to construct a mobile cart that could be programmed to move along a path. In the thirteenth century, Muslim mathematician and scientist Abu Al-'Iz Ibn Isma'il ibn Al-Razaz Al-Jazari created a set of programmable musicians. The drummer was operated by a rotating shaft that manipulated levers to produce rhythms. Around 1495, Italian painter and mathematician Leonardo da Vinci used his knowledge of the mathematics of anatomy and bodily movement

to sketch designs for a warrior robot outfitted in medieval armor.

Interest in robotics accelerated in the nineteenth century as early computer technology with punch cards began to be incorporated into systems such as that used for the Jacquard loom, named for Joseph Jacquard. Others, such as Pafnuty Chebyshev, studied the theoretical mathematics of linkages, inventing the Chebyshev linkage that converts rotating motion to approximate straight-line motion. Charles Babbage's mathematical engines were some of the first mechanical computers. These engines used finite differences to calculate the values of polynomials. Such inventions were forerunners of computer-controlled robot technology that quickly progressed in the mid-twentieth century to transistors and integrated circuits. Mathematician Norbert Weiner is often known as the "father of cybernetics," which is the science of self-regulating feedback systems, for his work and 1948 book *Cybernetics: Or Control and Communication in the Animal and Machine.*

Cybernetics is not synonymous with artificial intelligence or robotics, but this mathematical discipline is essential for environmentally responsive or adaptive robots. Some other areas of mathematics that have contributed to the development and implementation of robots included algebraic and differential geometry, which is used to help solve problems, such as orientation and movement in three dimensions; partial differential equations, which are used to model many aspects of behavior; optimization algorithms to help sequence tasks; combinatorics, which is used to investigate modular components and systems; and Bayesian statistical methods, named for Thomas Bayes, which can be employed in dynamic perception and machine learning.

Robotic Motion

In the twentieth and twenty-first centuries, many robots are complex, electromechanical devices that move and interact with physical objects, often replacing or augmenting human actions by carrying out certain tasks. Some mobile robots use articulated legs or wheels. Somewhat more common are stationary robotic arms with joints that allow for motion similar to the way joints allow human limbs to move. Having more joints increases the possible angles for movement and degrees of freedom, and hence increases fluid motion and accuracy. Articulated robots, used widely in various industries to perform tasks such as welding components or spray-painting parts, look much like human arms and have at least three joints. If the joints are slide-only, called "prismatic joints," then the robot arm can reach any position in a rectangular workspace by means of translations. If one joint is hinged, which is called a "revolute joint," then all points within a cylindrical workspace can be reached by a combination of rotation and translation. If two of the joints are hinged, a robot arm with a polar geometry is achieved. Inventor George Devol and engineer Joseph Engelberger developed one of the first modern-day programmable robots, Unimate, which began operation in 1961 at a General Motors plant. In 1969, Stanford University student Victor Scheinman created the predecessor for all robotic arms, the Stanford arm.

Mathematical programming and calibration for proper movement of robots depends on kinematics, which is the study of motion; and dynamics, which is the study of how force affects motion. With articulated or jointed robots, for example, the mathematics of kinematics is at the heart of positioning, collision avoidance, and redundancy. Direct kinematics makes use of given joint values to determine the end position that a robot arm may achieve. The mathematics of inverse kinematics is used to determine the required values for the joints when the end position of the robotic arm motion is known. Getting the robot arm to the right position is only half of the mathematical problem. The other half involves calculating forces using dynamics. For example, a robot designed to fight fires would need motors to move the robot and its arms. Calculations incorporated in determining which motors to use would involve dynamics. Inverse dynamics would help determine the required values

of forces to generate the desired acceleration of the robot or its components. The movement involved in robotics most often occurs in three-dimensional space, so geometry plays a role in the positioning and movement of robots. Matrices can be used to represent the points through which robots navigate. These algebraic representations are then reviewed and coordinated using sophisticated applications of basic calculus principles, like differentiation, to ensure maximum efficiency when designing and operating robots.

Movement and action in robots are driven by algorithms. Some robots respond to direct human input from keyboard commands or from haptic devices that respond to tactile or body motion. Others autonomously perform programmed tasks. Some robots are "smart" or "intelligent," meaning that they are able to sense and adapt to their surroundings while completing their tasks. Even then, these robots are able to

In November 2010, Robonaut 2 was brought to the International Space Station where it will remain as the first humanoid robot to work in space.

accomplish tasks only because they have been programmed to do so. For example, "smart" mobile robots make use of a variety of sensors with terrain-identification and obstacle-detection programs using input data and probabilistic models to guide trajectory and avoid collisions. Probabilistic robotics is increasingly of interest, with the goal of developing algorithms that facilitate accurate autonomous decision making in the face of real-work complexity and uncertainty, which would increase the reliability of automated behavior and more closely replicate the type of processing that occurs in the human brain.

Robots: Fiction and Fact

Robots are widely used in entertainment, especially science fiction. Mary Shelley's 1818 novel *Frankenstein* is cited by some as showing that scientific creations able to perform human tasks long preceded television and movies. Some well-known examples include C-3PO from the *Star Wars* series and Wall·E from the 2008 Pixar movie of the same name. Data, from the 1987–1994 television series *Star Trek: The Next Generation*, is an example of a fictional android. The Borg species from the *Star Trek* series and the Terminator robot from *The Terminator* movie series are examples of cyborg characters, usually hybrid humans whose biological capabilities are sustained or enhanced through robotic elements—though the Terminator may be thought of by some as a robot enhanced by biology. Enhancing human capabilities through robotic elements, like pacemakers and prosthetic devices, is common in the twenty-first century. However, the medical applications of robotics have not focused on humans achieving superhuman powers (as is done in fiction) but rather on helping those with medical conditions and disabilities.

Robots in Education

Robots are often used in schools to motivate learning of mathematics concepts, such as two- and three-dimensional coordinate geometry. The roBlocks construction system was developed by computational design scientists Mark Gross and Eric Schweikardt. Users can build robots using modular sensor, logic, and actuator blocks to study concepts like kinematics, feedback, and control. They can also create their own control programs to further explore robot mathematics and dynamics. The Lego Group produces a robotic construction and programming system called Mindstorms NXT that has been marketed for both education and entertainment.

Further Reading

Craig, John J. *Introduction to Robotics: Mechanics and Control.* 3rd ed. Upper Saddle River, NJ: Prentice Hall, 2004.

Murray, Richard M., Zexiang Li, and S. Shankar Sastry. *A Mathematical Introduction to Robotic Manipulation.* Boca Raton, FL: CRC Press, 1994.

Thrun, Sebastian, Wolfram Burgard, and Dieter Fox. *Probabilistic Robotics.* Cambridge, MA: MIT Press, 2005.

Deborah Moore-Russo
D. Keith Jones

See Also: Coordinate Geometry; Interplanetary Travel; Matrices; Nanotechnology; Neural Networks; Science Fiction; Surgery.

Roller Coasters

Category: Games, Sport, and Recreation.
Fields of Study: Algebra; Calculus; Geometry; Measurement.
Summary: Roller coasters are mathematically designed to provide safe and thrilling rides.

Roller coasters are entertainment rides designed to put the rider through loops, turns, and falls, inducing sudden gravitational forces. The rapid ascents and descents coupled with sharp turns create momentary sensations of weightlessness. One known precursor of roller coasters are seventeenth-century Russian ice slides, which sent riders down a tall, ice-covered incline of roughly 50 degrees. Modern roller coasters can be traced to the late 1800s. As of 2010, Ohio's Cedar Point held the record for most roller coasters (17) in a single amusement park.

Conservation of Energy

The law of conservation of energy states that energy can neither be created nor destroyed, but can only be converted from one form to another. Roller coasters exploit this law by converting the potential energy gained by the car as it ascends to the top of a hill into kinetic energy as it descends and goes through the turns and loops. The potential energy of the car at the top of the loop is given by

$$E = m \times g \times h$$

where E is the total potential energy (joules), m is the total mass of the car (kg), g is the acceleration due to gravity (9.8 m/s^2), and h is the height (m).

For example, consider a roller coaster car weighing 2200 pounds perched at the top of Cedar Point's Top Thrill Dragster, which is about 426 feet high. The car, at this point, has accumulated $1000 \times 9.8 \times 130 = 1{,}274{,}000$ joules or 1.2 megajoules of energy—the same amount of energy released by the explosion of a quarter kilogram of TNT. This potential energy is converted into kinetic energy as the car hurtles down the loops.

As the car expends potential energy, it is converted into kinetic energy, propelling it forward. In an ideal situation where there is no friction or air drag, the car would travel forever. However, because of friction and other resistive forces, the car decelerates and finally stops when it has expended all its potential energy.

Centripetal Force

Centripetal force is responsible for keeping the rider glued to the seat as the car executes turns and loops and even puts the rider upside down. Centripetal and centrifugal forces act on a body that is traveling on a curved path. Whereas centrifugal force is directed outwards, toward the center of curvature, centripetal force acts inward on the body.

G-Force and Loop Design

G-forces are non-gravitational forces, and can be measured using an accelerometer. Humans have the ability to sustain a few g's (a few times the force of gravity), but deleterious effects are a function of duration, amount, and location of the g-force. Many roller coasters accelerate briefly up to six g's, depending on the shapes, angles, and inclines of loops, turns, and hills. Early roller coaster loops were circles. To overcome gravity, the cars entered the circle hard and fast, which pushed riders' heads continually into their chests as the coaster changed direction. In the 1970s, coaster engineer Werner Stengel worked with National Aeronautics and Space Adminis-

tration (NASA) scientists to determine how much force riders could safely tolerate. As a result of this and other mathematical investigations, he began to use somewhat smoother clothoid loops, which are based on Euler spirals, named for Leonhard Euler. In 2010, using the same equations that describe how planets orbit the sun, mathematician Hanno Essén drew a new and unique series of potential rollercoaster loops. Riders would get the thrilling visual experience of a loop without any of the typical jolting and shaking, because the force that riders would feel pushing them into their seats would stay exactly the same all the way around the loop.

Further Reading

Alcom, S. *Theme Park Design: Behind The Scenes With An Engineer*. Orlando, FL: Theme Perks Press, 2010.

Koll, Hilary, Steve Mills, and Korey Kiepert. *Using Math to Design a Rollercoaster*. New York: Gareth Stevens Publishing, 2006.

Mason, Paul. *Roller Coaster!: Motion and Acceleration*. Chicago: Heinemann-Raintree, 2007.

Rutherford, Scott. *The American Rollercoaster*. Norwalk, CT: MBI, 2000.

ASHWIN MUDIGONDA

See Also: Energy; Gravity; Weightless Flight.

Roman Mathematics

Category: Government, Politics, and History.
Fields of Study: Connections; Number and Operations; Representations.
Summary: The ancient Romans, who are often remembered for their applied mathematics, made important contributions to surveying, time-keeping, and astronomy.

The Roman period for mathematics could be said to have started when a Roman soldier was sent to seize Archimedes during the capture of Syracuse. Told by Archimedes to wait as he finished his diagrams, the soldier lost patience with the old man and slew him. The popular stereotype of the Romans is that they did little to advance Greek discoveries in mathematics, instead merely applying Greek methods to practical problems. This conception is not entirely fair. The Roman Empire was not one homogenous zone, but was rather a collection of culturally diverse provinces. For this reason, many works produced during the time of Roman rule, like the books of Ptolemy, writing in Alexandria, Egypt, are written in ancient Greek rather than Latin. Therefore, these books could be considered Greek, Roman, or Greco-Roman depending on the context. However, despite this diversity, the Roman period led to the dominance of some mathematical practices that still have an influence in the twenty-first century.

Roman Numerals

One of the most distinctive remnants of Roman mathematics is the use of Roman numerals, which are letters that stand for specific values and usually work as additive values. The numerals are

I = 1	V = 5	X = 10
L = 50	C = 100	D = 500
M = 1000.		

So: $LXXVII = 50 + 2(10) + 5 + 2(1) = 77$.

The numerals are written with the largest values at the left, proceeding to the smaller values. They can also have subtractive constructions. I preceding subtracts one from a 10 to make nine. X before an L or C produces 40 or 90, and C before D or M produces 400 or 900. So

$$MCMXLVIII =$$

$$1000 + (1000 - 100) + (50 - 10) + 5 + 3(1) = 1948.$$

The origins of the system are unknown. It has been proposed that they were based on tally marks, with I being a notch, V being a double notch to mark five, and 10 as crossed-notches (though it could also be that X was formed from two V symbols). The number IV to represent 4 is a later addition based on medieval Latin and does not seem to have been used by the Romans, who instead used IIII.

This system is not very helpful for arithmetic, and so it is little surprise to find that the Romans developed the portable abacus to ease mathematical operations. This device was a tray with a number of columns etched into it that could hold pebbles. A pebble (in

Latin, the word "calculus") had a value depending on the column that held it. Moving a pebble a column to the left increased its value by a factor of 10. Such an abacus could be used by merchants in the city or by surveyors working for the military.

Survey

Roman surveyors employed geometry to divide the landscape and lay out cities with effects that can still be seen in the twenty-first century. The key to Roman survey was a tool called a *groma*, which was a tall staff with a beam, known as a *rostro*, at right-angles to the staff at the top. The rostro supported a wooden cross, and at each end of the cross-beams was hung a plumb line. Sighting across these lines allowed Roman surveyors to lay out grids of perpendicular lines in the landscape. Surveyors could then divide land for agricultural purposes, and some field systems in Europe are based on these ancient surveys. The *groma* also left an impression on modern cities. The Romans frequently built new cities in conquered territories, for either native inhabitants or new settlements of veteran soldiers. At the heart of a Roman settlement lay the forum, the central civic space, which usually lay at the intersection of the Cardo maximus (the main north-south street) and the Decumanus maximus (the main east-west street). This system created new cities with grid-plans in which the main intersection was laid out by a *groma*. These perpendicular grids were the origins of many European settlements and was adopted in the planning of many U.S. cities in the nineteenth century.

The Roman Calendar

The Roman calendar instituted by Julius Caesar made a radical change to time-reckoning in Europe. Before this development, European calendars outside Rome were usually luni-solar calendars. As such, each month was related to the lunar cycle, which is not commensurate with the solar year, and so periodically whole months, known as "inter-calary months" would be inserted into the year to keep the months in step with the seasons. Insertions would usually have to be done every two or three years. Even ancient authors recognized that this system was inefficient, including Herodotus, who wrote in the late fifth century B.C.E. that the Egyptians had a much more accurate solar calendar. In 45 B.C.E., Julius Caesar adapted the Egyptian method of time-keeping for Roman use.

Each month was counted as a period of days, usually 30 or 31 but with 28 or 29 in February. In addition, Julius Caesar laid down rules for when an inter-calary day would be added to February. The Egyptians corrected the calendar by adding a day every fourth year. Unfortunately, the Romans counted inclusively, meaning that the leap year was in the fourth year, rather than after the fourth year. For example, 2020 is a leap year. For the ancient Romans, the second year in the cycle is 2021 and the third is 2022. Therefore, 2023 is the fourth and the Romans of Julius Caesar's time would have made this a leap year, rather than 2024. Augustus Caesar corrected this error in the early years of the first century C.E.

This method of keeping the years remained until the reforms of Pope Gregory XIII in 1582, though Britain and the American colonies did not implement the Gregorian calendar until 1752. The difference between the two calendars is that years divisible by 100 are not leap years, unless the year is divisible by 400. Otherwise, years are marked by the same cycle of months as the ancient Romans did.

Mathematics and the Cosmos

Even though ancient mathematicians had a relatively small set of tools based in geometry and arithmetic, these could be used to create incredibly intricate models. Ptolemy proposed a model of the universe that contained circles rotating upon circles to reproduce the movement of the planets. The connections between mathematics and cosmology made mathematics attractive to philosophers of the Roman period. The assertion that mathematics could reveal truth became increasingly contentious in late antiquity. Pagan philosophers came into conflict with a new religious sect, Christianity, which was increasingly powerful. One notorious incident was the killing of Hypatia, a female mathematician philosopher, in the city of Alexandria by a Christian mob. For some ancient historians, her death marks the end of the period known as classical antiquity.

Further Reading

Cuomo, Serafina. *Ancient Mathematics*. London: Routledge, 2001.

Dilke, Oswald. *Roman Land Surveyors: Introduction to the Agrimensores*. Newton Abbot, England: David and Charles, 1971.

Hannah, R. *Time in Antiquity*. London: Routledge, 2009.

Jaeger, Mary. *Archimedes and the Roman Imagination*. Ann Arbor: University of Michigan Press, 2008.

ALUN SALT

See Also: Arabic/Islamic Mathematics; Archimedes; Calendars; Greek Mathematics; Sacred Geometry.

Ross, Mary G.

Category: Mathematics Culture and Identity.
Fields of Study: Algebra; Communication; Connections; Data Analysis and Probability.
Summary: Mary Ross was a prominent Native-American mathematician and engineer.

Mary G. Ross (1908–2008), a Native American of Cherokee heritage, had a distinguished career as a mathematician, space scientist, and engineer. She was the first female engineer to work at the Lockheed corporation and also the first female Native-American engineer. Ross was born in the Oklahoma territory and as a child lived with her grandparents in the Cherokee Nation of Tahlequah in order to pursue her education. She often credited a strong family and tribal focus on equal education for boys and girls as being crucial to her career. At age 16, she enrolled in Northeastern State Teachers College (Oklahoma), receiving her bachelor's degree in mathematics in 1928. Ross taught high school mathematics and science in Oklahoma for nine years before moving to Washington, D.C., to work as a statistical clerk in the U.S. Bureau of Indian Affairs. Her talent and education were quickly recognized and she was reassigned to work as an advisor (similar to a dean) for a coeducational Indian boarding school in Santa Fe, New Mexico (later to become the Institute of American Indian Art). At the same time, she pursued graduate studies in mathematics and astronomy, receiving her master's degree from Colorado State Teachers College in 1942. Ross received numerous awards during her lifetime.

Aeronautical Engineering

In 1942, Ross began working as a mathematician at the Lockheed Aircraft Corporation. She was given the opportunity to study aeronautical and mechanical engineering, taking evening classes at UCLA as well as an emergency war training course offered at Lockheed and, in 1949, received professional engineering classification as a mechanical engineer (there was no classification for aeronautical engineering at the time). As a research engineer at Lockheed, Ross worked on a number of projects related to transport and fighter aircraft and, in 1953, was chosen to be one of 40 engineers who became the nucleus of Lockheed Missiles and Space Company, now known as Lockheed Martin. In this group, she worked on a number of missile systems, including the Polaris ballistic missile, which required her to work in the new field of hydrodynamics because the Polaris missile was designed to be launched underwater from a submarine.

Ross continued to advance at Lockheed, becoming a research specialist in 1958, an advanced systems engineer in 1960, and a senior advanced systems engineer in 1961. She worked on the Agena series of rockets and the Polaris reentry vehicle. She also helped develop criteria for missions to Mars and Venus, designing orbital space systems and interplanetary expeditionary systems and writing a volume of the *NASA Planetary Flight Handbook*. About her career, she said, "I have always considered my work a joint effort. I was fortunate to have worked on great ideas and with very intelligent people. I may have developed a few equations no one had thought of before but that was nothing unusual—everybody did that . . . it has been an adventure all the way."

Other Accomplishments
Ross became an advocate of women's and Native-American education following her retirement from Lockheed in 1973. Her great-great-grandfather was principal chief of the Cherokee for 40 years, and she expressed the idea that, "there is a lot of ancient wisdom from Indian culture that would help solve the problems of today." She co-founded the Los Angeles section of the Society of Women Engineers and also worked to expand educational opportunities within the American Indian Science and Engineering Society and the Council of Energy Resource Tribes.

Further Reading
Briggs, Kara. "Cherokee Rocket Scientist Leaves Heavenly Gift." *Cherokee Phoenix* (December 18, 2008). http://www.cherokeephoenix.org/19913/Article.aspx.

Riddle, Larry. "Mary G. Ross." *Agnes Scott College Biographies of Women Mathematicians*. http://www.agnesscott.edu/lriddle/women/maryross.htm.
Sheppard, Laurel M. "An Interview with Mary Ross." *Lash Publications International*. http://www.lashpublications.com/maryross.htm.

SARAH BOSLAUGH

See Also: Airplanes/Flight; Interplanetary Travel; Minorities; Women.

Ruler and Compass Constructions

Category: History and Development of Curricular Concepts.
Fields of Study: Communication; Connections; Geometry; Measurement.
Summary: Ruler and compass constructions form the basis of geometry and have challenged mathematicians for thousands of years.

Ruler and compass constructions have long been important in mathematics. In geometry, a ruler and compass construction refers to a geometric construction that uses only an unmarked ruler and a compass. The ancient construction problems of squaring the circle, duplicating the cube, and trisecting the angle were unsolved until they were proved impossible by algebraic techniques. Early tile makers and architects were also interested in these constructions. Aside from historical considerations, limiting constructions to these two tools is important because the restrictions generate a variety of rich problems. In the twenty-first century, dynamic geometry software programs allow students, teachers, and researchers to explore, save, and share constructions.

Euclid
The most significant early compendium of ruler and compass constructions is Euclid's *Elements* written c. 300 B.C.E. In fact, Euclid's book organizes everything around these constructions in an attempt to build as

much geometry as possible starting with the most basic tools. Drawing a line using a ruler and a circle using a compass are seen as elementary in Euclid's tradition—hence, the title *Elements*—and it is preferred to reduce as much of geometry as possible to these elementary tools. *Elements* begins with five common notions and five "self evident" postulates. The first three postulates specify the rules for geometric constructions:

- A straight line segment can be drawn joining any two points.
- Any straight line segment can be extended indefinitely in a straight line.
- Given any straight line segment, a circle can be drawn having the segment as radius and one endpoint as center.

The final two postulates of Euclid are

- All right angles are congruent.
- If two lines are drawn which intersect a third in such a way that the sum of the inner angles on one side is less than two right angles, then the two lines inevitably must intersect each other on that side if extended far enough.

The last one is the famous fifth postulate and is equivalent to the more common parallel postulate: from a given point not on a given line, one can draw exactly one line parallel to the given line. Euclid based the whole edifice of rigorous geometry on these axioms, hence ruler and compass constructions are at the center of Euclidean geometry.

The Three Classical Problems

Three ancient construction problems captured the imagination of mathematicians for many centuries: doubling a cube, trisecting an angle, and squaring a circle.

- *Doubling a cube*: Given the side of a cube, can one construct, using an unmarked ruler and a compass, the side of another cube whose volume is twice the first one?
- *Trisecting an angle*: Given an arbitrary angle, can one draw a line, using an unmarked ruler and a compass, that trisects the angle?
- *Squaring a circle*: Given a line segment that is the radius of a circle, can one construct,

using an unmarked ruler and a compass, the side of a square that has the same area as the original circle?

None of these constructions are possible, but surprisingly, despite more than 2000 years of effort, a satisfactory answer to these three questions was given only in the nineteenth century.

Each of these classical problems has a long history. For example, the problem of doubling a cube was known to the Egyptians, Greeks, and Indians. In one version of the Greek legend, the citizens of Athens consulted the oracle of Apollo at Delos to put a stop to a plague in Athens. The oracle prescribed that the Athenians double the size of their altar. Efforts to find a way of doubling the volume of the cube failed, and it is claimed that Plato (427–347 B.C.E.) had remarked that the oracle really meant to "shame the Greeks for their neglect of mathematics and for their contempt of geometry." The original legend did not specify the tools to be used, and, in fact, solutions using a number of tools were found. However, a construction using the elementary tools of an unmarked ruler and a compass remained elusive.

Tool Variations

Variations on the tools are possible. For example, if one were allowed to make two marks on the ruler, then with the use of this marked ruler and a compass, one can trisect an arbitrary angle.

An interesting variation arose in the work of Abu'l Wafa Buzjani (940–997 C.E.). Abu'l Wafa in a work aimed at artisans (such as tile makers, designers of intricate patterns, and architects) limited the geometric tools to an unmarked ruler and a "rusty" compass. In other words, he wanted to only use a compass that had a fixed opening and could not be adjusted to draw different sized circles. He believed that working with such a fixed compass would be more accurate, less error-prone, and more useful for artisans. Abu'l Wafa constructs, among other polygons, regular pentagons, octagons, and decagons using a rusty compass. Since the opening of the compass used in Euclid's *Elements* could vary, Abu'l Wafa could not rely on the constructions in *Elements*. Hence, he constructed anew, using the rusty compass, all the needed basic results.

In Europe, the Danish mathematician Georg Mohr (1640–1697) showed, rather surprisingly, that all ruler and compass constructions can be done with a compass alone. In such constructions, one cannot draw a line segment, and a line segment is considered constructed as long as its two endpoints are found. This result is now known as the Mohr–Mascheroni theorem. The Italian Lorenzo Mascheroni (1750–1800) had independently found the same result. Georg Mohr also proved that all ruler and compass constructions can be done with a ruler and a rusty compass. Finally, the German mathematician Jacob Steiner (1796–1863) and the French mathematician Jean-Victor Poncelet (1788–1867) proved that all constructions using a ruler and a compass can be made with a ruler and only one use of the compass.

Proofs

Going back to the classical problems, the first rigorous proof of the impossibility of doubling the cube and trisecting an arbitrary angle using a ruler and a compass was given by the French mathematician Pierre Laurent Wantzel (1814–1848). In 1882, the German mathematician Ferdinand Lindemann (1852–1939) proved that π is transcendental. From this, it followed that one cannot square a circle using a ruler and a compass. In general, using only these tools, it is possible to construct line segments of any rational length as well as line segments whose length is the square root of the length of any already constructed segment. However, one can prove that it is impossible to construct other lengths using the theory of fields that was developed with the help of Niels Henrik Abel and Évariste Galois on the solvability of equations. The proof essentially boils down to the fact that, using a ruler and a compass, one can draw only straight lines and circles, and the only new points are the intersections of these lines and circles. Since lines have linear equations and circles have quadratic equations, finding the points of intersection of these shapes is the same as equating their equations and finding the solutions. These all can be achieved using the quadratic formula, which involves only square roots.

Polygons

Constructing regular polygons with a straightedge and compass is also an interesting ruler and compass construction problem. An n-gon is a regular polygon with n sides. Ancient Greeks could construct regular n-gons for $n = 3, 4, 5$, and 15 (triangles, squares, regular pentagons, and regular pentadecagons). They also knew that if one can construct a regular n-gon with a straightedge and compass, then one can also construct a regular $2n$-gon. Carl Friedrich Gauss (1777–1855) added to this knowledge, by constructing, when he was 19 years old, a regular heptadegon (a 17-gon).

A Fermat prime is a prime number of the form $2^{2^k} + 1$, where k is a non-negative integer. The only Fermat primes known are 3, 5, 17, 257, and 65537. It is not known whether there are any other Fermat primes or not. In any case, Gauss stated, and Wantzel gave a proof, that a regular n-gon is constructible with ruler and compass if and only if n is an integer greater than two such that the greatest odd factor of n is either one or a product of distinct Fermat primes.

Further Reading

Hadlock, Charles Robert. "Field Theory and Its Classical Problems." *Carus Mathematical Monographs*, 19 (1978).

Katz, Victor, ed. *The Mathematics of Egypt, Mesopotamia, China, India, and Islam. A Sourcebook*. Princeton, NJ: Princeton University Press, 2007.

Martin, George E. *Geometric Constructions*. New York: Springer-Verlag, 1998.

Sutton, Andrew. *Ruler and Compass: Practical Geometric Constructions*. New York: Walker & Co., 2009.

Shahriar Shahriari

See Also: Arabic/Islamic Mathematics; Greek Mathematics; Measurement, Systems of; Measurements, Area; Measurements, Length; Parallel Postulate; Pi; Squares and Square Roots.

S

Sacred Geometry

Category: Friendship, Romance, and Religion.
Fields of Study: Connections; Geometry; Number and Operations; Representations.
Summary: Cultures have long imbued various spaces, shapes, forms, ratios, and geometric concepts with special significance and ritual power.

Humanity has long attributed sacred meaning to certain geometric forms and concepts. The term "sacred geometry" was popularized during the twentieth century to represent the religious, philosophical, and spiritual beliefs surrounding geometry. The core of its teachings may be found in very ancient cultures, with varying metaphysical systems and worldviews. Some attribute the modern renaissance of the movement to artist Jay Hambridge. The image of a nautilus shell with overlaid golden rectangles is common in the twenty-first century, but when Hambridge investigated mathematical proportion and symmetry in Greek art and architectural design in the beginning of the twentieth century, his work on dynamic symmetry led to debate about definitions of dynamic versus static symmetry.

The development of sacred geometry led to more debate as some asserted that it showed the continuity and universality of mathematical concepts or forms, such as the golden proportion, the logarithmic spiral,

or the flower of life, across cultures, millennia, and the universe. In its most common conception, sacred geometry is then a metaphor for universal order—a metaphor found in the artistic expression of many cultures, especially in religious architecture. In its most ambitious conception, it is itself a practice for enlightenment or self-development, similar to meditation, prayer, or artistic techniques. The knowledge and exercise of geometrical skills can be taken to form a practice that awakens the practitioner to underlying order or truth. The movement has inspired its followers, who look for these forms in art, architecture, nature, and science. People like Drunvalo Melchizedek, who originally planned to major in physics and minor in mathematics but graduated with a fine arts degree, have organized spiritual workshops related to sacred geometry. Some attribute sacred geometry to people's needs to seek out connections. Astrophysicist Mario Livio found some of the analyses "rather contrived . . . with lines drawn conveniently at points that are not obvious terminals at all. Furthermore, some of the ratios obtained are too convoluted . . . to be credible."

Sacred diagrams and figures are omnipresent across ages and cultures. For example, the square has religious significance in Hindu architecture and design. The diagram known as the circular "mandala," for instance, symbolizes to some the cosmos through its symmetry and sectors, which represent elements,

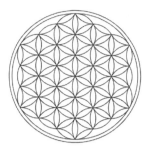

Flower of Life mandala

seasons, divinities, and various categories of religious and metaphysical interest. Practitioners believe that meditating on The Flower of Life icon, one example of a mandala, will reveal the mysteries of the universe. The Egyptians used regular geometric polygons and pyramids in important architectural structures and in representations of the gods. Geometric figures, such as the platonic solids, were assigned additional significance in ancient Greece.

For instance, Earth was associated with the cube, air with the octahedron, water with the icosahedron, fire with the tetrahedron, and the dodecahedron was a model for the universe. In his work *The Timeas*, Plato noted: "So their combinations with themselves and with each other give rise to endless complexities, which anyone who is to give a likely account of reality must survey." In the twentieth century, sacred geometry has become the universal language of nature, mastering shapes and patterns equally found in stars, snowflakes, and DNA, which ultimately represent a sort of blueprint of creation.

Golden Ratio

A common element in sacred geometry is the golden ratio. Many of the sacred geometry principles of the human body are found and subsumed into the famous "Vitruvian Man" drawing by Leonardo Da Vinci. "Vitruvian Man" was inspired by the work of Marcus Vitruvius Pollio, a first century Roman architect who wrote *De architectura*, or *The Ten Books on Architecture*. Vitruvius detailed systems of ratios he believed were found in the human body and that could be used to construct buildings, including temples, to achieve his three necessary criteria for structural perfection: beauty, durability, and utility. Da Vinci also lived and studied with the fifteenth-century mathematician Fra Luca Pacioli and drew the illustrations of the book *De Divina Proportione* (About Divine Proportion). In it, Pacioli explains and illustrates mathematical proportion in its direct relation of artistic patterns and forms and explores architecture and the vital proportion of the golden ratio, the ultimate divine proportion extensively.

Devotees of twentieth-century sacred geometry note the high occurrence of the golden ratio, such as its recursive occurrence in the Parthenon; the Notre Dame Cathedral; the great pyramid of Giza; the relations between platonic solids; the ratio of segments in a five-pointed star (called a *pentagram*); the ratio of adjacent terms of the famous Fibonacci Series, named after Leonardo Fibonacci; the symmetrical pattern of aperiodic tilings, thanks to which Roger Penrose discovered new aspects of quasicrystals; in movements of the stock market; and even in Erik Satie's compositions.

Further Reading

Lawlor, Robert. *Sacred Geometry: Philosophy & Practice.* London: Thames & Hudson, 1982.

Livio, Mario. *The Golden Ratio: The Story of PHI, the World's Most Astonishing Number.* New York: Broadway Books, 2003.

McWhinnie, H. J. "Influences of the Ideas of Jay Hambridge on Art and Design." *Journal of Computers & Mathematics with Applications* 17, no. 4–6 (1989).

Skinner, Stephen. *Sacred Geometry: Deciphering the Code.* London: Gaia Books, 2006.

Marilena Di Bucchianico

See Also: Houses Of Worship; Numbers and God; Religious Symbolism; Symmetry.

Sales Tax and Shipping Fees

Category: Business, Economics, and Marketing.
Field of Study: Number and Operations; Measurement.
Summary: Different types of sales taxes and shipping fees affect the final price of a purchase.

Benjamin Franklin famously noted, "Our Constitution is in actual operation; everything appears to promise that it will last; but in this world nothing is certain but death and taxes." When someone makes a purchase, often times there are extra charges added to the customer's bill. These costs may include a tax, shipping charges, or fees. These extra amounts, however, have a special purpose and they are each computed differently. For example, a sales tax is based on a percentage of the total amount of the sale and that percent is regulated by local and state governments. On the other hand, shipping is charged to cover the delivery of merchandise from the retailer to the customer's location. These fees are based on the policies of the company selling the goods as well as how quickly the customer would like their purchase delivered. Lastly, fees can be special charges; for example, insurance might be added to a purchase to cover the cost of the merchandise in the event it is lost or damaged during delivery. Albert Einstein commented that preparing a tax return "is too difficult for a mathematician. It takes a philosopher." The calculations to determine sales tax and shipping fees utilize percentages, multiplication, and addition, but Einstein may have been referring to the ever-changing instructions.

Both mathematicians and philosophers have long been involved in issues related to taxation. The *Jiuzhang suanshu* (*Nine Chapters on the Mathematical Art*) contains related problems. In the tenth century, astronomer and mathematician Abu'l-Wafa wrote a text on mathematics for scribes and businessmen, with part four of the book containing seven chapters devoted to various kinds of taxes and related calculations. In the seventeenth century, lawyer and amateur mathematician Étienne Pascal worked as a tax assessor and was appointed as the chief tax officer. In order to help his father in his tax work, mathematician and philosopher Blaise Pascal invented the Pascaline, which is reported to be the first digital calculator. In the twenty-first century, financial planners, mathematicians, and actuaries create mathematical models and investigate a variety of mathematical concepts related to taxes and fees, including the impact of flat rate, progressive, symmetric, or asymmetric taxation; and game theory applied to the interaction between taxpayers and tax collectors. They also investigate equilibrium states and how increasing or decreasing sales taxes or shipping and handling fees or using a nonlinear structure impacts consumer decisions about purchases and business sales.

Sales Tax
Many states, counties, and municipalities levy a sales tax as a way to increase revenues for their government or to balance their budget; however, not every state or local government charges a sales tax. The rate of the tax varies depending on the laws of the governmental unit. In other words, a purchaser will encounter different sales tax rates throughout the United States. The charges in 2010 varied from 0% in states like Alaska or Delaware to a high of 8.25% in California. This means that a person in Alaska who pays $100 for an mp3 player would not be required to pay any tax on the sale. However, a person buying that same mp3 player in California would be required to pay this tax. In other words, that $100.00 purchase would have an 8.25% tax added to the cost, meaning the new purchase price would be the original cost ($100.00) plus the sales tax ($8.25) for a total of $108.25.

Many localities exempt certain classifications of goods from their sales tax. Some common exceptions include groceries and prescriptions. On the other hand, special items such as gasoline, cigarettes, and alcohol have a significantly higher sales tax, as they have the potential to add sizeable revenue to a state's budget. A federal law called the Internet Tax Freedom Act (ITFA) specifically addresses sales over the Internet. The law provides that no governmental unit is allowed to add any special or additional tax on Internet purchases. This means that a sales tax may be charged on Internet purchases at the same rate as items purchased in person or by phone but no extra tax charge can be added.

Shipping and Handling Fees
Shipping and handling fees vary dramatically by seller as well as by the type of shipping the buyer requests. Common factors used to compute delivery costs include (1) how many items are being purchased, (2) how much the order weighs, and (3) how quickly the customer would like to receive their merchandise. However, common shipping types include free shipping, overnight delivery, two day or expedited delivery, and standard shipping, which may vary from three to seven days. In addition, the cost may change based on the number of items purchased or the weight of the merchandise. The following three examples illustrate different types of shipping options:

- *Flat fee*: The seller charges a flat shipping fee for all purchases regardless of price, weight, or number of items.
- *Progressive*: The seller charges a progressively larger shipping charge based on the cost of

the purchase. Shipping for a $50 purchase might cost $5, while shipping for a $100 purchase might cost $10.

- *Flat fee and item charge*: The seller charges a flat shipping rate plus an item charge (shipping + charge × number of items). Assume that the base shipping is $3.99, and there is a charge of $.99 for each item. A one item purchase would have a charge of $3.99 + $0.99 = $4.98. However, suppose the purchaser buys three items. In that case, the charge would be $3.99 + 3($0.99) = $6.96.

Shipping and fees are often grouped together as one charge; however, some vendors are known to charge each of these as separate and distinct charges. Vendors often add an additional charge to deliver a purchase. One example would be a package that requires special handling based on size or weight, such as a piece of furniture. Higher cost items such as jewelry might have an insurance charge added to the customer's total.

Further Reading

Anderson, Patrick. *Business Economics and Finance With MATLAB, GIS and Simulation Models*. Boca Raton, FL: CRC Press, 2000.

Consortium for Mathematics and Its Applications. *Mathematical Models with Applications*. New York: W. H. Freeman & Company, 2002.

Marks, Gene. "Don't Forget the Handling!" *Accounting Today* 23 (2009).

Scanlan, M. "Use Tax History and Its Implications for Electronic Commerce." *The Information Society* 25 (2009).

KONNIE G. KUSTRON

See Also: Income Tax; Money; Shipping.

Sample Surveys

Category: History and Development of Curricular Concepts.
Fields of Study: Communication; Connections; Data Analysis and Probability.

Summary: Mathematicians and statisticians help design sampling methods and techniques to better represent populations and account for biases and missing data.

A survey is a statistical process by which data are collected from a representative sample of some population of interest in order to determine the attitudes, opinions, or other facts about that population. A census is the special case where everyone in the population is surveyed.

For example, the Babylonians are known to have taken a population census around 3800 B.C.E. In one of the first modern surveys, the *Harrisburg Pennsylvanian* newspaper polled city residents about the 1824 presidential election. Polling continued to be largely a local phenomenon until a 1916 national survey by *Literary Digest* magazine, which predicted the winners of several presidential elections despite using highly unscientific survey methods. Their famously incorrect assertion that Alf Landon would beat Franklin Roosevelt in the 1936 election is cited as contributing to the magazine's failure. Journalist and market researcher George Gallup, who correctly predicted Roosevelt's 1936 victory, was a pioneer in statistical sampling in the early twentieth century, though at the time, many considered his ideas quite radical. A post–World War II boom in manufacturing led companies to survey consumers to tailor products to preferences and increase sales. In the twenty-first century, public opinion polls on all aspects of society are pervasive and surveys frequently shape society's opinions and actions in addition to simply measuring them.

Students begin learning how to collect survey data in the primary grades. Researchers in many disciplines also routinely rely on data gathered via surveys. Mathematicians and statisticians work on mathematically valid methods for selecting samples that are random and representative as well as methods to reduce bias in surveys, effectively analyze data, present results that adjust for random error, and account for the effects of missing data. Many of these individuals belong to the Survey Research Methods Section of the American Statistical Association. Leslie Kish, a recipient of the association's prestigious Samuel S. Wilks Award, was especially cited for his worldwide influence on sample survey practice and for being "a humanitarian and true citizen of the world . . . [whose] concern for those living in less fortunate circumstances and his use of the statistical profession to help is an inspiration for all statisticians."

History of Surveys

In practice, surveys are collections of questions administered to individuals. Organizations like Gallup (founded as the American Institute of Public Opinion in 1935) specialize in conducting scientifically valid surveys. In the early part of the twentieth century, surveys were mostly conducted door-to-door by trained surveyors, a procedure used by both Gallup and the U.S. Census. Frequently, surveyors used the mail, like in the case of *Literary Digest*. Telephone surveys increased notably in the 1960s, which was attributed in large part to the fact that the costs of in-person research were escalating and trends in non-response suggested that people were growing less willing to answer face-to-face surveys, which diminished their prior advantage over phone surveys. Around 1970, statisticians Warren Mitofsky and Joseph Waksberg developed an efficient method of random digit dialing that revolutionized telephone survey research. However, some major organizations, like Gallup, continued door-to-door surveys into the mid-1980s, at which point they determined that a statistically sufficient proportion of U.S. homes had at least one telephone.

In 2008, Gallup notably expanded its methodology to include cell phones, since an increasing proportion of people no longer use landlines. In the twenty-first century, surveys are increasingly conducted via the Internet, though the U.S. Census still uses a combination of mail and house-to-house surveys. Harris Interactive, which went public in 1999, is a company that specializes in interactive online polls like the Harris Interactive College Football Poll, which ranks the top 25 Bowl Conference Series football teams each week.

Bias

Each survey method has different implications for both response bias and nonresponse bias. It is unclear when mathematicians and pollsters first began to recognize the negative influences of these biases, though adjustments were made in the latter half of the twentieth century. Systematic investigations can perhaps be traced to the mid-twentieth century, coincident with similar concerns in experimental design, like the placebo effect and psychologist Henry Landsberger's

naming of the Hawthorne effect. Overall, these biases are problematic because they are non-random and cannot be accounted for by most traditional statistical methods. As a result, they may produce misleading results. Methods to combat these biases are the subject of a great deal of ongoing research and are typically addressed via incentives and proactive planning rather than adjustments after the fact.

Sampling

Randomness is a critical component of survey methodology. Statistical techniques commonly assume that the sample is a random subset of the population. When this is true, the results are more likely to be representative and informative of the population. Though random sampling is the standard in modern scientific polling, early pollsters like Gallup tended to use convenience or quote sampling—taking a sample of whomever was accessible or convenient, sometimes grouped according to other influential variables like political party, gender, or neighborhood. In some cases, this was simply an issue of practicality in terms of time and financial resources. Mathematical statistician Jerzy Neyman is credited with presenting the first developed notion regarding making inferences from random samples drawn from finite populations, what is now called "probability sampling," at a professional conference in 1934. He also contrasted probability sampling with non-random methods. The U.S. Department of Agriculture, in partnership with the statistical laboratory at Iowa State University, began researching probability sampling methods in the late 1930s, as did the U.S. Census Bureau. One of these influential survey researchers was William Cochran, who also helped build many academic statistics programs, including at Harvard. Through the 1940s and beyond, the formal methods of probability sampling and analysis sampling were developed, implemented, and refined in a wide variety of situations.

In the late 1970s and beyond, some researchers' attention turned to more advanced concepts like model-dependent sampling. In probability sampling, the characteristics of the population are wholly inferred from the sample. Model-dependent sampling, in contrast, assumes some probability model for the population beforehand and designs both a sampling and an analysis plan around this model. This method allows the researchers conducting the survey to optimally match the statistical properties of chosen estimators to the population. Statisticians Morris Hansen, William Madow, and Benjamin Tepping discussed many of the principal advantages and limitations of this method in a 1978 presentation and 1983 publication. Morris Hansen was an internationally known expert on survey research, an associate director for research and development at the Census Bureau, and later chairman of the board for polling company Westat, Inc. He also served as president of the American Statistical Association and Institute for Mathematical Statistics.

U.S. Census

Though the U.S. Constitution calls for a count of the population in the decennial census, the U.S. Census Bureau conducts other types of surveys and has been using sampling since 1937. In 1940, the bureau began asking a random sample of people counted in the decennial census extra questions to allow better characterization of population demographics as well as to estimate coverage errors. The ongoing American Community Survey helps determine how billions of federal and state dollars are distributed each year. In the late twentieth century, in large part because of substantial difficulties during the 1990 census, many statisticians proposed completely substituting sampling methods for the decennial counting process or at least substantially increasing the role of sampling. They felt that issues like undercoverage of certain subpopulations could be better addressed with increasingly sophisticated statistical methods. Cost was also considered. They had the support of many cities, states, civil rights groups, and members of Congress. The proposal was opposed by many other politicians and segments of the general population for both political reasons and because of skepticism regarding the sampling process. It ultimately required a ruling by the U.S. Supreme Court, which allowed supplemental sampling for some purposes but required a count to determine congressional apportionment.

Further Reading

Brick, J. Michael, and Clyde Tucker. "Mitofsky–Waksberg: Learning From the Past." *Public Opinion Quarterly* 71, no. 5 (2007). http://poq.oxfordjournals.org/content/71/5/703.full#ref-24.

Hansen, Morris. "Some History and Reminiscences on Survey Sampling." *Statistical Science* 2, no. 2 (1987).

http://projecteuclid.org/DPubS/Repository/1.0/Disseminate?view=body&id=pdf_1&handle=euclid.ss/1177013352.

GARETH HAGGER-JOHNSON

See Also: Census; Data Mining; Elections; Internet; Measurement in Society.

Satellites

Category: Communication and Computers.
Fields of Study: Algebra; Geometry; Measurement.
Summary: Mathematics is fundamental to the design, function, and launch of satellites.

Astronomy and mathematics have long developed together. Many early mathematicians studied the motion of celestial objects. The term "satellite" comes from the Latin *satelles* (meaning "companion"), which was used by mathematician and astronomer Johannes Kepler to describe the moons of Jupiter in the seventeenth century. Mathematician Giovanni Cassini correctly inferred that Saturn's rings were composed of many small satellites in the seventeenth century. Mathematicians Jean Delambre and Cassini Jacques both published books of astronomical tables, including planetary satellites, in the eighteenth century. When artificial satellites were developed, the term "satellite" largely came to refer to those in common speech, while "moon" was applied to natural bodies orbiting planets. Mathematicians like Michael Lighthill and engineers like John Pierce helped develop satellites in the 1960s.

By the first decade of the twenty-first century, there were several hundred operational satellites orbiting the Earth to facilitate communication, weather observation, research, and observation. The advantage of satellites for communication are that signals are not blocked by land features in the same manner as a lower-altitude signal would be, making long-distance communication possible without multiple ground-based relays. Early communication satellites simply reflected signals back to Earth to broaden reception. Modern satellites use many different kinds of orbits to facilitate complex functioning, including low Earth orbit; medium Earth orbit; geosynchronous orbit; highly elliptical orbit; and Lagrangian point orbit, named for mathematician Joseph Lagrange. Mathematics is involved in the creation and function of such satellites, as well as for solving problems related to launching satellites, guiding movable satellites, powering satellite systems, and protecting satellites from radiation in the Van Allen belt, named for physicist James Van Allen. For example, graph theory is useful in comparing satellite communication networks. Techniques of origami map folding, researched by mathematicians like Koryo Miura, have been used in satellite design. Chaos theory has been used to design highly fuel-efficient orbits, derived in part from mathematician Henri Poincaré's work in stable and unstable manifolds. Government agencies like the U.S. National Aeronautics and Space Administration (NASA) and private companies like GeoEye employ mathematicians for research and applications. The Union of Concerned Scientists (UCS) maintains a database of operational satellites.

Orbits

The orbit of a satellite about the Earth determines when it will pass over various points on the Earth's surface and how high it is above the Earth. In general, orbits are characterized by altitude, inclination, eccentricity, and synchronicity. As defined by NASA, low Earth orbits have altitudes of 80–2000 kilometers. This orbit includes the majority of satellites, the International Space Station, and the Hubble Space Telescope. Statistical estimates at the start of the twenty-first century suggest that the number of functional satellites and nonfunctional debris in low orbit ranges from a few thousand (tracked by the U.S. Joint Space Operations Center) to millions (including very small objects). Objects in low orbit must travel at speeds of several thousand kilometers per hour, so even a small object can cause damage in a collision. Medium Earth orbit extends to about 35,000 kilometers (21,000

miles), the altitude determined by Kepler's laws of planetary motion for geosynchronous orbits. Inclination is an angular measure with respect to the equator, while eccentricity refers to how elliptical an orbit is. Geosynchronous satellites rotate at the same rate as the Earth spins, so they appear stationary relative to Earth. They usually have inclination and eccentricity of zero; they circle the equator to balance gravitational forces. The Global Positioning System (GPS) is one example of satellites at this orbital level. Sun synchronous orbits are retrograde patterns that allow a satellite to pass over a section of the Earth at the same time every day. They have an inclination of 20–90 degrees and must shift by approximately one degree per day. These orbits are often used for satellites that require constant sunlight or darkness. The maximal inclination of 90 degrees denotes a polar orbit. A halo or Lagrangian orbit is a periodic, three-dimensional orbit near one of the Lagrange points in the three-body problem of orbital mechanics, which was used for the International Sun/Earth Explorer 3 (ISEE-3) satellite.

Signals

Antennas and satellite dishes are used to receive satellite signals on Earth. Most satellite dishes have a parabolic shape. A signal striking a planar surface reflects directly back to the source. If the surface is curved,

Antennas and satellite dishes generally have a parabolic shape and are used to receive satellite signals on Earth.

the reflection is in the plane tangent to the surface. A parabola is the locus of points equidistant from a fixed point and a plane, so a parabolic dish focuses all incoming signals to the same point at the same time, increasing the quality of the signal. Mathematics is used to compress, filter, interpret, and model vast amounts of data produced by satellites. Reed–Solomon codes, derived by mathematicians Irving Reed and Gustave Solomon, are widely used in digital storage and communication for satellites. Much of the data from satellites is images, which utilize mathematical algorithms for rendering and restoration. One notable case that necessitated mathematical correction is the Hubble Space Telescope. An incorrectly ground mirror was found to have a spherical aberration, which resulted in improperly focused images. Mathematical image analysis allowed scientists to deduce the degree of correction needed. Some of the mathematical concepts involved in these corrections include the Nyquist frequency, which is a function of the sampling frequency of a discrete signal system named for physicist Harry Nyquist, and the Strehl ratio, named for mathematician Karl Strehl, which quantifies optical quality as a fraction of a system's theoretical peak intensity.

Further Reading

Montenbruck, Oliver, and Gill Eberhard. *Satellite Orbits: Models, Methods and Applications.* Berlin: Springer, 2000.

Whiting, Jim. *John R. Pierce: Pioneer in Satellite Communication.* Hockessin, DE: Mitchell Lane Publishers, 2003.

Bill Kte'pi

See Also: Digital Storage; GPS; Interplanetary Travel; Planetary Orbits; Wireless Communication.

Scales

Category: Arts, Music, and Entertainment.
Fields of Study: Algebra; Measurement; Number and Operations; Representations.
Summary: Musical scales have distinct mathematical properties and patterns.

Western music is based on a system of 12 pitches within each octave. The interval between adjacent pitches in this 12-tone system is called a "half step" or "semitone." Pitches separated by two successive semitones are said to be at the interval of a "whole step," or a "tone." Based on a variety of theoretical underpinnings, the concept and sound of tones and semitones have evolved throughout the history of Western music. In modern music practice, a uniform division of the octave into 12 equally spaced pitches, known as "equal temperament," holds sway. Scales are arrangements of half and whole step intervals in the octave. Denoting a half step as *h* and a whole step as *w*, the familiar diatonic major scale is defined by the sequence *wwhwwwh*. The diatonic natural minor scale is *whwwhww*. Beginning these patterns from each of the 12 pitches results in 24 distinct diatonic scales. This suggests a set-theoretic description by which each major scale can be represented as a transposition (in algebra this would be called a "translation") of the set of pitches C, D, E, F, G, A, B, and C. In the twentieth century, such mathematical formalisms have led to the conceptualization of non-diatonic scales with special transposition properties.

Octave Equivalence

The concept of octave (the musical interval between notes with frequencies that differ by a factor of two) is fundamental to understanding musical scales. In Western music notation, pitches separated by an octave are given the same note name. The piano keyboard provides a visual representation of this phenomenon. Counting up the white keys from middle C as "1," the eighth key in the sequence is again called C. This eight-note distance explains the etymology of the word "octave." The perception and conceptualization of such pairs of pitches as higher or lower versions of the same essential pitch is called "octave equivalence." Octave equivalence is thought to be common to all systematic musical cultures. Evidence of octave equivalence is found in ancient Greek and Chinese music. Recent psycho-acoustic research suggests a neurological basis for octave equivalence in auditory perception.

The mathematical explanation of octave equivalence comes from the fact that the sound of a musical pitch is a combination of periodic waveforms that can be modeled as sinusoidal functions of time. In the two periodic functions, $f(t) = \sin(t)$ and $g(t) = \sin(2t)$, with frequencies 2π and π, every peak of the lower

Table 1. The diatonic scale in three intonation schemes, Pythagorean, just, and equal temperament.

	C	D	E	F	G	A	B	C
Pythagorean	1:1	9:8	81:64	4: 3	3:2	27:16	243:128	2:1
interval		9:8	9:8	256:243	9:8	9:8	9:8	256:243
Just	1:1	9:8	5:4	4:3	3:2	5:3	15:8	2:1
interval		9:8	10:9	16:15	9:8	9:8	9:8	16:15
Equal	1:1	1.1225:1	1.2600:1	1.335:1	1.4983:1	1.6818:1	1.8878:1	2:1
interval		$2^{1/6}$	$2^{1/6}$	$2^{1/12}$	$2^{1/6}$	$2^{1/6}$	$2^{1/6}$	$2^{1/6}$

frequency function coincides with a peak of the high-frequency function. In sonic terms, this is the highest degree of consonance possible for two pitches of different frequencies.

History of Scales

As Western music developed from the Middle Ages through the twentieth century, the central construct was the diatonic scale. This arrangement spans an octave with seven distinct pitches arranged in a combination of five whole steps and two half steps. Interestingly, the pattern of intervals (and not the absolute pitch of the starting note) was the only distinguishing feature of scales until the rise of tonal harmony in the seventeenth century. Pitch-specific examples help illustrate the interval patterns.

The diatonic scale traces its origins to the ancient Greek *genus* of the same name, referring to a particular tuning of the four-stringed lyre (tetrachord) consisting of two whole steps and one half step in descending succession. An example of this tuning can be constructed with the pitches A, G, F, and E. Concatenation of two diatonic tetrachords [A-G-F-{E}-D-C-B] produces the pitches of the diatonic scale (the piano white keys). In medieval European musical practice, the distinct Church Modes (such as Lydian or Phrygian) developed from the diatonic scale by the assignment of a tonal anchor or final tone. For example, the Dorian mode is characterized by the sequence of ascending half and whole steps in the diatonic scale *whwwwhw*; for example D-E-F-G-A-B-C-D, while the Phrygian mode is *hwwwhww*: E-F-G-A-B-C-D-E. The diatonic major scale *wwhwwwh* (C-D-E-F-G-A-B-C) came into widespread use in the seventeenth century. The diatonic natural minor scale is *whwwhww* (A-B-C-D-E-F-G-A).

Intervals, Ratios, and Equal Temperament

The simplest musical interval is the octave. The frequency ratio between pitches separated by an octave is 2:1. The interval of a perfect fifth has frequency ratio 3:2. Using these two ratios, pitches and corresponding intervals for the diatonic scale can be assigned according to Pythagorean tuning. Simpler diatonic scales based on ratios of small integers are known as "just tunings." Western music in the modern era uses a symmetric assignment of intervals known as "equal temperament." In equal temperament, the 12 half steps that comprise the frequency doubling octave each have frequency ratio $2^{1/12} \approx 1.0595$. For these three tuning schemes, frequency ratios relative to the starting pitch and intervals between adjacent scale notes are illustrated and compared in Table 1. For each intonation, the first row gives the frequency ratio from the tonic C to the given note. The second row in each case gives the frequency ratio between adjacent diatonic pitches.

Modern Scales

In contrast to the idiosyncratic pattern of intervals that comprise the diatonic scales, the chromatic scale *hhhhhhhhhhh* is perfectly symmetric. In particular, the set of pitches that form the chromatic scale is unchanged by transposition—there is only one set of pitches with this intervallic pattern. This set of pitches is referred to as having order one. The elements of the pitch set forming a diatonic scale, which generates 12 diatonic scales by transposition, has order 12. This point of view suggests other scales of interest with respect to transposition. The set of six pitches in a whole-tone scale *wwwwww* (for example, C-D-E-F♯-G♯-A♯-C) are unchanged by transposition by an even number of half steps. A transposition by an odd number of half steps results in the whole tone scale containing the remaining six pitches (for example, C♯-D♯-F-G-A-B-C♯). Thus, the set of pitches in the whole-tone scale has order two. Whole-tone scales are a characteristic feature in much of the music of Claude Debussy.

The twentieth-century composer and music theorist Olivier Messiaen codified a number of eight-tone "scales of limited transposition." Among these are the order three scales *hwhwhwhw* and *whwhwhwh*, which are called "octatonic scales" in the music of Stravinsky and sometimes referred to as "diminished scales" in jazz performance. It can be seen that transposition by one and two half steps produce new diminished scales, but transposition by three half steps leaves the original set of pitches unchanged.

Further Reading

Grout, Donald Jay. *A History of Western Music*. New York: Norton, 1980.

Hanson, Howard. *Harmonic Materials of Modern Music: Resources of the Tempered Scale*. New York: Appleton-Century-Crofts, 1960.

Johnson, Timothy. *Foundations of Diatonic Theory: A Mathematically Based Approach to Music Fundamentals*. Lanham, MD: Scarecrow Press, 2008.

Pope, Anthony. "Messiaen's Musical Language: An Introduction." In *The Messiaen Companion*. Edited by Peter Hill. Portland, ME: Amadeus Press, 1995.

Sundberg, Johan. *The Science of Musical Sounds*. San Diego, CA: Academic Press, 1991.

Eric Barth

See Also: Composing; Geometry of Music; Harmonics; Pythagorean and Fibonacci Tuning.

Scatterplots

Category: History and Development of Curricular Concepts.
Fields of Study: Communication; Connections; Data Analysis and Probability.
Summary: Scatterplots are useful tools for mathematicians and statisticians to graph and present data.

Human beings are constantly exploring the world around them to discover relationships that can be used to explain past and current events or phenomena and perhaps to predict future occurrences.

The colloquial expression "a picture is worth a thousand words" is traced back to many possible historical sources, including French leader and noted student of mathematics Napoleon Bonaparte, who purportedly said, "A good sketch is better than a long speech." In the twenty-first century, graphing is a fundamental first step in any exploratory data analysis, and graphical representations are common in the media. Scatterplots, which most often represent values of paired variables in a Cartesian plane, help data investigators identify relationships, describe patterns and correlation, fit linear and nonlinear functions using techniques like regression analysis, and locate points known as "outliers" that deviate from the predominant pattern. In the primary grades, students often use line graphs, which some consider to be a special case of scatterplots, while scatterplots for data may be explored beginning in the middle grades in both mathematics and science classes.

Early History

Mathematicians and others have long sought alternative methods of representation for researching, presenting, and connecting the mathematical concepts they studied. The Cartesian plane, named for René Descartes, facilitated graphing of algebraic equations and data beginning in the seventeenth century. Historians have traced scatterplots to 1686, though the term "scatter diagram" is attributed to early twentieth-century researchers such as statistician Karl Pearson, and "scatterplot" seems to have first appeared in a 1939 dictionary.

Examples of early pioneers of data graphing include "political arithmetician" Augustus Crome, who studied the relationships between nations' population sizes, land areas, and wealth; mathematician and sociologist Adolphe Quetelet, who conducted studies of body measurements that helped contribute to the measure now known as the Body Mass Index, which relates height and weight; and engineer and political scientist William Playfair, who called himself the "inventor of linear arithmetic," a term he used for graphs. He said: ". . . it gives a simple, accurate, and permanent idea, by giving form and shape to a number of separate

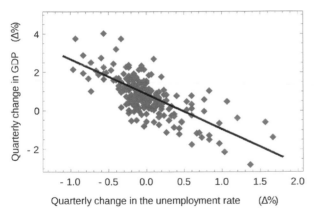

A scatterplot chart showing the relationship between gross domestic product growth and unemployment.

ideas, which are otherwise abstract and unconnected." Playfair's eighteenth-century graphical summaries of British trade across various years are perhaps the earliest example of what would now be referred to as "time series plots" (or in some cases "line graphs"), which may be considered a special case of scatterplots.

While Playfair plotted many economic variables as functions of time, the most extensive early use of scatterplots to relate two observed variables is probably the anthropometric and genetic research of Francis Galton, a cousin of scientist Charles Darwin. After studying medicine and mathematics in college, he became interested in the investigation and characterization of variability and deviations in many natural phenomena. He established a laboratory for the measurement and study of human mental and physical traits, focusing on empirical and statistical studies of heredity in the latter half of the nineteenth century. Many of Galton's scatterplots involved graphing parental characteristics on one axis, usually the X, and offspring characteristics on the other. Like scientist Gregor Mendel, some of his initial genetic experiments were conducted on peas; later, he investigated measurements of people. Scatterplots of height appeared in his 1886 publication *Regression Towards Mediocrity in Hereditary Stature*, which is the origination of the name for the statistical technique of regression analysis. The word "mediocrity" in this context was a reference to the mean or average height (not a qualitative judgment) and was used to describe a pattern observed in the data: very short parents tend to have taller children, and very tall

parents tend to have shorter children, in both cases closer to the mean.

Recent Developments

Prior to the development of computers and data analytic software, data had to be graphed by hand. In the twenty-first century, computers facilitate many types of scatterplots. In addition to the standard plots of two variables in the Cartesian plane, there are three-dimensional scatterplots that display point clouds to explore the ways in which three variables relate and interact. Symbols used to represent points on a two- or three-dimensional scatterplot may also be coded using different colors or shapes to indicate additional variables and uncover patterns. Matrix plots are square grids of scatterplots for a set of variables that plot all possible pairwise sets, usually arranged such that all of the plots in the same row share the same Y variable and all plots in the same column share the same X variable. Mathematicians, statisticians, computer scientists, and other types of researchers have explored the theoretical and methodological links between scatterplots and map surfaces for use in applications such as data mining and spatial analysis of geospatial information system (GIS) data.

While they are useful tools for exploration and representation, scatterplots are often subject to misinterpretations. For example, sometimes relationships or correlations shown in scatterplots are mistakenly taken as evidence of cause and effect, which must be inferred from the way in which the data were collected rather than from the strength of the association.

Further Reading

Few, Stephen. *Now You See It: Simple Visualization Techniques for Quantitative Analysis.* Oakland, CA: Analytics Press, 2009.

Friendly, M., and D. Denis. "The Early Origins and Development of the Scatterplot." *Journal of the History of the Behavioral Sciences* 41, no. 2 (2005).

Stigler, Stephen. *The History of Statistics: The Measurement of Uncertainty Before 1900.* Cambridge, MA: Belknap Press of Harvard University Press, 1990.

Gareth Hagger-Johnson

See Also: Coordinate Geometry; Forecasting; Graphs; Visualization.

Scheduling

Category: Business, Economics, and Marketing.
Fields of Study: Data Analysis and Probability;
Number and Operations.
Summary: Scheduling can be a complex
mathematical exercise and is necessary to keep
businesses and supply chains running efficiently.

Intense competitiveness forces companies to optimize
performance in terms of cost, time, and resources.
Scheduling is the process of developing and implementing optimal operational plans. Formal concepts
of scheduling date to the Industrial Revolution and
innovations like Henry Ford's assembly line, although
the basic ideas probably existed from antiquity in any
society where people manufactured goods.

In manufacturing, multiple tasks are carried out in
sequence to produce a final output from raw materials. Further, steps in a manufacturing process may be
performed on different machines that require variable time to deliver outputs and it is possible that
materials will be transported between facilities. A
mathematically determined schedule that takes into
account all relevant variables in the process serves to
optimally allocate resources with respect to demand
of the tasks, including shortening time intervals to
reduce unproductive time and minimizing costs from
wasted time and materials. Operations research is a
field of applied mathematics and science that uses
mathematical tools, such as simulation and modeling, linear programming, numerical analysis, graph
theory, and statistical analysis, to arrive at optimal
or near-optimal solutions to complex problems like
scheduling. It may also tackle problems in which the
resources are not materials but people. The scheduling of airplane crews is a highly constrained and difficult problem because of legal limits on work and rest
times as well as the need for crews to return to a home
base. Allocation of police, fire, and ambulance services
is also a widely used and very important application
of scheduling theory.

Production Management

As a part of production management, scheduling
interferes with many different aspects of business
such as the supply chain, inventory maintenance, and
accounting. For example, consider a paint company
that makes provisions of sales for the next month by
analyzing previous data. In light of these provisions,
schedulers determine the expected arrival time and
amount of different types of chemicals, which have
different delivery times.

The supply chain should be able to deliver the correct amounts of chemicals in time. In a similar way,
accounting of the cost of supply and inventory should
be accessible for the schedulers. Because of the number of operational parts of business that scheduling
is related with, it is apparent that scheduling is a very
complex process. It gets more complex with larger
variation in types of products and larger numbers of
machines varying in processing times. Thus, schedulers demand thorough knowledge of factors such as the
processing time of each machine, delivery time, the
amount of resources to allocate among machines, and
the size and flow of operations for each product.

Manufacturing

In many manufacturing processes, different machines
might share the same input, or inputs of a machine
might consist of outputs from multiple machines.
Scheduling operations in these type of cases requires
extensive mathematical modeling. Two basic types of
modeling for production scheduling are distinguished
by the presence of randomness within. Deterministic models do not include the probability of faults in
processes or critical changes in capacity or resource
availability. They are based on previous averages of
production figures and output rates, so they do not
easily adapt to changes in demand or capacity constraints. In these cases, rescheduling is needed, which
causes time and resource loss if repeated too many
times. They are best suited to manufacturing productions that involve less risk of defects. Stochastic
models, on the other hand, involve the probability of
unexpected malfunctions or critical changes by distributing probability analytically to individual steps
of the schedule. Usually, they are appropriate for processes consisting of many individual operations. For
example, these models examine machine failure rates
and aim to provide options for when a breakdown
occurs. Also, these models maintain an inventory of
materials, which may prove critical in maintaining
production. Simulations of models provide schedulers an environment to test possibilities that can
obstruct the flow of production.

Further Reading

Conway, Richard W., William L. Maxwell, and Louis W. Miller. *Theory of Scheduling.* New York: Dover Publications, 2003.

Pinedo, Michael. *Scheduling: Theory, Algorithms, and Systems.* New York: Springer, 2008.

Blazewicz, J., K. H. Ecker, E. Pesch, G. Schmidt, and J. Weglarz. *Scheduling Computer and Manufacturing Processes.* New York: Springer, 2001.

Kogan, K., and E. Khmelnitsky. *Scheduling: Control-Based Theory and Polynomial-Time Algorithms.* New York: Springer, 2000.

UGUR KAPLAN

See Also: Data Mining; Mathematical Modeling; Parallel Processing.

Schools

Category: Architecture and Engineering.
Fields of Study: Data Analysis and Probability; Geometry; Number and Operations.
Summary: Principles of geometry affect school design and mathematical models of risk may help identify safety issues.

When people think of mathematics in schools, most probably envision the teaching and learning of mathematics that occurs inside classrooms. However, there are many aspects of twenty-first century schools that depend on mathematics. For example, the transition of school design from one-room schoolhouses that were common in the nineteenth century, through the often rectangular and symmetric classroom buildings of the latter nineteenth and early twentieth centuries, to the open-plan schools initiated in the 1950s, to twenty-first century schools that consider contemporary concerns about renewable energy, technology, and safety. Changes in teaching philosophies over time, such as loop education and emphasis on science, technology, engineering, and mathematics (STEM) education principles in the lower grades, led to some of these changes, as did studies on tragedies like the shootings at Columbine High School and Virginia

Polytechnic Institute and State University, popularly known as Virginia Tech (VT). Mathematics principles can be used to map the flow of students to and from classes, optimize locker placement and access, build accommodations and accessibility for students with disabilities, and plan for athletic facilities and other non-classroom spaces. These applications are increasingly important as schools seek to educate students to live and work within the rapidly changing economies, technologies, and environments of the twenty-first-century global society. Other studies may determine whether to retrofit old buildings or construct new facilities using mathematical methods like cost-benefit analysis. There are many organizations and publications devoted to discussing the mathematics, engineering, and technical aspects of school design and construction.

Optimizing School Design

The notion of what constitutes "optimal school design" has markedly changed over time. There are some who consider the classic one-room schoolhouse to be the original open-plan design, since the teacher accommodated all students in all grades in a single space, dividing class time among the various grades. Famed Boston architect Gridley J. F. Bryant, who also studied engineering, is credited with revolutionizing the design of many public buildings. His Quincy School, which opened in 1847, was among the first multi-classroom schools. The school was three stories tall, with four identical and symmetrically arranged classrooms on each floor. This model was used for schools throughout the late nineteenth and early twentieth centuries and would be further evolved with movable desks and tables to allow for some flexibility within the "box in a box" construction, as it was called by some. This design led to other considerations such as optimal selection and placement of furniture such as desks, tables, chairs, and later computers, as well as features such as lockers and storage spaces, all of which must be fit into a limited amount of space yet be accessible and functional for a varying student body. Proper placements rely on mathematical concepts such as volume and are related to mathematical packing problems. Detractors often likened Bryant's school configuration to prisons, which he also designed. The evolution of open-plan schools of the latter twentieth century was motivated by cost and changes in teaching philosophies, derived in part from research in mathematics education. There was and continues to be controversy regarding the efficacy and desirability of open plan schools. Mathematicians, architects, facilities planners, and others continue to research effective strategies for design and construction. For example, architect Prakash Nair is internationally recognized as a leader in school design, and has been cited for using educational research as a basis for designs that optimize teaching and learning. He helped develop a "pattern language" that draws on geometric ideas and uses a modular set of design patterns, sub-patterns, and groupings to match school designs to goals and needs. It can be used to develop new schools and assess existing structures. Other education professionals like C. Kenneth Tanner, whose background includes work in design, mathematics, statistics, and operations research, have also used a combination of data-based research with mathematical techniques and tools to address a broad spectrum of school planning issues, such as technology integration. Organizations like the School Design and Planning Laboratory at the University of Georgia use data-driven methods and models for assessing school design and forecasting student populations and demographics, which may impact design, use, and sustainability.

Safety

The safety of children in U.S. schools has become a growing concern for parents, teachers, and society in general. The 1999 shootings at Columbine High School focused national attention on issues of school security, safety, and patterns of police response to such incidents. Even more debate occurred after the 2007 shootings at VT. The Secret Service, the Department of Education, and the Federal Bureau of Investigation (FBI) conducted broad studies into the causes and prevention of school violence. For example, the Secret Service and the Department of Education studied all 37 shootings in U.S. schools between 1974 and 2000. Data analysis revealed no identifiable statistical patterns; school shooters came from a variety of ethnic, economic, and social classes, and most had no history of violent behavior that would reliably predict later actions. Using statistical methods, profilers from the FBI also concluded that the "oddball" students that society commonly perceives to be potential trouble-

makers were not in fact more likely to commit violence, though such studies are limited by the relatively small number of incidents and data available for modeling. Probabilistic, predictive profiling is quite controversial, but many educators and others still advocate its use in risk assessment.

Another mathematically based strategy schools may employ for risk assessment is actuarial methods. Actuarial models for school risk statistically combine empirically chosen threat factors to produce probabilities for particular outcomes or behaviors, and sometimes they may be standardized for specific student populations. In some cases where there are sufficient data and the models can be validated, they have often performed better at identifying in-school threats than subjective human judgments. However, other model-based assessments of risk that are based on sparse data or with a short window for prediction have not been shown to be as reliable. Some researchers have tried to develop expert systems for school threat assessment and decision making, which are automated or semi-automated tools that use artificial intelligence and algorithms developed from data, achieving mixed success. Both actuarial models and expert systems for schools may be revised to incorporate new data as it is identified, making them flexible mathematical modeling tools.

Further Reading

Institute for the Development of Educational Activities, Inc. "The Open School Plan; Report of a National Seminar" (1970). http://archone.tamu.edu/CRS/engine/archive_files/efl/6000.0205.pdf.

Nair, Prakash, and Randall Fielding. *The Language of School Design: Design Patterns for 21st Century Schools.* 2nd ed. Minneapolis, MN: Designshare, Inc. 2005.

Reddy, Marissa, Randy Borum, John Berglund, Bryan Vossekuil, Robert Fein, and William Modzeleski. "Evaluating Risk for Targeted Violence in Schools: Comparing Risk Assessment, Threat Assessment, and Other Approaches." *Psychology in the Schools* 38, no. 2 (2001). http://www.secretservice.gov/ntac/ntac_threat_postpress.pdf.

Tanner, C. Kenneth, and Jeff Lackney. *Educational Facilities Planning: Leadership, Architecture, and Management.* Boston: Allyn and Bacon, 2006.

Sarah J. Greenwald
Jill E. Thomley

See Also: Engineering Design; Forecasting; Learning Models and Trajectories; Packing Problems; Risk Management.

Science Fiction

Category: Arts, Music, and Entertainment.
Fields of Study: Communication; Connections; Representations.
Summary: Mathematics plays many roles in science fiction, sometimes as content or characters, other times bringing elements to life on the screen.

Like mathematics, writing science fiction is a craft grounded in deduction and extrapolation. The writer begins with certain axioms: the world as we know it and the world as we believe it could be. He introduces certain new variables: a thinking robot, an alien invasion, human clones. Explicitly or implicitly, the story explores the consequences, the corollaries of these new things according to the implications of those initial axioms. Stories that do not do this are considered fantasy or sometimes science-fantasy or soft science fiction, if they otherwise contain the set-dressing of the science fiction genre.

The setting is often the future, an alternative history, or some alternate reality, and may include use of time travel. Alien characters frequently interact with humans in science fiction. Many books, movies, comics, graphic novels, computer games, and Internet applications use science fiction themes, sometimes as a context in which to explore deeper philosophical questions. Mathematics and science play a variety of roles within the science fiction genre. Sometimes, mathematics and science are written into the story to give validity and believability to the futuristic setting or to the technology. Mathematics is also used to bring fantastic science fiction elements to life on screen, such as in the groundbreaking *Star Wars* franchise or the 2010 film *Avatar*. At other times, characters in science fiction works are mathematicians or scientists who act as the primary heroes or villains, or who explain scientific elements to the audience. The inclusion of mathematically talented characters in science fiction is sometimes done to exploit commonly held stereotypes about mathematicians for narrative purposes, such as genius or aloofness. In other works, mathematics becomes the explicit subject of the story, and the mathematics of science fiction in both written and visual media has been explored in college courses and mathematics research. Mathematicians or individuals that have mathematical training often create science fiction, and science fiction may inform mathematical research. The widely noted "Big Three" authors of twentieth-century science fiction—Arthur C. Clark, Robert A. Heinlein, and Isaac Asimov—all had mathematical training or mathematically based science backgrounds and made nonfiction contributions to areas such as satellites, rocketry, robotics, and ethics.

Early History of Science Fiction
Because of the varying definitions of science fiction, it is difficult to determine exactly what might be the first science fiction story. The Mesopotamian epic poem *The Epic of Gilgamesh*, which is among the oldest surviving works of literature, is cited by some scholars as

Science fiction is often set in an alternate reality and may involve time travel or alien characters.

containing elements of science fiction. Some researchers note that the Bible, when examined as a work of literature, has stories that could be classified as science fiction, such as the ascension of the prophet Elijah to heaven in a fiery chariot. In the second century, the Greek satirist Lucien of Samosata wrote about interplanetary travel and alien life forms in his *True Histories* (or *True Tales)*. English lawyer and philosopher Thomas More's 1516 work *Utopia* described a perfect society, which became a common theme among later science-fiction writers. Some scholars argue that such early works cannot be claimed as the first science fiction because neither the audience nor the authors likely knew enough about the underlying science. Correspondingly, they might claim that the origin of science fiction coincided with the post-medieval scientific revolution and discoveries in science and mathematics made by people such as Isaac Newton and Galileo Galilei. Mathematician and astronomer Johannes Kepler wrote a story in 1634 called *Somnium*, which imagined that a student of astronomer Tycho Brahe had been transported to the moon and described how Earth might look when viewed from that location. It contained mathematical computations and is considered by some to be a scientific treatise, while others cite it as the first science fiction, including both Asimov and astronomer and author Carl Sagan. Author Brian Aldiss asserts that science fiction derives many of its structure and conventions from the Gothic horror genre, which suggests that Mary Shelley's 1818 novel *Frankenstein* is "the first seminal work to which the label SF can be logically attached." This labeling is perhaps because of its introduction of science fiction themes like a mad scientist, the potential misuse of technology, and the presence of an non-human being as a main character.

The Foundations of Twentieth-Century Science Fiction

Jules Verne and Herbert George (H. G.) Wells are often jointly known as the "fathers of science fiction" for their creative influence on the development of twentieth-century science fiction. Jules Verne consistently incorporated the newest technological discoveries and experiments of his lifetime into his work. Many of his most popular novels, like *A Journey to the Center of the Earth* (1864), *From the Earth to the Moon* (1865), *Twenty Thousand Leagues Under the Sea* (1869), and *Around the World in Eighty Days* (1873), have been widely translated into other languages and adapted into plays, movies, television shows, and cartoons. Some scholars have called Verne's books visionary and even prophetic for describing mathematical and scientific phenomena such as weightlessness and heavier-than-air flight before they were well-known or understood. His attention to realistic scientific principles and detailed descriptions of problems and solutions would later challenge many real-life mathematicians, scientists, and engineers. Physicist and engineer Hermann Oberth and scientist Konstantin Tsiolkovsky, who are known as the "fathers of rocketry and astronautics" along with physicist Robert H. Goddard, reported being inspired by Verne's books.

Like Verne's work, the novels of H. G. Wells have been widely adapted into various other media, and the 1895 novel *The Time Machine*, in particular, is cited as inspiring many other works of fiction. The invention of the now commonly used term "time machine" is attributed to Wells, as is the notion of time being the fourth dimension. In the 1897 novel *The Invisible Man*, a scientist named Griffin makes himself invisible by changing the refractive index of his body so that it neither absorbs or reflects light. Some of Wells's books were considered to be exceptionally bold and compelling. His 1898 novel *The War of the Worlds* is well-grounded in mathematical and scientific theories from the time it was written, like mathematician Pierre-Simon Laplace's formulation of the nebular hypothesis. It shared a vision of space travel common to late nineteenth-century novels, including Verne's *From the Earth to the Moon*. Large cylinders were fired from cannons on the Mars surface to transport the aliens to Earth. Later mathematical models and calculations necessary to send people into Earth's orbit and to the moon, as well as to guide probes to Mars and the far reaches of the solar system, demonstrated that the parabolic trajectories were often quite complex and that the forces required to propel a cylinder from Mars to the Earth would likely be lethal to passengers. Wells was also a science teacher and political activist who recognized and asserted the importance of quantitative knowledge, noting: "Statistical thinking will one day be as necessary for efficient citizenship as the ability to read and write."

Mathematical Science Fiction

While mathematics is widely used to help build or validate the setting or technology of a science fiction story, such as in the works of Verne and Wells, in some cases it is a central component of the plot. There are many mathematical science fiction novels that have been written about a variety of themes. The 1946 short story *No-Sided Professor*, written by mathematician and author Martin Gardner, disusses the Möbius strip, a one-sided figure named for mathematician August Möbius. It addresses the possibility of a zero-sided figure and other concepts in topology. *Occam's Razor*, by author David Duncan in 1956, posits the notion of discontinuous time, which can be bridged by minimal

Mathematical Characters and Stereotypes

Science fiction authors often include mathematicians or mathematically talented individuals as characters in order to explain scientific elements to the audience or to exploit commonly held associations and stereotypes, which can be a shortcut for characterization, including intelligence, logic, emotional coldness, eccentricity, arrogance, or general strangeness or differences between mathematicians and supposedly "normal" people. For example, in Michael Crichton's 1990 novel *Jurassic Park,* mathematician and chaos theorist Ian Malcolm, sometimes cited as having been modeled in part on Ian Stewart, expounds with some arrogance on the mathematics that shape the increasingly dangerous situations in which the characters find themselves. However, he otherwise defies many of the stereotypes associated with mathematicians, such as social ineptness. The notion of logic and mathematical reasoning as male modes of thinking and understanding, versus understanding via female emotion and intuition, is also pervasive in older science fiction. Some point to lessening trends in this theme in the latter twentieth century, and a few works like Chiang's *Division by Zero* contain female characters who are coldly logical and distant rather than emotional. It is an issue of debate whether this should be seen as a positive or negative shift.

surfaces in certain topologies. Asimov's 1957 novel *The Feeling of Power* addresses scientific computing in a futuristic society in which people have lost the ability to perform basic arithmetic calculations. The rediscovery of hand-multiplication therefore becomes a new "secret weapon" for the society's military. Author Stanislaw Lem discusses countably infinite sets in his 1968 novel *The Extraordinary Hotel*, while author and mathematician Greg Egan's 1991 work *The Infinite Assassin* includes a discussion of the Cantor set, named for mathematician Georg Cantor, an important concept in topology and some other mathematical fields. Other mathematical science fiction urges appreciation of mathematics as if it is a form of poetry. Examples include author Kathryn Cramer's 1987 work *Forbidden Knowledge*, author Normal Kagan's 1964 work *The Mathenauts*, and multiple stories by author Eliot Fintushel. Mathematician and author Vernor Vinge often addressed the mathematical themes of superhuman artificial intelligence and a predicted technological singularity: a point in time where the exponential growth of technology results in essentially instantaneous change. These themes are also found Clark's *2001: A Space Odyssey* and its sequels. The term "technological singularity" is credited to mathematician Irving Good and is also linked to Moore's law, named for Intel co-founder Gordon Moore, which mathematically models the trend in the evolution of computer processor speeds.

Several science fiction novels challenge the foundations of mathematics itself or the commonly proposed notion of mathematics as a universal language. Author Ted Chiang's *Division by Zero*, a 1991 short story, discusses the discovery of a proof that mathematics is inconsistent, which may be possible according to Gödel's Incompleteness Theorems, named for mathematician Kurt Gödel. Chiang's later 1998 work *Story of Your Life* involves humans trying to communicate with aliens whose mathematics is based on variational formulations rather than algebra. In the same year, mathematician and author David Reulle's *Conversations on Mathematics With a Visitor from Outer Space*, which was published in a collection of nonfiction mathematical essays, argued that mathematics on Earth is essentially human in nature, so humans should not expect aliens to share human's unique mathematical language. Sagan's 1985 novel *Contact* alternatively suggested that humans and aliens may communicate via mathematics, but rather than the typical mode of receiving radio waves containing messages from space, communications are instead embedded within the very framework of mathematics itself.

Since Wells introduced the notion of the fourth dimension in *The Time Machine*, dimensionality in many forms has been a widely used theme in science fiction, including mathematical science fiction. In the 1940 novel *And He Built a Crooked House* by Heinlein, a mathematical architect designs a house that is constructed as an "inverted double cross" representation of an unfolded tesseract net in three-dimensional space. Following an earthquake, the structure spontaneously shifts and folds itself into an actual tesseract, whose four-dimensional properties are explored and described by characters. The satirical Edwin Abbot novel *Flatland: A Romance of Many Dimensions,* which was written largely as a social commentary on Victorian norms and mores, may also be considered science fiction because it depicts an alternate two-dimensional world inhabited by polygonal creatures, which is visited by three-dimensional creatures in a manner that resembles twentieth- and twenty-first-century depictions of human-alien interactions. More than a century after its initial publication, *Flatland* remains popular in the mathematical community because of its entertaining and enlightening discussions of what some people consider to be an abstract mathematical concept, and it was once described by Asimov as, "The best introduction one can find into the manner of perceiving dimensions."

Other authors have used the novel as inspiration. Mathematician and author Ian Stewart's 2001 work *Flatterland: Like Flatland, Only More So,* explores several mathematical topics such as Feynman diagrams, named for physicist Richard Feynman, superstring theory, quantum mechanics, fractal geometry, and the recurring science fiction theme of time travel. He includes mathematical jokes and puns such as a one-sided cow named Moobius to make concepts relatable to a broader audience. Stewart also co-authored the semi-fictional *Science of Discworld* series, which compares mathematically and scientifically the natural laws of sperical planets or "roundworlds" like Earth to the created or imagined physical laws of the flat, disc-shaped setting of author Terry Pratchet's *Discworld* novels. Some other works that are commonly cited as extensions of ideas found in *Flatland* include mathematician and author Dionys Burger's 1953 novel *Sphereland: A Fantasy About Curved Spaces and an Expanding Universe* and two works by mathematician and author Rudy Rucker: the 1983 short story *Message Found in a Copy of Flatland* and the 2002 novel *Spaceland*.

Mathematics is a living discipline that is constantly evolving, and mathematical science fiction sometimes underscores this point. Gardner's 1952 story *The Island of Five Colors* is the sequel to the *No-Sided Professor.* The characters in the story attempt to solve the Four Color theorem, which was unproven at the time. It illustrates the inherent time dependence of some elements of science fiction, since imagined creations and the mathematics on which they are based frequently become reality later. Gardner stated: "the true four-color theorem, unproved when I wrote my story, has since been established by computer programs, though not very elegantly. As science fiction, the tale is now as dated as a story about Martians or about the twilight zone of Mercury." At the same time, others argue that the themes of such novels are still useful and relevant when considering the nature of mathematics and that these stories do not automatically lose value as entertainment or inspiration simply because the mathematical or scienctific frameworks become somewhat out of date.

Mathematicians as Science Fiction Authors

Mathematicians and mathematically trained individuals such as Martin Gardner, Isaac Asimov, Greg Egan, Ian Stewart, and Vernor Vinge often contribute to both science fiction writing and mathematical or scientific research, and several mathematicians have won the Hugo Award, the premier prize in science fiction and fantasy literature. Perhaps one of the most well-known of these is Rudy Rucker, who is considered among the founding fathers of the science fiction subgenre of cyberpunk, a style that draws inspiration from Gothic horror like *Frankenstein*, film noir, punk, computer science, and cybernetics, a discipline whose twentieth-century development is attributed to mathematician Norbert Weiner. Rucker credits his mathematical background for influencing not only the content of what he writes but also the way in which he writes: "I think of the writing process itself as a fractal. I have the big arc of the plot, the short-story-like chapters, the scenes within the chapter, the actions that make up the scenes, and nicely formed sentences to describe the actions, the carefully chosen words in the sentence. And hidden beneath each word is another fractal, the entire language with all my ramifying mental associations."

He also notes that both mathematics and science fiction writing can be thought of as ways of exploring the consequences of imposed constraints or assumptions, and that science fiction writing provides a structure for carrying out interesting "thought experiments" about concepts such as alternative mathematical structures to explain the nature of reality.

Visual Media

Science fiction has long been translated to other forms of media, and mathematics plays a dual role as a subject and a technique for bringing both realistic and fantastic images to life. Further, mathematicians are often involved as writers or consultants. Stanley Kubrick's *2001: A Space Odyssey* is one of the most well-known science fiction films. Mathematician Irving Good consulted on the film, as did novelist Clark, and it is praised for its scientific realism and pioneering special effects. The *Star Wars* franchise, launched in 1977, now includes books, comics, movies, video games, and Web media. It was groundbreaking in its use of mathematically based special effects techniques, including extensive stop motion animation and then later computer animation for backgrounds, props, costumes, and even entire characters. Effects that were once limited to big-budget films have now made their way onto television. The *Star Trek* franchise is notable not just for its visual imagery but also for references to real-world mathematical concepts including π and Fermat's last theorem, named for mathematician Pierre de Fermat. Other examples of shows that contain frequent real-world mathematical include SyFy's *Eureka* and the animated series *Futurama*. Producer and writer Ken Keeler has a mathematics Ph.D. from Harvard. Along with other mathematically trained writers, he co-creates many of the mathematical references found on *Futurama*, and once notably constructed a new mathematical proof to validate an episode's plot twist.

Further Reading

Bly, Robert. *The Science in Science Fiction: 83 SF Predictions that Became Scientific Reality*. Dallas, TX: BenBella Books, 2005.

Chartier, Timothy, and Dan Goldman. "Mathematical Movie Magic." *Math Horizons* 11 (April 2004).

Gouvêa, Fernando. "As Others See Us: Four Science Fictional Mathematicians." *Math Horizons* 11 (April 2004).

Kasman, Alex. "Mathematical Fiction." http://kasmana.people.cofc.edu/MATHFICT/.

———, "Mathematics in Science Fiction." *Math Horizons* 11 (April 2004) http://kasmana.people.cofc.edu/MATHFICT/sf-mathhoriz.pdf.

Raham, Richard. *Teaching Science Fact With Science Fiction*. Santa Barbara, CA: Libraries Unlimited, 2004.

Shubin, Tatiana. "A Conversation With Rudy Rucker." *Math Horizons* 11(April 2004).

SIMONE GYORFI

See Also: Geometry of the Universe; Interplanetary Travel; Literature; Mathematics, Elegant; Movies, Mathematics in; Universal Language; Writers, Producers, and Actors.

Sculpture

Category: Arts, Music, and Entertainment.
Fields of Study: Geometry; Representations.
Summary: Mathematics may be necessary to assure the stability of a sculpture and sculptures can represent mathematical concepts in three dimensions.

The word "sculpture" comes from Latin *sculpere*, meaning "to carve." Sculptures can be made from variety of materials, including wood, metal, glass, clay, textiles, or plastic that is carved, cast, welded, cut, or otherwise formed into shapes. Topiary and bonsai are living sculptures. Modern sculptors even experiment with light and sound. Additionally, sculptures may be free-standing objects or appear as reliefs on surfaces like walls.

The Taj Mahal, one of the most recognizable structures on Earth, includes many geometric reliefs. Sculptures can be static or kinetic, like Rube Goldberg contraptions, and projection sculptures change appearance when viewed from different sides. The outdoor Penrose tribar sculpture in East Perth, Australia, appears to be the illusory figure developed by Roger Penrose when viewed from the correct angle. While mathematical forms have long been used to create sculpture, mathematicians have come to embrace this incredibly flexible art form to investigate many mathematical concepts that might otherwise be difficult to visualize. Many mathematical sculptures are quite aesthetically pleasing in addition to being highly functional in clarifying and representing mathematical ideas. Displays of mathematical sculptures are now a regular part of many art exhibitions and mathematics conferences.

Mathematical Sculptures

Researchers who explore higher degrees of dimensionality often find it challenging to represent these concepts to people whose everyday perception is three-dimensional. Mathematician Adrian Ocneanu's work includes modeling regular solids mathematically and physically. His "Octatube" sculpture, on display in Pennsylvania State University's mathematics building, maps a four-dimensional space into three dimensions using triangular pieces bent into spherical shapes. "Octatube" is conformal; the angles between faces and the way the faces meet are uniform. It was sponsored by Jill Grashof Anderson, whose husband was killed on September 11, 2001. Both graduated with mathematics degrees in 1965. Mathematician Nigel Higson said, "For professionals the sculpture is very rich in meaning, but it also has an aesthetic appeal that anyone can appreciate. In addition, it helps to start conversations about abstract mathematical concepts—something that is generally hard to do with anyone other than another expert."

Other concepts explored by mathematical sculptures include minimum variation surfaces, such as spheres, toruses, and cones, which humans tend to judge to be aesthetically pleasing because of their constant curvature; zonohedra, a class of convex polyhedra with faces that are point-symmetrical polygons, such as parallelograms; and Möbius loops, Klein bottles, and Boy's surfaces, named for mathematicians August Möbius, Felix Klein, and Werner Boy. Sculptures on exhibit at the Fermi National Laboratory, like "Monkey-Saddle Hexagon," focus in part on saddle-shaped minimal surfaces.

Mathematicians Who Sculpt

Art and mathematics have been intertwined for centuries and many historical sculptors such as Leonardo DaVinci were also mathematicians. Cubist sculptors explored many new perspectives on dimension and geometry. Spouses Helaman and Claire Ferguson have created and written extensively about mathematical sculpture. Helaman developed the PLSQ algorithm for finding integer relationships, considered by many to be among the most important algorithms of the twentieth century. He creates his award-winning sculptures to represent mathematical discoveries, and the pair's worldwide presentations have been praised for their accessibility and for initiating dialogue among multiple disciplines.

George Hart, another mathematician-sculptor, has worked in fields like dimensional analysis. He regularly hosts "sculptural barn raisings," where people are invited to help assemble large mathematical sculptures and discuss their properties. This includes a traveling sculpture for use at schools and conferences. Hart also uses rapid prototyping technology for mathematics and sculpture work. In 2010, he left Stony Brook University to be chief of content at the interactive Museum of Mathematics, with an opening date of 2012.

An "impossible" Penrose triangle sculpture built in 2008 in Gotschuchen, Austria, by a physics association.

Computer-Generated Sculpture

Self-taught artist and mathematician Brent Collins and computer scientist Carlo Séquin created their Fermi mathematical sculpture exhibit as part of a prolific ongoing collaboration. Séquin started researching geometric modeling in the early 1980s and Collins created saddle-form sculptures during the same period, though he only later learned their mathematical names. The Séquin-Collins Sculpture Generator combines the aesthetics of sculpture, mathematical theory, and computer visualization to allow sculptors to rapidly prototype and refine ideas electronically before beginning to work in their chosen medium. A designer can move around and through the model as well as slice and transform it. Some consider the computer images themselves to be "virtual sculpture." In contrast, some sculptors see computer modeling as too restrictive on the symbiotic processes of design and implementation. Some directions of mathematical sculpture include knots, three-dimensional tessellations, surfaces defined by parametric equations, fractal structures, and models of complex natural entities such as organic molecules.

Other Representations and Projects

The Hyperbolic Crochet Coral Reef project combined mathematics and marine biology to call attention to global warming and other environmental issues using three-dimensional crocheted sculptures of reef lifeforms. Artists create reef components using iterative patterns, which can be permuted to produce a broad variety of lifelike designs. The project is an extension of the hyperbolic crochet work pioneered by mathematician Daina Taimina, who demonstrated that hyperbolic surfaces can be modeled physically.

Some mathematically themed sculptures represent the connections between mathematics and other aspects of society rather than trying to model explicit mathematical concepts. Oakland University's Department of Mathematics and Statistics has a sculpted ceramic mural called *Equation*, which was created to explain the development of mathematics and its relationship to the universe and humanity. Though not a mathematician, artist Richard Ulrish stated that he has fond memories of the mathematics courses he took at Oakland.

Further Reading

Abouaf, Jeffrey. "Variations on Perfection: The Sequin-Collins Sculpture Generator." *IEEE Computer Graphics and Applications* 18 (November/December 1998).

Ferguson, Claire, and Helaman Ferguson. *Helaman Ferguson: Mathematics in Stone and Bronze.* Erie, PA: Meridian Creative Group, 1994.

George W. Hart. "Geometric Sculpture." http://www.georgehart.com/sculpture/sculpture.html.

Hyperbolic Crochet Coral Reef. http://crochetcoral reef.org.

Peterson, Ivars. *Fragments of Infinity: A Kaleidoscope of Math and Art.* Hoboken, NJ: Wiley, 2001.

MARIA DROUJKOVA

See Also: Crochet and Knitting; Escher, M.C.; Painting; Surfaces; Visualization.

Search Engines

Category: Communication and Computers.
Fields of Study: Algebra; Data Analysis and Probability; Number and Operations.
Summary: Using complex and sometimes proprietary algorithms, search engines locate and rank requested information, usually on the Internet or in a database.

Search engines are used for finding information from digitally stored data. Based on a search criterion like a word or phrase, search engines find information from the Internet and personal computers and present search results appropriately. A search engine is a very efficient tool for effortless finding information from millions of Web sites and their Webpages. For example, information on movies or weather forecast from the Internet can be easily found using search engines. To sort through vast amount of data, search engines use statistics, probability, mathematics, and data analysis.

Types of Search Engines

Different types of search engines are developed for different purposes. The simplest one is a desktop search engine, which is used for finding information stored within a computer. An enterprise search engine searches for digitally stored information within only one organization. A Web search engine looks for information on the World Wide Web (WWW). Sometimes, federated search engines are used for searching online databases or related items. Though there are different types, the term "search engine" generally refers to Web search engines.

Search Mechanisms

Searching for a word or phrase in a document in a computer is very simple and sophisticated search engines are not needed for this. A program simply reads the whole or selected part of the document, looks for where the intended word or phrase is located, and highlights the locations in the document.

Desktop search engines perform more complicated searches. These engines read all files and folders kept in the computer to collect information and index them. Indexing is a method of storing information about files and folders considering several factors like file names, contents, types, authors, and locations of files. It uses mathematical manipulations involving numbers, operations, and data mining. Once indexing is finished, the engine follows that index for searching. For example, if the word *algebra* is searched in a computer, the engine reads the index and tries to find out where the word *algebra* is located (if anywhere), and it shows the resulting files or folders.

The most complicated and interesting search engines are Web search engines. The Web contains billions of Web pages, and each page contains information. These search engines search for information from almost all of them. These engines generally work in three major steps: (1) collecting information from the Web, (2) indexing, and (3) presenting search results.

For reading Webpages and collecting information, almost all Web search engines have their own computer program, often called a "crawler." A Web search engine may have one or more crawlers. The information collected by crawlers contains subject matters, hyperlinks, images, and other information. Next, the search engines index the collected data and store them for future retrieval. The index is like a giant catalogue and involves huge mathematical applications to prepare. When a search criterion is given for searching, search engines follow this index; they find which Webpages contain the information and present results as lists of links to those pages.

A challenging task for Web search engines is to present the search results properly and quickly. While showing the results, it is expected that the more relevant pages corresponding to the search criterion should appear earlier than less relevant pages. Different search engines have different algorithms for arranging pages based on relevance. For example, the Google search engine uses an algorithm called PageRank for this purpose. It uses probability, data analysis, matrix algebra, and related fields.

Examples of Search Engines

Web search engines began to be developed in the 1990s and are constantly improving to handle the increasing size and content of the Web. Many of the individuals who develop and refine search engines have degrees in mathematics. Popular search engines like AltaVista (launched in 1995), Google (1998), Yahoo Search (2004), and Bing (2010) are only a few examples. Google Desktop, GNOME Storage, Windows Search, and Easyfind are among the most popular desktop search engines, while OpenSearchServer and DataparkSearch are good examples of enterprise search engines.

Further Reading

Levene, Mark. *An Introduction to Search Engines and Web Navigation*. London: Pearson, 2005.

Voorhees, E. M. *Natural Language Processing and Information Retrieval*. New York: Oxford University Press, 2000.

SUKANTADEV BAG

See Also: Data Analysis and Probability in Society; Data Mining; Internet; Matrices; Number and Operations; Probability; Randomness; Statistics Education.

Segway

Category: Travel and Transportation.
Fields of Study: Algebra; Calculus; Geometry; Measurement; Number and Operations.
Summary: The Segway is a personal transporter built on the principle of dynamic equilibrium.

The Segway is an electric, two-wheeled personal transportation device that utilizes principles of balance and equilibrium both to create and control its motion. The Segway transporter was developed in part to combat the congestion and pollution caused by automobiles. In many cities, Segway tours are now alternatives to walking or bus tours. The Segway is often cited as an it will stand upright on top of the cart. However, this condition is unstable; if the pole is moved away from this resting position, it falls.

An interesting property about the inverted pendulum (or cart-and-pole) problem is that as long as the base, or cart, is resting, the upright position is unstable. However, if the base or cart is in motion, oscillating at the right frequency, the upright position becomes stable. Imagine that the cart is moving forward and backward ever so slightly and very rapidly; in this case, the pole can remain upright. Now, the pole is in a dynamically stable position. This type of motion-induced stability is similar to what happens as humans walk. If an individual leans forward with his or her feet firmly planted on the ground, the individual will fall. However, if the feet are allowed to move, the individual will not fall but instead will move forward (or backward, depending on the direction of the lean). Allowing the feet to move has made the leaning position dynamically stable. With the feet moving, it is much harder for the individual to fall.

Dynamic Equilibrium

The Segway transporter operates on this principle of dynamic equilibrium. Riders lean forward to cause the wheels to move forward and lean back to cause the Segway to stop or reverse. The wheels and base are dynamically moving to keep the rider in an upright position instead of falling to the ground. Balance sensors in the base of the Segway regulate and control the motion by incorporating the pitch angle (or tilt) of the rider, the change in pitch angle, the wheel speed, and the wheel position. Mathematicians, physicists, and engineers relate all these variables through differential equations describing motion; these equations have long been studied in each of these fields. The Segway transporter is one example of a project resulting from the interplay of all three fields.

Riders lean forward to make the wheels move forward and lean back to stop the Segway.

application of a classical dynamical systems problem: the inverted pendulum problem. It can also be referenced in illustrating the phenomenon of dynamic stability that also occurs in human walking.

Inverted Pendulum

In a traditional pendulum problem, the pendulum is composed of a mass attached to a string that is itself attached to a pivot point. In this case, the mass hangs below the pivot point. The position in which the mass hangs below the pivot point is stable—the pendulum eventually returns to that position even if pushed away from that position. In fact, it is relatively easy for the pendulum to rest in this equilibrium position. In an inverted pendulum problem, the situation in which the mass is above the pivot point is considered. Frequently, one can visualize this scenario as a "cart and pole." With the cart at rest, if the pole is perfectly positioned,

Further Reading

Kalmus, Henry P. "The Inverted Pendulum." *Journal of Physics* 38, no. 7 (1970).

Kemper, Steve. *Code Name Ginger: The Story Behind Segway and Dean Kamen's Quest to Invent a New World.* Cambridge, MA: Harvard Business School Publishing, 2003.

Tweney, Dylan. "Dec. 3, 2001: Segway Starts Rolling." *Wired* (December 3, 2009). http://www.wired.com/thisdayintech/2009/12/1203segway-unveiled.

Vasilash, Gary S. "Learning From Segway: Innovation in Action." *Automotive Design & Production* (January 2006).

Angela Gallegos

See Also: Mathematical Modeling; Mathematics, Applied; Trigonometry.

Sequences and Series

Category: History and Development of Curricular Concepts.
Fields of Study: Algebra; Communication; Connections; Number and Operations.
Summary: Sequences and series are important mathematical representations with numerous, interesting applications.

A sequence is a list of objects, called "terms," arranged in a fixed pattern such as 1, 3, 5, 7, 9, . . . or Monday, Tuesday, Wednesday, Thursday, Friday, ... In a series, the terms of a sequence are typically added together. Series have a long history of being used to approximate functions or represent geometric quantities. For example, in the seventeenth century, James Gregory showed how the areas of a circle and hyperbola could be obtained using series. In the early days of calculus, series represented geometric quantities and were manipulated using methods extended from finite procedures. Mathematicians like Niels Abel critiqued the rigor of series and expressed concerns with the foundations of calculus. The theory of series was later made rigorous within the field of analysis. Series are important to many areas in science and engineering. Sequences are explored in the primary and middle grades, while series are introduced in high school.

Famous Sequences

One very famous sequence emerges when considering the reproductive habits of rabbits. Consider two rabbits that are too young to reproduce after their first month of life but can and do reproduce after their second month of life. That pair of rabbits produces another pair after its second month and for each month there-after. If one assumes that none of the rabbits die and that each pair reproduces in the same manner as the first, the number of pairs of rabbits at the end of each month corresponds to the elements of the sequence 1, 1, 2, 3, 5, 8, 13, 21, 34, ... This sequence is known as the "Fibonacci sequence." It is named after Leonardo de Pisa who was called Fibonnaci, a nickname meaning "son (*filius*) of Bonaccio." He wrote about it in his 1202 book *Liber Abaci*. With the exception of the first two terms, each successive term is found by adding the two terms prior to it. This sequence appears in nature in other situations, including the arrangement of leaves on the stems of certain plants, the fruitlets of a pineapple and the spirals of shells. Some mathematical historians suggest that a Fibonacci-like sequence of integers is also represented in stone balance weights excavated in the 1960s that originated in the eastern Mediterranean during the Late Bronze Age.

Other specific types of sequences have been explored. In 1940, Pavel Aleksandrov introduced a concept called "exact sequences," which found relevance in a wide variety of mathematical fields. In 1954, Jean-Pierre Serre was awarded a Fields Medal, the most prestigious award in mathematics, in part because of his work on spectral sequences.

Series

A series is often the sum of the terms of a sequence. Series originate as early as the Indian mathematician and astronomer Brahmagupta who gave rules for summing series in his 628 c.e. work *Brahmasphutasiddanta* (*The Opening of the Universe*). The sum of the terms of an arithmetic sequence is called an *arithmetic series*. The arithmetic series $1 + 2 + 3 + 4 + \cdots + 97 + 98 + 99 + 100$ is also a well known one, as it is related to mathematician Carl Friedrich Gauss (1777–1855). At a very young age (around 6 years old), Gauss found the sum of the natural numbers (1, 2, 3, 4, …) from 1 to 100. That is, the sum given by the series $1 + 2 + 3 + 4 + \cdots + 97 + 98 + 99 + 100$. He was given this task by his teacher to keep him busy while the teacher worked with the other students in the class who were not as mathematically gifted as Gauss. After a relatively short time, Gauss returned to the teacher with the sum 5050. Gauss's method was to pair up the terms of the series. Taking the sum of the first and last term $(1 + 100)$ yields 101. This is the same as the sum of the second and second to last $(2 + 99 = 101)$, the third and third to last $(3+98+101)$, and so forth. In all, there are 50 such pairs, each of which sums to 101. Thus,

$$1+2+3+4+\cdots+97+98+99+100$$
$$=(50)(101)=5050.$$

Many mathematicians advanced the theory of important series such as power series, trigonometric series, Fourier series, and time series. For example, Nicholas Mercator represented the function $\log(1+x)$ as a series in 1651. Taylor series, named after Brook Taylor, is a representation of a function as an infinite sum of terms calculated from the values of its derivatives at a single point. From the early history of analysis, these power series were important in the study of transcendental functions. Data given as a sequence of data points over time led Wilhelm Lexis to develop time series in 1879.

Applications of Series

Some other series arose in the context of questions related to physics and sparked controversy. The mathematics and physics of a vibrating string and solutions of the wave equation led to trigonometric series. Daniell Bernoulli, Jean Le Rond d'Alembert, Leonhard Euler, and Joseph-Louis Lagrange debated the nature of trigonometric series in the eighteenth century. Joseph Fourier developed Fourier series for the heat equation in the nineteenth century, which was criticized at the time because it contradicted a theorem by Augustin-Louis Cauchy but was explored more rigorously by Johann Dirichlet. An overshoot or ringing in Fourier series was first observed by H. Wilbraham and later explored by Josiah Gibbs. The Gibbs phenomenon has implications in signal processing. The three-body problem, which investigates the behavior and stability of three mutually attracting orbiting bodies in the solar system, was solved by Delaunay in 1860 via representing the longitude, latitude, and parallax of the moon as an infinite series.

However, in 1892, Jules Henri Poincaré showed that these and similar solutions were not in general uniformly convergent, and this criticism created doubt about proofs of the stability of the solar system and eventually led to the formation of the field of deterministic chaos. A prize was offered by King Oscar II of Sweden for a solution to the extension of the three-body problems to n bodies. It has since been proven

that no general solution is possible, but the n-body problem was also connected to series in Quidong Wang's 1991 work.

Series were also important as mathematicians searched for efficient ways to represent π and find its digits. Keralese mathematician Madhava of Sangamagramam may have been the first when he used 21 terms of a series and stated π correctly to 11 places. In the 1800s, William Shanks used a series to calculate digits of π in the morning and check them in the evening. He calculated 707 digits of π using this method. However, there was a suspicious lack of the number "7" in the last digits, and it was later found that only the first 527 digits were correct. Johann Lambert used the same series to show in 1761 that π must be irrational—it cannot be expressed as a ratio of whole numbers and has an infinite, non-repeating decimal expansion. Srinivasa Ramanujan found series that converged more rapidly than others, and these efficient series were used as the foundations of computer algorithms.

Binary Series

A very famous series is the binary series that consists of powers of 2: $2^0 + 2^1 + 2^2 + 2^3 + 2^4 + 2^5 + \cdots$. It is theorized that the King of Persia, finding himself very bored, asked that a game be invented for his amusement. The inventor of the game the king found most enjoyable would be given a reward. A servant of the king created the game of chess that was most pleasing to the king. When asked what prize he would like, the servant replied that he wanted grains of rice. The chessboard consists of 64 small squares. As a reward the servant asked for 1 grain of rice for the first square, 2 for the second square, 4 for the third square, 8 for the fourth square and so forth, until all 64 squares had been accounted for. The number of grains of rice requested is the sum $2^0 + 2^1 + 2^2 + 2^3 + 2^4 + 2^5 + \cdots 2^{63}$, and it amounts to 274,877,906,944 tons of rice, which is more rice than has been cultivated on Earth since recorded time. The story goes that the king grew furious at the servant once he knew what was requested. The servant was taking the rice as he received it and distributing it among the poor. At some point, the king indicated that he did not have the rice to pay the servant. The servant indicated that he was content with the amount that he had already received and that it was the king who offered a reward not he who made the initial request. Both parties were pleased.

Applications in Economics

Series appear in many other contexts as well. For example, the future value of an ordinary annuity can be found using a series. An ordinary annuity is an account where an individual makes identical deposits on a regular schedule. The money in the account earns interest that is compounded with the same frequency as the deposits. Suppose an individual deposits $100 every year into an account that earns 6% interest annually. Three years later, the first year's deposit has earned interest over two years, the second account over one year, and the last deposit not at all. The money in the account after three years is given by: $100(1.06)^0 + 100(1.06)^1 + 100(1.06)^2$. The general series can be expressed as a single number

$$A = P\left(\frac{(1+i)^n - 1}{i}\right)$$

where A is the future value of the annuity, P is the payment made at the end of each period, i is the interest rate per period, and n is the number of periods.

Limits

Though infinite sequences consist of infinitely many terms, it may be the case that the sum of the terms of such sequences converges on a given value. Such is the case of the geometric series $.9 + .09 + .009 + .0009 + \cdots$. In this series, the first term is .9 and the common ratio is

$$\frac{1}{10}$$

Applying the formula for the sum of the first n terms of the series yield

$$S_n = \frac{.9\left(1 - \frac{1}{10}^n\right)}{1 - \frac{1}{10}} = \left(1 - \frac{1}{10}^n\right).$$

As the number of terms approaches infinity (as $n \to \infty$), the fraction

$$\left(\frac{1}{10}\right)^n$$

becomes so small that one may consider it zero. Therefore,

$$S_n = \left(1 - \frac{1}{10}^n\right) = 1 - 0 = 1$$

as n grows infinitely large. Since

$$S_n = .9 + .09 + .009 + .0009 + \cdots = .\overline{9}$$

one arrives at the very famous result that $.\overline{9} = 1$.

Further Reading

Ferraro, Giovanni, and Marco Panza. "Developing Into Series and Returning From Series: A Note on the Foundations of Eighteenth-Century Analysis." *Historia Mathematica* 30, no. 1 (2003).

Klein, Judy. *Statistical Visions in Time: A History of Time Series Analysis, 1662–1938*. Cambridge, England: Cambridge University Press, 1997.

Kline, Morris. *Mathematical Thought From Ancient to Modern Times*. New York: Oxford University Press, 1972.

Laugwitz, Detlef. *Bernhard Riemann, 1826–1866: Turning Points in the Conception of Mathematics*. Boston: Birkhäuser Boston, 2008.

LIDIA GONZALEZ

See Also: Archimedes; Functions; Limits and Continuity; Numbers, Complex.

Servers

Category: Communication and Computers.
Fields of Study: Algebra; Communication; Number and Operations.
Summary: Servers help users connect to networks, including the Internet.

ARPANET, the first network of time-sharing computers, was connected in 1969. In subsequent decades, technology developments and the increasing benefits of distributed, shared access spurred network growth, ultimately resulting in the Internet and World Wide

Web. Most local, national, and global networks rely on servers, which manage network resources for client computers that are connected to it. A server may be a physical computer, a program, or a combination of hardware and software. In some cases, a system is a dedicated server. In other cases, software servers operate on multipurpose systems. A distributed server is a scalable grouping in which several computers act as one entity and share the work. In general, a network server manages overall network traffic, while specialty servers handle other tasks. CERN httpd (or W3C httpd), which debuted in 1990, is considered to be the first Web server. It was developed by scientists Tim Berners-Lee, Ari Luotonen, and Henrik Frystyk Nielsen at the European Organization for Nuclear Research (CERN). Servers and clients use communication protocols to exchange information to carry out tasks. There are server-to-server and client-server variations. Mathematicians, computer scientists, and others work to create technology and algorithms that make servers possible and increase their efficiency. They also study the properties of networks and servers, which facilitates advances in both mathematics and computers. For example, in a system with multiple parallel servers, jobs may be assigned to any server. Often, jobs are modeled with an exponentially distributed processing time or some other probabilistic distribution with some resource cost per unit of time. Mathematical methods may be used to find the optimal strategy for allocating jobs to servers to minimize costs.

Function

The term "server" does not describe a specific type of computer in the same sense that "desktop" or "Windows machine" does. When used in reference to hardware, a server is any computer running a server program, which can—and in practice does—include all configurations and operating systems. Since the 1990s and the increased demand for Internet services, there have been more and more computers that have been designed specifically to be used as Internet servers. Because they need to run for long periods of time without interruption, they must be durable, reliable, and have uninterruptible power supplies. Typically, hardware redundancy is incorporated, so that if a hard drive fails, another one is automatically put on line—a feature rarely found in personal computers. There is also a great deal of server-specific hardware, such as

water cooling systems, which help reduce heat, and Error-Correcting Code (ECC) memory, which corrects memory errors as they happen, preventing data corruption. Many components are designed to be hot-swappable, meaning that they can be replaced while the server runs—without needing to power it down. Furthermore, ordinary server operations including turning the power on or off can often be conducted remotely; for example, from a home computer. Some system operators maintain watch over multiple servers in multiple locations and physically visit the site only when necessary because of a crisis.

Communication

Sockets are the primary means by which network computers in a network communicate. They are the endpoints of the flow of interprocess communication (IPC) and provide application services. They are also the place where many security breaches take place. Mathematicians and computer scientists study the different socket types and their states to understand how they work and to improve function and security. Servers create sockets on start-up that are in listening state, waiting for contact to be made by client programs. For instance, a Web browser, like Firefox, is a client program used to access content from Web servers. Most servers connected to the Internet use a protocol known as Transmission Control Protocol (TCP), developed by computer scientists Vinton Cerf and Robert Kahn for ARPANET. An Internet socket is referred to by its socket number, a unique integer that includes Internet Protocol (IP) address and socket number. Listening sockets using TCP are usually assigned the remote address 0.0.0.0 and the remote port number 0. TCP servers can serve multiple concurrent clients by creating what is called a "child process" associated with each client and establishing TCP connections between child processes and clients. Each connection uses a unique dedicated socket. Two communicating sockets—the local socket created by the server and the remote socket of the client—are called a "socket pair," and their activity is referred to as a "TCP session."

A common feature of Web servers is server-side scripting, which allows Web pages to be created in response to client activity. For instance, a search for a book on Amazon.com results in a unique search results page. Without this capacity, every possible search would need to be conducted in anticipation of client needs.

Further Reading

Chevance, Rene. *Server Architectures: Multiprocessors, Clusters, Parallel Systems, Web Servers, and Storage Solutions.* Oxford, England: Elsevier, 2005.

Dshalalow, Jewgeni H. *Frontiers in Queueing Models and Applications in Science and Engineering.* Boca Raton, FL: CRC Press, 1997.

Gray, Neil A. B. *Web Server Programming.* Hoboken, NJ: Wiley, 2003.

BILL KTE'PI

See Also: Cerf, Vinton; Parallel Processing; Personal Computers; Wireless Communication.

Shipping

Category: Business, Economics, and Marketing.
Fields of Study: Geometry; Measurement; Number and Operations.
Summary: A variety of mathematical concepts, including packing, routing, and tracking, are necessary to make the process of shipping goods more efficient.

The shipping and delivery industry is a vast global business that is responsible for delivering packages, postal mail, and commercial cargo all over the world. In 2009 alone, express delivery companies made $130 billion in revenue worldwide, the U.S. Postal Service delivered 177 billion pieces of mail, and ocean liners transported more than $4.6 trillion worth of goods between nations. With so many items being delivered to so many different places, there is a need for mathematics to help manage the complex delivery network and ensure that deliveries are made correctly, safely, cheaply, and quickly. Mathematics has had a significant impact in three key areas of the shipping industry: container packing, vehicle routing, and package tracking.

Container Packing

To minimize transportation costs and maximize profit, a shipper would naturally prefer to pack cargo into as few shipping containers as possible. Determining the optimal way to arrange items in a container is a decep-tively difficult problem. Given a set of differently sized objects, the Bin-Packing Problem is to find the order in which to place the objects so that they fill the minimum number of bins. Testing every permutation of packing the objects would be too time-consuming, so an efficient and simple algorithm is required.

A common packing procedure is the First-fit algorithm, where the objects are ordered from largest to smallest, and each object is placed in the first available bin that will hold it. It can be proven mathematically that this algorithm is not guaranteed to produce the optimal packing. In the worst case, the result can be far from optimal and require the use of more bins than a more sophisticated packing. The First-fit algorithm is an example of an approximation algorithm, which means it produces a good approximate answer but not necessarily the optimal arrangement of objects. Other more sophisticated bin-packing algorithms have been developed, but as of 2010, no efficient algorithm was known that always produced the optimal packing.

In practice, there are more considerations to packing shipping containers. Some packages will be irregularly shaped and do not stack well. Some cargo is fragile

NP-Complete Problems

The NP-complete set is a list of mathematical problems for which there is no known fast algorithm for solving the problem exactly. The Traveling Salesman Problem is an example of a NP-complete problem. While there are fast algorithms for finding a good answer, the only known algorithm for finding the single shortest route is extremely slow. However, just because there is no known fast algorithm for solving these problems, it does not mean that such an algorithm does not exist. In 2000, the Clay Mathematics Institute offered a $1 million prize to anyone who could devise an algorithm that would solve an NP-complete problem quickly or prove that no such algorithm exists. While not technically an NP-complete problem, the Bin-Packing Problem is in a related category of problems known as "NP-hard."

and must be secured separately. Sometimes, a delivery vehicle will make several stops, so the packages that are delivered first should be packed into a container last to make them easily accessible.

Through World War II, most cargo was shipped in wooden crates of various sizes. A big step forward came in 1956, when trucker Malcolm McLean patented the modern shipping container made of corrugated steel. This sturdy container was easier to move between truck, rail, and ocean liner. More importantly, having a standard-size container meant that packing procedures could be standardized. Prior to 1956, it was estimated that loose cargo cost $5.86 per ton to load. After the standardized container was introduced, it was estimated the loading cost dropped to 16 cents per ton, a 3600% improvement.

Vehicle Routing

Cargo travels by a variety of transportation modes, including truck, rail, air freight, and ocean liners. The goal of routing is to determine a vehicle for each piece of cargo to be delivered and then find the shortest delivery route for each of the vehicles. The Traveling Salesman Problem is a simple mathematical example of a routing problem. In practice, the value of a route is not determined by just the distance. The problem is complicated by considerations such as personnel, fuel costs, traffic, tolls, and tariffs.

Mathematical analysis of delivery routes can lead to huge improvements in shipping efficiency. As the first Postmaster General of the United States, Benjamin Franklin ordered careful surveying of delivery routes, refined the post office accounting practices, and increased public access to mail. Under this new system, the U.S. Postal Service became profitable for the first time, and it is estimated that the mail delivery time between major cities was cut in half.

The routing problem is an example of a problem studied in operations research, the branch of mathematics that studies the cost-effectiveness of decisions made by corporate management such as scheduling and personnel assignments. The field of operations research has its origins in World War II, when the Allied Forces were interested in coordinating the manufacturing and organization needed to mobilize the military. One of the early researchers in operations research was Tjalling Koopmans, who proposed a mathematical model for the routing problem for shippers.

Tjalling Koopmans (1910–1985)

Tjalling Koopmans was a Dutch economist who helped develop the mathematical field of operations research. Working for the British Merchant Shipping Mission in the 1940s, Koopmans derived a mathematical model for finding the most cost-effective shipping routes. Later, he became a professor of economics at University of Chicago and then at Yale University. In 1975, Koopmans received the Nobel Prize for Economics for developing mathematical tools for the analysis of corporate management and efficiency.

Package Tracking

It is important for a shipper to carefully track a package until it reaches its destination. A common system for identifying a package is the barcode. By encoding the destination as a sequence of black and white bars, the packages can be sorted quickly by automated sorting machines equipped with laser scanners. The U.S. Postal Service has developed a special barcode that encodes the address as a sequence of short and tall black bars. The mail is first read by an Optical Character Recognition (OCR) program, which translates the handwritten address into a barcode. The barcode is stamped onto the package and then automatically sorted to be sent to the next distribution center.

Radio-frequency identification (RFID) is a tracking technology that could potentially have a large impact on the shipping industry. A small electronic tag that emits a radio signal would be placed on each item to be shipped. Generally, this tag is a microchip just a few millimeters on a side. Potentially, this microchip would allow a shipper to determine the entire contents of a shipping container without ever opening the container. However, the technology still needs to be refined to make RFID a cheaper alternative to the barcode. Furthermore, since an item could theoretically still be tracked after the delivery is made, RFID technology is somewhat controversial because of privacy concerns.

Further Reading

Hillier, Frederick. *Introduction to Operations Research.* New York: McGraw-Hill, 2009.

Lodi, Andrea, Silvano Martello, and Daniele Vigo. "Recent Advances on Two-Dimensional Bin Packing Problems." *Discrete Applied Mathematics* 123 (2002).

Palmer, Roger. *The Bar Code Book.* Peterborough, NH: Helmers, 2007.

Roberti, Mark. "The History of RFID Technology." *RFID Journal* (2005).

TODD WITTMAN

See Also: Bar Codes; Scheduling; Traveling Salesman Problem.

Similarity

Category: History and Development of Curricular Concepts.
Fields of Study: Communication; Connections; Geometry; Measurement.
Summary: The concept of mathematical similarity has been studied since antiquity.

The concept of "similarity" is universal, playing a particularly large role in the field of geometry. In general, objects may be called "similar" if they share features that look alike, such as shape, color, or value. However, it is a much stronger statement to say that two objects are "mathematically similar." Similarity can be a powerful simplifying assumption in modeling situations. Scaling an object appears in many applications, such as in architecture. Scaling notions can also explain the speed of a hummingbird's heartbeat as compared to a human heart, and why certain insects would collapse under their own weight if they were scaled to a large size. Julian Huxley asserted that the evolutionary struggle to maintain similar surface-to-volume relationships is important in anatomy. Recognizing a similar object is also important. Logician and philosopher Willard Van Orman Quine felt that learning, knowledge, and thought all require similarity so that humans can order objects into categories with similar meaning. Similarity is often connected to triangles in mathematics, starting in grades three through five, but there are many other mathematical situations where it is also useful, such as in the definition of trigonometric functions, in axiomatic arguments, in matrices, in analysis of differential equations, and in fractals.

Early History

Distance calculations contributed to the development of similarity. Thales of Miletus is said to have measured the height of a pyramid using its shadow, but historians are unsure of the method that he used. A method that makes use of similar triangles is attributed to Thales by Plutarch of Chaeronea. In classrooms in the twentieth and twenty-first century, similar experiments are conducted. By measuring the length of the shadow of a tall object, like a pyramid, tree, or building, at the same time as measuring the length of a shadow of a known meter or other stick, a proportion with similar right triangles can be formed. The method assumes that light rays are parallel. In ancient China, instruments such as the L-shaped set-square or gnomon also needed similar triangles. In chapter nine of the *Nine Chapters on the Mathematical Art*, problems were posed and solved using similarity concepts. One of the problems has been translated as

> There is a square town of unknown dimensions. There is a gate in the middle of each side. Twenty paces outside the North Gate is a tree. If one leaves the town by the South Gate, walks 14 paces due South, then walks due West for 1775 paces, the tree will just come into view. What are the dimensions of the town?

Many other mathematicians have worked on a variety of similarity concepts and applications. In Euclid of Alexandria's *Elements*, the various definitions of similarity depend on the figure being examined. Apollonius of Perga explored the similarity of conic sections. During the seventeenth and eighteenth centuries in China, the proportionality of corresponding sides of similar triangles in the plane was quite useful in solving problems in spherical trigonometry. In some twenty-first-century college classrooms, students explore the reason why spherical triangles with shortest distance paths and the same angles must be congruent—there is no concept of similarity on a sphere. Mathematics educators also study the conceptual difficulties in teaching and learning similarity.

Other concepts of similarity arose from mechanics concerns. In his work on the equilibrium of the plane, Archimedes of Alexandria postulated that plane figures that are similar must have similarly placed centers of gravity. Galileo Galilei tried to generalize the notion of geometric similarity to mechanics. Isaac Newton, Hermann von Helmholtz, Joseph Fourier, James Froude, Osborne Reynolds, Lord Rayleigh (John Strutt), and others also worked on variations of similarity in physical situations. Building on their work, and motivated by the lack of a theoretical foundation for flight research, Edgar Buckingham articulated a formal basis for mechanical similarity in 1914. Aside from physical applications, in computer graphics, transformations that preserve similarity can be used to scale mechanical and dynamical behavior in addition to static images.

Further Reading

Fried, Michael. "Similarity and Equality in Greek Mathematics." *For the Learning of Mathematics* 29, no. 1 (2009).

Lodder, Jerry. "Proportionality in Similar Triangles: A Cross-Cultural Comparison." *Convergence*, 2008. http://mathdl.maa.org/mathDL/46/.

Sterrett, Susan. *Wittgenstein Flies A Kite: A Story of Models of Wings and Models of the World*. New York: Pi Press, 2005.

SARAH J. GREENWALD
JILL E. THOMLEY

See Also: Archimedes; Digital Images; Matrices; Transformations.

Six Degrees of Kevin Bacon

Category: Friendship, Romance, and Religion.
Fields of Study: Algebra; Geometry; Number and Operations.
Summary: Concepts from graph theory help explain the idea that people, including actor Kevin Bacon, are surprisingly closely connected with each other.

Six degrees of Kevin Bacon is an example of a network showing a high level of interconnection, known as the "small world" phenomenon. In the language of graph theory applied to films, nodes are film actors, and two nodes are connected by an edge if the corresponding actors have appeared together in a film. It is also a game that tests cinematic knowledge. The task is to find the shortest connection between a given actor and Kevin Bacon. For example, John Wayne is two connections from Kevin Bacon. They were never in a film together, so the distance is greater than one. John Wayne starred with Eli Wallach in *How the West Was Won*, and Eli Wallach starred with Kevin Bacon in *Mystic River*, establishing a shortest distance of at most length two.

The idea of quantifying distance by interpersonal connections dates at least to a 1929 short story called *Chain-Links* by the Hungarian writer Frigyes Karinthy, wherein the narrator determines a five-step connection between a riveter at the Ford Motor Company and himself. Almost 40 years later, the social psychologist Stanley Milgram, best known for his experiments on obedience to authority, devised an experiment to quantify interpersonal connections empirically. Letters were given to some 300 participants, each charged with forwarding the letter to an acquaintance who should move the letter toward the intended recipient. Writing in 1969 with Jeffrey Travers, Milgram stated, "The mean number of intermediaries observed in this study was somewhat greater than five; additional research (by Korte and Milgram) indicates that this value is quite stable." Rounding up, this value became the popular notion "six degrees of separation"—that any two people on the planet are connected by six links. It served as the title of John Guare's 1990 play and 1993 movie about the confidence man David Hampton. In the play, a character speaks to the audience, "Six degrees of separation. Between us and everybody else on this planet. The President of the United States. A gondolier in Venice. Fill in the names. I find that A) tremendously comforting that we're so close and B) like Chinese water torture that we're so close." Exactly how close people are is something sociologists continue to debate, since the nodes and edges of this network are not precisely known.

Mathematics Networks

There are large networks where the nodes and connections are exactly known, allowing for precise analysis. In a collaboration network, nodes are researchers, and two nodes are connected by an edge if the corresponding researchers worked together on a published paper. As early as 1957, mathematicians determined their "Erdös numbers," the collaboration distance from Paul Erdös, the most prolific mathematician of recent years, with some 1500 published research papers and more than 500 collaborators. For instance, the author never wrote a paper with Erdös, but Robin Wilson wrote a paper with Erdös in 1977, and the author wrote a paper with Robin Wilson in 2004, so the author's Erdös number is two. The American Mathematical Society's MathSciNet electronic publication computes the "collaboration distance" between any two authors in its database of some 500,000 authors and 2.5 million publications.

Kevin Bacon

Film Networks

Of more interest to the general public than mathematicians and their papers, the Internet Movie Database (IMDb, found at imdb.com) includes over 1 million actors around the world and some 250,000 films from the 1890s to titles in production. The Web site Oracle OfBacon.org accesses the IMDb and determines the shortest link between any two actors. The network is very tightly connected; it is surprisingly difficult to name any pair of actors even four apart. Consider Kevin Bacon, who has been in over 60 films with over 2200 total co-stars. That is a very small percentage of the total number of actors in the database, but there are over 225,000 actors who, like John Wayne, are co-stars of co-stars of Kevin Bacon. Actors within four links of Kevin Bacon comprise approximately 98% of the database. About 99% of the actors in the IMDb all connect to one another. Finding actors within the last 1% who are five or more from Kevin Bacon is another entertaining part of the game. As of 2010, there are 17 actors with a distance of eight from Kevin Bacon, so that "six degrees" is a misnomer.

Another variant of the game is to determine the actor who is best connected on average. The average every actor's Kevin Bacon number is 2.980. This number means, roughly, that a randomly chosen actor is within three links of Kevin Bacon. It is interesting to consider which sorts of actors have the lowest averages. John Wayne, with significantly more movies and co-stars than Kevin Bacon, has an average of 3.026 links to the rest of the connected actors. The best-connected actor, as of 2010, is Dennis Hopper, with an average distance of 2.772. The IMDb is regularly updated with new actors and films, and the connection data change accordingly.

Why is it six degrees of Kevin Bacon, and not some other actor? The game was created by students at Albright College in January 1994; they had watched *Footloose* earlier in the day, then saw a commercial for another Kevin Bacon film, *The Air Up There*, and a pop culture phenomenon was born. There are similar games based on other large databases, such as baseball players connected by teams, and "six degrees of" remains a very common phrase in society. Kevin Bacon himself used the notion to build a Web-based charity fundraiser, SixDegrees.org. The notion of "small world" networks is being used by scientists in applications as diverse as neural networks of worms, the interconnection of power grids, analysis of the World Wide Web, and genealogical connections.

Further Reading

Grossman, Eric. "The Erdös Number Project." *Oakland University*. http://www.oakland.edu/enp/.

Hopkins, Brian. "Kevin Bacon and Graph Theory." *PRIMUS* 14, no. 1 (2004).

Watts, Duncan. *Six Degrees: The Science of a Connected Age*. New York: Norton, 2003.

BRIAN HOPKINS

See Also: Mathematics Genealogy Project; Movies, Mathematics in; Social Networks.

Skating, Figure

Category: Games, Sport, and Recreation.
Fields of Study: Algebra; Geometry.
Summary: The elements, equipment, and scoring system of figure skating all involve a mathematical framework.

Figure skating is a winter Olympic competitive sport, which involves artistically gliding on ice using metal blades. Ice skating rinks are generally shaped in the form of rectangles with rounded corners. The patterns skaters form on the ice can be explained in geometric terms. Physical principles are observed when watching figure skating. The scoring system used to judge figure skating involves algebraic computations.

Patterns

The bottoms of ice skating blades are not flat, but rather slightly curved, like arcs taken from the edge of a circle about seven to nine feet in radius. This enables the skater to angle and tilt to form patterns on the ice. These patterns can be represented geometrically. For instance, the most famous geometric pattern on ice is a figure eight, which can be formed by two circles of equal size tangential to each other. A skater could start the first circle of the figure eight on the right forward outside edge and skate the second circle on the left forward outside edge. The possible edge combinations include using the left or right foot, traveling forward or backwards, and using the inside or outside edges.

Mathematical Principles of Spinning and Jumping

In addition to basic compulsory figures, modern skating requires participants to execute increasingly difficult jumps and spins. In a jump, the skater's center of gravity follows a parabolic arc with respect to the ice, and a jump is frequently measured in terms of its vertical displacement (the height off the ice) as well as horizontal displacement (the distance). Both are a function of many variables, such as the takeoff angle and velocity immediately prior to the jump.

Spinning, whether in the air as part of a jump or on the ice, is also a complex function of many variables. Factors include the skater's body mass and speed when entering the spin, as well as the extension of the arms or legs from the body. For example, a spinning skater rotates more slowly with extended arms than when the arms are tucked in because as the radius between the body and the arms decreases, the angular velocity increases.

Judging

Four disciplines of figure skating are competitive at the Olympic level: singles (ladies' and mens'), pairs, and ice dance. In each of these disciplines, a choreographed program is skated to music in competition and is judged according to the International Skating Union's International Judging System. The International Judging System awards points for technical difficulty and artistry.

There are many types of skating elements. Jumps vary from their takeoff edges as well as numbers of rotations between one and four. Throw jumps are also performed by the pair teams. A variety of spins are possible, but there are three basic spin positions: upright, camel, and sit. Some spins involve a change of foot, change of position, flying entrance, or difficult variation. Footwork is an element in every program and requires steps and turns that fully cover the ice surface in a circular, straight line, or serpentine pattern. For pairs and ice dance skaters, combination spins, lifts, and other elements requiring two skaters are also scored.

Each of the skating elements performed in a program is assigned a numerical base value, which varies according to difficulty. For example, in the 2010–2011 skating season, the base value of a triple toe loop was 4.1 points, and the base value of the single toe loop was 0.4 points, indicating that the triple toe loop was a much harder jump. Judges add to or subtract from the base value of each element depending upon its execution. For instance, a poorly performed toe loop would receive fewer than 0.4 points. The sum of the values given for each element is called the "technical score."

In addition to a technical score for performance on the individual elements, overall scores for artistic aspects of the program, such as choreography, interpretations, transitions, and skating skills, are awarded as the program components score, which is added to the technical score for a total overall score. The skaters with the highest scores earn the highest rankings.

Further Reading

Carroll, Maureen, Elyn Rykken, and Jody Sorensen. "The Canadians Should Have Won!?" *Math Horizons* 10 (February 2003).

Kerrigan, Nancy, and Mary Spencer. *Artistry on Ice: Figure Skating Skills and Style*. Champaign, IL: Human Kinetics, 2002.

Schulman, Carole. *The Complete Book of Figure Skating*. Champaign, IL: Human Kinetics, 2002.

DIANA CHENG

See Also: Arenas, Sports; Ballet; Connections in Society; Hockey.

Skydiving

Category: Games, Sport, and Recreation.
Fields of Study: Algebra; Calculus.
Summary: Principles of calculus can be used to model a sky dive and to calculate the effect of the parachute on velocity.

Skydiving is the act of leaping out of an airplane at a sufficient altitude and placing your life in the hands of a piece of cloth—although a fairly large piece of cloth. Leonardo da Vinci left drawings of parachutists in his *Codex Atlanticus* circa 1485. The modern parachute was invented by Louis-Sébastien Lenormand in France, making the first public jump in 1783. In 1797, André Garnerin was the first to use a silk parachute, earlier versions being made of linen. The first parachute jump from an airplane was in Venice Beach, California, in 1911. The parachute was held in the arms and thrown out as the jumper left the plane. The soft-pack parachute was developed in 1924. There are two types of parachutes used for skydiving: round, and ram-air (square). The U.S. Army uses the round 35-foot diameter parachute to train its paratroopers because they are reliable and give the jumper a terminal velocity of about 15 feet per second. Most skydivers in the United States started using a 28-foot round canopy. They produced a terminal velocity of about 17–18 feet per second—a somewhat hard landing. The switch to ram-air types came in the 1970s; these give more comfortable landings and maneuverability. The rates of descent vary from canopy to canopy, but terminal velocities usually run from eight feet per second (5.5 mph) to 14 feet per second (9.5 mph).

A canopy's performance is determined by its wing-load, which helps determine the terminal velocity and speed at landing. Most canopies are flown with a wing-load between 0.8 and 2.8 pounds per square foot. To compute the right size of canopy, take the total weight (W) of the jumper and equipment divided by the assigned wing-load factor (WLF):

$$Area_{canopy} = \frac{W_{jumper} + W_{equipment}}{WLF}.$$

To model the parachute jump itself is much more complicated. It involves a first order differential equation to find the speed. The forces on a skydiver are the gravitational force, F_g, and the drag force, F_d, of air resistance and buoyancy. There are two factors to the drag: the time before and the time after the canopy deploys. If x is the distance above the Earth's surface, then $a = dv/dt$ is acceleration and $v = dx/dt$ is velocity. For most jumps, the gravitational force stays essentially constant.

In a first approximation to the problem, take the drag force to be proportional to the velocity. The coefficient of drag has one value when the skydiver is falling and a second value when the parachute is fully deployed. During the fall, the velocity satisfies the initial value problem:

$$m\frac{dv}{dt} = -mg - k_1 v \qquad v(0) = 0.$$

This is a separable ordinary differential equation. Its solution can be found by most students in a calculus class. The jumper's position then is found by integrating the velocity with initial condition that at time $t = 0$ the jumper is at the jump altitude. After the chute deploys, the velocity and position can be found exactly as above, except that the drag coefficient and initial conditions change.

A second approach is to assume that the drag force is proportional to the square of the speed. Then, a falling object reaches a terminal velocity:

$$V_T = \sqrt{\frac{2mg}{\rho A C_d}}$$

where V_T is the terminal velocity, m is the mass of the falling object, g is the acceleration due to gravity, C_d is the drag coefficient, ρ is the density of the fluid through which the object is falling, and A is the projected area of the object.

Based on air resistance, the terminal velocity of a skydiver in a belly-to-Earth free-fall position is about 122 miles per hour (179 feet per second). A jumper reaches 50% of terminal velocity after about three seconds and reaches 99% in about 15 seconds. Skydivers reach higher speeds by pulling in limbs and flying head down, reaching speeds close to 200 miles per hour. The parachute reduces the terminal velocity to the five to 10 miles per hour range. This is achieved by increasing the cross-sectional area and the drag coefficient, lowering the terminal speed.

Further Reading

Meade, Doug. "Maple and the Parachute Problem: Modeling With an Impact." *MapleTech* 4, no. 1 (1997).

Meade, Doug, and Allan Struthers. "Differential Equations in the New Millennium: The Parachute Problem." *International Journal of Engineering Education* 15, no. 6 (1999).

Poynter, Dan, and Mike Turoff. *Parachuting*. 10th ed. Santa Barbara, CA: Para Publishing, 2007.

The United States Parachute Association. http://www.uspa.org/.

DAVID ROYSTER

See Also: Airplanes/Flight; Calculus and Calculus Education; Calculus in Society.

Skyscrapers

Category: Architecture and Engineering.
Fields of Study: Algebra; Geometry; Measurement.
Summary: Mathematicians and engineers work together to design and build skyscrapers.

A skyscraper is a building noteworthy for its great height. As the name suggests, the building appears to touch the sky. There is no agreed-upon minimum height that classifies a building as a "skyscraper"; the term is used for any building that commands attention because of its height. Many people are fascinated by building, visiting, and measuring skyscrapers. The Eiffel Tower, designed by engineer Gustave Eiffel, revolutionized civil engineering and architectural design. In the design of a skyscraper, architects and engineers must consider load distribution and the impact of the wind and earthquakes. Scientists and mathematicians also investigate how to improve features such as seismic dampers. Many skyscrapers resemble rectangles or pyramids, but they may have other geometries, like the plan for the Helicoidal Skyscraper in New York or the sail-shaped skyscraper in Dubai—the Burj al-Arab Hotel. In Tokyo, St. Mary's Cathedral incorporates eight hyperbolic parabolas, and the HSB Turning Torso in Sweden uses five-story cubes that twist as they rise, with the top cube 90 degrees from the bottom cube. Buckminster Fuller proposed a city consisting of huge floating spheres, which he called Cloud Nine. The Wing Tower in Scotland was designed to rotate at the base in order to respond to changes in the direction of the wind. Proposed dynamic skyscrapers allow each floor to rotate independently, creating changing shapes, and using turbines to harness the power of the wind. There are various ways of ranking skyscrapers by height, and these buildings have other characteristics that can be quantified as well. Mathematician Shizuo Kakutani invented a mathematical skyscraper in ergodic theory called a "Kakutani skyscraper," so named because the mathematical process resembles the floors of a skyscraper. Students in some mathematics classrooms play a multiplication skyscraper game.

History

Throughout history, there have been buildings that were considered unusually tall, including pyramids, towers, and religious structures. The 10-story Home Insurance Building in Chicago, designed by William Le Baron Jenney and completed in 1885, is considered by many to be the world's first skyscraper. A variety of technological developments made the first skyscrapers possible. These included the mass production of steel, the invention of the elevator, the ability to achieve water pressure at altitude, the fireproofing of flooring and walls, and the development of reinforced concrete. The 792-foot Woolworth Building in New York City, completed in 1913, was typical of how skyscrapers would be constructed for the rest of the twentieth century. It had a steel skeleton and a foundation of concrete. Modern skyscrapers typically have a frame that supports the building's weight, with walls suspended from the frame. This feature distinguishes them from smaller buildings in which the walls are usually weight-bearing.

The Empire State Building in New York City reigned for 41 years as the world's tallest skyscraper and entered the public consciousness when the 1933 film *King Kong* depicted a giant ape that climbed the building. The movie had innovative special effects, including the use of scale modeling. In the twenty-first century, numerous television and FM radio stations transmit their signals from atop the Empire State Building and from skyscrapers in other cities.

Measurement

There are many different ways to measure the height of a skyscraper. It can be measured by the number of floors, highest occupied floor, spire height, or total height including such things as an antenna. Consequently, different figures can be found for the height of a single skyscraper. When lists of the world's tallest skyscrapers are published, a single skyscraper often ranks in different places on lists that use different rules of measurement. For example the Willis Tower in Chicago, formerly known as the Sears Tower, is the world's second tallest building when ranked by number of floors or when antennae are included, but it places seventh worldwide when spires are counted, but antennae are not.

Since 1998, a number of skyscrapers in Asia have surpassed the tallest American buildings in height. The Burj Khalifa, which opened in 2010 in Dubai, United Arab Emirates, is the world's tallest skyscraper as of 2010, whether ranked by its 163 floors, its 2,093-foot highest floor, or its spire height of 2,717 feet. The progression of record skyscraper heights over time can be graphed and modeled by a regression equation.

Other Aspects

Skyscrapers are noteworthy for other quantities besides their heights. When known geometric solids are used to model a skyscraper's shape, the building's surface area can be estimated. Because of differences in elevation, a skyscraper often experiences measurably different weather conditions at its top and bottom. In addition to its noteworthy height measurements, the Burj Khalifa contains over 20 acres of glass, has over 5 million square feet of floor space, and has elevators that travel over 26 miles per hour. Tall buildings are known to sway slightly in windy conditions. A rule of thumb for estimating a building's sway is to divide its height by 500 to arrive at the amount of horizontal sway near the top of the building. In many skyscrapers, steel tubes, or bundles of tubes, give the building strength against this swaying. The distance one can see from the top of a skyscraper can be computed. When the curvature of Earth is considered, the sight line is tangent to Earth's surface. On a clear day it is possible to see over 100 miles from atop the world's highest skyscrapers.

Further Reading

Koll, Hillary, Steve Mills, and William Baker. *Using Math to Build a Skyscraper*. Milwaukee, WI: Gareth Stevens Publishing, 2007.

Wells, Matthew. *Skyscrapers: Structure and Design*. New Haven, CT: Yale University Press, 2005.

David I. Kennedy

See Also: City Planning; Clouds; Elevation; Elevators; Engineering Design; Measurement in Society; Movies, Making of; Radio.

SMART Board

Category: Communication and Computers.
Fields of Study: Geometry; Representations.
Summary: Interactive whiteboards use a touch-sensitive display to mimic the functionality of a whiteboard while enhancing the user's options.

The SMART Board is a brand of interactive whiteboard. Unlike traditional whiteboards and chalkboards, the SMART board does not require markers, chalk, or erasers. Instead, the SMART Board utilizes a projector and a touch sensitive display. The projector displays computer images onto the screen. The screen itself allows the user to interact directly with applications similar to a large touch screen. For instance, touching the screen is equivalent to left-clicking with a mouse. Typically, SMART Boards come with four digital pens and a digital eraser. These digital devices allow the user to write on the screen using digital ink. SMART Board interfaces are available for Windows and for the Mac operating system.

Development

David Martin and Nancy Knowlton, inventors of the SMART Board, initially devised the idea in 1986 and began promoting it in 1991. Knowlton previously taught accounting and computer science, while Martin has a bachelor's degree in applied mathematics and began his career working on computer simulations. The SMART Board was the first interactive whiteboard that gave users touch control of computer applications. In 2003, their company developed and later patented Digital Vision Touch (DViT) technology, which relies on concepts of three-dimensional geometry, such as projection, reflections, and parallel lines to effectively display information and allow the user to interact with the board. It uses digital cameras and sophisticated recognition algorithms to determine the position of the user's fingertip and to make a distinction between single clicking, double clicking, and drag and drop. These recognition algorithms differentiate it from other touch technologies, like tablet personal computers. As of 2010, SMART Board was the most popular interactive whiteboard on the market in the United States.

Advantages

There are several advantages to SMART Board technology in the mathematics classroom. First, lectures done using the SMART Board can be saved, which allows instructors to access information written minutes, weeks, or even years earlier. By exporting these files as a pdf or a similar universal format, the instructor can post classroom notes on their course Web page, allowing students to review notes from previous classes, either to prepare for a test or to catch up on material that was covered when they were absent. In addition, the digital images saved by the SMART Board can more easily be read and transcribed for students with disabilities. Further, images on the SMART Board can be individually selected and copied to additional pages, which allows complex mathematical formulas and diagrams to be reproduced accurately and quickly. SMART board systems are typically connected to computers, meaning any application that is accessible on the computer is available on the SMART Board. Instructors may access spreadsheets, word processors, and the Internet. For these reasons, SMART Boards can greatly enhance the educational experience for both the instructor and the student. SMART Board–type lectures can also be accomplished using a tablet computer installed with the appropriate software and a projector system.

Since its introduction in 1991, SMART Boards have been incorporated into classrooms of all levels from kindergarten to college. In addition, many corporate boardrooms feature SMART Boards allowing for interactive presentations. As of 2010, over 1 million SMART Board systems have been installed across the world. It is likely that SMART Boards and similar systems will continue to replace or supplement the more traditional whiteboards and chalkboards found in current classrooms.

Further Reading

Bitter, Gary G. *Using Computer Technology in the Classroom.* Boston: Allyn and Bacon, 1999.

Ellwood, Heather. "Practice Makes Perfect: Building Creative Thinking Skills in High School Mathematics." *EdCompass Newsletter* (March 2009).

Gooden, Andrea R. *Computers in the Classroom: How Teachers and Students Are Using Technology to Transform Learning.* San Francisco, CA: Apple Press, 1996.

McAndrews, Alyson. "Improving STEM Engagement. New Program Uses SMART Products to Help Students Collaborate and Connect." *EdCompass Newsletter* (December 2009).

Morrison, Gary R. *Integrating Computer Technology Into the Classroom.* Upper Saddle River, NJ: Merrill, 1999.

Peters, Laurence. *Global Learning: Using Technology to Bring the World to Your Students.* Eugene, OR: International Society for Technology in Education, 2009.

Price, Amber. "Ten Ways to Get Smart With SMARTboard." *Tech & Learning* (August 2008).

Robert A. Beeler

See Also: Calculus and Calculus Education; Curricula, International; Curriculum, College; Curriculum, K-12; Digital Images; File Downloading and Sharing; Geometry and Geometry Education; Internet; Personal Computers; Schools; Statistics Education.

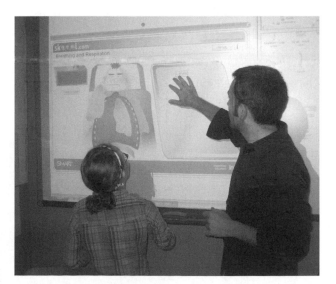

A projector displays video output on the SMART board, which acts as a large touch screen.

Smart Cars

Category: Travel and Transportation.
Fields of Study: Algebra; Geometry; Measurement.
Summary: A smart car is able to respond to the conditions it detects, such as sounding an alarm if it detects that a driver is becoming drowsy.

A smart car is also sometimes referred to as a "biometric car." The overall design and technology of such vehicles should incorporate many functions: protection of the driver and passengers, reliable and easy navigation, and better mechanical and fuel efficiency. Mathematicians, engineers, and many others are involved in the development of improved vehicle technology, including aerodynamics and computerized systems that use mathematical techniques from geometry; mathematical and computer modeling; and statistical analyses of data regarding safety, ergonomics, and consumer preferences. Methods from artificial intelligence, such as cellular automata, are also very useful. According to mathematician John von Neumann, cellular automata can be thought of as "cells" or agents that behave according to relatively simple sets of mathematical rules or algorithms. These rules include responses to neighboring cells' behaviors, making them useful in modeling many biological processes, like flocking birds or traffic.

Ideal Functions

In many peoples' minds, the primary purpose of a smart car should be to help a driver in ways that prevent accidents and encourage safe driving. For example, many car accidents occur because drivers do not realize that they are drowsy, so they consequently fall asleep at the wheel. A biometric smart car could alert drivers to such conditions by measuring eye movements relative to typical alert driver behavior to detect inattention and lack of scanning of the instruments and the road. Drivers that deviated too far from established safety norms would then be alerted. Other systems may involve a steering detector that responds to angular movements of the steering wheel that exceed a specified degree or a system that measures the angles of a driver's head and sound an alert if the head nods too far forward. In 2010, a Japanese company launched a system designed for commercial truck drivers that analyzes a driver's unique patterns and variability taking into account variables such as time. It then uses mathematical algorithms to proactively recommend rest breaks and measures to increase alertness and safety.

But What Actually Makes a Car Smart?

In addition to reactive systems like driver alertness warnings, some feel that a truly smart car should anticipate conditions to be avoided. Speeding when road conditions are poor or attempting to pass another car in low visibility could be predicted and avoided. Smart car systems would not only anticipate but also correct any anomaly so that a driver has time to recover. Further, they might suggest actions to a driver in advance of adverse conditions by monitoring the road and weather. Aspects of these features are present in many models of cars at the start of the twenty-first century facilitated by the introduction of real-time technology, such as interactive maps and global positioning systems (GPS), which depend on external communication with the environment to provide data beyond the drivers' senses. For example, many agencies provide data on road grade and surface, work zones, hazards, or speed restrictions. A smart car also monitors its internal state, taking measures of aspects like tire pressure and fluid levels using electronic sensors—functions that used to have to be performed by hand.

Advanced instrumentation, once found mostly in luxury cars, is becoming commonplace in vehicles. These systems may include smart starting that relies on electronics embedded in the car's keys; biometric features, like fingerprint scans; or keyless entry that may also require a computer chip, code, or fingerprint to activate. Many hybrid gas–electric vehicles balance energy usage to obtain maximum performance in mileage. Future smart cars may automatically sense variables like weight distribution and suggest load adjustments for better balance and braking. There are even notions that future smart cars will be able to dynamically reshape their surfaces for maximum aerodynamic efficiency. There is work being done on systems such as neural networks that may monitor and analyze all driver decisions in order to better provide feedback for safety and performance for particular geographic regions. Networks within smart cars may also interact with other cars and "smart roads," which could use computer technologies and mathematical modeling or algorithms, coupled with control and communications features, to improve issues like road safety and traffic capacity by directing traffic and helping drivers make better and safer decisions.

Further Reading

Scientific American Frontiers. "Inventing the Future Teaching Guide: Smart Car." http://www.pbs.org/safarchive/4_class/45_pguides/pguide_701/4571_smartcar.html.

Volti, Rudi. *Cars and Culture*. Baltimore, MD: Johns Hopkins University Press, 2006.

Whelan, Richard. *Smart Highways, Smart Cars*. Norwood, MA: Artech House, 1995.

JULIAN PALMORE

See Also: GPS; Highways; HOV Lane Management; Neural Networks; Street Maintenance; Traffic.

Soccer

Category: Games, Sport, and Recreation.
Fields of Study: Data Analysis and Probability; Geometry; Measurement.
Summary: Mathematical modeling and statistical analysis can help inform individual techniques and team tactics in soccer.

Soccer is a sport that has been enjoyed worldwide for more than a century by both players and spectators. In the early part of the twentieth century, mathematician Harald Bohr, founder of the field of almost periodic functions and brother of famed physicist Niels Bohr, was a skilled player and a silver medalist on the 1908 Danish Olympic soccer team. He was reported to be so popular that his doctoral dissertation defense was attended by more soccer fans than mathematicians. In general, the sport is often cited for its equal emphasis on individual skills and team tactics. As in other sports, statistics are frequently cited by sports commentators. In addition, technically demanding individual actions, as well as masterfully executed plays, can all be described and analyzed using statistics and mathematics, which is done worldwide by numerous sports scientists. One could even say that the players,

perhaps unconsciously, use or display "mathematics in motion."

Individual Technique

The effectiveness of any of the various moves a player uses (kicking, heading, or dribbling) depends on a combination of physical qualities and technical skills. This idea can be demonstrated using the instep kick as an example; the instep kick, with the aim to kick the ball as hard as possible, is by far the most studied soccer movement by sport scientists. In order to maximize the forward swinging speed of the shank, physical qualities (such as strength and speed of contraction) of the knee extensor muscles and the hip flexor muscles are important. However, research has shown that technical skills are equally important. The specific technical skill required for optimal kicking is coordination—how the shank moves relative to the thigh.

Coordination is one of the topics studied in the scientific field of biomechanics, which relies heavily on

The skill required for optimal kicking is coordination—how the shank moves in relation to the thigh.

mathematics. Biomechanics researchers use high-speed cameras in their laboratories to record kicking performance from top level players. From the video footage, the researchers can obtain the three-dimensional position in space of selected points on the kicking leg. Using mathematical concepts from vector algebra and trigonometry, joint and segment angles can subsequently be calculated. These data, in turn, allow calculations of a number of kinematic parameters of the foot, shank, and thigh, comprising linear velocity and acceleration and angular velocity and acceleration.

In mathematics, the most common way to calculate velocities and accelerations from position data is to use calculus. This method, however, requires the position data to be specified as a mathematical function. This is not the case with position data obtained from video footage, which are discrete in nature—they consist of thousands of numbers, specifying the three-dimensional position of numerous points on each video frame. From the cameras' frame rate, the elapsed time between frames can be calculated, which instead allows numerical differentiation of the position data using a computer. Finally, by combining the kinematic data with data for each segment's mass and moment of inertia (a measure of a segment's inertia when rotating) and using the principles from Newtonian mechanics, the researchers can calculate how the movement of the thigh affects the movement of the shank and vice versa. The forward swing of the thigh generates a force at the knee that causes the shank to swing faster forwards. The force is larger, the faster the thigh moves, while the effect of the shank is larger, the closer the knee angle is to 90 degrees. Top players instinctively coordinate thigh and shank movements in order to take maximum advantage of these intersegmental forces, although science so far has failed to determine precisely what optimal coordination is.

Team Tactics

When a midfielder executes a beautiful play that a forward picks up between defending opponents and scores, a lot of "hidden" mathematics is occurring. The midfielder's team members and opponents are all moving simultaneously in different directions with different speeds, yet the midfielder still manages to precisely calculate the required ball speed and direction to execute his play, so the ball and forward meet at the intended spot out of reach of defending opponents. Situations like

this are analyzed by sport scientists and coaches using the methods of notational analysis. With video footage and specialized software, the various actions (sprinting, moving sideways, tackling, or heading) of each player from both teams can be registered. Statistical calculations can reveal which situations are most likely to lead to a certain outcome, such as scoring a goal, and which general tactics lead to most of these situations. Digital representations have also been used to assist with tactics and analysis. Researchers from the University of Sheffield digitized a soccer ball (including even the stitching) and computed airflow around the ball. They found that the specific shape and surface of the ball, and its initial orientation, are significant in determining the ball's trajectory through the air. Measurements on actual balls in a wind tunnel at the University of Tsukuba verified these mathematical simulations.

Further Reading

Chartier, Tim. "Math Bends It Like Beckham." *Math Horizons* 14 (February 2007).

Putnam, C. A. "Sequential Motions of Body Segments in Striking and Throwing Skills: Descriptions and Explanations." *Journal of Biomechanics* 26 (1993).

Reilly, T., and M. Williams. *Science and Soccer.* New York: Routledge, 2003.

HENRIK SØRENSEN

See Also: Arenas, Sports; Connections in Society; Hockey; Kicking a Field Goal.

Social Networks

Category: Friendship, Romance, and Religion.
Fields of Study: Geometry; Number and Operation; Representations.
Summary: Social networks can be described and analyzed using graph theory.

A social network is a set of actors and the relationships that connect them. The actors are usually people, but may be other individual or collective actors, such as organizations, gangs, clubs, municipalities, nations, or social animals. Social network analysis is a cross-disciplinary method for analyzing social networks that integrates techniques from science, social science, mathematics, computer science, communication, and business. In keeping with its diverse origins, various types of social relationships have been studied using social network analysis, such as friendship, sexual relationships, kinship and genealogy, competitions, collaboration, and disease spread.

Sociogram, Sociomatrix, Graph, and Network

Modern social network analysis can be traced to Austro-American psychiatrist Jacob Levy Moreno, though many of the methods he employed in his work had been used before in a more piecemeal fashion. For example, French probabilist Irénée-Jules Bienaymé modeled the disappearance of closed families (for example, aristocrats) and family names in the nineteenth century. In his 1934 book *Who Shall Survive*, Moreno used diagrams he called "sociograms" to analyze friendships among girls in a training school in New York State. The girls were represented by points, and pairs of girls who were friends were connected by a line. In sociograms of relationships such as liking, which are not necessarily reciprocated, an arrowhead indicates the direction.

While very simple social networks can be analyzed by visual inspection, the power of social network analysis arises from the conceptualization of the sociogram as a mathematical graph, which can be analyzed using the concepts and methods of graph theory. Moreover, a graph can be represented by a square adjacency matrix in which each row and column represent a point, and the cell entries represent the presence or absence of lines between points. A graph can be generalized in several ways. Lines can have numerical values representing, for example, the strength, intensity, or frequency of a relationship. There can be multiple types of lines between pairs of actors, each representing one type of relationship. Actors can have various attributes with numerical values or qualitative labels. In social network analysis, real-life social networks are modeled by mathematical networks, then the properties of the networks are analyzed mathematically in order to draw conclusions about the structure of the social relationships.

Social Cohesion

Social cohesion is a fundamental issue in the social sciences; it is the "glue" or bond that holds a social group together. According to social network analysis, it is the network of social ties among members of the group. Therefore, to measure the level of social cohesion in a social group or subgroup, one must measure the extent of ties among the members. The density of ties among members is the simplest measure of connectedness. It is defined as the ratio of the number of actual ties to the number of possible ties and ranges from 0 to 1. In a network with one symmetric (undirected) type of tie, and k members, the total possible number of ties is

$$\frac{k(k-1)}{2}.$$

A network in which every actor is connected is called a "complete" graph, or a "clique."

It is easy to imagine four people all being friends with one another but less realistic to postulate a clique with a large number of members. For example, in a clique with 30 members, each would have to maintain ties with the 29 other members—an onerous task. Limits on human beings' time, energy, and memory constrain the number of people with whom they can maintain social ties. Therefore, social networks tend to become more sparse (the ties become less dense) as they become larger. Residents of a small village may know all the other residents, but this is impossible for city-dwellers. Thus, the village will tend to be more socially cohesive than the city. Density of ties has also been used to study social cohesion in such areas of social life as marriage, the family, small groups in laboratories, community elites, intercorporate relationships such as share ownership and interlocking directorates, scientific communities, and the spread of ideas and diseases.

The overall density of ties is a rather crude measure of connectivity and cohesion in a social network, because it is insensitive to local variations. Real-life social networks tend to contain islands of actors tied relatively densely to one another but disconnected or only loosely connected by sparse ties to other such islands. In the friendship network of a high school, there are likely to be a number of small cliques, perhaps loosely connected into larger subgroups that are in turn perhaps totally disconnected from one another. Detection of relatively cohesive subgroups in a network and delineation of their articulation into larger, less cohesive groups are a major theme in social network analysis.

Centrality

The centrality of an actor in a network is an important attribute, because centrality is associated with power, prestige, prominence, and popularity. In a network of ties representing flows or potential flows of valued social goods, such as information, a central actor is in a privileged position for both reception and transmission. The centrality of an actor may be intuitively evident from visual inspection of the drawing of a graph, especially if the graph is small or highly centralized. In larger graphs, a precise definition and formula are needed. The four main definitions of centrality are degree, closeness, betweenness, and power (or "eigenvector") centrality.

Degree centrality is the proportion of the other actors to which an actor is directly connected. The closeness centrality of an actor is based on how close the actor is to each of the other actors in the network and is the inverse of distance. The betweenness centrality of an actor is the extent to which the actor is "between" other actors; in other words, how often the shortest paths between pairs of other actors pass through the actor. Power centrality is defined recursively taking into account the power centrality of the actors to which an actor is adjacent.

Applications of Social Networks

The popular party game Six Degrees of Kevin Bacon tries to connect any movie actor to actor Kevin Bacon via costars in movies using the shortest number of steps. That value is an actor's Bacon Number. The Web site "The Oracle of Bacon," originally implemented in 1996, can be used to find the shortest path for any actor that can be linked to Kevin Bacon. The average path length as of September, 2010, was about three. It also allows a user to find a measure of centrality for the Hollywood network based around any actor in the database in terms of the average path length.

On a more personal level, the social network Web site Facebook includes an application called Friend Wheel that lets users visualize the interconnections among their friends as nodes and ties. Further, it selectively arranges the friends' names around the circumference of the wheel so that closely-knit groups or cliques are placed together and color-coded. Thomas Fletcher, a computer science and mathematics student at Bath University, developed the application and made it available in 2007.

Harkening back to Moreno's study, in 1995 a team of sociologists was the first to map the romantic and sexual relationships of an entire high school. Unlike similar adult networks, which tend to have several highly interconnected cores with loose interconnections (like airline hubs), the students were connected via long chains, more like a rural phone network. One chain linked 288 of the 573 romantically active students, though there were also many unconnected dyads or triads. Researchers attributed this finding in part to the often-elaborate teenage social rules about who may date. The surprising finding had important implications for educational practices like sex education programs.

Further Reading

Bearman, P. S., J. Moody, and K. Stovel. "Chains of Affection: The Structure of Adolescent Romantic and Sexual Networks." *American Journal of Sociology* 110, no. 1 (2004).

Furht, Borko. *Handbook of Social Network Technologies and Applications*. New York: Springer, 2010.

Moreno, Jacob L. *Who Shall Survive?* Washington, DC: Nervous and Mental Disease Publishing, 1934.

Wasserman, Stanley, and Katherine Faust. *Social Network Analysis*. New York: Cambridge University Press, 1994.

PETER J. CARRINGTON
SARAH J. GREENWALD
JILL E. THOMLEY

See Also: Connections in Society; Graphs; Matrices; Six Degrees of Kevin Bacon; Visualization.

Social Security

See *Pensions, IRAs, and Social Security*

Software, Mathematics

Category: Communication and Computers.
Fields of Study: Algebra; Geometry.

Summary: Mathematics software has long been used as a teaching aid and has become an important tool in applied mathematics.

Mathematics software refers to a wide variety of computer programs designed to manipulate, graph, or calculate numeric, symbolic, or geometrical data. Along with the development of computer technology and wider access to personal computers, these types of programs gained popularity at end of the twentieth century. Within the mathematics community, it has influenced instruction, applications, and research. Instruction has changed so that mathematics is more accessible to larger numbers of students; it is more engaging, more visual, and more focused on conceptual understanding rather than on computational facility.

Mathematics software has changed research and the nature of mathematical proof so that computers are now tools for exploration, for applications, and for performing repetitive tasks. There are numerous journals devoted to the development, use, or implementation of software in research and teaching, such as the *Transactions on Mathematical Software* journal. Computer software has influenced what mathematics is being taught, how it is being taught, the nature of its applications, and the way mathematics is explored. Computer software provides society a different modality for learning, understanding and applying mathematics.

What is Computer Software?

"Computer software" is a general term reserved for a collection of computer programs that provide step-by-step instructions for a computer to perform specific tasks. There are four major types of software:

- *Operating systems*: System software, often called the computer "platform" (for example, Microsoft Windows, Mac OS, and Linux);
- *Computer languages*: The code and syntax used for developing software (for example, Java, C/C++, Visual BASIC, Pascal, and Fortran)
- *General applications software*: Software designed for general purposes (for example, word processors, database systems, spreadsheets, and communications software)
- *Specific applications software*: Software designed for performing content-based

tasks (for example, MATLAB, Mathematica, Geometer's Sketchpad, SPSS, and MINITAB)

Software for the subject of mathematics falls in the category of Specific Applications Software.

Mathematics Software

The term "mathematics software" refers to computer programs designed to manipulate, graph, or calculate numeric, symbolic, or geometrical data. The journal *Transactions on Mathematical Software* (TOMS), produced by the Association of Computing Machinery (ACM), provides current information on available mathematics software. Through TOMS, the reader can gain access to large indexed mathematics software repositories. The majority of the software is written in Fortran or C++ for solving mathematics problems occurring in the sciences and engineering. Research scientists are invited to use these modules in developing their own software.

Software for Applied Mathematics

According to the National Research Council (NRC), computer software has had a major impact on applied mathematics and has illuminated new areas for mathematical research. The use of computer software in research in applied mathematics is prevalent, especially when repetitive computations are necessary. The most prominent computer software packages used in college-level instruction in the early twenty-first century are MATLAB, Mathematica, and Maple. These are computer algebra systems (CAS) that perform both symbolic and numeric computations. Software is used for statistics applications by professionals in mathematics, sciences, education, and social sciences, such as SPSS, SAS, BMDP and SPlus. These allow users to easily explore and visualize data and automate the computational aspects of many commonly used statistical procedures, which can be significantly difficult for larger data sets. It also facilitates more complex modeling and computer-intensive methods like exact tests, resampling techniques like bootstrapping, and many types of Bayesian statistical procedures, which are named for mathematician Thomas Bayes.

Software for Mathematics Research

Mathematics software is also gaining prominence in the fields of pure mathematics such as number theory, abstract algebra, and topology. An outstanding example of the impact of software on topological research occurred in 1976 when a computer program was used to check all of the possible cases in the Four-Color Map conjecture.

To understand the Four-Color Map conjecture, consider a map of the United States. Suppose the task is to color the individual states so that no two contiguous states are the same color. How many different colors are necessary to complete the task? Such a question arose in 1852. The Four-Color Map conjecture states that, at most, four colors are needed to color the map. In 1976, Kenneth Appel and Wolfgang Haken finally proved this conjecture (thus establishing it as a theorem) by using a computer program, representing the first time computer software was used in the proof of a mathematics theorem.

This computer-based proof led to considerable controversy within the mathematics community. The controversy centered on the nontraditional nature of the proof, which required a computer program for testing all of the possible cases, namely 1936 maps. Some mathematicians argued that this procedure did not constitute a formal mathematical proof, which is typically based on deductive logic and mathematical principles (such as definitions, axioms, and theorems). Instead, it was an exhaustive test of all possible cases, made possible by a computer program. Thus, neither deductive logic nor mathematical structure was required. Regardless of the controversies surrounding the proof of the Four-Color Map theorem, the result was to alter the attitudes of mathematicians toward the role of computer software in formal mathematical proof. Consequently, since the 1970s, computer software has become a major research tool for both pure and applied mathematicians.

A further consequence of the use of computer software in mathematical research is a trend for mathematicians to use open-source software, rather than proprietary software. Many commercial or proprietary software programs were originally developed and sometimes freely distributed as part of grant-funded projects or by individual mathematicians, computer scientists, and others to meet specific research or teaching needs. Some of these programs were also developed in conjunction with educators and students. With proprietary software, the user is denied access to the algorithms used in solving problems and thus cannot

have complete confidence in the fidelity of the mathematical results obtained by the programs. On the other hand, open-source software provides the source code to its users so they can modify and apply it with confidence to their research endeavors. Sage is an important example of open-source software that contains one of the world's largest collections of computational algorithms. For this reason, it is gaining in popularity among contemporary research mathematicians.

Software for Mathematics Education

Since the 1980s, computer software has been utilized regularly in the research of both pure and applied mathematics and it has made its way into mathematics classrooms. However, the adoption of mathematics software in teaching has not been without controversy. For instance, in 1993, students at the University

Algebra, Trigonometry, and Calculus Software

Mathematics instruction at all levels has changed considerably because of the profusion of graphing calculators in schools. These so-called calculators are actually handheld computers that have numerous built-in mathematical functions and programming capability. As a consequence, the mathematics curriculum is now more focused on conceptual development rather than building computational facility. Additionally, classroom computers and the use of interactive whiteboards (large interactive computer panels) have served to make mathematics instruction far more interactive and engaging for students.

Popular commercial software packages for college instruction are MATLAB, Mathematica, and Maple. A powerful piece of free-ware for algebra instruction at the high school level is Winplot (for Windows platforms only), which is available in 14 languages. It is a virtual graphing utility that can plot and animate functions, relations, and three-dimensional surfaces in a variety of formats.

of Pennsylvania complained about frustrations with Maple in calculus classes, citing a lack of support and faculty expertise. Some students even wore shirts printed with vulgarities about Maple, which attracted national attention. The use and implementation of software in classes has continued to generate debate regarding the balance between students exploring concepts and solving problems using traditional methods and computers. There are also questions regarding how much teaching time should be focused on instructing students in software use versus addressing concepts.

More recently, mathematics instruction in grades K–12 has benefited from computer software. This trend is due in part to the recommendations of major professional educational organizations and from federal programs and legislation. In 2000, the National Council of Teachers of Mathematics predicted that technology would enhance the learning of mathematics, support mathematics teaching, and influence the content that is taught. Educators have also praised the advantages of interactive software on student motivation and for providing a different modality for instruction—a modality that is visual, concrete, and interactive. Thus, anticipated impacts of computer technology on student achievement are encouraging. In 2002, the No Child Left Behind Act provided $15 million for research on the effects of computer technology on K–12 instruction.

Geometry Software

Computer software for teaching geometry is prevalent in American schools. The software of choice is dynamic software, which allows students to construct geometric shapes and actively explore their properties on the computer screen by (1) dragging vertices, (2) measuring component parts, (3) transforming them in the coordinate plan, (4) animating them, and (5) tracing points, and so on. Examples of dynamic geometry software are Cabri II Plus, The Geometer's Sketchpad (GSP), and GeoGebra.

When using dynamic geometry software, high school students have been able to make new discoveries in Euclidean geometry. For example, in 1994, Ryan Morgan, a sophomore at Patapsco High School in Baltimore, used GSP to discover a generalization to Marion Walter's theorem.

First, consider Marion Walter's theorem: If the trisection points of the sides of any triangle are connected to the opposite vertices, the resulting hexagon has area one-tenth the area of the original triangle.

Based on the prior theorem, Morgan discovered the following: If the sides of the triangle are instead partitioned into n equal segments (for $n =$ an odd integer) and each division point is connected to the opposite vertex, a central hexagon is still formed.

Morgan's theorem states that this hexagon has an area

$$A = \frac{8}{(3n+2)(3n-1)}$$

relative to the original triangle.

Discoveries by high school students, such as Morgan's theorem, lend credence to using dynamic software for geometry instruction in the nation's high schools.

Further Reading

American Mathematical Society. "Mathematics on the Web." http://www.ams.org/mathweb/mi-software.html.

Bailey, David, and Borwein, Jonathan. "Experimental Mathematics: Examples, Methods and Implications." *Notices of the American Mathematical Society* 52, no. 5 (2005).

DeLoughry, Thomas. "A Revolt Over Software: Penn Students Call Calculus Program Frustrating and Say Faculty Didn't Know How to Use It." *The Chronicle of Higher Education* 40, no. 14 (November 24, 1993).

Lutus, Paul. "Exploring Mathematics With Sage." http://www.arachnoid.com/sage/.

Quesada, Antonio. "New Mathematical Findings by Secondary Students." *Universitas Scientiarum* 6, no. 2 (2001). http://www.javeriana.edu.co/universitas_scientiarum/universitas_docs/vol6n2/ART1.htm.

Sarama, Julie, and Doug Clements. "Linking Research and Software Development." *Research on Technology and the Teaching and Learning of Mathematics* 2 (2008).

SHARON WHITTON

See Also: Calculators in Society; Curriculum, College; Educational Manipulatives; Geometry and Geometry Education; Personal Computers; Statistics Education; Telephones.

Solar Panels

Category: Architecture and Engineering.
Fields of Study: Algebra; Geometry; Measurement.
Summary: The angle of inclination of a solar panel array is key to its efficiency, among other factors.

Solar panels are interconnected assemblies of photovoltaic cells that collect solar energy as part of a solar power system, either on Earth or in space. Typically, several solar panels will be used together in a photovoltaic array along with an inverter and batteries to store collected energy. Photovoltaic cells convert the energy of sunlight into electricity via the photovoltaic effect (the creation of electric current in a material when it is exposed to electromagnetic radiation), which was observed by French physicist Alexandre-Edmond Becquerel in 1839. Prior to that time, many scientists and mathematicians built and researched parabolic burning mirrors, which are another way to focus solar energy. Diocles of Carystus showed that a parabola will focus the rays of the sun most efficiently. Archimedes of Syracuse may have built burning mirrors that set ships on fire. George LeClerc, Comte de Buffon, apparently tested the feasibility of such a mirror by using 168 adjustable mirrors in order to vary the focal length to ignite objects that were 150 feet away. It was also investigated experimentally in the early twenty-first century on the television program *Mythbusters*. Mathematics

teacher Augustin Mouchot investigated solar energy in the nineteenth century and designed a steam engine that ran on sun rays. Some consider this invention to be the start of solar energy history. The first working solar cells were built by the American inventor Charles Fritts, in 1883, using selenium with a very thin layer of gold. The energy loss of Fritts's cells was enormous—less than 1% of the energy was successfully converted to electricity—but they demonstrated the viability of light as an energy source. Engineer Russell Ohl's semiconductor research led to a patent for what are considered the first modern solar cells, and Daryl Chapin, Calvin Fuller, and Gerald Pearson, working at Bell Labs in the 1950s, developed the silicon-based Bell solar battery. There were fewer than a single watt of solar cells worldwide capable of running electrical equipment at that time. Roughly 50 years later, solar panels generated a billion watts of electricity to power technology on Earth, satellites, and space probes headed to the far reaches of the galaxy. Scientists and mathematicians continue to collaborate to improve solar panel technology. One such focus is creating scalable systems that are increasingly efficient and economically competitive with various other energy technologies.

Physics and Mathematics of Solar Panels

In 1905, Albert Einstein published both a paper on the photoelectric effect and a paper on his theory of relativity. His mathematical description of photons (or "light quanta") and the way in which they produce the photoelectric effect earned him the Nobel Prize in Physics in 1921. In general, the photons or light particles in sunlight that are absorbed by semiconducting materials in the solar panel transfer energy to electrons—though some is lost in other forms, such as heat. Added energy causes the electrons to break free of atoms and move through the semiconductor. Solar cells are constructed so that the electrons can move in only one direction, producing electrical flow. A solar panel or array of connected solar panels

The solar field at Nellis Air Base in Nevada has more than 72,000 panels and supplies the base with more than 30 million kilowatt-hours of power.

produces direct current, like chemical batteries, which can be stored. An inverter can convert the direct current to alternating current for household use.

Mathematics is involved in many aspects of solar panel design, operation, and installation. For example, the perimeter of an array of multiple solar panels may change with rearrangement of the panels, but the area stays the same. Since area is one critical variable for power collection, this suggests different optimal arrangements for surfaces where solar panels might be arranged, like walls and roofs. Satellites often use folding arrays of solar panels that deploy after launch, and folding portable solar panel arrays have been designed for applications like camping and remote or automated research and monitoring stations. Space scientist Koryo Miura developed the Muria–Ori map folding technique, which involves mathematical ideas of flexible polygonal structures and tessellations. It has been incorporated into satellite solar panels that can be unfolded into a rectangular shape by pulling on only one corner.

Arrays

A solar panel array may be fixed, adjustable, or tracking. Each method has trade-offs in installation cost versus efficiency and energy over the lifetime of the installation, which can be analyzed mathematically in order to optimize an individual setup. Fixed arrays are solar panels that stay in one position. Optimal positioning of such arrays usually involves facing the equator (true south, not magnetic south, when in the northern hemisphere), with an angle of inclination roughly equal to their latitude. Using an angle of inclination slightly higher than the latitude has been shown in some studies to improve energy collection in the winter, which can help balance shorter days or increased heating energy needs. Setting the inclination slightly less than the latitude optimizes collection for the summer. Adjustable panels can have their tilt manually adjusted throughout the year. Tracking panels follow the path of the sun during the day, on either one or two axes: a single-axis tracker tracks the sun east to west only, while a double-axis tracker also adjusts for the seasonal declination movement of the sun. Tracking panels may lead to a gain in power, but for some users, the cost trade-off might suggest adding additional fixed panels for some applications instead. Solar power companies and other entities provide maps showing the yearly average daily sunshine in kilowatt hours per square meter of solar panel. Combined with the expected energy consumption of a building, this data helps determine how many solar panels and batteries will be needed for an installation. Science and mathematics teachers often have students build solar panels and collect data to facilitate mathematical understanding and critical thinking, as well as make mathematics, science, and technology connections.

Further Reading

Anderson, E. E. *Fundamentals of Solar Energy Conversion.* Reading, MA: Addison Wesley Longman, 1982.
Hull, Thomas. "In Search of a Practical Map Fold." *Math Horizons* 9 (February 2002).
Kryza, F. *The Power of Light: The Epic Story of Man's Quest to Harness the Sun.* New York: McGraw-Hill, 2003.

BILL KTE'PI

See Also: Electricity; Light; Origami; Satellites.

South America

Category: Mathematics Around the World.
Fields of Study: All.
Summary: Long before European settlement, mathematics flourished in South America.

South America includes Argentina, Bolivia, Brazil, Chile, Colombia, Ecuador, French Guiana, Guyana, Paraguay, Peru, Suriname, Uruguay, and Venezuela. The history of South American mathematics begins with pre-Columbian developments like the Nazca lines and *quipus* ("KEE-poos") and continues through the astronomy boom of the colonial period to work by modern mathematicians and ethnomathematics studies in Brazil.

Quipus

The Incan empire, with its capital in Cuzco, Peru, dominated pre-Columbian South America. The Incan civilization emerged from the highlands in the early thirteenth century and extended over an area from what is now the northern border of Ecuador, Peru, western and south central Bolivia, northwest Argentina, northern and central Chile, and southern Colombia. The Incas reached a high level of sophistication with remarkable systems of agriculture, textile design, pottery, and administration. Since the Incas had no written records, the *quipu* (or *khipu*) played a pivotal role in keeping numerical information about the population, lands, produce, animals, and weapons.

Quipus were knotted tally cords that consisted of a main cord from which hung a variable number of pendant cords containing clusters of knots. These knots and their clusters conveyed numerical information in base-10 representation. For example, if the number 365 was to be recorded on the string, then five touching knots were placed near the free end of the string followed by a space, then six touching knots for the 10s, another space, and finally three touching knots for the 100s. Specific information was conveyed via the number and type of knots, cluster spacing, color of cord, and pendant array. Inca administrators and accountants employed this complex system for numerical storage and communication. Quipus were mathematically efficient and portable. Unfortunately, the Spanish destroyed many quipus, potentially hiding clues to understanding Incan architectural processes, irrigation, and road systems.

Nazca Lines

The Nazca lines are a set of figures that appear engraved in the surface of the Nazca desert in southern Peru. The lines include hundreds of geometric shapes and renderings of animals and plants, including birds, a spider, a monkey, flowers, geometric figures, and lines—some of them miles long. The Nazca lines, best appreciated from an airplane, are one of the world's enduring mysteries. It is hard to explain how the ancient people of Nazca (900 B.C.E.–600 C.E.) achieved such geometrical precision in an area over 300 square miles. German-born mathematician and archaeologist Maria Reiche spent five decades studying and preserving these lines. She, like many other scientists, believed that the Nazca lines represented an astronomical calendar and observatory, while other theories suggest that they map areas of fertile land.

Mathematics in the Colonial Era

The accidental arrival of navigator Christopher Columbus in the Americas in 1492 marked the beginning of a 300-year period of Spanish and Portuguese colonial rule in South America that ended in the early nineteenth century. Under the Treaty of Tordesillas (1494), Portugal claimed what is now Brazil, and Spanish claims were established throughout the rest of the continent with the exception of Guyana, Suriname, and French Guiana. Roman Catholicism and an Iberian culture were imposed throughout the region, and mathematical systems and practices of ancient cultures were replaced by the Hindu-Arabic decimal system used by the Spanish.

Mathematical activity in Spain between the sixteenth and nineteenth century decisively influenced mathematical thinking and practices in South America. In sixteenth-century Spain, two lines of mathematical thought existed: the arithmeticians (calculators, interested in the uses of mathematics) and the algebraists (abstract or pure mathematicians). Because the European countries used the colonies to enhance their trade and economic resources, the emphasis in South America was on applied mathematics.

Later, the Spanish and the Portuguese established schools—mostly run by Catholic religious orders—which concentrated mathematics teaching on economic applications related to trade. There was also an interest on mathematics related to astronomical observations. The first nonreligious book published in the Americas was an arithmetic book related to gold and silver mining printed in 1556.

Astronomy was a major area of interest in South America in the seventeenth century. In Brazil, research on comets was of major importance, as exemplified by the work of Valentin Stancel (1621–1705), a Jesuit mathematician from Prague who lived in Brazil from 1663 until his death (his astronomical measurements are mentioned in Newton's *Principia*). As in many cultures, most astronomical interpretations attempted to explain divine messages to humankind. Other developments in Brazil included the first aircraft known to fly: the Passarola, invented by Bartolomeu de Gusmão, a Brazilian priest and scientist from Sao Paulo. De Gusmão, also known as the "Flying Priest," studied mathematics and physics at the Universidade de Coimbra in Portugal. The Passarola was an aerostat heated with hot air and flew in Lisbon, Portugal, in 1709.

Mathematics in the Era of Independence

In the first quarter of the nineteenth century, many successful revolutions resulted in the creation of independent countries in South America. Mathematical activity increased throughout Latin America in the twentieth century. For instance, Argentinian mathematician Alberto P. Calderon (1920–1998) developed new theories and techniques in classical and functional analysis. Professor Calderon worked at the University of Chicago for many years. He was awarded the National Medal of Science in the United States.

Research by Professor Ubiritan D'Ambrosio and his students in the slums and indigenous communities in Brazil focused on ethnomathematics—a sub-field of mathematics history and mathematics education. The goal of ethnomathematics is to understand connections between culture and the development of mathematical processes and ideas. Other researchers have explored specific mathematical habits and methods in South American cultures. In the 1980s, Terezinha Nunes and her collaborators studied differences between street mathematics and school mathematics in Brazil by comparing how street vendors (including children) and farmers solve problems compared to those who encounter similar problems in formal school situations.

For example, in their study of young street vendors in Recife, the interviewers acted as customers and asked questions that required the use of arithmetic skills (such as making change). The children did much better in this "real" situation than on a formal test given a week later that used similar numbers and operations. One possible explanation is that the children were better able to keep the meaning of the problem in mind in the "real" situation. Many others, such as Geoffrey Saxe, have found similar results. An implication of these studies is that the essence of school mathematics, which the Recife children were not as successful at, is highly symbolic and possibly devoid of meaning. These studies have been important in advancing the goal of mathematics education that students must initially construct appropriate meanings for the various concepts and methods they encounter.

Further Reading

Ascher, Marcia. "Before the Conquest." *Mathematics Magazine*. 65, no. 4 (1992).

D'Ambrosio, Ubiritan. "Ethnomathematics and Its Place in the History and Pedagogy of Mathematics." *For the Learning of Mathematics* 5, no. 1 (1985).

Nunes, Terezinha, et al. *Street Mathematics and School Mathematics*. New York: Cambridge University Press, 1993.

Ortiz-Franco, Luis, Norma Hernandez, and Yolanda De La Cruz, eds. *Changing the Faces of Mathematics: Perspectives on Latinos*. Vol. 4. Reston, VA: National Council of Teachers of Mathematics, 1999.

Gisela Ernst-Slavit
David Slavit

See Also: Astronomy; Calendars; Incan and Mayan Mathematics; Knots.

Space Travel

See *Interplanetary Travel*

Spaceships

Category: Travel and Transportation.
Fields of Study: Algebra; Geometry; Measurement.
Summary: Every task involving spaceships, from their design to their launch to effective collision avoidance and communication, is mathematically intensive.

Spaceships, also called "spacecraft," are manned or automatic vehicles for flying beyond planet atmospheres. Different types of spaceships serve different purposes, including scientific or applied observations and data collection, exploration of celestial bodies, communication, and recreation.

According to the routes they take, spaceships can be classified as suborbital, orbital, interplanetary, and interstellar. According to the type of propulsion used, spacecraft engines can be designated as reaction engines, including rockets; electromagnetic, such as ion thrusters; and engines using fields, such as solar sails or gravitational slingshots. Mathematics is fundamental for spaceship design, operation, and evaluation. For example, mathematics is used to plan efficient trajectories, avoid collisions, communicate with satellites, transmit data over vast interplanetary distances, and solve complex problems like those that occurred in the famous Apollo 13 mission.

Mathematics in Spaceship Systems

Propulsion of a spaceship poses scientific and engineering problems that involve balancing forces and computing sufficient fuel, energy, work, and fluid mechanics. For any type of engine, the impulse it gives to the craft has to be calculated and compared to the craft's tasks such as leaving the gravity well of a planet or maintaining an orbit. For example, calculations for rocket engines involve variables including the changing mass of the craft as its fuel is spent, the efficiency of the engine, and the velocity of the rocket's exhaust. Solar sail theories involve such variables as radiation pressure of the light, the area of the sail, and the weight of the craft.

Mechanics and material sciences problems involved in the structure of spacecraft include withstanding the forces, temperatures, and electromagnetic fields involved in moving through space. For example, moving through a planetary atmosphere at speeds necessary to leave the planet's gravity well involves high temperatures from friction.

The guidance and navigation systems of a spaceship collect data and then compute position, speed, and the necessary velocity and acceleration to reach the destination. These systems also determine the relative position of the spaceship to nearby celestial bodies, which influence the craft's motion by their gravitational and electromagnetic fields. For example, mathematical description of a craft orbiting a planet includes the six Keplerian elements (for example, inclination and eccentricity) defining the shape, the size, and the orientation of the orbit, named for Johannes Kepler.

Most twenty-first-century spacecraft do not carry living organisms, but when they do, life support systems are necessary. Life support systems protect people, animals, or plants in the spaceship from harmful environments and provide air, water, and food. The design of life support systems involves biology, physiology, medical sciences, plant sciences, ecology, and bioengineering. Mathematical models for life support typically include calculations of safety margins, such as maximum allowable radiation doses. All organisms need some inputs (such as food, water, or oxygen) and produce some outputs depending on a variety of variables, such as activity levels. Spaceship ecosystem designers strive to produce waste-free, closed systems where water is reclaimed and plants are used to purify the air. Because of the complexity of the closed eco-

Escaping a Planet's Gravity

The problem of escaping the gravitational field of a large celestial body, such as Earth, is different from the problem of flight in space far from large bodies. For example, a certain velocity, called *escape velocity*, is required to leave any given planet. At the sea level of Earth, the escape velocity is about 11 kilometers per second (km/s) or 7 miles per second (mi/s). However, spaceships usually fly slower at first. The escape velocity is inversely proportional to the square root of the distance from the planet's center of gravity. Spaceships leaving the Earth reach these lower escape velocity levels at some distance from the surface. For comparison, the escape velocity from the Sun is about 600 km/s (373 mi/s) and the speed record as of 2010 for a spacecraft leaving the Earth is about 16 km/s (10 mi/s). This means that flights near the Sun are not technologically possible in the early twenty-first century. The escape velocity of a black hole is greater than the speed of light (over 300,000 km/s or 186,000 mi/s), which is the highest theoretical speed possible.

system problem, most current flights employ simpler, machine-driven life support systems.

Atmospheric Flight

Flight within an atmosphere presents very different problems compared to flight in a vacuum. The problems solved by applied mathematicians who study atmospheric flight include friction, turbulence, wing lift, aerodynamic shapes, and control of temperature. Spaceships launching or landing on planets have to be equipped for atmospheric flight. Because of differences in the vacuum and atmosphere flight requirements, many spaceships are designed to change their configuration when they cross atmospheric boundaries. For example, mathematical theories originally developed for origami are used to fold and unfold solar batteries, which can be used only in a vacuum because of their large area.

Science Fiction and Computer Game Mathematics

Space travel frequently appears in science fiction, where plots deal with various existing engineering or physics limitations. Hard science fiction is the more scientifically oriented subgenre, and it frequently includes extensions, discussions, and speculations dealing with the current scientific research. This tradition of blending science and literature started in the late nineteenth century with the works of Jules Verne; many of his then-fantastic devices and ideas (for example, televisions and submarines) were implemented relatively soon after.

As an example of experiments with scientific limits in literature, science-fiction spaceships may travel at superluminal (faster than light) speeds, often through non-physical spaces such as "hyperspace," "subspace," or "another dimension." These are terms from existing mathematical theories, which hard science fiction sometimes discusses.

Sci-fi spaceships may also be living organisms, completely or partially. This idea is a reflection of the current interest in bioengineering and has connections with exciting research in ecology, genetics, cybernetics, and artificial intelligence, as well as social sciences such as philosophy and bioethics.

Computer games and movies about space flight created a demand for applied mathematicians who can model fantastic situations with passable realism. The physics and mathematics of three-dimensional modeling is a fast-growing area, with new courses and programs opening in universities and an expanding job market. What started in the nineteenth century as an exotic occupation for very few writers has become a profession for many programmers and applied mathematicians.

Further Reading:

Battin, Richard. *An Introduction to the Mathematics and Methods of Astrodynamics*. New York: American Institute of Aeronautics and Astronautics, 1987.

National Aeronautics and Space Administration. "Design a Spaceship." http://www.nasa.gov/centers/langley/news/factsheets/Design-Spaceship.html.

Osserman, Robert. "Mathematics Awareness Month: Space Exploration." http://www.mathaware.org/mam/05/space.exploration.html.

Maria Droujkova

See Also: Airplanes/Flight; Elevation; Energy; Fuel Consumption; Interplanetary Travel; Origami; Radiation; Satellites; Solar Panels; Vectors; Weightless Flight.

Spam Filters

Category: Communication and Computers.
Fields of Study: Number and Operations; Data Analysis and Probability; Problem Solving.
Summary: Spam filters use probability and Bayesian filtering to sort spam from legitimate e-mails.

Most people with an e-mail address receive unsolicited commercial e-mail, also known as spam, on a regular basis. Spam is an electronic version of junk mail, and has been around since the introduction of the Internet. The senders of spam (called spammers) are usually attempting to sell products or services. Sometimes, their intent is more sinister—they may be trying to defraud their message recipients. Since the cost of sending spam is negligible to spammers, it has been bombarding e-mail servers at a tremendous rate. Some estimate that as much as 40% to 50% of all e-mails are spam. The cost to the message recipients and businesses can be considerable in terms of decreased productivity and unwelcome exposure to inappropriate content and scams. As frustrating and potentially damaging as spam e-mail is, fortunately, much of it does not reach recipients thanks to spam filters. Spam filters are computer programs that screen e-mail messages as they are received. Any e-mail suspected to be spam will be redirected to a junk mail folder so that it does not clutter up a user's inbox. How does the filter decide which messages are suspect? Spam filters are implementations of statistical models that predict the probability that a message is spam given its characteristics. The filter classifies messages with large predicted probabilities of being spam, as spam.

Filters

Primitive filters simply classified a message as spam if it contained a word or phrase that frequently appeared in spam messages. However, spammers only need to adjust their messages slightly to outsmart the filter, and all legitimate messages containing these words would automatically be classified as spam. Modern spam filters are designed using a branch of statistics known as "classification." Bayesian filtering is a particularly effective probability modeling approach in the war on spam. Bayesian methods are named for eighteenth-century mathematician and minister Thomas Bayes. He formulated Bayes' theorem, which relates the conditional probability of two events, A and B, such that one can find both the probability of A given that one already knows B (for example, the probability that a specific word occurs in the text of an e-mail given that the e-mail is known to be spam); the reverse, the probability of B given that one knows A (for example, the probability that an e-mail is spam given that a specific word is known to appear in the text of the e-mail).

The underlying logic for this type of filter is that if a combination of message features occur more or less often in spam than in legitimate messages, then it would be reasonable to suspect a message with these features as being or not being spam. An extensive collection of e-mail messages is used to build a prediction model via data analysis. The data consist of a comprehensive collection of message characteristics, some of which may include the number of capital letters in the subject line, the number of special characters (for example, "$", "*", "!") in the message, the number of occurrences of the word "free," the length of the message, the presence of html in the body of the message, and the specific words in the subject line and body of the message. Each of these messages will also have the true spam classification recorded. These e-mail messages are split into a large training set and a test set. The filter will first be developed using the training set, and then its performance will be assessed using the test set. A list of characteristics is refined based on the messages in the training set so that each of the characteristics provides information about the chance the message is spam.

However, no spam filter is perfect. Even the best filter will likely misclassify spam from time to time. False positives are legitimate e-mails that are mistakenly classified as spam, and false negatives are spam that appear to be legitimate e-mails so they slip through the filter unnoticed. An effective spam filter will correctly classify spam and legitimate e-mail messages most of the time. In other words, the misclassification rates will be small. The spam filter developer will set tolerance levels on these rates based on the relative seriousness of missing legitimate messages and allowing spam in user inboxes.

Spam filters need to be customized for different organizations because some spam features may vary from organization to organization. For instance, the word "mortgage" in an e-mail subject line would be quite typical for e-mails circulating within a banking institution, but may be somewhat unusual for other businesses or personal e-mails. Filters should also be updated frequently. Spammers are becoming more sophisticated and are figuring out creative ways to design messages that will filter though unnoticed. Spam filters must constantly adapt to meet this challenge.

Further Reading

Madigan, D. "Statistics and the War on Spam." In *Statistics: A Guide to the Unknown*. 4th ed. Belmont, CA: Thompson Higher Education, 2006.

Zdziarski, J. *Ending Spam: Bayesian Content Filtering and the Art of Statistical Language Classification*. San Francisco, CA: No Starch Press, 2005.

BETHANY WHITE

See Also: Internet; Predicting Preferences; Search Engines; Social Networks; Software, Mathematics.

Sports Arenas

See *Arenas, Sports*

Sport Handicapping

Category: Games, Sport, and Recreation.
Fields of Study: Algebra; Data Analysis and Probability; Number and Operations.
Summary: Various calculations are used to set fair, competitive handicaps in sports.

Sport handicapping is an important methodology that affects millions of people worldwide and potentially impacts billions of dollars worth of bets. In many sports, handicaps are calculated for individuals or teams and are used as a way of "equalizing" performance by giving a scoring advantage or other in-game compensation to some players. This process allows lower skilled players to compete with higher skilled players while preserving perceived fairness. The term "handicap" refers to both the adjusted scores and the process of determining them, and may also be used for whole tournaments that rely extensively on the method. Handicap in this context derives from a seventeenth century lottery game called "hand-in-cap," where players put their bets in a literal cap. A point spread, frequently used in sports betting, is a related idea for computing or estimating relative advantage to equalize teams in competitive sports. Examples of sports using handicapping at various levels include bowling, golf, horse racing, and track and field.

Handicapping

In sport, a handicap is usually imposed to enable a more equal competition to take place. The handicap is calculated according to specific criteria set down for each of the sports that use the technique, meaning that some are much more complex than others. To understand why a handicap may be used, consider one of the most well-known sports that employ a handicap system, golf.

If a recreational golf player were to compete against the best golfer in the world in a round of golf, then the outcome would almost certainly be a win for the better golfer. A win by a large margin would also have been very likely. If a handicap were applied that was based on each player's average scores, then the outcome would be much less certain. There would have been a distinct possibility that, if the recreational player had played well, they would have had the opportunity to beat the better golfer—or at least not loose by many shots—after the handicap was applied.

In most sports when professionals compete against each another, the events are usually free from handicapping. A professional golf tournament will usually engage those who play with a scratch (or zero) handicap.

One of the primary reasons for using a handicap is to make an event more competitive. In many respects, this makes the given sport more enjoyable and can help to make it more appealing and increase the number of those wishing to participate.

Tenpin bowling is a sport that has more participants worldwide than most other sports. The overwhelming majority of players are recreational, although many take part in annual league competitions. Most leagues are not scratch based (on actual total pin fall) but are handicapped. In tenpin bowling handicap leagues, the scores that are used to determine who has won are a combination of the total pins actually knocked down and the handicap value. This method allows players (and teams) with lower averages to compete against players (and teams) that have much higher averages.

The handicap in Tenpin Bowling is usually of the form: Handicap value (per game) = 80% of the difference between the player's average and 200 pins.

If a bowler averages 100 pins, then the bowler would, using the handicapping system, gain a handicap value of 80 pins: $(200 - 100) \times 0.80$. The total pinfall for a game would be 80 plus whatever number of pins the bowler actually knocked down.

This handicap system is versatile in that the two values used (the 80% and the 200 pins, in the example above) can be manipulated to suit the particular league. For instance, if there are a number of players who average over 200, for example 210 or 220, then the handicap may be 80% of the difference between each bowler's average and 220 pins. Alternatively, if the players are grouped quite closely together, then the handicap may be 66% of the difference between each bowler's average and 200 pins.

Athletics

Athletics, or track and field, is another mass participation sport, but one in which, at the highest level, age is intrinsically linked to performance—few athletes compete internationally in their late 30s and beyond. There is still huge participation in the sport by people older than 30, and there are obvious health benefits to doing so.

There is a scoring system that takes age into account by comparing race time to that of the world record holder in each age group. It is often known as a World Association of Veteran Athletes (WAVA) Rating and is expressed as a percentage between zero and 100. If one gets a WAVA rating of 50%, it means that the competitor is half the pace of the world record holder. WAVA rating is a useful way to make comparisons between runners of all ages and can form the basis of a handicap league.

Horse Racing

A further important application of handicapping is that seen in horse racing, a sport on which billions of dollars worth of bets are made each year. In a handicapped race, the horse must carry a certain additional weight, which when added to the weight of the jokey gives it an assigned impost (or total weight). These weights are held in saddle pads with pockets.

The calculation for the weight a horse is required to carry is based on a number of factors. A great deal of work is done with past data to create and then ensure that the handicaps are as fair as possible. These handicaps allow for horses of differing abilities to race against each other over a given distance.

Further Reading

Mullen, Michelle. *Bowling Fundamentals.* Champaign, IL: Human Kinetics, 2003.

Tuttle, Joeseph J. *The Ultimate Guide to Handicapping the Horses.* Self published: Createspace, 2008.

Wright, Nick. *Lower Your Golf Handicap: Under 10 in 10 Weeks.* London: Hamlyn, 2006.

STEPHEN LEE

See Also: Algebra in Society; Betting and Fairness; Data Analysis and Probability in Society.

Squares and Square Roots

Category: History and Development of Curricular Concepts.
Fields of Study: Algebra; Communication; Connections; Geometry.
Summary: Squares and square roots have long challenged mathematicians and have led to various expansions of the number system and developments in number theory.

The square of a number x, denoted x^2, is the number $x \times x$. The inverse operation is called the *square root*: the number x is a square root of y if $y = x^2$, the notation used being $x = \sqrt{y}$. Historically, these operations have been a major source of new problems, ideas, and systems of numbers in the early and modern development of mathematics. Square roots have also

appeared in many applications, such as computing the standard deviation of a data set, and have often presented a challenge to scientists and mathematicians in the days before readily available calculating technology. Middle-grade students in the twenty-first century continue to use squares and square roots to simplify computations and solve problems, as do carpenters and engineers.

Definition

Geometrically, the "square" of a number x measures the area of a square whose side has length x. This idea explains the name and is likely the way that ancient civilizations were first confronted with the operation. The Pythagorean theorem is an equality between sums of areas of squares constructed on the sides of a right triangles, namely $a^2 + b^2 = c^2$ if a and b are the two legs and c is the hypotenuse. Applied to the triangle obtained by halving a square of side length one along one of its diagonals, it shows that such a diagonal has length equal to $\sqrt{2}$.

A member of the Pythagorean School sometimes identified as Hippasus of Metapontum (c. fifth century B.C.E.) discovered that this number cannot be expressed as the ratio of two integers—it is irrational. The discovery was a sensation amid the Pythagorean School where it was preached that all numbers were rational and called for an extension of the number system.

In the centuries that followed, extensions of the number system would include all numbers expressible with an infinite number of decimal digits, so that each positive number has a square root (for example, $\sqrt{2} = 1.4142136 \ldots$) and negative numbers, which can be multiplied according to the usual associative rules, and the following additional ones governing signs: $-1 \times x = -x$; $(-1) \times (-1) = 1$, which implies that $(-1)^2 = 1$ so that -1 should also be counted as a square root of 1.

More generally, both extensions can be combined to yield the system of real numbers, which are the numbers with sign and infinite decimal expansions. In this system, each square of a number is a positive number (or zero), and each positive number has exactly two square roots, which differ by a sign. For example, 2 has as square roots the numbers $1.4142136 \ldots$ and $-1.4142136 \ldots$, a fact denoted by the expression $\sqrt{2} = \pm 1.4142136 \ldots$.

Computation

Square roots can be computed by hand, by calculator, or by computer (up to the desired numerical approximation) by several methods, including those using sequences, exponentials, logarithms, or continued fractions. Mathematicians in ancient Egypt and Babylonia are some of the first who are thought to have extracted square roots. Early Chinese, Indian, and Greek mathematicians also contributed to this area. According to some historians, the first method to be introduced in Europe was that of Aryabhata the Elder, a Hindu mathematician and astronomer. One of the oldest ones, still at the basis of many currently used algorithms, is the so called Babylonian method (which is also an instance of the modern Newton–Raphson method for solving general equations in one variable). Given a positive number S and choosing an initial "guess" x_0, the method produces a sequence of numbers x_n converging to the square root of S by the rule

$$x_{n+1} = \frac{1}{2}\left(x_n - \frac{S}{x_n}\right).$$

For example, the first approximations to $\sqrt{2}$ starting from $x_0 = 1$ are $x_1 = 15$, $x_2 = 1.416...$, $x_3 = 1.414215...$, $x_4 = 1.4142135623746...$, the last one already having 11 correct decimal digits.

Solving the Quadratic Equation

Square roots are used to solve the general quadratic equation $ax^2 + bx + c = 0$, where a, b, and c are parameters, and a is not zero. The formula, at least partially known to the ancient Greek, Babylonian, Chinese, and Indian mathematicians, is

$$x = \frac{-b \pm \sqrt{b^2 - 4ac}}{2a}$$

provided that the so-called discriminant of the equation, the number $b^2 - 4ac$, is not negative.

Imaginary Numbers

The Italian mathematician Rafael Bombelli, in his book *L'Algebra* written in 1569, proposed the introduction of a new number i, which should denote the square root of -1. Multiplying the number i by real numbers would yield square roots of negative real numbers. The new numbers so obtained are called "imaginary numbers,"

a name introduced by René Descartes (who meant it to bear a derogatory connotation). A new number system is obtained with the numbers formed by adding a real and an imaginary number; such numbers are called "complex numbers." Complex numbers can be added, multiplied, and divided, and the preceding quadratic formula shows that any quadratic equation has two solutions that are complex numbers; this remains true even if the parameters a, b, c are allowed to be complex number themselves. Actually, a stronger result holds true: Carl Friedrich Gauss (1777–1855) discovered that any equation of the form $a_0 x^d + a_1 x^{d-1} + \cdots + a_{d-1} x + a_d$ has d solutions in complex numbers, an important theorem known as the fundamental theorem of algebra. Partly thanks to this property, complex numbers are of fundamental importance in modern mathematics and in many fields of science and engineering, such as telecommunications.

Implications in Number Theory

Questions regarding squares, square roots, and quadratic forms have played a particularly important role in number theory, often giving rise to the simplest instances of rich theories. Numbers that are squares of integers are called "perfect squares," the first examples being 1, 4, 9, 16, 25, ... Galileo Galilei examined perfect squares in the attempt to understand infinity. Leonardo Fibonacci wrote a number theory book called *Liber Qudratorum*, the book of squares.

The problem of representing integers as sums of perfect squares has also received much attention. Pierre de Fermat (c. 1607–1665) proved that the odd prime numbers that are sums of two perfect squares are exactly those that have remainder 1 when divided by 4, an example being $13 = 2^2 + 3^2$ (whereas, for example, the prime number 7 has no such representation). Joseph Louis Lagrange (1736–1813) proved that every positive integer can be written as the sum of at most four perfect squares (for example, $15 = 9 + 4 + 1 + 1$); three squares suffice only for those numbers which are not of the form $4^k(8m + 7)$, as was later proved by Adrien-Marie Legendre.

In his 1801 masterpiece *Disquisitiones Arithmeticae*, written at the age of 21, Gauss investigated two problems whose generalizations are still major topics of current research. The first one is related to the question of representing integers as the sum of squares and asks for a classification of binary quadratic forms, which are functions of two variables x and y of the shape $f(x, y) = ax^2 + 2bxy + cy^2$, where a, b, and c are integer parameters, in terms of the set of integers they represent—the set of possible values of $f(x, y)$ as x and y range among the integers. The second problem considered by Gauss is the following: given two odd prime numbers p and q, is it possible to write p as the difference of a perfect square and a multiple of q (in symbols $p = n^2 - mq$)? Conversely, is it possible to write q as the difference of a perfect square and a multiple of p? Gauss proved that if at least one of p, q leaves remainder 1 when divided by four, then the two questions have the same answer; and that if p and q both leave remainder 3 when divided by 4, then the answer to the second question is "no" whenever the answer to the first question is "yes" and vice versa.

As a consequence of this result (known as the "quadratic reciprocity law") he was able to give an efficient method for answering the question. In fact, Gauss found not one but eight different proofs of this fact, which is so central in modern number theory that about 200 more proofs were later found.

Further Reading

Conway, John H., and Francis Y. C. Fung. *The Sensual (Quadratic) Form*. Washington, DC: Mathematical Association of America, 1991.

Mazur, Barry. *Imagining Numbers: (Particularly the Square Root of Minus Fifteen)*. New York: Farrar, Straus and Giroux, 2003.

Nahin, Paul J. *An Imaginary Tale: The Story of i*. Princeton NJ: Princeton University Press, 2010.

DANIEL DISEGNI

See Also: Babylonian Mathematics; Carpentry; Chinese Mathematics; Numbers, Complex; Numbers, Rational and Irrational; Pythagorean School.

Stalactites and Stalagmites

Category: Weather, Nature, and Environment.
Fields of Study: Algebra; Geometry.

Summary: The growth, age, and shape of stalactites and stalagmites can be mathematically calculated, depending on a variety of variables.

Stalactites and stalagmites are secondary minerals, also called "speleothems," formed as calcium carbonate, calcium oxide, and other minerals first dissolved in water and are then precipitated as water drips. Stalactites hang from cavern ceilings and concrete structures, and stalagmites rise from floors, sometimes meeting to create columns. Mathematicians, statisticians, geologists, and other scientists involved in studying stalactites and stalagmites develop complex, interdisciplinary theories and models as well extensions and applications. This work draws from many areas of mathematics, chemistry, and physics, especially fluid dynamics.

Growth and Dating

Stalactites and stalagmites form from chemical reactions involving ground water and minerals in the earth and the open areas of caves. The reactions typically consist of dissolving, precipitating, and—sometimes—evaporation. Chemical reactions of minerals first dissolving in water and then precipitating out of water are directly opposite to one another. The mathematical analogy of this relationship is an inverse function, and in either case, these processes may be quantified mathematically using standard chemical notation and formulas. Some stalactites and stalagmites are slow-forming, such as those made of calcium carbonate. Concrete or gypsum stalactites, which are made from more water-soluble materials, form much faster. For example, calcium hydroxide, which originates concrete stalactites, is about 100 times more soluble than calcium carbonate. Gypsum stalactites are formed by simple evaporation.

Dating of stalactites and stalagmites is complex because fluctuations in temperature or humidity can affect the pace of growth in such ways that length is not directly proportional to age. In some caves, because of minerals dissolving in water seasonally, stalactites and stalagmites may have annual bands, much as trees have rings, visible by the naked eye or under ultraviolet light. Dating with such direct methods, when available, can then be used to mathematically estimate and reconstruct temperature and humidity variation patterns in ancient times. However, the process is currently not reliable for anything less than very drastic climate changes.

Another method of dating involves collecting data on stalactite and stalagmite growth over several years. Then, data are used to determine the relationship between the size and the age, with approximations such as the method of least squares.

Dating with radioactive isotopes measures the ratio between a radioactive element, usually uranium, and the product of its radioactive decay. Electron spin resonance (ESR) dating is based on measuring radiation damage on calcium that forms stalactites and stalagmites.

These three methods of dating consistently produce average growth rates of about 0.1 millimeters (0.004 inches) per year in lime cave stalactites, with several times slower rates for stalagmites. Gypsum and concrete stalactites, formed by different reactions, grow several hundred times faster.

Unique, Optimal Shape

Plato supported the notion that there are true or ideal forms in nature, many of which may be expressed geometrically. While stalactites vary widely in size, they all tend to have a distinct, uniform shape that varies only by scale or magnification. Physicist Raymond Goldstein, part of an interdisciplinary team that investigated the mathematics of stalactite shape, said, "Although any particular stalactite may have some bumps and ridges that deform it, one might say that within all stalactites is an idealized form trying to get out." Using equations from fluid dynamics and other information about stalactite growth, the team developed a simulation and grew virtual stalactites under a variety of conditions, which they compared to real stalactites. The broad range of initial conditions for the mathematical model as well as for situations in real caves produced the same shape, though in caves, shapes can be distorted by impurities or breaks. The findings relate to other natural growth situations, including thermal vents and mollusk shells. To measure stalactites' shapes exactly without destroying them, the researchers use high-resolution digital cameras and scaled photography. This work also facilitates the mathematical study of stalactites' rippled patterns.

Further Reading

Ford, Derek, and Paul Williams. *Karst Hydrogeology and Geomorphology.* Hoboken, NJ: Wiley, 2007.
Pickover, Clifford. *The Math Book: From Pythagoras to the 57th Dimension: 250 Milestones in the History of Mathematics.* New York: Sterling Publishing, 2009.
Short, Martin, et al. "Stalactite Growth as a Free-Boundary Problem: A Geometric Law and Its Platonic Ideal." *Physical Review Letters* 94 (2005).

Maria Droujkova

See Also: Carbon Dating; Caves and Caverns; Probability; Transformations.

State Legislation

See *Government and State Legislation*

Statistics Education

Category: History and Development of Curricular Concepts.
Fields of Study: Communication; Connections; Data Analysis and Probability.
Summary: Statistics education has grown and adapted since the nineteenth century.

At the start of the twentieth century, science fiction author H. G. Wells asserted, "Statistical thinking will one day be as necessary for efficient citizenship as the ability to read and write." While use of statistical methods dates to earlier times, the first college statistics departments were founded in the early twentieth century, and many textbooks were written on statistical subjects like the design of experiments. A century after Wells's prediction, the notion of statistical thinking permeates all levels of education from kindergarten through college. In the early twenty-first century, there are increasing calls for statistical literacy in the United States and abroad in order to help people manage an increasingly complex and data-driven world.

Etymology

The word "statistics" derives from the term "state arithmetic," which refers to the various counting and calculating operations necessary for governments to operate effectively. The ancient Babylonians, Egyptians, Greeks,

Romans, Chinese, and others appear to have used various kinds of mathematics for activities like partitioning land and determining army sizes. The eleventh-century *Domesday Book*, a survey of England ordered by William the Conquerer, is another example of such state arithmetic. Statistician Maurice Kendall cites the first possible occurrence of the term "statistics" in the sixteenth-century work of Italian historian Girolamo Ghilini, who wrote about "*civile, politica,* [and] *statistica e militare scienza.*" However, he also traces the conceptual beginnings of the field to the "political arithmetic" of the seventeenth century and the work of researchers like pioneer demographers John Graunt and William Petty, who examined population growth and commerce in London versus Rome and Paris; and mathematician Edmond Halley, who some consider to be the founder of actuarial science for his work on life expectancy tables and insurance calculations. German historian and economist Gottfried Achenwall is frequently credited with inventing the German form of the word "statistics" in the eighteenth century and the related term *Staatswissenschaft* for political science. Their shared root *staats* means "state." Scottish politician John Sinclair appears to have been the first to use the term "statistics" in English in his *Statistical Accounts of Scotland*, a late eighteenth-century work addressing people, geography, and economics. He said: "I thought a new word might attract more public attention, I resolved to use it."

Historical Applications

In the nineteenth century, the ideas of statistical counting and calculating began to spread into a wider variety of political, social, scientific, and financial applications. For example, British physician William Farr received statistical training in France and applied statistics to medicine and models of epidemic diseases, calling his methods "hygiology" after the word "hygiene." He is credited as the founder of the field of epidemiology. Another pioneering epidemiologist was physician John Snow, who famously used statistical methods to trace the source of an 1854 cholera outbreak in London. His conclusions were politically controversial. In approximately the same period in the United States, self-taught statistician Lemuel Shattuck was appointed to plan a census of Boston in 1845 and later helped plan national census activities. He ultimately helped implement many local and state public health measures. Governments, businesses, and academic institutions increasingly used data and statistical methods to inform decisions. During this period, countless mathematicians, statisticians, economists, scientists, and others contributed to the development of statistical methods and the mathematical foundations of statistics, as well as the related field of probability. Many of them addressed both the theory and application of statistics.

Historical Education

Universities had existed in Europe since the Middle Ages. In other parts of the world, there were centers of learning at which scholars gathered to exchange ideas and teach. However, education in many academic subjects was often accomplished through mentorships or private tutoring. For example, nineteenth-century statistician and nurse Florence Nightingale was tutored in arithmetic, algebra, and geometry. She, in turn, tutored others before becoming involved in nursing. One of her tutors was the well-known mathematician of the period, James Sylvester. She was also influenced by the work of Farr and corresponded with mathematician Adolphe Quetelet, who was a pioneer in the use of statistics for anthropometry and criminology. She called him "the founder of the most important science in the world."

Other statisticians formed relationships with universities for research. For example, Karl Pearson, Francis Galton, and Walter Weldon worked at University College London. Pearson gave statistics lectures starting in 1894, and the trio founded the journal *Biometrika* in 1901 "as a means not only of collecting or publishing under one title biological data of a kind not systematically collected or published elsewhere in any other periodical, but also of spreading a knowledge of such statistical theory as may be requisite for their scientific treatment." Upon his death in 1911, Galton bequeathed the university a large endowment. Pearson became the first Galton Professor of Eugenics, sometimes called Galton Professor of Applied Statistics, perhaps because of the controversial nature of eugenics. That same year, Pearson was instrumental in creating the university's Applied Statistics department, now the Department of Statistical Science, which was recognized as the world's first college statistics department. It merged biometrics and eugenics (genetics) laboratories that had been founded by Pearson and Galton—though the Galton Laboratory later moved to the Department of Biology. Some other statisticians who worked or studied at University College London in the early nineteenth century include William "Student" Gossett, who is credited with the development of the Student's *t* distribution; Karl Pearson's son, Egon Pearson, who became the head of the Applied Statistics department when it split with the Department of Eugenics; Ronald Fisher, who was the first head of the Department of Eugenics and is referred to by some as the "father of modern statistics;" and Jerzy Neyman, who co-developed what is often called "Fisher–Neyman–Pearson inferential methods" or "classical" methods of statistical inference. These techniques typically use what is known as the "frequentist approach" to statistical analysis, which is based on defining probabilities of events as the limits of their relative frequencies over a large number of trials or experiments. It is perceived by many as being wholly objective and therefore "scientific." This approach is in contrast to Bayesian methods, named for mathematician Thomas Bayes. Bayesian statistical methods allow for subjective or belief-driven probabilities that may or may not be derived from observation or experimentation. The Applied Statistics department at University College London temporarily relocated during World War II; the war was to have a broad impact on mathematics and statistics in Europe and the United States.

Education in the United States

The Unites States was also developing its own college-level education programs at the beginning of the twentieth century. Similar to the department at University College London, many programs and other efforts started with individuals offering courses and partnerships between researchers and universities. One often-cited example is Iowa State University. George Snedecor, a professor in the Department of Mathematics, taught courses that included statistics content starting in 1914. He often focused on agriculture problems, a significant research area at the university. In 1924, he co-wrote a worldwide publication about computational statistical methods with Henry Wallace, who would later become Secretary of Agriculture and vice president of the United States. Iowa State created a statistical consulting and computing service in 1927, which was available to researchers in many disciplines. This service led to Iowa State's creation of the first recognized statistical laboratory in the United States, in 1933, and its Department of Statistics, in 1947. However, statistics degrees were offered before that time, beginning with Gertrude Cox's master's degree in 1931.

Cox went on to help found the Department of Statistics at North Carolina State University, one of the oldest statistics departments in the United States. She was the first female full professor and first female department head at the school and went on to start other college programs as well. An anecdote about her hiring at North Carolina State reports that, when Snedecor was asked to recommended five men for the job, he added to his letter: "...if you would consider a woman for this position I would recommend Gertrude Cox."

European statisticians also proved influential on U.S. statistics education and, in some cases, on government policy. Fisher visited Iowa State in the 1930s, and his agricultural work at the Rothamsted Experimental Station made a great impact on Snedecor. William Cochran, who was born in Scotland, also worked at the Rothamsted Experimental Station and taught at Iowa State. He went on to help create many statistics departments, including the one at Harvard, and he served on the committee that produced the 1960 *Surgeon General's Report on Smoking and Health*. Statistics proliferated, and similar efforts took place elsewhere, such as at the University of California, Berkeley. Neyman, who was born in Poland and also studied in England, France, and Russia, started working at Berkeley in 1938. Like many mathematicians and statisticians of the time, he was fleeing the growing Nazi influence in Europe. Prior to World War II, colleges sometimes offered a few undergraduate and graduate statistics courses but entire departments were still fairly rare. Thanks largely to Neyman's efforts, Berkeley had a department by 1955. He would also contribute significantly to experimental design, including some methods used by the United States Food and Drug Administration to test new medicines. Berkeley would become a center for mathematical statistics and was chaired for a time by statistician and mathematician David Blackwell, the first tenured African-American professor at Berkeley.

Post–World War II Statistics Education

Statistics and statistics education exploded after World War II, influenced by developments that occurred during the war and the subsequent Cold War. Statisticians had contributed significantly to the war effort in both the United States and Europe. For example, Hungarian mathematician and statistician Abraham Wald, who had suffered persecution for being Jewish, helped solve the problem of where to armor British bombers against antiaircraft fire. Others, like French-German Wolfgang Doeblin, would die as a result of the war. Later studies of Doeblin's works showed that he was an early pioneer of Markov chains, named for Andrei Markov. John Tukey was one of the most influential statisticians working in the mid- and later twentieth century. According to statistician Frederick Mosteller, the first chair of Harvard's statistics department and an influential force in statistics education: "He probably made more original contributions to statistics than anyone else since World War II." Tukey worked at the government's Fire Control Research Office during World War II, among his many roles. At the same time, he was often praised for his teaching. Mathematician Robert Gunning called him a "very lively presence on campus" and "a good and energetic teacher," who also helped schedule class and exam times in his head. As a member of Princeton's mathematics department, Tukey helped found the school's Department of Statistics in 1966, following earlier work by statistician Samuel Wilks, who had worked for the Office of Naval Research and profoundly influenced the application of statistics to military planning. The American Statistical Association's Samuel S. Wilks Award was named in his honor. Later, the department became the Committee for Statistical Studies, which encourages cross-disciplinary study of statistics and coordinates courses in many departments and programs.

The post-war extension of statistics into areas like clinical trials (pioneered by statistician Austin Bradford Hill), business, manufacturing (influenced by statisticians like W. Edwards Deming), and financial economics (for which economists Harry Markowitz, Merton Miller, and William Sharpe won a Nobel Prize), as well as the revival of Bayesian methods, meant that statistics was reaching a broader audience. It also meant that, more often, statistics courses were taught outside traditional mathematics and statistics departments. The debate over who should teach statistics was not new. Given that the discipline had been developed within so many fields—agriculture, psychology, biology, sociology, business, just to name a few—it was only natural that teaching would occur within these fields. Statistician John Wishart, who had worked with Pearson at University College London and with Fisher at Rothamsted, asserted that non-statisticians were not equipped to teach statistics or supervise statistical research. Fisher took a different approach, citing statistics' basis in research and applications and arguing for focused statistics offerings in departments in which statistics were often used, like psychology and biology. Around 1940, Harold Hotelling, who taught at Stanford University, Columbia University, and the University of North Carolina Chapel Hill, presented the idea that being a strong mathematician is not sufficient for teaching statistics, so mathematicians and statisticians were not always superior instructors versus individuals in other disciplines. He asserted that a statistics teacher must meld quantitative skills with "a really intimate acquaintance with the problems of one or more empirical subjects in which statistical methods are taught." Hotelling recognized that in typical academic structures, there might be some reluctance among faculty to teach courses that lay outside their specialty areas and that keeping current with statistics might be a daunting task for non-specialists. These issues remain matters of debate at the start of the twenty-first century. A study published in 2000, funded by the National Science Foundation, suggested that students were more likely to receive statistics education from instructors outside mathematics or statistics departments.

Employment

Through the 1970s, universities in the United States and elsewhere produced many statisticians or statistically trained practitioners in other disciplines, many of them to meet growing industry demands. However, employers were showing increasing concern that their new employees did not know how to practice statistics on the job, even if they had been instructed in current applied methods and practices in their academic programs. The American Statistical Association (ASA), which was founded in 1839, created a committee in the late 1940s to consider matters related to the training of statisticians. In 1980, the ASA Committee on the Training of Statisticians for Industry presented guidelines for programs that train industrial statisticians. One conclusion that spurred further debate stated: ". . . it is generally agreed that the MS degree is a minimum requirement for the professional statistician . . . it is recommended that someone interested in statistics as a profession obtain solid foundations in science or engineering and mathematics." Some discussion centered on balancing theory, applications, and employer-desired skills such as communication and teamwork. In Great Britain, the 1986 report *Supply of and Demand for Statisticians* cited both teaching factors and unrealistic expectations on the part of employers. Overall, in the 1980s, there were many general calls from statisticians to increase both the number and quality of programs, with mixed success. In the 1990s, there were also calls to increase the quality of undergraduate education and provide more interdisciplinary opportunities to graduate-level statisticians to "modernize" statistics for the twenty-first century. This call hearkened back to statistics' inherently interdisciplinary roots in previous centuries.

New Emphasis

The hallmarks of statistics education in the latter twentieth century and into the twenty-first century would be an increased focus on concepts over computation, statistical literacy, statistical thinking, use of real data, use of technology for both data analysis and conceptual understanding, and assessment to gauge student learning and understanding. Reports by several professional mathematical and statistical organizations contributed to this shifting educational emphasis. For example, the 1991 Focus Group on Statistics Education, part of the Curriculum Action Project of the Mathematical Association of America, produced *Heeding the Call for Change*. Later, the ASA Undergraduate Statistics Education Initiative (1999) focused on many aspects of education. One concern they noted was that many students were having a negative first experience in introductory statistics. In 2005, the Guidelines for Assessment and Instruction in Statistics Education (GAISE) committee, sponsored by ASA, produced

Impact of Computers

The advent of computers contributed to changes in statistical practice and new debates related to statistics teaching. Until then, many statistics courses had, of necessity, focused on teaching computational formulas, and statistical practice relied on techniques that were computationally tractable for researchers analyzing data by hand. Larger and larger data sets were becoming more common, requiring computer assistance for analyses. In the late 1960s, social scientist Norman Nie and computer scientists C. Hadlai Hull and Dale Bent developed the Statistical Package for the Social Sciences (SPSS) for mainframe computers. Academic use of the program soared when McGraw-Hill published a user's manual in 1970. By 1984, SPSS was the first statistical package offered for disk operating system (DOS) personal computers. Also in the 1970s, Numerical Algorithms Group (NAG) introduced its Algol 60 and Fortran algorithm libraries for mainframe systems.

The Statistical Analysis System (SAS) Institute emerged in 1976 from roughly a decade of work, starting with the University Statisticians Southern Experiment Stations, a consortium of universities funded by the U.S. Department of Agriculture and National Institutes of Health. SPSS, NAG, and SAS continue in the twenty-first century to offer a breadth of statistical software. Many other software packages and algorithms to graph and analyze data also emerged, some for general purposes and others for specific applications. One example is LISREL (an abbreviation of "linear structural relations"), which is used for structural equation modeling. It was developed by statisticians Karl Jöreskog and Dag Sörbom in the 1970s. Computers also revitalized interest in computationally intensive exact tests, iterative methods such as bootstrapping, and Bayesian analyses. Instructors debated the role of computers in the classroom. Many argued for statistical programming in languages such as Fortran or C, rather than point-and-click packaged routines, believing that statisticians should understand what the computer was doing. On the other hand, some classroom instructors advocated for the pedagogical utility of programs that computed statistics in a quick and easy manner, leaving the students free to focus on interpretation of results and "statistical thinking." The debate is ongoing. In the twenty-first century, many statistical programs contain both programming and menu-driven options, such as S-PLUS and its freeware version R.

K–12 and undergraduate reports focusing on instructional practice and assessment. There have also been recurring meetings, such as the International Conference on Teaching Statistics (ICOTS), which allow statistics instructors to address and debate issues, including the place of "classical" statistical methods versus Bayesian or computationally intensive exact methods in introductory classrooms; how best to meet the needs of non-majors taking statistics courses in mathematics and statistics departments; the "best" structure for introductory statistics textbooks; or the role of online tools and distance education.

The 2000 edition of the *National Council of Teachers of Mathematics Principles and Standards for School Mathematics* outlined standards for mathematics education that included statistics threaded from kindergarten through the last year of high school. Previously, statistics had been offered in various forms in high schools, though it presented some difficulty because many did not think it fit neatly into the traditional algebra, geometry, trigonometry, calculus sequencing used by many schools. The Advanced Placement (AP) Statistics exam was first offered in 1997. More than 7000 students took the exam, the most for a first offering of any AP exam as of 2010, and between 1996 and 2010 the rate of enrollment increased more quickly than any other course offered by AP.

Further Reading

Aliaga, Martha, Carolyn Cuff, Joan Garfield, Robin Lock, Jessica Utts, and Jeff Whitmer. "Guidelines for Assessment and Instruction in Statistics Education (GAISE) College Report." Washington, DC: American Statistical Association, 2005. http://www.amstat.org/education/gaise/.

Anderon, C. W., and R. M. Loynes. *Teaching of Practical Statistics*. Hoboken, NJ: Wiley, 1987.

Fienberg, Steinberg. "When did Bayesian Inference Become 'Bayesian?'" *Bayesian Analysis* 1, no. 1 (2006).

Gargield, Joan, ed. *Innovations in Teaching Statistics (MAA Notes #65)*. Washington, DC: The Mathematical Association of America, 2005.

Hulsizer, Michael, and Linda M. Woolf. *A Guide to Teaching Statistics: Innovations and Best Practices*. Hoboken, NJ: Wiley-Blackwell, 2009.

Salsburg, David. *The Lady Tasting Tea: How Statistics Revolutionized Science in the Twentieth Century*. New York: Holt Paperbacks, 2002.

Stigler, Stephen. "A Historical View of Statistical Concepts in Psychology and Educational Research." *American Journal of Education* 101, no. 1 (1992).

SARAH J. GREENWALD
JILL E. THOMLEY

See Also: Blackwell, David; Data Analysis and Probability in Society; Expected Values; Measures of Center; Normal Distribution; Permutations and Combinations; Probability; Randomness; Scatterplots.

Step and Tap Dancing

Category: Arts, Music, and Entertainment.
Fields of Study: Communication; Geometry; Representations.
Summary: Step and tap dancing each involve rhythms and combinations that can be analyzed mathematically.

Step dance is the type of dance focusing on feet movements. It de-emphasizes the other two spatial dance aspects—hand and body movement—and repositions dancers relative to the ground to form movement patterns. There are forms of step dancing in several cultural traditions, such as Malambo from Argentina, Irish stepdance, African-American stepping, and traditional Cherokee dancing. Related forms include clog and tap dancing.

The movements of these styles of percussive dance may be performed by a single dancer or choreographed among several dancers. Tony Award–winning choreographer and dancer Danny Daniels noted that, while an individual dancer may improvise, groups must be coordinated. The rhythms and counts for the dances he designed or performed on Broadway could be organized and detailed using mathematically based musical notation. Dance theorist Rudolf Laban used ideas from various fields, including crystallography, when he modeled dance dynamics. Scientists and dancers continue to develop notation and models to express human movement in tap and other dances. Dance algorithms may help create natural robotic movement. Dancer Gregory Hines said: "My style is part choreography, part improvisation. That gives me a chance to show people the possibilities of tap dancing, which, at its heart, is mathematics with endless possibilities."

Ratio and Proportion

There are several ratios related to music and choreography that determine movement in step dancing. Music time signature is written as a fraction with the denominator signifying the size of the notes used, and the numerator signifying the total length—in such notes—of a bar, which is the unit of music. For example, traditional music for Irish slip-jig has 9/8 time signature in the note pattern: quarter, eighth, quarter, eighth, dotted quarter (three-eighth). The five notes in the time signature correspond to two-and-a-half dance steps per bar, with long graceful slides between the steps.

The formula for a dance includes the number of bars in each repeating cycle (sometimes performed symmetrically) first for one starting foot and then the other. For example, a song that has 40 bars may be choreographed to include five step cycles, each spanning eight bars. Another ratio important for step dancing is the tempo of music, measured in beats per minute (bpm). Dancing competitions specify the tempo range for each type of dance. For example, single jig must be 112–120 beats per minute. Tap dancers of the past used their signature "time steps" (particular combinations of taps) to communicate the tempo to the accompanying band.

Patterns and Improvisation

In step dances, themes are expressed using sequences of the basic elements or steps. For example, common elements in tap dancing include shuffles, flaps, pullbacks, wings, and stomps. These sequences may be strictly choreographed from beginning to end, sometimes with repeating patterns or permutations of shorter elements, which can be repeated by any dancer who has learned the sequence. Improvisation allows the dancer to take basic elements and rearrange them in ways that may appear to be random to the casual observer.

Some step dance music has built-in departures from the standard bar structures. For example, Irish stepdance "crooked tunes" may have seven-and-one-half bar parts in addition to eight bar parts. Step dance patterns have multiple levels: steps within a bar, combinations of steps spanning multiple bars, and patterns of these step combinations. Order and perceived randomness can be manifested at all levels.

Dance-Dance Revolution

Dance-Dance Revolution (DDR) is a step dancing video game. The goal of the game is to match the pattern of steps on the screen and their rhythm on the special gaming pad with four or eight foot positions. The combination of visual, audio, and kinesthetic representations of the same rhythm have kept versions of the game popular around the world since its release in 1998.

Later versions of DDR use a mathematical visualization of multi-dimensional data, called radar diagrams, to rate the difficulty of individual dances. The variables describe different characteristics of the dance, such as steam (the density of steps) and chaos (the amount of steps that do not occur on beat).

Further Reading

Apostolos, M. K., M. Littman, S. Lane, D. Handelman, and J. Gelfand. "Robot Choreography: An Artistic-Scientific Connection." *Computers & Mathematics with Applications* 32, no. 1 (1996).

Maletic, Vera. *Dance Dynamics: Effort and Phrasing.* Columbus, OH: Grade A Notes, 2005.

Sethares, William. *Rhythm and Transforms.* New York: Springer, 2007.

MARIA DROUJKOVA

See Also: Ballroom Dancing; Contra and Square Dancing; Permutations and Combinations; Video Games.

Stethoscopes

Category: Medicine and Health.
Fields of Study: Algebra; Geometry.
Summary: Some modern stethoscope designs digitize sound waves, which can be modeled and analyzed.

The stethoscope is perhaps one of the most iconic pieces of medical equipment and is used by doctors in nearly every area of clinical practice around the world. From its beginnings as a simple tube to amplify sound, in the twenty-first century the stethoscope is evolving into a highly mathematical and computerized tool. It can record, analyze, and display diagnostic information using software and algorithms developed from clinical data using a variety of concepts and techniques from statistics, signal processing, spectral analysis, and related sciences. Further, mathematical models and simulations are increasingly used to support and validate clinical results.

History and Development

French physician René Laennec is credited with the invention of the "stethoscope" in 1816. The name comes from the Greek words meaning "chest" and "to examine." Knowing that solid bodies conduct and amplify sound, Laennec used tightly rolled and glued sheets of paper to hear patients' heartbeats. Experimenting with cylinders of various materials, he observed that an aperture maximized magnification of internal body sounds. His ultimate design was a straight, eight-inch wooden tube with a conical chest piece and a funnel-shaped stopper. Later physicians developed stethoscopes from materials like rosewood, papier-mâché, and even glass. The binaural form was popularized in the United States in the early 1900s by William Osler.

In the twenty-first century, the binaural acoustic stethoscope consists of a chest piece with a plastic disc (called a "diaphragm") on one side and a hollow cup (called a "bell") on the other. The bell transmits low frequency sounds and the diaphragm transmits high frequency sounds. A majority of clinicopathological correlations and diagnostic techniques used today result from patient data acquired by physicians listening with stethoscopes or a bare ear. Refinements in design and the increasingly widespread use of stethoscopes—coupled with training—improved observations. With respect to the heart, these included better precision in timing cardiovascular sounds, focusing on segments of the cardiac cycle in turn, and devising quantitative symbols to describe sounds. On the other hand, stethoscopes have also been investigated as a vector of disease transmission in busy clinical settings like emergency rooms.

Mathematical Modeling

Electronic systems of collecting and analyzing data have begun to supplement or even supplant the use of the stethoscope. Some predict that before 2020, manual stethoscopes will become obsolete. Electronic stethoscopes convert acoustic sound waves into electrical signals, which can be amplified and enhanced, producing both visual and audio output. Software can then represent cardiopulmonary sounds graphically and interpret them using mathematical algorithms. Signals may also be recorded or transmitted, facilitating remote diagnosis and teaching. Some research suggests that mathematical methods improve accuracy in diagnosing conditions, such as heart murmurs, but some methods have not yet shown clinical usefulness. Mathematicians and physicians continue to investigate and model cardiac sounds from murmurs and prosthetic valves, as well as other types of hemodynamic data, using techniques from spectral waveform analysis and physics concepts like damped oscillations of viscoelastic systems. They have also sought to quantify pulmonary sounds, like wheezing and crackles, and address signal processing issues, such as noise reduction, amplification, and filtration.

Measuring Blood Pressure

Blood pressure is the amount of pressure exerted by the blood upon the arterial walls. A clinician uses a device known as a "sphygmometer"—a device that pumps air into a cuff wrapped around a patient's arm—and listens for pulse sounds with a stethoscope, observing the height in millimeters of a column of mercury supported by the blood pressure. The sounds are known as "Korotkoff sounds," named for Russian physician Nikolai Korotkoff. A contraction of the heart that causes a pulse beat that supports a column of mercury 120 millimeters high is called a "systolic reading of 120." The reading in the period between contractions of the heart or pulses is called the "diastolic blood pressure." If the diastolic reading is 80 millimeters, the blood pressure is recorded as 120/80 and is read as "120 over 80." These numbers represent a ratio rather than a true fraction. The U.S. National Heart, Lung and Blood Institute defines normal blood pressure to be <120 for systolic *and* <80 for diastolic pressure and defines hypertension to be >140 *or* >90 for systolic and diastolic, respectively. These values are derived in part from statistical studies of typical human variation in blood pressure and associations with medical conditions like stroke and heart disease. Early diagnosis and appropriate treatment of hypertension is recognized as one of the most significant advances of modern medicine in reducing morbidity and mortality.

Further Reading

Bishop, P. J. "Evolution of the Stethoscope." *Journal of the Royal Society of Medicine* 73 (1980). http://www.ncbi .nlm.nih.gov/pmc/articles/PMC1437614.

Pullan, Andrew, Leo Cheng, and Martin Buist. *Mathematically Modeling the Electrical Activity of the Heart: From Cell to Body Surface and Back Again.* Singapore: World Scientific Publishing, 2005.

KAREN DOYLE WALTON

See Also: Diagnostic Testing; EEG/EKG; Mathematical Modeling.

Stock Market Indices

Category: Business, Economics, and Marketing.
Fields of Study: Algebra; Data Analysis and Probability; Measurement.
Summary: Stock market indices use sophisticated mathematical formulas to track the performance of the stock market and to help inform investors.

Mathematical stock market indices are used for a variety of purposes: as indicators of overall market health and activity, as measures of specific corporate profitability and activity, as performance metrics against which institutional investors (such as mutual fund managers) are measured, and for individual portfolio optimization and risk assessment. Some mathematicians and economists were developing price-based indices as early as the nineteenth century as well as analyzing pricing trends for explanation and prediction of market behavior. The Dow Jones Industrial Average (DJIA), named for journalist Charles Dow and statistician Edward Jones, appeared in 1896. Initially, it was a simple sum or average of the stock prices from 12 large companies. Since then, stock market indices have increased in their variety and mathematical complexity. For example, technical analysts use Fibonacci retracement levels, named after mathematician Leonardo Pisano Fibonacci, in order to model support and resistance levels in the currency market. Mathematicians and statisticians are instrumental in producing these indices. They also conduct theoretical and applied studies of market performance using these indices as data. In 1999, French-American mathematician Benoit Mandelbrot showed that market volatility can be modeled by fractal geometry, which contradicted

some aspects of modern portfolio theory. Author and mathematician John Allen Paulos addressed many mathematical stock market issues in his popular book *A Mathematician Plays the Stock Market.*

Definition and Examples

When describing the performance of the stock market as a whole (or a segment of the market, such as selected large-company stocks, or all small-company stocks, or stocks of all companies belonging to a particular industry), one is usually referring to a stock market index. Such an index is a representation of a hypothetical portfolio that contains a certain quantity of each of the stocks in the market (or market segment). The quantity of each stock in the fictitious portfolio depends upon the "weighting" technique employed.

Some of the more commonly encountered stock indexes include the following:

- S&P 500, comprised of 500 large-company U.S. stocks that cover about 75% of U.S. equities
- DJIA, comprised of 30 large-company U.S. stocks
- Wilshire 5000, comprised of the most common stocks in the United States (although not necessarily exactly 5000 of them)
- Nikkei 225, an index of Japanese equities
- FTSE, a collection of indices of British stocks

Building a Stock Market Index

The wide variety of stock market indices fall into several weighting categories, each involving a different mathematical approach to combining stocks within a hypothetical portfolio. One can imagine a potentially unlimited number of ways of creating a portfolio that includes numerous company stocks: for example, a portfolio comprised of one share of each stock, a portfolio comprised of the same dollar amount of each stock, and so on. The most common methods of weighting stocks within an index are price-weighting and market-value-weighting. (To simplify, stock performance is treated as only a function of changes in the stock price over time—as capital gains and losses. In reality, dividends, stock splits, and a variety of other issues must be taken into account, which makes the specific mathematical applications more complex than represented in this entry.)

Price-Weighted Indices

A price-weighted stock index represents a theoretical portfolio that includes one share of each stock comprising the index. The price or value of the index is then equal to the average of individual stock prices. Therefore, the relative impact of a given company stock on the index is a function of the company's stock price per share: larger prices per share imply greater influence on the index.

Suppose that $S_i(t)$ represents the per-share price of stock i at time t, and let $S_I(t)$ be the value of the index at time t. Then, the price of a price-weighted index could be defined as simply the arithmetic average of the stock prices in the index:

$$S_I(t) = \frac{\sum_{i=1}^{n} S_i(t)}{n}(t)$$

where n is the number of stocks comprising the index.

While the value of a price-weighted index is simple to calculate, typically the measure of most interest to an investor is not the actual price of the index, but rather the percentage change (the rate of return) in the index over a period of time. Let $r_i(t,t+1)$ be the rate of return on stock i during the period from time t to time $t+1$, and let $r_I(t,t+1)$ be the return on the index between times t and $t+1$ (assume an annual return period for purposes of this discussion, but returns can also be calculated daily, monthly, quarterly, or over any other period of time).

Then, the return on a price-weighted index is

$$r_I(t,t+1) = \frac{S_I(t+1)}{S_I(t)} - 1 = \frac{\sum_{i=1}^{n}\left[r_i(t,t+1) \times S_i(t)\right]}{\sum_{i=1}^{n} S_i(t)}.$$

Multiplying this value by 100 yields the return expressed as a percentage change. The DJIA and other Dow Jones averages are examples of price-weighted indices.

Market-Value-Weighted Indices

A market-value-weighted (also called "value-weighted") stock index is one that weights the individual per-share stock prices according to the relative market values, or market capitalizations (called "market cap" for short), of the component stocks. A company's market cap is simply the totally value of its outstanding equity and is calculated as the per share stock price multiplied by the number of stock shares outstanding. Thus, an individual company's influence on a value-weighted index is a function of the overall equity value, or size, of the company—larger companies have greater influence on the movement of the index.

Using the notation introduced above, and letting N_i represent the number of shares of stock i outstanding, the rate of return on a value-weighted stock index would be

$$r_I(t,t+1) = \frac{\sum_{i=1}^{n}\left[r_i(t,t+1) \times S_i(t) \times N_i\right]}{\sum_{i=1}^{n}\left[S_i(t) \times N_i\right]}.$$

The S&P 500 and other Standard & Poor's indices are examples of market-value-weighted indices.

Other Types of Index Weightings

While price-weighted and value-weighted indices are common, there are other weighting techniques that can be used. For example, it is possible to create an index that gives equal weight to the return of each stock comprising the index. In such a case, the return on the index would be calculated as

$$r_I(t,t+1) = \frac{\sum_{i=1}^{n} r_i(t,t+1)}{n}.$$

With such an index, the performance of each stock has the same impact on the overall index return as every other stock.

Another possibility in creating an index would be to use geometric, as opposed to arithmetic, averaging. A geometric average is calculated by multiplying n numbers together and taking the n-th root of the product (as opposed to summing the numbers and dividing by n, as with an arithmetic average).

The key in interpreting the various types of stock market indices is to know their underlying construction and to understand and interpret them appropriately. Price-weighting and equal-weighting, for example, can result in very different index performance indications than value-weighting, even relative to the same underlying stock return data. The appropriate index to use in a given situation depends upon the specific purpose in mind. If one wants a measure of market performance that is more influenced by the price movements in the stocks of larger companies, for example, a value-weighted index may be most appropriate. If the sizes of companies are not relevant for analytical purposes, or if the companies that comprise an index are very similar in size and other attributes, a price-weighted or value-weighted index may be appropriate.

Further Reading

Bodie, Zvi, Alex Kane, and Alan Marcus. *Investments.* New York: McGraw-Hill/Irwin, 2008.

Paulos, John Allen. *A Mathematician Plays the Stock Market.* New York: Basic Books, 2003.

RICK GORVETT

See Also: Money; Mutual Funds; Pensions, IRAs, and Social Security; Probability.

Strategy and Tactics

Category: Government, Politics, and History.
Fields of Study: Geometry; Measurement; Problem Solving; Representations.
Summary: Mathematical concepts and processes can be used to analyze optimal strategies in a variety of situations.

In a competitive situation, such as businesses selling similar products, armies engaged in battle, opponents playing games, oil companies deciding where to drill, and employees bargaining for better salaries, successful outcomes depend on choosing the best plan of action from among a set of strategies to achieve a specific outcome. In many cases, mathematics can be used to analyze the situation and help to choose the best strategy. Mathematical techniques have been—and will continue to be—developed to address a wide range of problems in areas such as military logistics, intelligence, and counterintelligence.

The first step in the process is to determine the objective. That goal may be to maximize profit, beat the opposing army, or win the game. Next, the possible strategies to choose from and the limitations or constraints that may affect the choice of strategy need to be identified.

In competitive situations, the opponent's choice of strategy must be taken into consideration as well. While there are many examples of systematically analyzing and selecting the "best" strategies throughout history, the twentieth century—especially the World War II era—saw the emergence of operations research as the discipline that explores and develops systematic techniques for making decisions that are the "best" in some sense, usually maximizing profits/benefits or minimizing costs/liabilities.

Decision making can be approached mathematically in a number of ways depending upon the situation involved and the information available.

Linear Programming: Choosing the Best Option When Resources are Limited

Many decision problems arose out of troop supply needs during World War II. With a war on several fronts, deciding how to ship the limited troops and supplies to maximize their effectiveness was daunting. Many of the situations had the following characteristics:

- There were resources needed in specific combinations by a number of end users, and the amount of each resource was limited.
- The resources were used proportionally for each combination (in other words, to assemble whole units from raw materials, the number of raw materials needed was the same for each unit produced).
- The goal was to maximize the benefit or minimize the cost, and the cost or benefit was proportionally related to the number of units produced (in other words, the more produced, the higher the benefit or cost).

These characteristics yield a mathematical structure that is linear. Each resource corresponds to an equation or inequality that is a linear combination of unknown quantities representing the units to be combined or produced The objective function is also a linear combination of the number of units. See Example 1 for a very simple, classic example that involves deciding how to prepare a "balanced" meal.

Example 1. A linear programming problem.

A dietician wishes to prepare a salad meal that has a minimum amount of calories but still satisfies nutritional requirements. In particular, it must have at least 30 grams of protein and at most 9 grams of fat. The foods available are an ounce of lettuce with 4 calories, no fat, and 1 gram of protein; and slices of roast beef with 90 calories, 3 grams of fat, and 16 grams of protein. What amounts of lettuce and beef should the dietician serve with a diet salad dressing? Minimize calories $= 4L + 90B$, where $1L + 16B \geq 30$ and $3B \leq 9$.

These problems are easy to solve when they are small, like the problem in Example 1. The problems that arise in practice—such as those under consideration during World War II—are usually much larger and can involve hundreds of unknowns. During World War II, British and U.S. mathematicians looked for an approach that could make use of computers, which were being developed at that time and offered the possibility of performing many simple calculations quickly. In 1947, too late for the war effort, U.S. mathematician George Danzig (1914–2005) developed the simplex algorithm for solving linear programming problems. The simplex algorithm is an efficient recipe for solving linear programming problems of any size and is very easy to program on a computer. In the decades since the development of the simplex algorithm, many industries have used this procedure to solve problems in fields as diverse as banking, natural resources, manufacturing, and farming.

Linear programming problems are usually used to model static situations in that the final solution is essentially the result of one decision made under a clear set of assumptions. Many decision problems are more complicated, with a number of intermediate decisions to be made. These more dynamic problems often involve a probabilistic component as well, with uncertainty playing a complicating role in each decision.

Game Theory

Often, people are faced with a decision in which the resulting payoff will depend on external forces that are hard to predict (like natural forces). One option may always be best, but it is more likely that the best choice will simply "depend" on other factors. For example, when deciding which crop to plant, a farmer can list seed costs and profits based upon yield, but the yield will depend on the weather. A table can be made for each crop choice based upon several different weather scenarios, with past experience used to assign a probability to each possible weather scenario. Example 2 provides a standard format, usually called the "payoff matrix."

Example 2. A payoff matrix.

	List the possible states of the external forces
List the possible actions to choose from in making the decision	List the gain (profit, benefit, etc.) for each combination of actions and states.

Many decisions can be similarly structured, including determining what stocks to buy, what products to market, and what wars to wage. Different people will make different decisions depending upon their comfort level with risk.

Strategies for systematic decision making can be placed in four categories:

1. *Optimist strategy*: "MaxiMax" (Maximize the maximum gain). Find the best gain for each possible action and choose the largest of these maximums. Of course, that action may have the most risk associated with it, since the maximum gain may also coincide with the least likely state for the external force. In this case, the farmer may plant something that would have huge profits but only in the most unlikely weather conditions.
2. *Pessimist strategy*: "MaxiMin" (Maximize the minimum gain). Find the smallest gain for each action, and choose the largest of these minimums. This is a safe choice because it yields the minimum guaranteed gain regardless of external forces. In this case, the farmer may choose a "safe" crop to plant. If weather is really good, another crop would have been a better choice.
3. *Balanced strategy*: "MiniMax Regret." Calculate the "regret" for each possible action by determining the cost of choosing that

action compared to benefits of the best state of the external forces. Find the worst (largest) regret for each action and pick the action with the smallest worst-case regret.

4. *Averaging strategy*: "Expected Value." Use the probabilities governing the external forces to determine the expected gain for each action and choose the highest one. Expected gain or payoff is calculated as a weighted average of the gain for each state of the external force where the weight for each state is the probability of that state occurring. This strategy can be thought of as determining the action that, when chosen repeatedly, provides the best average benefit over the long term. For the farmer, this may not seem reasonable, since the decision under consideration is what to plant in a single, given year.

When the external force is an opponent with choices to make rather than a natural phenomenon with a random component, these decision situations can be examined as mathematical games. Two-person games can be represented with a payoff matrix as in Figure 2. The "row" player lists strategies on the left and the "column" player lists strategies across the top. The entries of the matrix are pairs of numbers, the row player's payoff, and the column player's payoff, respectively. In situations where the row player's winnings are equal to the column player's losses, and vice versa, the payoff matrix entries can be completely defined with one number, conventionally the row player's payoff. These games are called "zero-sum games" because for a particular pair of strategies, the row player's payoff and the column player's payoff, being negatives of each other, sum to zero.

Example 3. The prisoner's dilemma.

Two suspects are arrested by the police. They are each offered the same deal: Confess and receive a reduced sentence. If one confesses and the other does not, the confesser goes free and the other gets a 10-year sentence. If both confess, each gets a five-year sentence. If neither confesses, both get a one-year sentence on reduced charges. Neither prisoner knows what the other will say. What should they do?

	Confess	Refuse
Confess	$(-5,-5)$	$(0,-10)$
Refuse	$(-10,0)$	$(-1,-1)$

While mathematicians have been studying decision making and games of strategy systematically for several centuries, game theory emerged as a recognized mathematical approach to analyzing these decision processes in the 1930s and 1940s through research published by John von Neumann (1903–1957). The "prisoner's dilemma" (Example 3) was investigated in the 1950s and led to additional interest in the field.

The prisoner's dilemma captures many interesting features of competitive situations. Analysis shows that the intelligent prisoner should always confess, since the "best" outcome will occur no matter what the other prisoner decides to do: -5 is better than -10 if the other prisoner confesses; 0 is better than -1 if the other prisoner refuses to talk. However, this individual "best choice" results in each prisoner confessing and getting

Figure 1. A "decision tree."

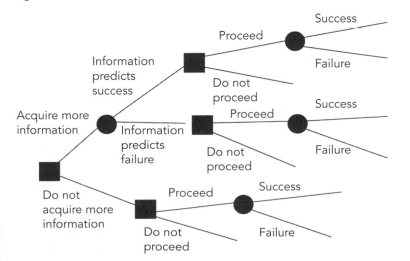

a five-year sentence, whereas if neither confesses, they only get one-year sentences. This feature of competitive behavior and strategies can be thought of as the friction between basing strategic decisions on individual goals or on the common good.

With appropriate choices for the values in the table, these games could model a number of competitive situations, such as two companies trying to determine what price to set for competing products or two armies determining how to wage war.

Decision Trees

In situations where the ultimate decision depends on an intermediate choice, a decision tree can help to organize the information and facilitate a systematic analysis. A company may be ready to bring a product to market and needs to decide whether or not to invest funds up front in a test market exercise. The test market may bring in better information about how to market the product on a larger scale, thus increasing profit, but the cost of the test market exercise would also take away from the profit. An oil drilling company could choose to invest funding in test wells before determining the final drilling location. A university may be trying to hire a senior administrator and could choose to invest funds in a head-hunter search firm.

In all of these situations, the outcomes can be organized into a tree diagram like the one in Figure 1. Each "decision fork" is represented by a square, and each event fork—governed by external, possibly random forces—is represented by a circle. The branches leading from the event forks have probabilities assigned based upon the likelihood that an outcome will occur. Typically, acquiring additional information will result in an increased probability of success (or failure), and so the probabilities of success and failure will be different for different event forks.

Each terminal branch represents a final outcome. If current assets, the cost of the information acquisition, and the gains or losses under success and failure are known, then each terminal branch can be labeled with the net gain (or loss) for that option. Once those values are determined, the tree can be "folded back" through calculating the expected outcomes from the probabilities to determine which decisions to make to maximize the gain.

The decision points and events may include more than two options or outcomes, and there may be more

than two decisions to be made before the final outcome, so the tree may have more forks and branches than the one in Figure 1 but the analysis process is the same.

From these trees, the value of the additional information acquired can be calculated. This calculation can assist companies in determining how much they should be willing to pay for that information. Also, the amount of risk a company is willing to assume can be incorporated into the process, allowing companies that are willing to shoulder a larger risk for the (slimmer) chance of a larger gain to include that information into the analysis.

Further Reading

Mesterton-Gibbons, M. *An Introduction to Game-Theoretic Modeling*. Redwood City, CA: Addison-Wesley, 1992.

Raifa, H. *Decision Analysis: Introductory Lectures on Choices Under Uncertainty*. Reading MA: Addison-Wesley 1968.

Winston, W. L. *Operations Research: Applications and Algorithms*. 4th ed. Belmont, CA: Brooks Cole-Thompson Learning, 2004.

HOLLY HIRST

See Also: Coding and Encryption; Intelligence and Counterintelligence; Predicting Attacks; Risk Management; Scheduling.

Street Maintenance

Category: Travel and Transportation.
Fields of Study: Data Analysis and Probability; Geometry; Measurement.
Summary: Street maintenance requires planning, preparedness, and risk assessment, all of which involve mathematics.

Stone paved roads date back thousands of years and mathematicians and architects have long investigated ways to lay paving stones. Another connection between mathematics and streets dates to when Hermann Minkowski proposed numerous metric spaces, one of which is referred to in the twenty-first century as "taxicab geometry." Some streets are laid out on a grid system, leading to mathematical investigations in taxicab geometry. The surface curvature of roads is also mathematically interesting and important in drainage and safety issues. Street maintenance is a combination of services that includes resurfacing of streets and curbs, pothole patching, sweeping, snow removal, and maintenance of drains. Mathematical problems that arise within street maintenance have to do with engineering, applied physics and chemistry, logistics, budgets, and communication.

Types of Streets

Different types of roads call for different maintenance. Civil engineers can use tools such as falling weight deflectometers to measure properties of street coverings—in this case, deformation under dropped weight. A heavily loaded truck can damage the street surface approximately 10,000 times more than a small passenger car. This fact explains why streets with industrial traffic require more frequent maintenance, owners of trucks pay more taxes, and trucks are not allowed on most streets.

There are many materials used to cover streets, and the choice of material provides interesting mathematical optimization problems. For example, rubberized asphalt contains recycled tires, which is an environmental bonus and can reduce the noise of the road by about 10 decibels, which is valuable for nearby homes. However, it can only be laid in certain temperatures. Concrete is more durable than asphalt but is more expensive and harder to repair. Brick and cobblestone coverings do not form potholes and can hold heavy loads. However, they are noisy, they require manual installation and maintenance, and they can damage cars.

Potholes and Fatigue

Most potholes happen because of what is known in the materials science as "fatigue" of the surface. Fatigue occurs when materials are subject to periodic forces, such as heavy cars passing through. Small cracks start to appear, which then aggregate into networks of cracks, which then give way to a pothole. Calculus, differential equations, and statistics models are used to test road surface materials for resistance to fatigue and to predict fatigue's time through the statistically derived fatigue curves (S-N curves). Cycles of heating and cooling can quickly extend existing cracks and make potholes larger as well as freezing water that has seeped into cracks.

Cleaning

The mathematics of cleaning schedules involves balance among many random variables, such as traffic or seasonal leaves removal. In a typical city, urban streets with heavy pedestrian traffic are swept daily, and other streets are swept every week or two. Statistical data on street use determines where to place garbage cans and how often to empty them, when to send heavy sweep machines for cleaning, and how to avoid disrupting regular street use and events with cleaning activities.

Some street maintenance measures prevent street dirt. Highly visible trash cans can drastically reduce littering. In many communities, residents are invited to participate in street cleaning and maintenance to some degree, from sites where they can report potholes to street cleaning celebrations on holidays or weekends. Birds can be attracted to appropriate places, and dog owners are guided to special parks and runs. Mathematical models behind such measures come from studies of human and animal behavior.

Accidents and disasters—from dust storms to spilled poisons—may require special cleaning activities. Because such events are rare but require special knowledge and equipment, it usually makes sense to maintain tools and specialists for these special events only in large cities and to send teams to smaller places that need help.

Snow Maintenance

Streets under snow require special maintenance, including mechanical removal of the snow by snowplows, snow-blowers, or shovels; inert surface treatment for traction with sand or sawdust; and chemical surface treatment. The mathematics of dealing with snow includes economical and environmental factors. When snow immobilizes traffic, productivity and sales are lost. However, snow-removal measures cost money and take time. In cities where it snows infrequently, it is usually cheaper to wait for the snow to melt rather than to maintain a fleet of removal machines.

Most of the chemical treatment of snow is done with sodium chloride (table salt). Salt makes snow melt at about 10 degrees Fahrenheit less than usual (freezing-point depression). Switzerland uses more than a pound of salt a year for every square yard of its roads. Chemical treatments can damage plants and animals throughout the watershed. Safe amounts of chemicals can be determined based on ecological models. Chemicals also cause vehicle damage and faster road deterioration. These costs are part of the decision of which type of snow maintenance is more economically sound.

Further Reading

Kelly, James, and William Park. *The Roadbuilders.* Reading, MA: Addison-Wesley, 1973.

Krause, Eugene. *Taxicab Geometry: An Adventure in Non-Euclidean Geometry.* New York: Dover Publications, 1987.

Perrier, Nathalie, Andrew Langevin, and James Campbell. "A Survey of Models and Algorithms for Winter Road Maintenance. Part IV: Vehicle Routing and Fleet Sizing for Plowing and Snow Disposal." *Computers & Operations Research* 34, no. 1 (2007).

MARIA DROUJKOVA

See Also: Bicycles; Green Design; Traffic.

String Instruments

Category: Arts, Music, and Entertainment.
Field of Study: Geometry; Number and Operations; Representations.
Summary: The harmonics and timbre of wind instruments are described and computed using mathematics.

All stringed instruments exhibit a fundamental property of physics in that when impacted, they vibrate at numerous frequencies. The vibration of the string displaces the air around it, which—when impacted on the human eardrums—creates the sensation of sound. Some of the common instruments in the string family are violin, guitar, harp, mandolin, cello, and banjo. A modern violin has about 70 parts, and the overall design of such complex string instruments is inherently mathematical. Features such as string tension, area, and shape of the top plate, and spacing of frets all have mathematical properties that influence sound.

For any string, at a given tension, only one note will be produced. To generate multiple notes from the instrument, many strings may be used to span the desired frequency spectrum (for example, harps) or the string may be forced to vibrate at different lengths, thereby changing the frequency (for example, guitars). On an equally tempered instrument like a guitar, the spacings of the frets, which help a player adjust string length, have to be scaled by the ratio $2^{1/12}$. This problem is mathematically equivalent to duplicating a cube, which is one of the classic problems of antiquity. Mathematician Jim Woodhouse has studied violin acoustics using linear systems theory and mathematically modeled "virtual violins," as well as related vibration problems like vehicle brake squeal.

Harmonic Series and Fundamental Frequency
When a string is plucked, struck, or bowed, it resonates at numerous frequencies simultaneously. The waves travel up and down the string. These waves reinforce and annul each other, which results in standing waves. The one-dimensional wave equation is used to model string instruments. A harmonic series is composed of frequencies that are an integer multiple of the lowest frequency. Fundamental frequency is the lowest frequency in a harmonic series. The musical pitch of a note is usually perceived as the fundamental frequency. The fundamental frequency (f) of a string can be computed as

$$f = \frac{\sqrt{\frac{T}{\frac{m}{L}}}}{2L}$$

where T is the string tension in newtons, m is the string mass in kilograms, and L is the string length in meters. The fundamental frequency is also known as the "first harmonic."

Timbre
Timbre is the quality of a musical note and is what defines the character of a musical instrument. When two different instruments play the same note, the note could have the same frequency. The human ear distinguishes the source of the note because of timbre. Hermann Helmholtz was the first to describe timbre as a property of sound. When an instrument plays a certain note, the outputted sound consists of the fundamental frequency

and its harmonics. These harmonics differ from instrument to instrument—what is known as "timbre."

Further Reading

Hall, Rachel W., and Kresimir Josic. "The Mathematics of Musical Instruments." *American Mathematical Monthly* 108, no. 4 (2001).

Mottola, R. M. "Liutaio Mottola Lutherie Information Website: Technical Design Information." http://liutaiomottola.com/formulae.htm.

Rossing, Thomas. *The Science of String Instruments.* New York: Springer, 2010.

Ashwin Mudigonda

See Also: Harmonics; Percussion Instruments; Pythagorean and Fibonacci Tuning; Wind Instruments.

Stylometry

Category: Arts, Music, and Entertainment.
Fields of Study: Data Analysis and Probability; Measurement.
Summary: Stylometry is a descriptive science that uses statistical techniques to identify authorship of written materials.

Stylometry is a descriptive science that uses statistical techniques to identify authorship or written materials. In addition to comparing simple frequency patterns of words, stylometry focuses on the groupings of words and the position of these words in sentences. Using stylometry, scholars have tried to determine if Homer wrote the last book of the *Odyssey*, if the Apostle Paul wrote the *Letter to the Ephesians*, and if Shakespeare wrote the first act of the play *The Booke of Sir Thomas Moore*. Because of the successful use of stylometry, its techniques have been expanded to help identify composers from their musical compositions and analyze artists from their paintings.

Beginnings of Stylometry

In 1851, August de Morgan, an English mathematician, initiated the field of stylometry when he suggested that authors could be identified by the average number of letters in their written words. Because de Morgan's suggestion was simplistic and often misleading, stylometry did not gain validity until 1944, when Udny Yule published his pioneering work that suggested that an author's vocabulary usage did not depend on sample size. Analyzing Paul's Epistles and the words of the physician Hippocrates in 1957, W. C. Wake was the first to produce an acceptable test of authorship using distributions, sampling methods, and periodic effects within distributions. In 1961, A. Q. Morton and others used computer technology to both extend and verify Wake's approach.

Uses of Stylometry

Stylometry has other constructive uses, such as the use of statistical techniques to examine concordances (13 million words) of the works of Saint Thomas Aquinas (illustrated above). As a result, scholars not only identified spurious additions introduced by editors of Aquinas's works, but also successfully reconstructed lost passages. Scholars used a similar approach to examine stylistics differences among the three Greek tragedians—Euripides, Aeschylus, and Sophocles—trying to also establish chronological progressions across a single author's works in terms of vocabulary, themes, and use of iambic trimester.

A specific example of scholars' use of stylometry involves *The Booke of Sir Thomas Moore*, a play about a martyred Englishman in 1535. Scholars first concluded that the play was a composite effort of five authors, with handwriting analyses accepted as proof that William Shakespeare was the sole author of two of the play's sections. Then, computer analyst Thomas Merriam created computer databases of the play in question and three other Shakespearean plays—*Julius Caesar*, *Pericles*, and *Titus Andronicus*.

The concordances generated for all four plays revealed significant similar frequencies of "word habits" or repeated combinations of words and phrases. Though Merriam concluded that Shakespeare was the sole author of *The Booke of Sir Thomas Moore*, his stylometric data did not convince all scholars. Skeptics such as these claim that Merriam's techniques are at best informative, being suspect because the three comparison plays are not the best representatives of Shakespeare's style.

Modern Applications

Stylometry has been used in court cases to identify "fraudulent" wills and "false" criminal confessions. In the late 1970s, defense attorneys for kidnap victim and accused bank robber Patty Hearst tried to introduce stylometric evidence that "proved" the tape-recorded "communiqués" read by Hearst were not her own words.

Their evidence was based on concordances built from previous essays by Hearst, oral conversations, her confession, and materials produced by the Symbionese Liberation Army. The attorneys carefully analyzed these concordances using statistical discrimination, cluster analysis, and *t*-test comparisons to examine factors such as average sentence lengths, parsing patterns involving conjunctions, and linguistic habits. Despite the defense's protests, the trial judge and the appeals court both ruled that the stylometric evidence was not admissible and thus was never used.

Donald Foster, a Vasser College English Professor, used stylometry to identify with 99% confidence the "anonymous" author of the political text, *Primary Colors*. Though *Newsweek* columnist Joe Klein originally denied being the suspected author, he eventually admitted to the deed. Since that time, Foster has helped confirm Ted Kaczynski's authorship of the *Unabomb Manifesto* and identify Eric Rudolph as a suspect in the 1996 Atlanta Olympics bombing.

Further Reading

Juola, Patrick. "Authorship Attribution." *Foundations and Trends in Information Retrieval* 1, no. 3 (2006). http://www.mathcs.duq.edu/~juola/papers.d/fnt-aa.pdf (Accessed February 2011),

Michaelson, S., A. Q. Morton, and N. Hamilton-Smith. "Fingerprinting the Mind." *Endeavor* 3, no. 4 (1979).

Morton, A. Q. Literary Detection: *How to Prove Authorship and Fraud in Literature and Documents*. New York: Charles Scribner's Sons, 1979.

Roberts, David. "Don Foster Has a Way With Words." *Smithsonian* (September 1, 2001).

Schwartz, Lillian. "The Art Historian's Computer." *Scientific American* (April 1995).

Yule, Udny. *The Statistical Study of Literary Vocabulary*. Cambridge, England: Cambridge University Press, 1944.

JERRY JOHNSON
WESTERN WASHINGTON UNIVERSITY

See Also: Diagnostic Testing; Literature; Probability.

Submarines

See *Deep Submergence Vehicles*

Succeeding in Mathematics

Category: School and Society.
Fields of Study: Communication; Connections; Problem Solving.
Summary: Poor mathematics performance can be attributed to a variety of factors and numerous organizations and strategies are believed to help students achieve mathematics success.

Many educational initiatives are designed to motivate U.S. students to excel in science and mathematics, with the goal of building the strong science, technology, engineering, and mathematics (STEM) workforce needed to meet twenty-first century challenges. Three overarching goals of the 2009 federal Educate to Innovate program are increasing STEM literacy for everyone; improving teaching so that American students meet or exceed those in other nations; and expanding STEM education and career opportunities for underrepresented groups. However, success in mathematics, or even literacy, can be difficult to define. Some see it as some minimum skill set, number of courses, or type of courses taken. Others conceptualize it by what sorts of problems students are able to solve or by their ability to manage real-world mathematical problems, such as budgets or loans.

Measuring Success

There are many barriers to achieving success. Broad application of standardized testing in mathematics education sometimes reduces the measure of success to a single score or change in scores over time. Modern educational approaches and programs at all levels increasingly emphasize problem solving, which the National Council of Teachers of Mathematics asserts is not well measured by standardized tests, since problem solving reaches beyond simply remembering some encapsulated set of concepts, formulas, and skills to include broader applications, novel situations, and mathematical thinking and reasoning. There are calls for innovative assessment alongside changes in educational practice in order to attempt to capture what it means to know, do, and be successful in mathematics at home, school, and work. Among professional mathematicians, measures of success vary as well, with ongoing debate about various aspects of teaching and scholarship, including how and whether to assess measures like the number of publications, the number of citations to an author, the quality of a journal, or letters from peers, and the role of other measures like student evaluations. Educational researcher Christopher Jett, who examined mathematics success among African-American men, uniquely defined mathematics success as, "being able to use mathematics as an analytical tool to educate, stimulate, and liberate the (my) people."

Failure and Anxiety

Albert Einstein was once quoted as saying, "Do not worry about your difficulties in mathematics. I can assure you mine still are greater." This assertion seems contrary to many people's belief that success in mathematics is binary: people are successful or not, with no middle ground. Popular culture portrayals of mathematicians as geniuses often inadvertently support the mistaken belief that one must be gifted to succeed in mathematics. Further, while mathematicians are portrayed as wizards, mathematics itself is often shown as a sort of mysticism—a secret and arcane knowledge accessible only by a select few. In reality, mathematics encompasses a diversity of fields and professional mathematicians have varying sets of competencies, personalities, and working styles. Likewise, students at any level may have command of a wide variety of skills and concepts, and while those concepts are related and may build on one another, competence is not uniform. A student who struggles all through algebra may still be successful in geometry. A student who labors over constructing a proof may have a flair for data analysis and statistics.

Mathematics is inherently cumulative in nature. The feeling of failure at mathematics, especially given the common binary view, can seemingly be caused by a small problem that actually immediately impacts only one small area. This partial temporary failure can result in a long-lasting loss of confidence. Mathematics anxiety is an increasingly recognized phenomenon that interferes with students' ability to learn mathematics and perform at the best of their abilities, regardless of their actual skill. Many people who are perfectly capable of learning and using mathematics feel anxiety about it and will avoid using mathematics whenever they have the option. Over time, this can lead to degradation of their abilities as they fall out of practice, which cyclically reinforces the anxiety. In addition to avoidance, mathematics anxiety can sometimes negatively affect working memory. As the anxiety grows, the student has more trouble keeping track of tasks, leading to poor performance and yet again reinforcing the anxiety. Some believe that mathematics anxiety is caused in part by poor performance on mathematics achievement tests and in part by early difficulty in mathematical skill development. The anxiety remains even after actual performance has improved and may be related to a belief that the earlier difficulties reflect some inherent character trait rather than a situational difficulty. People who later in life describe themselves as "terrible" at mathematics but who display education-appropriate competence in mathematics may not remember the original event that inculcated in them this belief that they are poor performers. Beyond performance, some cite teacher or classroom practices as contributing to mathematics anxiety. Like other mathematicians, mathematics teachers do not possess equal skills and levels of comfort in all areas of mathematics. This may be especially true of elementary school teachers who must teach a wide array of subjects on a daily basis. Classroom practices such as emphasizing the "right" answer, which also frequently occurs on standardized tests, can increase anxiety because some students attach great significance to being "wrong." Other students may feel anxiety over being asked to "show their work," because they are less confident in their mathematical thinking than in their ability to produce a correct answer.

The Mathematics Anxiety Rating Scale (MARS) was developed by psychologists in the early 1970s and exists in several versions, including foreign language adaptations. Researchers using this scale and other measures have identified other situations more likely to trigger anxiety. For example, tests where problems become progressively more difficult appear to trigger mathematics anxiety more often than tests in which the distribution of problems by difficulty is more random, which is true even when all the problems on the test are well within the skill level of the test-taker. Timed tests and the possibility for public embarrassment, such as working at a board in the front of the class, are also factors that can induce anxiety. Many studies suggest an association between mathematical anxiety and gender; female students are more often anxious about mathematics, perhaps because they have embraced the belief that women are not as good at mathematics as men and thus have difficulty building self-confidence in their abilities. These stereotype effects can also extend to other underrepresented groups.

Stereotype Threat

One widely studied phenomenon regarding success on standardized tests is known as the "stereotype threat" in which the stereotyping of groups in society affects an individual. Researchers found that proficient white males performed more poorly on a difficult mathematics test when researchers induced the threat of superior performance by Asians as compared to a control group. The impact of stereotype threat for many groups of students has been researched under a wide variety of

Neuroscience

Research including verbal descriptions of problem solving, observation, and biological data from tools such as magnetic resonance imaging has confirmed that spatial ordering, temporal-sequential ordering, higher order cognition, memory, language, and attention are among the functions of the mind that are at work when children think with numbers—a wide array of functions, not a single "mathematics center" of the brain. Sequential ordering may be used to solve multi-step problems. Spatial ordering lets a child recognize symbols and geometric forms, among others. Higher order cognition lets a child "think about thinking," considering different problem-solving strategies, being aware of what they are doing as they do it, and generalizing and applying old skills to new problems. Language skills affect a student's ability to understand instructions given to them and articulate their thinking. They are also critical because of the discipline-specific vocabulary at work in mathematics. Some terms, like "exponent" have no common, everyday usage. Other words like "times" and "multiply" are generally used interchangeably. Words like "normal" and "random" overlap with everyday non-mathematical vocabulary and may have broader colloquial meanings. Given the number of interacting components involved in the process of thinking about, working with, and learning mathematics, and the fact that a temporary "stall" in one area can seemingly magnify a negative, reflexive "I just don't get it" feeling in students, the roots of mathematics success or failure can be difficult to identify and address.

conditions. For instance, Asian women performed more poorly on mathematical tests in which they were cued as women, while they performed better when cued as Asians as compared to control groups. Some researchers theorize that students must contend with a subconscious whisper of inferiority when their abilities are most taxed. Whether they consciously or unconsciously accept the stereotype or not, they may still work harder in order to avoid confirming it, because failure has a more devastating meaning. The extra burden may be enough to impact performance. The stereotype cues may be subtle, like self-identification of gender, race, or culture before an exam. Researchers have also found that removing the cues can positively impact test performance. For example, in a 2009 meta-analysis of 18,976 students from five countries who were matched by researchers using past performance, stereotyped students performed better under conditions that reduced the threat.

Research and Strategies

In the 1970s, mathematics education researchers Elizabeth Fennema and Julia Sherman developed the Fennema–Sherman mathematics attitude scales to examine eight components considered critical for success in mathematics: attitude toward success in mathematics, mathematics as a male versus female domain, parent support, teacher support, confidence in learning mathematics, mathematics anxiety, motivation for challenge in mathematics, and mathematics usefulness. Their work has been cited among the most quoted social science and educational research studies of the latter twentieth century, and many versions of their scales exist.

The cumulative body of research suggests several strategies that may help students succeed in mathematics, though one point of general agreement seems to be that the key to mathematics success is active participation, active study, and engaging the material. For example, younger students can be encouraged to ask questions in and out of class and can be given mathematical exercises that are interactive rather than requiring them to only passively listen to explanations of the material. Hands-on activities with even simple objects like buttons, dried beans, or animal counters can help children develop number, counting, and arithmetic skills. More sophisticated tools, like tangrams and algebra tiles, develop geometric concepts and thinking about functions. This method has come to include computer-based virtual manipulatives. Asking questions and engaging in hands-on learning is also valuable in the later grades and college. With regard to attitudes, educators frequently encourage students to recognize that the act of learning and doing mathematics is likely to be different than other school subjects, particularly with regard to its cumulative nature and fact that working with a variety of mathematics problems is usually the only way to learn mathematics. Some instructors include explicit problem-solving and test-taking strategies in their instruction in addition to concepts. Students may also benefit from instruction in methods of note-taking, reviewing, and reading mathematics textbooks that encourage them to think reflectively about mathematics content rather than simply summarizing. At the same time, teachers may use a variety of presentation and engagement methods, including using real-world problems, considering different learning styles, being aware of anxiety and stereotypes that may affect students, and engaging parents in an ongoing dialogue about mathematics education to gain support and make them partners in their children's success.

Organizations

Beyond the classroom, clubs, professional organizations, and scholarship programs have been shown to contribute to success and some are particularly targeted toward groups that may be more at-risk, such as women and minorities. The Meyerhoff Scholars Program is a notable example of such a program. It was initially created in 1988 to target African-American men, though admission is no longer restricted by gender or ethnicity. A 2010 statistic noted that program participants were 5.3 times more likely to be attending or have graduated from a STEM Ph.D. or M.D./Ph.D. program than others who were invited to join but declined and attended another school. The program's success is attributed in large part to its emphasis on mentorship, particularly since women and minorities interested in mathematics may never have met a woman or a minority mathematician or have not been exposed to research and challenges to excel rather than to imply succeed. The Hypatia Scholarship program for women at the University of South Australia, founded in 1997 and named for woman mathematician Hypatia of Alexandria, awards not only financial support but also provides women with shared office space and computer resources in close proximity to faculty to encourage interaction and build confidence. It funds summer employment to encourage the participants to use their mathematical training in industry or academia. Feedback from students indicated that the women valued the social network more than the financial support, saying it motivated them and helped reduce anxiety. Other organizations offer financial scholarships and some opportunities for networking, such as the American Statistical Association's Gertrude Cox Scholarship. Mathematics clubs and honor societies, like Pi Mu Epsilon, provide social and academic opportunities for students with mathematics interests, and researchers have found some evidence that participation is associated with increases in retention, positive attitudes about mathematics, and higher grade point averages.

The successful Upward Bound program targets disabled, low-income, homeless, and foster care youth, as well as those who would be first-generation college students to encourage comprehensive success in secondary and higher education. It provides instruction in mathematics, laboratory sciences, composition, literature, and foreign languages, along with support like counseling, academic tutoring, and assistance with college admission and financial aid.

Further Reading

Baumann, Caroline. *Success in Mathematics Education.* Hauppauge, NY: Nova Science Publishers, 2009.

Cooke, Heather. *Success with Mathematics.* New York: Routledge, 2002.

Leinwand, Steven. *Accessible Mathematics: Ten Instructional Shifts That Raise Student Achievement.* Portsmouth, NH: Heinemann, 2009.

Martin, Danny. *Mathematics Success and Failure Among African-American Youth.* Mahwah, NJ: Lawrence Erlbaum Associates, 2006.

BILL KTE'PI

See Also: Curriculum, K–12; Learning Models and Trajectories; Mathematicians, Amateur; Minorities; Women.

Sudoku

Category: Games, Sport, and Recreation.
Fields of Study: Algebra; Number and Operations.
Summary: The game of Sudoku is explained by and informs graph theory and randomness.

Sudoku is a number puzzle based upon the mathematical concept of a "Latin square." Latin squares are arrays of numbers in which each number is listed only once in any row or column. Leonhard Euler originated the term, calling them "Latin squares" because he used Latin letters rather than numbers in his investigations. Completed Sudoku grids are Latin squares that are further subdivided into subgrids in which the numbers also appear only once in the subgrid. The graphic below shows a completed 9-by-9 Sudoku puzzle.

7	3	1	4	9	5	8	6	2
9	8	4	6	2	1	3	5	7
5	2	6	3	7	8	9	4	1
4	1	9	5	3	2	7	8	6
8	7	5	1	6	9	4	2	3
3	6	2	7	8	4	1	9	5
1	9	8	2	5	7	6	3	4
6	5	7	8	4	3	2	1	9
2	4	3	9	1	6	5	7	8

Originally published in the 1970s in Europe and the United States, Sudoku surged to popularity in Japan in the late 1980s and reappeared in Europe and America in the mid-1990s, becoming popular among puzzlers. The popularity of the puzzle has continued to grow in the twenty-first century, leading the puzzle to become the subject of mathematical scrutiny.

Sudoku is usually based on a Latin square with nine rows and columns; puzzles of other sizes are possible such as 4-by-4, 16-by-16, and 25-by-25. Some of the numbers are filled in to start. The goal is to quickly and accurately complete the puzzle, with the digits 1–9 placed once in each row, column, and subgrid. Below is an unsolved Sudoku puzzle:

	1	3	5		9	2		
2				8				
9	8		6		7		1	
3	5						8	
		6	8		1	7		
	2						9	1
	6		3		4		2	5
				5				9
		2	9		6	8	4	

Graph Theory

There are a number of interesting mathematical questions associated with Sudoku puzzles, in particular the conditions under which they have one solution. In 2007, Agnes Herzberg and M. Ram Murty showed that Sudoku puzzles can be recast as graph coloring problems allowing the broad, well-developed theory of graphs to be applied to the solution question. In particular, they showed that a standard Sudoku puzzle can be thought of as a graph where each cell in the puzzle is represented by a vertex with 20 edges, each edge connecting the cell to another cell in the row, column, or sub-grid. Graphs for which all vertices have the same number of edges are called "regular," so Sudoku graphs made in this way are "20 regular" graphs.

Since each digit can appear only once in any row, column, or subgrid, putting the nine digits into the cells is equivalent to coloring the vertices of the graph with nine colors such that no vertices connected by an edge are the same color—or in graph theory terminology, finding a "proper 9-coloring" of the graph. The number of ways to color a regular graph with n colors is a well-known formula that is a function of the number of colors and the number of vertices. Using this and other ideas about coloring graphs, Herzberg and Murty proved that 9-by-9 puzzles must have at least eight different digits shown in the starting configuration to have a unique solution.

There are still many unanswered questions about when Sudoku puzzles have one solution. Assuming that eight different digits are used in the starting configuration, how many numbers total must be shown in the starting configuration to ensure a unique solution? The answer is not known; a small number of distinct puzzles with 17 entries in the starting configuration are known to have a unique solution. There are no known puzzles with 16 or fewer entries that have unique solutions. Does one exist? Would the answer be different if all nine digits are used in the starting configuration? Mathematicians and puzzlers are investigating these and other interesting questions.

For example, in 2010, mathematicians Paul Newton and Stephen DeSalvo demonstrated that the arrangement of numbers in Sudoku puzzles is more random (by some definitions of randomness) than 9-by-9 matrices produced by random generators, since Sudoku rules excludes some of the possible arrangements that have innate symmetry.

Further Reading

Chevron Corporation. "Sudoku Daily: History of Sudoku." http://www.sudokudaily.net/history.php.

Herzberg, Agnes, and Ram M. Murty. "Sudoku Squares and Chromatic Polynomials." *Notices of the American Mathematical Society* 54, no. 6 (2007).

HOLLY HIRST

See Also: Acrostics, Word Squares, and Crosswords; Mathematical Puzzles; Matrices; Puzzles.

Sunspots

Category: Weather, Nature, and Environment.
Fields of Study: Algebra; Data Analysis and Probability.
Summary: Sunspots have long been observed and mathematicians and scientists continue to try to understand them and their effects.

Sunspots are a not yet fully explained phenomenon tied to solar activity. The sun is Earth's richest source of heat and light. Furious eruptions of energy take place on the surface of the sun. In the core, nuclear reactions occur because of the immense temperature and pressure. Through a process known as "convection," millions of tons of hydrogen are converted into helium every second and are then expelled at the surface of the sun as light and heat. Sunspots have a magnetic field strength that is thousands of times stronger than Earth's magnetic field. These magnetic fields inhibit convection to create relatively cooler areas, which appear as dark spots on the surface of the sun. Scientists and mathematicians have long attempted to understand their behavior and oscillations and have used mathematical tools like differential equations, hexagonal planforms, and time series analyses. They also count the number of sunspots and examine possible relationships between this number and factors on Earth, like radio disruptions, land temperature, and weather phenomena.

History

Direct observation of the sun is very dangerous, which historically made sunspots hard to study and quantify.

In ancient times, Chinese astronomers recorded solar activity. Mathematician and astronomer Thomas Harriot, noted for his work on algebra, is also credited as the discoverer of sunspots. Increased understanding of the nature of sunspots, including the observation that they often occurred in groups and that they moved relative to one another as the sun rotated, is tied to the development of the telescope in the seventeenth century. One of Galileo Galilei's works on sunspots offered evidence for the heliocentric system of Nicolaus Copernicus, and this led to debate about sunspots, as evidenced in astronomer, mathematician, and Jesuit Christoph Scheiner's views and works.

In the eighteenth century, Alexander Wilson used a geometric argument to show that sunspots were depressions. In the nineteenth century, pharmacist and amateur astronomer Heinrich Schwabe collected data on the periodicity of sunspots. Systematic observations, such as the approximately 11-year cycle, were made by Rudolph Wolf starting in 1848, who also measured the number of sunspots present on the surface of the sun. Wolf was primarily an astronomer but he also taught mathematics and physics. His observations were disputed by other astronomers, but his methods, which were based on statistical analyses, were eventually accepted as correct. Wolf's formula continues to be used in the twenty-first century as one of the sunspot indices. The International Sunspot Number is compiled worldwide by the Solar Influences Data Analysis Center in Belgium and by the U.S. National Oceanic and Atmospheric Administration. In the twenty-first century, sunspots are observed with solar telescopes, which use various filters, and specialized tools such as spectroscopes and spectrohelioscopes. Amateurs generally observe sunspots using projected images.

Waxing and Waning

Scientists know that the sun had a period of relative inactivity in the seventeenth century, which corresponds to a climatic period called the "Little Ice Age." Evidence suggests that similar periods existed in the distant past, which means there might be a connection between solar activity and terrestrial climate. The magnetic activity that accompanies the sunspots can change the ultraviolet and soft X-ray emission levels, affecting Earth's upper atmosphere. Some researchers have proposed that sunspots and solar activity are the main cause of global warming rather than carbon dioxide greenhouse gas emissions.

Further Reading

Izenman, Alan. "J R Wolf and the Zurich Sunspot Relative Numbers." *The Mathematical Intelligencer* 7 (1985). http://astro.ocis.temple.edu/~alan/WolfMathIntel.pdf.

National Aeronautics and Space Administration. "The Sunspot Cycle." http://solarscience.msfc.nasa.gov/SunspotCycle.shtml.

Spaceweather.com. "The Sunspot Number." http://spaceweather.com/glossary/sunspotnumber.html.

SIMONE GYORFI

See Also: Telescopes; Temperature; Weather Forecasting.

Subtraction

See *Addition and Subtraction*

Surfaces

Category: History and Development of Curricular Concepts.
Field of Study: Communication; Connections; Geometry.
Summary: Surfaces are two-dimensional manifolds, some of which have been studied for their special properties.

Living beings interact with much of the world through surfaces. Humans walk on surfaces and eat and sleep on them. Surfaces like the one-sided Klein bottle, named for Felix Klein, stretch the imagination and are the subject of mathematical investigations. They are often represented using physical models as well as computer models, including sculptures and computer animations. In twenty-first-century classrooms, students investigate a variety of surfaces and their properties, including area and volume. The National Council of Teachers of Mathematics recommends an understanding of the area and volume of rectangular solids for primary school students; of prisms, pyramids, and cylinders for middle grades students; and of cones, spheres, and cylinders for high school students. The parametrization and volume of surfaces is further explored in a multivariable calculus course.

History of Study

Mathematicians have long developed the theory of surfaces, and they continue to investigate their properties. In addition to the plane, polyhedra, such as the surface of a cube or an icosahedron, are among the first surfaces studied by the ancient Greeks in geometry. Their view of surfaces was entirely different from the functional description used in investigations of surfaces in the twenty-first century. The Greeks also had a good knowledge of surfaces of revolution and pyramids. With the introduction of analytic geometry in the seventeenth century, the study of surfaces developed into one of the most studied branches of mathematics. Mathematicians like Carl Friedrich Gauss, Pierre Bonnet, Barnhard Riemann, Gaspard Monge, and their followers firmly established surfaces on a rigorous basis during the eighteenth and nineteenth centuries.

One of the greatest achievements of the theory of surfaces is the Gauss–Bonnet theorem. Versions of the theorem were explored by Gauss in the 1820s and Bonnet and Jacques Binet in the 1840s. The form of the theorem that is standard in undergraduate differential geometry courses is attributed to Walther von Dyck in 1888. Smooth surfaces are defined as those surfaces in which each point has a neighborhood diffeomorphic to some open set in the plane. This added structure allows the use of analytic tools. The parametric functions for a smooth surface define two quadratic differential forms: the first and second fundamental forms, which are local invariants defined as functions of the arc length. The Gaussian curvature of the surface is an isometric invariant; hence, an intrinsic property of a surface, which is known as the Theorema Egregium of Gauss. The Gaussian curvature measures the deviance of the surface from being flat at each point. The parametric equations of a surface determine all six coefficients of its first and second fundamental forms; conversely, the fundamental theorem of surfaces states that given six functions satisfying certain compatibility conditions, then there

exists a unique surface (up to its location in space). A geodesic curve on a smooth surface is characterized as a locally minimizing path. In a sense, geodesics are the straight lines of surfaces, which are essential for defining the notions of distance, area, and angle on a surface. Bonnet investigated the geodesic curvature, which measures the deviance of a curve on the surface from being a geodesic. The Gauss–Bonnet theorem states that for an orientable compact surface, the total Gaussian curvature is 2π times the Euler characteristic of the surface, named for Leonhard Euler. A consequence of this theorem is that the sum of the interior angles of a geodesic triangle is greater than, less than, or equal to π, depending on if the Gaussian curvature of the surface is positive, like on a sphere; negative, like on a hyperboloid; or zero, like in the plane.

Types of Surfaces

The classification of surfaces is another topic that is explored in undergraduate geometry or topology classes. In 1890, Felix Klein asked what surfaces locally look like the plane. In Klein's Erlangen Program, a space was understood by its transformations. Heinz Hopf published a rigorous solution in 1925 that arose from groups of isometries acting on the plane without fixed points. The surfaces are the plane; the cylinder; the infinite Möbius band, named for August Möbius; the flat Clifford torus or donut, named for William Clifford; and the flat Klein bottle. Intuitively, surfaces seem to always have two sides; however, the Möbius band and the Klein bottle have only one side. Other surfaces like the projective plane resemble the sphere, and many-holed donuts

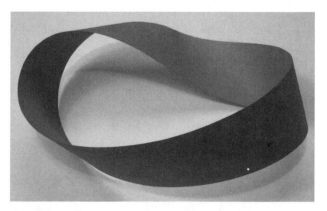

A paper Möbius band is a developable surface (it has zero Gaussian curvature).

resemble hyberbolic space. The Euler characteristic is used to classify the topology of a surface.

Some important types of surfaces that are studied intensively in geometry and analysis are minimal surfaces with zero mean curvature, such as catenoids and helicoids; developable surfaces with zero Gaussian curvature, like the plane, the cylinder, the cone, or a tangent surface; and ruled surfaces that can be generated by the motion of a straight line, like the cylinder and the hyperboloid of one sheet. While some of these surfaces date to antiquity, others are more recent. In the eighteenth century, Euler described the catenoid and Jean Meusnier the helicoid. Discoveries in 1835 by Heinrich Scherk and in 1864 by Alfred Enneper included minimal surfaces that are now named for each of them. In the 1840s, Joseph Plateau's experiments indicated that dipping a wire ring into soapy water will create a minimal surface. Jesse Douglas won a Field's Medal in 1936 for his solution to Plateau's problem in minimal surfaces. A minimal surface that originated at the end of the twentieth century is because of Celso Coasta in 1982.

Representations and Investigations

Algebraic geometers investigate algebraic surfaces, such as cubic or quartic surfaces that can be represented by polynomials. These led to rich mathematical investigations in the nineteenth and twentieth centuries. For instance, in 1849, Arthur Cayley and George Salmon showed that there were 27 lines on a smooth cubic surface. Quartic surfaces were of interest in optics, and mathematicians such as Ernst Kummer studied them.

While mathematicians had long built physical models of surfaces, which were typically housed in universities and museums, in the late twentieth and early twenty-first centuries, computer-generated surfaces revolutionized the visualization and construction of surfaces and led to many interesting mathematical questions. For instance, numerous mathematicians and computer scientists have explored the method of subdivision of surfaces, including Tony DeRose, and Jos Stam, who won a Technical Achievement Award in 2005 from the Academy of Motion Picture Arts and Sciences. This method often takes advantage of the similarity between the local structure of a surface and a small piece of a plane. For instance, surfaces may be represented using small flat triangle or quadrilateral mesh representations. These representations are easier to manipulate, but they can still appear smooth to the eye.

Further Reading

Andersson, Lars-Erik, and Neil Stewart. *Introduction to the Mathematics of Subdivision Surfaces.* Philadelphia, PA: Society for Industrial and Applied Mathematics, 2010.

Farmer, David, and Theodore Stanford. *Knots and Surfaces: A Guide to Discovering Mathematics.* Providence, RI: American Mathematical Society, 1996.

Fischer, Gerd. *Mathematical Models: Photograph Volume and Commentary.* Braunschweig, Germany: Friedrick Vieweg, 1986.

Gottlieb, Daniel. "All the Way With Gauss-Bonnet and the Sociology of Mathematics." *American Mathematical Monthly* 104, no. 6 (1996).

Henderson, David W. *Differential Geometry: A Geometric Introduction.* Upper Saddle River, NJ; Prentice Hall, 1998.

Pickover, Clifford. *The Math Book: From Pythagoras to the 57th Dimension: 250 Milestones in the History of Mathematics.* New York: Sterling, 2009.

Dogan Comez
Sarah J. Greenwald
Jill E. Thomley

See Also: Animation and CGI; Crochet and Knitting; Geometry in Society; Measurements, Volume; Polyhedra; Sculpture.

Surgery

Category: Medicine and Health.
Fields of Study: Data Analysis and Probability; Number and Operations; Representations.
Summary: Mathematical models can be used for various aspects of surgical operations in order to predict effects and improve recovery.

Surgery is the branch of science that typically involves medical treatment through an operation. There are a variety of reasons for why a surgery is performed. Diseases such as cancer and various forms of heart disease may be treated through surgical procedures.

When cancer is found, surgery may be performed to remove a tumor in order to reduce the likelihood of the cancer spreading or to alleviate pressure caused by a tumor pressing against another organ. Heart surgeries include a heart transplant, a coronary artery bypass, or a heart-valve repair or replacement. Injury is usually treated with surgery when the body is unable to repair itself. Torn ligaments or tendons can be surgically treated through reattachment or replacement. Burn victims can be treated with a skin graft as a permanent replacement for the damaged skin. Deformities can also be treated with surgery. Spinal fusion surgery can be used to treat spinal deformities like scoliosis. A cleft lip and palate is a fetal deformation that can be corrected with surgery soon after birth. The surgery usually involves an incision with medical instruments. Surgical incisions can vary in size from large incisions, such as in some open-heart and brain surgeries, to tiny incisions, such as in the case of laparoscopic procedures. The advantage to smaller incisions is that the wound heals faster, leaves a smaller scar, and reduces the likelihood of infection.

Mathematically Modeling Surgery

Improving surgical techniques and devising new surgical procedures is an active area of research. Multidisciplinary approaches are required for developing successful techniques and procedures. The National Institutes of Health (NIH) has stated that these approaches should include computational and mathematical simulations to facilitate this biomedical research. Simulations can be accomplished, in part, through mathematical modeling.

Mathematical modeling is the process of using a mathematical language in order to describe, in this case, a biological phenomenon treated with a biomedical procedure. While any mathematical model is a simplification of reality, computational solutions of the mathematical model may provide useful insights for researchers and clinicians when the model has been formulated under biologically and physically sound principles with realistic treatment strategies.

In order to develop a mathematical model for surgical treatment, it is common to have a team of researchers work closely on a given problem because of the different areas of expertise needed to address what is likely a complex biomedical problem. The first step is for the researchers or clinicians to define, as clearly as possible, the problem or question that they want the modeler to analyze and identify the benefits of the modeling project. From there, they should work to determine the

appropriate scales with which to study the problem. Is the most appropriate scale at the molecular, cellular, or tissue level or is it more of a systemic problem? Is the time scale (if there is assumed to be temporal variation) on the order of minutes, days, or years? Answers to these questions will help determine if the model is best described in terms of discrete units or continuous variables, whether temporal or spatial variation should be included, and if a deterministic or statistical approach is more appropriate. This will also help determine what computational platform and method might best be used to analyze the question at hand. Developing a model diagram can help visualize the model formulation and process. From there, the research team needs to decide how best to access the quality of the model. From the model formulation, what are the assumptions and model limitations? Are the assumptions biologically reasonable? Are there data that can be used to quantify some of the model parameters? Are there data that can be used to quantitatively or qualitatively compare to the initial model simulations? If the research team is at the beginning stages of a new surgical treatment, can the model be developed to put forth hypotheses tested for animal experiments or clinical trials? The mathematical model works best as an iterative process when the modeler and experimentalist or clinician exchange ideas. The first set of suggested guidelines for biomedical research teams with mathematical modelers was proposed for the mathematical modeling of acute illness.

Applications

In connection with surgical procedures, mathematical models can be used in a variety of ways. Mathematical models may be used to predict the likelihood of a surgical procedure's success. For example, in reconstructive microsurgery (where skin tissue is moved from one location to another), a mathematical model was developed to predict successful tissue transfer based on oxygen delivery, tissue volume, and blood-vessel diameter. Mathematical models can be used to explore ways of making existing surgical procedures more successful. Stents are tubes inserted into a blood vessel (or another tubular body part) to keep open the vessel but are associated with a higher risk of a heart attack or a blood clot. A mathematical model was developed to analyze drug delivery to stent locations where two or more arteries meet.

Mathematical models can be developed to analyze controversial questions. For example, a model was developed to analyze whether a more liberal or constrictive allowance of fluid level allows for a more successful recovery from abdominal surgery. Mathematical models can be used to predict changes following surgery. A model was developed to predict changes in the knee joint following a wedge osteotomy, which is the removal of a wedge region of bone around the knee. The model was validated by predicting the results of 30 patients undergoing the surgical procedure, then the results were compared to actual measurements 14 months after surgery. In spite of these efforts, mathematical modeling of surgical procedures (and questions related to the surgical procedures) is still a relatively new concept.

Ideally, mathematical models can be used on individual patients to predict a likely and optimal outcome when considering surgical treatment. With advances and improvements in imaging techniques and computer software, this may be possible for treatment of some diseases and injuries. For example, when dealing with more complex arterial geometries near stent locations, researchers may be able to predict appropriate drug treatment strategies. However, in the absence of patient-specific models, mathematical models may be used to help clinicians make decisions based on patient variability. Patient variability implies that although there are differences in individual patients, there may be common characteristics in subpopulations of patients with a similar disease or injury. These common characteristics might be measured in common biomarkers from urine or blood analysis or similarities in imaging analysis. Mathematical models can be used to investigate surgical treatment strategies for patients with similar characteristics.

Mathematical models can also help with the development of new treatment surgical protocols or the analysis of existing treatment strategies. When exploring these questions, it is common for researchers to conduct experimental trials on animal models. However, animal experiments can be time consuming and costly. Mathematical models used to analyze a given question can provide a significant cost savings. For example, computer simulations of the model can be used to initially screen different experimental trials in order to decide which ones are worth pursuing and which are not. Furthermore, successful experimental results on animals do not guarantee the same level of success in clinical trials on humans. Mathematical models can not only give an idea as to how experimental trials on animals translate into surgical treatment on humans but also provide necessary insights when animal experiments are not possible or clinical trials on humans are unethical.

To help address the many questions that arise from current surgical procedures and the development of new surgical methods, an interdisciplinary team of researchers is required to formulate and analyze the problem at hand. It has been suggested that the team include mathematical modelers. Mathematical modeling can potentially provide a way to investigate novel treatment strategies and predict possible problems that may arise for a given surgical procedure. Furthermore, there exists the possibility of significant cost savings, in part, by reducing the number of animal experiments or clinical trials performed. In these ways, the mathematics underlying the description of the biology can be beneficial to the surgeon or biomedical researcher.

Further Reading

An, Gary, et al. "Translational Systems Biology: Introduction of an Engineering Approach to the Pathophysiology of the Burn Patient." *Journal of Burn Care & Research* 29, no. 2 (2008).

Geris, Liesbet, et al., "*In silico* Biology of Bone Modeling and Remodeling: Regeneration." *Philosophical Transactions of The Royal Society A: Mathematical, Physical, & Engineering Sciences* 367, no. 1895 (2009).

———. "*In silico* Design of Treatment Strategies in Wound Healing and Bone Fracture Healing." *Philosophical Transactions of the Royal Society A: Mathematical, Physical, & Engineering Sciences* 368, no. 1920 (2010).

Kolachalama, Vijaya, et al. "Luminal Flow Amplifies Stent-based Drug Deposition in Arterial Bifurcations." *PLoS One* 4, no. 12 (2009).

Matzavinos, Anastosios, et al. "Modeling Oxygen Transport in Surgical Tissue Transfer." *Proceedings of the National Academy of Sciences USA* 206, no. 29 (2009).

National Institutes of Health. "New Roadmap Emphasis Areas for 2008." http://nihroadmap.nih .gov/2008initiatives.asp.

Sariali, Elhadi, and Yves Catonne, "Modification of Tibial Slope After Medial Opening Wedge High Tibial Osteotomy: Clinical Study and Mathematical

Modeling." *Knee Surgery, Sports Traumatology, Arthorscopy* 17, no. 10 (2009).

Tartara, Tsuneo, et al. "The Effect of Duration of Surgery on Fluid Balance During Abdominal Surgery." *Anestheia & Analgesia* 109, no. 1 (2009).

Vodovotz, Yoram, et al. "Evidence-Based Modeling of Critical Illness: An Initial Consensus From the Society for Complexity in Acute Illness." *Journal of Critical Care* 22, no. 1 (2007).

Washington University in St. Louis. "Plastic Surgery To the 'Nines.'" *Science Daily* 23 (July 2002). http://www.sciencedaily.com/releases/2002/07/020723080514.htm.

RICHARD SCHUGART

See Also: Mathematical Modeling; Medical Imaging; Transplantation.

Swimming

Category: Games, Sport, and Recreation.
Fields of Study: Data Analysis and Probability; Measurement; Number and Operations.
Summary: Swimming performance can be modeled and improved mathematically.

Mathematical modeling and statistical analysis have been applied to swimming in a variety of ways. Modeling the properties of fluids in motion is the subject of fluid dynamics, a sub-branch of mechanics. Placing objects in the fluid complicates the physics enormously. The interaction of the fluid and object at the point where the object meets the fluid (called the "boundary") is of particular interest. Problems studied in this way include why flags "flap" in a breeze and how fish swim. Statistical analysis has been applied to a number of questions about swimming performance, including the prediction of future world record times, the modeling of deterioration in swimming performance as a function of age, and the evaluation of whether triathlons are fair to swimmers.

Improving Human Performance in Swimming
Modeling human swimming presents serious challenges for researchers. The use of arms and legs to propel the swimmer through the water adds complexity to the fluid dynamics models. Because the human swimmer is not completely immersed in the water but keeps part of the body above the surface, the interaction between the swimmer and the surface is particularly difficult to model. Researchers have applied smoothed particle hydrodynamics to the study of human swimming performance which, unlike traditional fluid dynamics, treats fluid flow as the motion of individual particles. This method enables researchers to more accurately model and simulate the interactions of the swimmer at the surface. The goal of this research is to help individual swimmers improve their performance in competition.

Predicting World Record Swim Times
Statistical analysis of human swimming performance encompasses a number of different approaches and methods. An analysis of world records in swimming from 1960 to 2010 shows a nearly steady decrease in times, resulting in between 15% to slightly more than 25% improvement, depending on the event. The question remains how long times can continue to decrease, how much is because of increased participation in swimming (especially women's swimming), and how much is because of advances in technique and conditioning.

Predicting the Swimming Performance of Aging Swimmers
On a different tack, Ray Fair modeled the performance of elite swimmers of different age groups and modeled the performance at various swimming distances by age. For example, he predicted that a 60 year old will swim a time about 10% slower than the swimmer had done at age 35, while a 70-year-old will be 25% slower.

Are Triathlons Fair to Swimmers?
Richard De Veaux and H. Wainer investigated the relative disadvantage of swimmers to runners and cyclists in a triathlon. Because the times taken for the three events are so different, they argued that the standard triathlon proportions (including the Iron Man and Olympic triathlons) are grossly unfair to swimmers. The best marathon runners in the world take about two hours, seven minutes to run the 26.2-mile marathon (with variation due to course and weather). An elite cyclist can cover about 60 miles in the same time,

and an elite swimmer can travel 7.5 miles. Thus, to be fair in terms of average time taken, a triathlon based on a marathon should also contain a 60-mile bike leg and a 7.5 mile swim. In reality, the Iron Man is a 26.2-mile run, a 112-mile bike leg, and only a 2.4-mile swim, and thus disadvantages swimmers enormously.

Further Reading

Cohen, R. C. Z., P. W. Cleary, and B. Mason. "Simulations of Human Swimming Using Smoothed Particle Hydrodynamics." In *Proceedings of the Seventh International Conference on CFD in the Minerals and Process Industries.* Melbourne, Australia: CSIRO, 2009.

———. "Improving Understanding of Human Swimming Using Smoothed Particle Hydrodynamics." *6th World Congress on Biomechanics* (2010).

Fair, Ray C. "How Fast Do Old Men Slow Down?" *Review of Economics and Statistics* 76, no. 1 (1994).

Wainer, H., and R. D. De Veaux. "Resizing Triathlons for Fairness." *Chance* 7 (1994).

———. "Making Triathlons Fair: The Ultimate Triathlon." *Swim Magazine* 10, no. 6 (1994).

Richard De Veaux

See Also: Data Analysis and Probability in Society; Mathematical Modeling; Tides and Waves.

Symmetry

Category: Architecture and Engineering.
Fields of Study: Geometry; Measurement.
Summary: An ancient mathematical concept, there are various forms of symmetry.

"Symmetry," which comes from the Greek word roots meaning "same" and "measure," describes a picture, shape, or other object that looks the same when viewed from another perspective or that can be transformed in some way without changing its important properties. The word "symmetry" can refer to this property, to the transformation itself, or more holistically to an aesthetically pleasing sense of balance. Eighteenth-century mathematician Adrien-Marie Legendre revolutionized the concept of symmetry when he connected it to transformations. There are a wide variety of uses of the word "symmetry" in different domains, including art, architecture, and science, and many of these have existed from antiquity. The concept of symmetry is inherent to modern science and architecture, and its evolution reflects in many ways the dynamic nature of these fields.

Visual Symmetry

In the context of geometric figures drawn in the plane, there are three fundamental types of symmetry:

1. A figure has "reflection" symmetry if it coincides with its own mirror image across some line. The capital letters M and W have a single reflection symmetry, while the letter H has two symmetries, horizontal and vertical.
2. A figure has "rotational symmetry" if it can be rotated around a fixed point, leaving the figure unchanged. For example, the capital letters N, Z, and S are unchanged when rotated 180 degrees. The pattern of black squares in traditional crossword puzzles also has this half-turn symmetry.
3. A figure has "translational symmetry" if it can be slid or moved without changing. A typical example is a repeating pattern on wallpaper.

Construed in the broadest terms, symmetry plays a role in almost all art and is related to balance and harmony. One of the many ways in which the narrower geometric notion of symmetry applies to art is tessellations. A tessellation is a covering of the plane by copies of a limited set of tiles. Such figures are often highly symmetric. Tilings by squares, hexagons, and triangles are common enough, both in art and on kitchen floors, and more fanciful tessellations involving animal and plant shapes are also possible. Tessellations, dynamic symmetry, and mathematical sophistication are especially evident, for example, in the art of M.C. Escher (1898–1972).

Abstract Symmetries

Symmetry is not just a geometric concept. Any structure or object can have symmetry. Abstractly, a symmetry is any transformation of an object resulting in an object that is "the same" in the sense of having all the same properties that are important in context. Often,

Many plants are radially symmetric with almost identical petals or leaves growing at regular intervals.

the object is a geometric figure, and the relevant properties are length, angle, and area, but it need not be so.

Consider the game rock-paper-scissors. Renaming the scissors gesture to "paper," renaming paper to "rock," and renaming rock to "scissors" would leave the rules of the game unaltered. This is an abstract symmetry of the game. Then, there are enough symmetries to identify any move with any other, so all three options are intrinsically "equally good." In this example, there is symmetry but no geometry whatsoever.

Symmetry and Groups

In higher mathematics, notions of symmetry are expressed in the language of group theory. A "group" is a set (G) of objects that can be composed together (in other words, if x and y are elements in a group, $x \times y$ is also an object in the group), subject to three conditions: associativity, identity, and inverse criteria. The salient feature of this definition is that the set of all the symmetries of any object satisfies these conditions. The associativity property is automatic from function composition; but what about the other two? These are restatements of the convention that the transformation that does nothing is a symmetry and the idea that symmetries are "undo-able." Symmetries leave an object "structurally the same as it was," so there will always be another symmetry to undo any given symmetry.

The symmetries of any object that preserve any desired features form a group, called the "symmetry group" of the object. Often, one can understand a com-

plicated object much better by studying the size and structure of its symmetry group.

Klein and the Erlangen Program

Felix Klein (1849–1925) greatly strengthened the connection between geometry and group theory. His insight was that, if one really wants to understand a geometric structure, then one should study the group of symmetries that preserve the structure. This philosophy has proved very fruitful and is now known as the Erlangen program.

For example, in ordinary Euclidean plane geometry, the focus is on lengths and angles. The group of symmetries that preserve lengths and angles consists of translations, rotations, reflections, and combinations of these. Given any two points, each with an arrow pointing away from it in a given direction, one can always translate and rotate the plane so that the image of the first point lies on the second point, and the arrows are pointed in the same direction. This is the sophisticated way to understand the notion that every point and direction in the plane are functionally the same as every other point and direction.

The Erlangen program has played a fundamental role in the development of nineteenth- and twentieth-century geometric thinking, clarifying the relationships and distinctions between geometry and topology; projective and affine geometry; and Euclidean, hyperbolic, and spherical geometry.

Symmetry and the Universe

Those who study the shape of space are greatly concerned with symmetry. Consider the question of whether the universe is "homogeneous." That is, do the laws of physics treat every place the same as every other place? Is every direction physically like every other direction? What answers to those questions are believed to be correct determines what shapes, structures, and geometries are viable candidates to model the universe.

Time symmetry is another issue of importance in physics research; one wants to know to what extent the physical laws of the universe treat the past and future symmetrically. On a small enough scale, particle interactions have time symmetry. If one watched a "movie" of particle interactions on a small enough time-scale, it would be impossible to tell whether the movie was playing forward or backward. On the other hand, the large-scale events observed in everyday life do not possess such past-future symmetry; for example, eggshells break but do not spontaneously assemble, people age but do not become more youthful. This discrepancy between small-scale symmetry and large-scale asymmetry is rather mysterious, and one can hope that reconciling the two will lead to greater understanding of physics.

Symmetry and Architecture

Symmetry has long been connected with architecture. In Greek and Latin, symmetry was used to indicate a common measure or a notion of something well-proportioned, rather than as a reflection. However, reflection symmetries can be found in many buildings from different cultures, where the left side is a mirror image of the right side. Architects have also used symmetry in external views, layout, stability, or building details, such as stairs or windows. Some authors claim that the first recorded instance of the use of symmetry as a mirror reflection was in 1665, when Gian Lorenzo Bernini was asked to design an altar for the church of Val-de-Grace, while others assert that it was first found in Claude Perrault's 1673 treatise on columns. Perrault is best known as the architect of the east wing of the Louvre.

Concepts such as the symmetry groups of the plane also originate in architecture. Beginning with mathematician Edith Muller's 1944 analysis, experts continue to debate how many of the 17 groups can be found in the mosaics of the Alhambra at Granada, a fourteenth-century Moorish palace. Some assert that all 17 can be found there and in many other examples in Islamic art and architecture. A formal mathematical proof that there are no additional symmetry groups was proven independently by Evgraf Fedorov in 1891 and George Pólya in 1924. Partly because of a prohibition against using anthropomorphic forms, symmetry appears in many instances of Islamic-influenced architecture, such as the Taj Mahal.

The connections between symmetry and architecture continue into the twenty-first century. In numerous texts in the twentieth and twenty-first centuries, mathematicians such as Hermann Weyl illustrate concepts using architectural references. Architects and engineers also frequently use symmetry, though architects working in the modernist aesthetic reject symmetry in their designs.

Further Reading

Cohen, Preston Scott. *Contested Symmetries and Other Predicaments in Architecture*. Princeton, NJ: Princeton Architectural Press, 2001.

Gardner, Martin. *The Ambidextrous Universe: Symmetry and Asymmetry From Mirror Reflections to Superstrings*. 3rd ed. New York: W. H. Freeman, 1990.

Hon, Giora, and Bernard Goldstein. *From Summetria to Symmetry: The Making of a Revolutionary Scientific Concept*. New York: Springer, 2008.

Weyl, Hermann. *Symmetry*. Princeton, NJ: Princeton University Press, 1952.

MICHAEL "CAP" KHOURY

See Also: Escher, M.C.; Geometry and Geometry Education; Geometry of the Universe; Puzzles; Sudoku.

Synchrony and Spontaneous Order

Category: Weather, Nature, and Environment.
Fields of Study: Algebra; Connections; Data Analysis and Probability.
Summary: The world is filled with examples of spontaneously emerging order.

Humans are familiar with order: people order homes by placing belongings in one place; people also watch football games with players who follow orders given by a quarterback who directs the play. There are many examples of order in nature. Birds and fish order themselves by flying in flocks and swimming in schools. How is order created in a complex system with many parts? Experience indicates that order emerges from the actions or directions of a leader, just as the quarterback is the leader of a football team. It is possible, however, for a system to be ordered without the help of a single leader—an attribute that occurs in a spontaneously ordered system. Systems have a global (group) level and a local (individual) level. A school of fish is made up of thousands of individual fish, and a laser is a collection of particles of light (photons) that are emitted from trillions of atoms. When a system is spontaneously ordered, the order occurs because of local level interactions without global level direction. Imagine a spontaneously ordered football team. The quarterback on this team does not need to direct or call a play. This team is able to organize and execute plays simply by communicating with each other (individually) as each play unfolds. There will likely never be a team like this, but spontaneously ordered phenomena are all around if one knows where to look.

When multiple events are ordered in time, the result is "synchrony." Without synchrony, life would be very different. People would not enjoy watching a football game with unsynchronized players who run in different directions after—or before—the ball is snapped. Many of the technological devices that people use, including GPS, cell phones, and lasers, rely on synchrony to work properly. Scientists have even published evidence of synchrony in cloud patterns. When spontaneous order occurs, the result is often synchrony. Mathematicians and statisticians are involved in the collection of data that help define important variables related to synchrony and fuel the development of theories and models, as well as the formulation of mathematical models to describe and explain synchrony. This work draws from many areas of mathematics, including logic, probability, decision theory, geometry, and statistics, as well as related scientific fields.

Examples of Synchrony and Spontaneous Order

In some regions of Southeast Asia, large numbers of male fireflies flash on and off at the same time, creating a spectacular array of synchronized lights. It is believed that the males are flashing in unison to attract females. Physiologically, these fireflies have an internal firing mechanism that can generate a rhythmic flashing sequence. Experiments with individual fireflies demonstrate that the timing of their flashes can be altered to mimic that of an external stimulus, which is flashing rhythmically. This suggests that synchronized firefly flashing is the result of a spontaneously ordered process. To test this hypothesis, mathematicians Renato Mirollo and Steven Strogatz created a simple mathematical model by using an equation to describe an individual firefly as a biological oscillator (just as a plucked guitar string is a mechanical oscillator). They coupled multiple, identical oscillators together to form a system. Their mathematical model is a system of coupled differential equations. Mirollo and Strogatz analyzed the system and proved that in almost all cases, no matter how many oscillators there are or how the oscillations are started, synchrony is the result.

Fish often travel in schools. One advantage of this behavior is to allow fish to better avoid predators by performing highly synchronized, evasive maneuvers. Experimental data suggests that schooling fish have a preferred distance, elevation, and orientation relative to their nearest neighbor. Scientists Andreas Huth and Christian Wissel have modeled fish schooling as a spontaneously ordered system. They assume that schooling originates not because of a particular fish directing the group's movements but because of simple behavioral rules for individual fish. Their assumptions include that each fish desires to be close (but not too close) to another fish, each fish moves according to its perception of the position and orientation of neighboring fish, and individual fish movement is random. Huth and Wissel tested different movement rules for their model since there are no data that supports specific movement rules for schooling fish. They used the data generated from computer simulations of their model to determine the average direction of movement as a group and the average angular deviation by individual fish from the group's direction, which is defined as the "polarization" of the school. The polarization is a way to quantify the synchrony of the school because the larger the polarization, the more disoriented the school is. Since polarization depends on the movement rules, they used polarization to find movement rules for which their model best simulated synchronized schooling.

A fluorescent light bulb consists of a long tube filled with an inert gas. The light that we observe originates from the atoms in the gas. Each atom has multiple electrons that exist at specific energy levels. Electricity forces electrons through the tube and these electrons collide with the atoms in the gas. The collision raises the energy level of the atom's electrons, which then spontaneously revert back to a state of lower energy. This loss of energy causes a light particle (photon) to be emitted and the light that we see is from the emission of millions upon millions of photons. The light from a fluorescent light bulb consists of many different wavelengths and is scattered in many directions. Alternatively, the light from a laser, which stands for "light amplification by stimulated emission of radiation," is highly synchronized with a single frequency, direction, and phase. The first laser was constructed in 1960, but in 1917, Albert Einstein developed the quantum physics that predicted how a laser is able to synchronize the photons. When lasers were invented no one knew what to use them for.

Today, laser light is used for everything from grocery store checkout scanners to eye surgery. Just as with fluorescent light, raising and lowering the energy levels of individual electrons generates the light from a laser. An external energy source (such as electricity) continually stimulates electrons and raises them from lower energy states to higher energy states. Initially, when the laser is turned on and some electrons spontaneously fall back to their lower energy states, the emitted photons move in random directions. But a laser has mirrors at both ends and the photons are trapped between the mirrors for a long period of time before they can escape. Furthermore, a laser is constructed so that the photons will perfectly synchronize and amplify a light wave with a specific frequency and direction while filtering out the other light waves. One of the mirrors allows some of the light to escape in the form of a laser beam, an example of synchrony that we encounter each day.

Further Reading

Camazine, Scott, Jean-Louis Deneubourg, Nigel R. Franks, James Sneyd, Guy Theraulaz, and Eric Bonabeau. *Self-Organization in Biological Systems.* Princeton, NJ: Princeton University Press, 2001.

Haken, Hermann. *The Science of Structure: Synergetics.* New York: Van Nostrand Reinhold, 1981.

Strogatz, Steven. *SYNC: The Emerging Science of Spontaneous Order.* New York: Hyperion, 2003.

JOHN G. ALFORD

See Also: Clouds; Light; Mathematical Modeling; Mathematics Research, Interdisciplinary.

T

Tao, Terence

Category: Mathematics Culture and Identity.
Fields of Study: Number and Operations;
Representations.
Summary: One of the most accomplished
contemporary mathematicians, Terence Tao is a
groundbreaking number theorist as well as a
popular blogger.

Terence Chi-Shen "Terry" Tao (1975–) is a South
Australian–born mathematician. Some rank Tao
among the greatest living mathematicians in the early
part of the twenty-first century. A child prodigy, he
was taking college-level mathematics classes as early
as 9 years old and was awarded a Ph.D. from Princeton University at the age of 20. He is, as of 2010, a
professor at the University of California, Los Angeles. As a mathematician and writer, he is extremely
productive and has contributed elegant solutions to
difficult problems in diverse areas in mathematics.
His primary research interests are analytic number
theory, harmonic analysis, combinatorics, and partial
differential equations.

Honors and Contributions

Dr. Tao's contributions to mathematics, and his awards
for them, are numerous. In 2006, Terence Tao was

awarded the Fields Medal. The Fields Medal is sometimes called the "Nobel Prize of mathematics" and is
generally regarded as the most prestigious award in
mathematics. It is awarded once a year for superlative
achievement by a mathematician up to the age of 40.
At the age of 13, Tao won a gold medal at the International Mathematics Olympiad, an annual competition
intended to challenge the world's brightest students of
high-school age; as of 2010 he remained the youngest
person to ever win such a gold medal. His accolades
also include the Salem Prize, the Clay Research Award,
the SASTRA Ramanujan Award, the Australian Mathematical Society Medal, and the King Faisal International Prize. Tao has been pleased with his success, but
he would like to continue focusing on mathematics
research rather than reflect on his achievements.

One particularly significant contribution by Terence Tao to number theory came in 2004. In joint
work with Ben Green, he proved a remarkable result
about arithmetic progressions of prime numbers. An
arithmetic progression is a sequence of numbers with
a constant difference between them. For example,
$5, 11, 17, 23, 29$ is an arithmetic progression with length
5 and constant difference 6. The five numbers in the
sequence are prime. Green and Tao proved that it is
possible to find arithmetic progressions of five primes,
or 50 primes, or 50,000 primes. Indeed, they showed
that arithmetic progressions of primes exist that are as

long as desired. Understanding the distribution of the prime numbers is of paramount importance in number theory, and results of this type are often notoriously difficult to establish.

Communication and Strategy

In addition to all his papers and books, Terence Tao is a very well-respected and prolific blogger. On the "What's New" blog, Terence Tao frequently posts remarks on his ongoing projects, links to and commentary on current articles, and other mathematical topics. There are numerous active mathematical blogs at all levels of sophistication, but many consider "What's New" to be the "grandfather" of mathematical blogging. "What's New" is considered by many active mathematicians to be an important and influential source of information. As of 2010, the American Mathematics Society had published two books of excerpts from his blog.

While he has been described as the "Mozart of Math" because of his creativity and the mathematics that seems to flow out of him, Tao attributes his success to strategies that enable him to break up difficult problems into easier ones. Often, he focuses on one question at a time and tries a variety of techniques. He stated: "When I was a kid, I had a romanticized notion of mathematics—that hard problems were solved in Eureka moments of inspiration. With me, it's always, 'let's try this that gets me part of the way. Or, that doesn't work, so now let's try this. Oh, there's a little shortcut here.'"

Further Reading

Green, Ben, and Terence Tao. "The Primes Contain Arbitrarily Long Arithmetic Progressions." *Annals of Math* 167 (2008).

"Mathematical Minds: Terence Tao." Interview with the *Gazette of the Australian Mathematical Society* 36, no. 5 (2009).

Tao, Terence. *Poincaré's legacies: Pages From Year Two of a Mathematical Blog.* Providence, RI: American Mathematical Society, 2009.

———. *Solving Mathematical Problems.* 2nd ed. Oxford, England: Oxford University Press, 2006.

———. *Structure and Randomness: Pages From Year One of a Mathematical Blog.* Providence, RI: American Mathematical Society, 2008.

———. "What's New." http://terrytao.wordpress.com.

UCLA College of Letters and Science. "News: Terence Tao: The 'Mozart of Math.'" http://www.college.ucla.edu/news/05/terencetaomath.html.

Michael "Cap" Khoury

See Also: Number Theory; Mathematics, Elegant; Mathematics, Theoretical.

Tax

See *Income Tax; Sales Tax and Shipping Fees*

Telephones

Category: Communication and Computers.
Fields of Study: Geometry; Measurement; Number and Operations; Representations.
Summary: Mathematicians have played key roles in efficiently managing telecommunication networks and developing newer and more powerful phones and wireless networks.

Inventors Elisha Gray and Alexander Graham Bell independently designed devices to electrically transmit speech in the 1870s; however, Bell patented his device first. The American Bell Telephone Company created the first telephone exchange in 1877. A subsidiary company, American Telephone and Telegraph (AT&T), was incorporated in 1885 to develop and implement long-distance telephone service, and Bell Laboratories was founded in 1925 for research and development. Later, the labs would be managed by both AT&T and Lucent Technologies. The telecommunications industry has long relied on the contributions of mathematicians for its success and hundreds of companies continue to employ mathematicians to address the increasingly complex problems of twenty-first-century communication. They develop technology and algorithms for wired and wireless communication, which facilitate speed and efficiency for a variety of applications. They also research ways to increase security and prevent unauthorized listening or wiretaps. Some create business models or study issues such as customer satisfaction.

Finite Phone Numbers in an Expanding Communication Network

One notable mathematical problem of the early twenty-first century is assignment of phone numbers. In 2007, the Federal Communications Commission stated that 582 million of 1.3 billion available phone numbers in the United States were already assigned, increasingly to cell phones. Some mathematical models have suggested that exponential growth would exhaust the supply. Similar concerns have been raised about social security numbers and Internet addresses, since the number of digit permutations for any given string length is finite.

Cell Phones and Smartphones

Mathematical methods were also important in the development of cell phones and smartphones. In 1947, Bell Labs engineer W. Rae Young suggested a hexagonal tower arrangement for cellular mobile telephone phone systems, which was expanded upon by engineer Douglas Ring—though the technology did not exist to implement the idea until the 1970s. The Motorola DynaTAC 8000x, released in 1983, was the first truly portable cell phone. Cellular technology proliferated rapidly, and society has widely embraced smartphone technology. Described as a new generation of telephone, smart phones are, essentially, computers small enough to fit in a palm or pocket. The IBM Simon Personal Communicator, created by IBM and Bell-South and sold beginning in 1994, is cited as the first smartphone, while at the start of the twenty-first century, Apple's iPhone and the Motorola Droid are very popular. The Android open-source operating system, which forms the basis for the Droid smartphones, was invented by computer scientist Andrew Rubin. It has been compared to Lego system building blocks because of the structure of its software solution stack, which many consider to be more compatible than the discretely packaged and isolated programs of some other operating systems.

Mathematics has played an increasingly large role in cell phone and smartphone service. Smartphones not only serve as cell phones for verbal or textual communication, they also play media, provide access to the Internet, serve as GPS and navigation devices, and run various other software. Electronic signals from smartphones carry digitized speech and data, requiring mathematical algorithms to construct and compress information, as well as to correct errors. Mathematicians and information theorists, such as David Huffman and Jorma Rissanen, developed compression techniques using concepts from probability theory and entropy. Mathematical methods from signal processing and graph theory prevent interference between multiple callers and help to establish networks that provide uninterrupted coverage. The International

Contributions of Mathematicians

Many mathematicians have contributed to telephone systems. For example, George Boole, whose Boolean algebra was used in switching systems; Oliver Heaviside, who adapted complex numbers to study electrical circuits and worked on long-distance systems; and Agner Erlang, who modeled phone call waiting time using probability theory, in collaboration with the Copenhagen Telephone Company. Mathematical modeling has been used to design and study telecommunications systems since the beginning of the twentieth century. Many mathematical and scientific advances, both theoretical and applied, were developed by the multidisciplinary working groups at Bell Labs, AT&T, and Lucent, which included notable mathematicians and at least one Nobel Prize winner. Examples of significant advances with applications both within and beyond telephones include transistors, solar cells, lasers, satellites, the Unix operating system, the C programming language, and digital signal processing chips. Mathematician Claude Shannon is often referred to as the "father of information theory," which he developed while at Bell Labs. William Massey created performance models for telecommunication systems using queuing, stochastic methods, and special functions, and he has cited Bell Labs as especially supportive to minority mathematicians and scientists.

Mobile Telecommunications-2000 or 3G (third generation) is a global standard for mobile telecommunications introduced in 2000. It addresses critical issues such as data rates, bandwidth, frequencies, broadband compatibility, and issues of authentication, confidentiality, and privacy. As of 2010, scientists and mathematicians were developing further standards for mobile networks and devices, including a next generation 4G network. The Open Handset Alliance is a group of companies that develops and advocates for open standards for mobile devices.

Apps

As smartphone popularity booms, so do the tools developed for smartphones by computer scientists and others. Downloadable applications (commonly called "apps") are readily accessible for free or for purchase.

Many of them are aimed at education or academic subject areas, including mathematics. One set of apps offers the opportunity to practice with mathematics concepts and skills, like Math Flash Cards and Advanced Mental Math. Gamer-style apps like Math Ninja require players to answer challenge questions to advance. Other apps are electronic versions of mathematically based board games, like Mancala and Dominoes, while the popular game Tetris involves performing geometric transformations quickly to stack variously shaped objects, which is related to classical packing problems.

Further Reading

Horak, Ray. *Telecommunications and Data Communications Handbook.* 2nd ed. Hoboken, NJ: Wiley-Interscience, 2008.

Mercer, David. *The Telephone: The Life Story of a Technology.* Westport, CT: Greenwood Press, 2006.

Thompson, Richard. *Telephone Switching Systems.* Norwood, MA: Artech House Publishers, 2000.

Whitfield, Diffie, and Susan Landau. *Privacy on the Line: The Politics of Wiretapping and Encryption, Updated and Expanded Edition.* Cambridge, MA: MIT Press, 2010.

NORMA BOAKES

See Also: Cell Phone Networks; Fax Machines; MP3 Players; Software, Mathematics; Solar Panels; Wireless Communication.

Telescopes

Category: Space, Time, and Distance.
Fields of Study: Algebra; Geometry; Number and Operations.
Summary: Image clarity in telescopes is achieved through extremely precise measurements and mathematics.

In 1608, the Dutch lensmaker Hans Lippershey applied for a patent on what was soon named a "telescope." It is not clear if Lippershey was the true inventor; at least two other Dutch lensmakers also claimed credit. The news of this new invention quickly spread. In 1609, Galileo Galilei in Italy started using telescopes to observe heavenly objects. Among other findings, he discovered the rotation of the sun, the phases of Venus, and the first four satellites of Jupiter. A mathematician as well as a physicist and astronomer, Galileo also used geometry to measure the heights of lunar mountains by determining how long they remained illuminated after the lunar sunset.

Other mathematicians and physicists helped develop the modern telescope. Isaac Newton determined that lenses acted like prisms in spreading out the spectrum of visible light (a phenomenon known as *chromatic aberration*). Newton and the mathematician James Gregory independently invented the reflecting telescope, which does not have this problem. Leonhard Euler made a mathematical analysis of chromatic aberration, and in England so-called achromatic lenses (a combination of two lenses that together bring light of different colors to a focus) were invented in the early eighteenth century.

Optics

A telescope is an optical device for seeing objects that are either far away, or very dim, or both. Consider a typical magnifying glass, as shown in Figure 1, which is a piece of glass or other transparent substance shaped so that both sides are sections of spheres. Light rays from an object (such as a candle) come to a focus on Screen 1. In other words, light rays from any given point on the candle converge onto a single point of Screen 1, forming an image. Screen 2 is at the wrong distance, meaning the light rays do not converge properly on Screen 2. Screen 1 is said to be "in focus," and Screen 2 is "out of focus."

Figure 1.

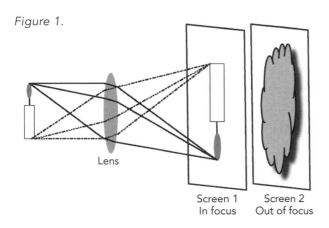

Lens

Screen 1
In focus

Screen 2
Out of focus

Figure 1a.

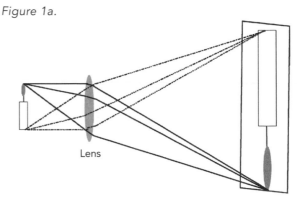

Lens

Figure 1 repeated, but with a lens with twice the focal length.

The first lens the light goes through is the called the "objective lens." The plane (Screen 1) where the image is in focus is called the "focal plane." The "focal length" is the distance to the focal plane for a source at infinity (incoming parallel rays).

Magnification is measured in diameters. If the image is twice as tall and also twice as wide as the original, then there is a magnification of two diameters. Some optical devices, however, are measured in power, which is the square of the magnification in diameters; for example, a microscope advertised as "100 power" actually magnifies 10 diameters.

Figure 1a shows the same configuration as Figure 1, but the focal length is twice as long. The image on the focal plane is thus twice as high and twice as wide—it is magnified twice as many diameters. Since the image is spread out over four times the area, it

is only one-fourth as bright. Conversely, for a given focal length, doubling the size of the objective lens lets in four times as much light, hence the image is four times as bright.

In Figure 1, if one were to put a light-tight box around screen 1, set up a shutter to control when light enters the box, and replace screen 1 with photographic film, the result is a camera. Replace the photographic film with an electronic light-sensitive screen, and the result is a digital camera. If the camera is used to take pictures of far-away or dim objects, then it qualifies as a telescope.

Astronomical Telescopes

Since astronomers are interested in dim celestial objects, a big objective is necessary for astronomical telescopes. Amateur astronomers frequently use a 6-inch (15-cm)

Figure 2. Eye plus magnifying glass.

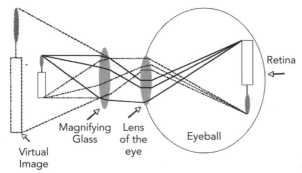

Virtual
Image

Magnifying
Glass

Lens
of the
eye

Eyeball

Retina

Figure 3. How a telescope delivers an enlarged image to the eye.

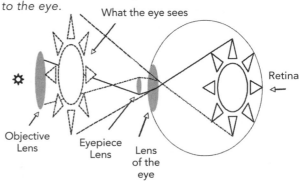

What the eye sees

Objective
Lens

Eyepiece
Lens

Lens
of the
eye

Retina

Figure 4. Grinding a mirror by hand (curvature greatly exaggerated).

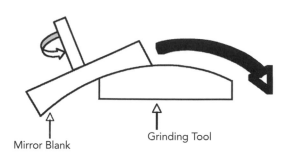

Figure 5. Four designs of reflecting telescopes (Gray bar shows focal plane).

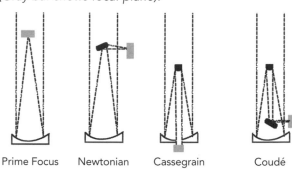

Prime Focus   Newtonian   Cassegrain   Coudé

objective as a good compromise between light-gathering power and cost. Professional astronomers rarely use objectives less than about half a meter (1.5 feet) in diameter. The largest objective lens in the world as of 2010 is 40 inches (1.106 meters) at Yerkes Observatory in Wisconsin.

The eye has its own lens, and the telescope has two lenses (or sets of lenses): the objective and the eyepiece. Figure 3 shows how the two-lens telescope delivers a greatly magnified image to the eye. The magnification in diameters is equal to the focal length of the objective divided by the focal length of the eyepiece. For example, the 40-inch telescope at Yerkes has a focal length of 744 inches. With a one-inch eyepiece, this telescope magnifies 744 diameters.

A microscope operates in the same way, except that the object being viewed, instead of distant and dim, is well lit and close to the objective lens.

Diffraction and Refraction

The useful magnification of a telescope is limited by diffraction. Light rays at the edge of the objective lens are diffracted—they are bent around the edge of the lens. These diffracted light rays cause a pattern of light and dark circles around bright images, which will blur adjacent images together. An empirical formula traditionally used to specify the limit of useful magnification is the Dawes Limit (also called the Rayleigh Limit): the resolution in arc-seconds is $4.56/D$, where D is the diameter of the objective in inches; or $11.6/D$, where

Non-Optical Telescopes

G ravitational fields bend light, as predicted by Albert Einstein's general theory of relativity. Hence, large gravitational fields act as lenses. The first test of general relativity was during a solar eclipse in 1919, when the effect of the sun's gravity was to make stars very near the sun's edge appear to be at a small—but measurable—angle further away from the sun than when they are viewed when the sun is not almost in front of them. In effect, the sun acted as a lens and magnified the image of the area around the sun. There are no lenses for radio waves, but radio telescopes that observe radio waves from astronomical objects, such as quasars, do exist. Most radio telescopes use a metal parabolic mirror to reflect the astronomical radio waves to a receiver at the focus of the parabola.

There also exist what might be called *sound telescopes*. One variety, for picking up sounds from a distance, uses a parabolic dish to reflect sound waves to a microphone at the focus of the parabola. Ultrasound machines, used for monitoring pregnant women, use the woman's own bladder to focus the ultrasound waves onto the receiver.

D is in centimeters. For example, the diameter of the pupil of the human eye when dark-adapted is approximately 8 mm. By the Dawes Limit, the eye can resolve 11.6/.8 (14.5 arc-seconds), or about 1/125 of the diameter of the full moon. The Yerkes telescope can resolve about 0.1 arc-seconds.

A telescope using a lens as its objective is called a *refracting telescope*, since light is "refracted" (bent) by the lens. As of 2010, the 40-inch Yerkes instrument is the largest refracting telescope. A lens that size has to be thick to stand up to gravity, and thick lenses absorb so much light that beyond the size of Yerkes, absorption begins to outweigh the increased light gathered by a wider lens. Hence all current telescopes with objectives greater than 40 inches are "reflecting telescopes" in which the objective is a mirror rather than a lens.

Observer Placement

Unlike a lens, an objective mirror has a parabolic rather than a spherical surface. There is also the mechanical problem of where to place the observer or camera. There are several possibilities, some of which are shown in Figure 5.

One method, called "prime focus," places the photographic film (or other astronomical instrument) inside the path of the incoming light. A few very large reflecting telescopes, such as the 200-inch Hale Telescope at Mount Palomar, actually allow for a human observer to ride in a cage at the prime focus.

A more common arrangement, invented by Isaac Newton and called the "Newtonian," consists of a small flat mirror at an angle, which moves the focal plane to the side of the telescope. Two other common arrangements have a convex mirror at the prime focus, reflecting the light back down the length of the incoming light and also increasing the focal length. In the Cassegrain arrangement, a hole is cut in the middle of the mirror for the light to pass through. In the *coudé* arrangement, the light is reflected one more time into the mounting of the telescope, allowing the use of stationary instruments too heavy to be loaded onto the tube of the telescope.

Further Reading

Alloin, D. M., and Jean-Marie Mariotti. *Diffraction-Limited Imaging With Very Large Telescopes*. Berlin: Springer, 1989.

Edgerton, Samuel. *The Mirror, the Window, and the Telescope: How Renaissance Linear Perspective Changed Our Vision of the Universe*. Ithaca, NY: Cornell University Press, 2009.

Gates, Evalyn. *Einstein's Telescope: The Hunt for Dark Matter and Dark Energy in the Universe*. New York: W. W. Norton, 2009.

Maran, Stephen. *Galileo's New Universe: The Revolution in Our Understanding of the Cosmos*. New York: BenBella Books, 2009.

JAMES LANDAU

See also: Conic Sections; Digital Cameras; Planetary Orbits; Relativity.

Television, Mathematics in

Category: Arts, Music, and Entertainment.
Fields of Study: Communication; Problem Solving.
Summary: Television shows routinely help shape the public's view of mathematics and mathematicians.

Like many other academic disciplines, mathematics has found its way to the small screen in the form of children's educational programming, various puzzle challenges on reality television and other game shows, and mathematically talented characters on a variety of scripted shows. These categories of programming and their attendant themes help shape and reflect the public's image of mathematics and mathematicians at different times. It is important to note that television viewership is determined though the statistically based Nielsen ratings, which networks use to calculate advertising revenue. As a result, the fate of a show is often tied to its Nielsen ratings.

Some of these programs promote mathematics as an exciting learning area (often in children's educational programming) or as a technical skill, which can give characters power and control. Problematic stereotypes persist, especially the still-common portrayal of mathematicians predominantly as white men. The stereotype of the mathematically talented character as a "nerd" is also prevalent and suggests that popular television representations of mathematics reflect both respect for the technical knowledge and fear about an expertise sometimes portrayed as mystifying or as the exclusive domain of obsessive "geeks."

Children's Educational Programming

The focus in children's educational programming that addresses mathematics is often on encouraging children to be excited about the subject area, along with helping them master skills and gain understanding. Most notably, the Children's Television Workshop (CTW), founded in 1967, ultimately created or inspired much of children's educational programming. Funded by federal and private sources, CTW designed *Sesame Street* to teach letter and number skills, as well as foundations of critical thinking, to preschoolers. The program revolutionized children's programming when it premiered in 1969 and has been broadcast continually ever since. Its core focus is on educational content that is presented using attention-getting and retaining tactics, such as fast movement, humor, puppets, and animation. The Count, for example, is a flamboyant Dracula-like character who loves to count. A popular animated segment, "Pinball Countdown," taught children to count using an elaborate pinball machine. Mathematics is also contextualized in segments involving real-life skills, like going to the grocery store.

Studies suggest that *Sesame Street* is viewed by almost half of all U.S. preschoolers on a weekly basis, and there are at least 10 foreign-language versions that have been broadcast in more than 40 countries. Not only is mathematics presented in the show, but it has also been used in shaping decisions about content and presentation. A multidisciplinary team, including Edward L. Palmer, who held a Ph.D. in educational measurement and research design, systematically studied early episodes of the show using data collection and statistical methodology to address both appeal and content comprehension. Other researchers in the early 1970s, including the Educational Testing Services, found both gains in learning and improvements in attitudes toward school in children who watched *Sesame Street*, but at the time it did not help close the gap between some groups of children as had been originally hoped. A longitudinal study found that exposure in the preschool years was significantly associated with better grades in English, mathematics, and science in secondary school, though one cannot infer direct causality from such a study.

In 1973, ABC premiered *Schoolhouse Rock!* as short, musical cartoons aired in between full-length shows on Saturday morning. The show was reportedly inspired by David McCall, the chairman of a public relations firm, whose son had difficulty with multiplication tables but could easily recall song lyrics. In the "Multiplication Rock" series, multiplication of numbers was set to music. Though there is no song about 10, "My Hero, Zero" discusses powers of 10 and the importance of zero. "Little Twelvetoes" examines the base-ten numeral system by imagining a world in which humankind is born with 12 fingers and toes instead of 10. The series "Money Rock" and "Computer Rock" also included applied mathematical concepts. Teachers often show "Multiplication Rock" in their classrooms, and the series is available as both audio and video recordings.

Other shows featured mathematical content as well. In the late 1980s, Children's Television Workshop created the mathematics show *Square One*, which featured guest stars and explored mathematical concepts through segments that parodied aspects in popular culture. In 2002, PBS premiered *Cyberchase*, in which three children and their bird use mathematics to prevail against evil schemes to destroy Cyberspace. Other examples of mathematics in educational programming for various levels of students are *The Metric Marvels*, *Math Can Take You Places*, *Bill Nye the Science Guy*, and *Blue's Clues*. One notable addition to the family of education mathematics programs in the twenty-first century is Nickelodeon's *Team Umizoomi*. This series, which premiered in early 2010, mixes 2-D and 3-D animation with live action to create a virtual world in which a team of characters helps children solve problems. Like many modern television programs, *Team Umizoomi* has an accompanying Web site. According to Nickelodeon executive Brown Johnson, "Math surrounds us everywhere we go, which is why we wanted to create a fun, adventure-filled, interactive series that engages preschoolers and encourages them to practice and refine their mathematical thinking skills."

Reality Television and Game Shows

This spirit of mathematics as an adventurous challenge also appears in other programming, especially on reality television and game shows, which include it as a key test of skill. Often, players must solve a puzzle that falls under the umbrella of some classical problem from the fields of game theory or probability. Instances of mathematics in *Survivor*, *The Mole*, and *The Real World/Road Rules Challenge* have been examined and catalogued. *The Price is Right* has been used to study probability in the classroom, and *Friend or Foe* has been used to analyze and study the Prisoner's Dilemma in the classroom.

Another way in which mathematics is applied to reality television is through its application of voting theory. Many reality shows use formulas to calculate voting results. The fall 2010 season of *Dancing with the Stars* was marked by a controversy in which contestant Bristol Palin, daughter of 2008 Republican vice presidential candidate and former Alaska governor Sarah Palin, consistently received low scores from the judges and yet escaped elimination week after week. The controversy prompted the ABC network to, for the first time, specify its voting scheme on the show and explain it on its Website. Under the system, the judges' scores for each couple are recalculated as a percentage of the judges' total scores for that night. Then, the votes each couple receives from home viewers are calculated as a percentage of the total number of votes received for that week. These two percentages are added together, and the couple with the lowest combined total is eliminated. Palin's high percentage of the popular vote meant that her combined share was rarely the lowest. While reality programs typically refuse to reveal the exact number of votes contestants receive, as in highly popular shows such as *American Idol*, in the case of *Dancing with the Stars*, viewer curiosity and voting controversy prompted an unusually detailed discussion of the mathematics involved.

Nerd-Genius

Moving beyond such simple tests of skill, scripted series sometimes treat mathematics on a deeper thematic level. One common theme is the "nerd-genius." Since the late 1990s, mathematicians and scientists have more frequently been appearing as the unlikely heroes of shows ranging from police procedurals (*NUMB3RS*) to sitcoms (*The Big Bang Theory*), from animated shows (*Futurama*) to reality gamedocs (*Beauty and the Geek*). The increasingly positive portrayal of "nerd-genius" may reflect a greater acceptance of the Information Age and of technical expertise and knowledge as positive attributes.

The popularity of *The Big Bang Theory*, which premiered in 2007 on CBS, speaks to this larger fascination with the nerd-genius. The sitcom follows four young scientists, two of whom are physicists at the California Institute of Technology (one in experimental physics, the other in theoretical physics), a third who is a Caltech astrophysicist, and the fourth, who is an aerospace engineer at a NASA field center. By the show's third season, it was drawing over 14 million viewers per week and ranked in the top 15 shows. In 2010, it won a People's Choice Award for Favorite TV Comedy, and star Jim Parsons won an Emmy Award for Outstanding Lead Actor in a Comedy Series.

The characters are all teased for being socially awkward and obsessive about mathematics and science, as is typical for the stereotype. However, they are also lauded for their intellect and the program presents their thought processes as both humorous and fascinating. The program takes the scientific content seriously, retaining a UCLA physics and astronomy professor, David Saltzberg, to review scripts for accuracy and provide mathematical equations and diagrams. The show has addressed such topics as string theory, loop quantum gravity, and dark matter.

Mathematics is featured in educational shows, reality and game shows, and with mathematically talented characters.

Women and Minority Mathematicians

Whenever mathematicians are depicted on screen, some audience members may form (sometimes prejudicial) opinions about what mathematicians look like or how they act. Mathematicians are often presented as nerdy white men. There are possible downfalls of such limited portrayals. For example, Ron Eglash describes how the dearth of African-American geek characters in popular culture reflects and somewhat reinforces the stereotype that white male nerds are the gatekeepers to full participation in science and technology. But, to their credit, some television shows have made an effort to broaden the demographic range of their mathematical characters, including women and African Americans among their number.

There have been a few female characters with mathematical ability on television. Early examples include three characters from the *Star Trek: Voyager* (1995–2001) series: captain Kathryn Janeway, chief engineer B'Elanna Torres, and Seven of Nine, who was rescued from the Borg (and thus joining the series) in season four. Often, these characters discuss intricacies of twenty-fourth-century physics, including warp speed travel and altering the time line. The show situates these three women (and the Vulcan Tuvok) as leaders among their shipmates in terms of knowledge of and ability in physics, engineering, and mathematics.

Another woman character with mathematical talent is Winifred "Fred" Burkle on the show *Angel* (1999–2004), created by Joss Whedon as a spinoff of his popular *Buffy the Vampire Slayer* series. Though a physicist by training, Fred displays her talents in mathematics, engineering, and invention on the show. Moreover, her character is supported by most of the other characters on the show—she is seen as a key player on the team. The Fred character has also been used as a case study of how Hollywood representations impact girls' mathematical education.

The show *NUMB3RS* (2005–2010) contains another mathematically talented female character: Amita Ramanujan, a Southern Californian of Indian origin. Throughout the series, Amita was a Ph.D. student, then colleague and fiancée, of mathematician Charlie Eppes. Charlie's brother, Don, works for the FBI and uses Charlie's mathematical skills to help solve crimes. Amita and physicist Larry Fleinhardt form Charlie's problem-solving team and inner social circle. As with Fred from *Angel*, Amita is supported by the other series characters who value her mathematical talents. However, the role of Amita has also been controversial because of her romantic relationship with her thesis adviser, Charlie.

Lisa Simpson, from the long-running animated show *The Simpsons* (1989–), also displays mathematical ability (among other nerdish qualities) at various times throughout the series. For instance, in the episode "Girls Just Want to Have Sums," which originally aired April 30, 2006, on FOX, Principal Skinner makes disparaging remarks about girls' mathematical abilities. As a result, the school is split into two single-sex schools. Upset by the lack of rigor in her mathematics class, Lisa is forced to dress as a boy, Jake, in order to attend the boys' mathematics class and learn "real" mathematics. When Jake wins an award for mathematical achievement, Lisa reveals her true identity, to which her brother Bart claims that she did so well in mathematics only because she learned to think like a boy. In the 2010 episode "MoneyBart," Lisa used the statistical methodology of Sabermetrics to manage Bart's baseball team.

Though African-American characters possessing mathematical talent are admittedly not common, two notable exceptions aired on television shows in the late 1980s. *A Different World* (1987–1993), a spinoff of the popular *Cosby Show* (1984–1992), featured Dwayne Wayne as a lead character. At different points throughout the series, Dwayne was a mathematics major and a calculus teacher. Known for his flip-up glasses, Dwayne was involved in romantic relationships with several of the female characters on the show. By contrast, Steve Urkel, on *Family Matters* (1989–1997) was the stereotypical geeky character, depicted in thick glasses, suspenders, and with a high-pitched voice. Whereas Dwayne was portrayed as popular with the opposite sex, Urkel was portrayed as an annoying neighbor of the Winslows who was grimly tolerated from week to week, though even he ultimately gained the audience's sympathy and became engaged to the Winslows's daughter Laura near the end of the show's run.

Other black characters with mathematical talent include Geordi LaForge of *Star Trek: The Next Generation* (1987–1994) and Turkov, of *Star Trek: Voyager*. Geordi eventually became the chief engineer on the Enterprise and a close friend of the android character Data. Often the two of them would discuss various details of twenty-fourth-century physics. Tuvok, though chief security officer of Voyager, also displayed a deep knowledge of science and mathematics. Both Geordi and Tuvok were valued members of their respective crews and were portrayed as scientific experts.

Such depictions of mathematicians and diversity offer great promise for the future, as television shows continue to reflect how society views mathematics and also impact those views themselves. In the twenty-first century, some have noted an increase in the portrayals of mathematics and mathematically talented individuals on television. Examples include mathematical discussions by the main characters on Bones (2005–); an intern on *House* (2004–) named Martha Masters has a Ph.D. in applied mathematics, who joined the cast in 2010; and forensic pathologist Dr. Maura Isles on *Rizzoli and Isles* (2010–), who often discusses mathematical concepts.

Further Reading

ABC.com. "ABC.com—Dancing with the Stars—How Voting Works." http://abc.go.com/shows/dancing-with-the-stars/about-voting.

Coe, Paul R., and William T. Butterworth. "Come On Down…The Prize Is Right in Your Classroom." *PRIMUS: Problems, Resources, and Issues in Mathematics Undergraduate Studies* 14, no. 1 (2004).

Devlin, Keith, and Gary Lorden. *The Numbers behind NUMB3RS: Solving Crime with Mathematics.* New York: Plume, 2007.

Eglash, R. "Race, Sex and Nerds: From Black Geeks to Asian-American Hipsters." *Social Text* 20, no. 2 (Summer 2002).

Greenwald, Sarah, and Andrew Nestler. "Mathematics and Mathematicians on The Simpsons." http://SimpsonsMath.com.

MacLean, Mark. "Math in the Real World: Mathematical Moments From Reality Television." *Math Horizons* 12, no. 4 (2005).

Pegg, Ed, Jr. "The Math Behind Numb3rs." http://numb3rs.wolfram.com/.

Polster, Burkard, and Marty Ross. "Mathematics Goes To the Movies." http://www.qedcat.com/moviemath/index.html.

Leigh H. Edwards
Christopher D. Goff
Sarah J. Greenwald
Jill E. Thomley

See Also: Movies, Mathematics in; Nielsen Ratings; Plays; Science Fiction; Televisions; Writers, Producers, and Actors.

Televisions

Category: Architecture and Engineering.
Fields of Study: Data Analysis and Probability; Geometry; Measurement; Number and Operations.
Summary: Innovations in television technology rely upon a sophisticated use of mathematics, physics, and engineering.

Humans process reality by initially recording light and sound waves through the eyes and ears and then transmitting these data to the brain where they are transformed and synthesized into intelligible matter. In a similar manner, the engineering challenge of television from its conception has been to record data, transmit them (via electricity), and then reconstitute them at a physical distance from its origin. Television is a relatively recent invention. The first appearance of the word (a combination of Greek and Latin words, meaning "far-seeing") occurred in 1900 at the International Electricity Congress at the Paris Exhibition. It was not one single person who invented television, but a number of scientists, engineers, and visionaries working independently in different countries who devised the necessary technology and mathematics. Television has changed dramatically from its first appearance as an electromechanical system to electronic systems, including cathode ray tube (CRT), liquid crystal display (LCD), plasma, and three-dimensional (3D) television.

Image Scanning and Aspect Ratio
Scanning the image required it to be disassembled into discrete pieces of picture that could then be transmitted separately and reassembled as a sequence of images on a screen, with each image recomposed from those smaller pieces of the picture. If the sequence of images could be displayed on the receiver's end rapidly enough, they would appear to the human eye as a continuous whole of moving images. This approach makes use of the fact that the human eye can distinguish two parallel lines

only if they are about one-thirtieth of a degree apart and will blend 12 images per second into a moving whole. In the 1920s, the transmission of images went from an unacceptably choppy five per second to 12.5 and more.

The earliest scanning mechanism is known as the "Nipkow disk," named for the German physicist Paul Nipkow, and versions and refinements of this were used as late as the 1930s. It consisted of a disk with a spiral of small holes in it and a photosensitive cell made of selenium on the other side of the plate from the image. One revolution of the disk corresponded to one complete image, with the holes as they rotated capturing the image in a series of lines. The number of such lines depended on the number of holes, which thus determined the degree of resolution of the image. A second disk was then rotated at the receiving end, playing back the captured image. One drawback of the Nipkow disk was that the scanned lines were not linear, which changed the geometry.

Historians debate why Thomas Edison chose to represent the geometry of television using the rectangular 4:3 aspect ratio, which indicates the ratio of the width to the height of the image. Some hypothesize that Edison chose this because the ratio approximates the golden mean while others assert that his motivation was to save money by cutting 70 mm film stock in half. The Society of Motion Picture Engineers adopted this ratio in 1917 and it was standard for many years. The international standard for high-definition television was devised mathematically in 1980 by electrical engineer Kerns H. Powers. Powers analyzed the common aspect ratios in use at the time and normalized them to a constant area to fit them in a rectangle. When overlapped via their centers, they shared a common inner rectangle. He computed the geometric mean to obtain the 16:9 aspect ratio that continues to be the standard for televisions in the twenty-first century.

A uniform aspect ratio for television created another problem of how to capture the ratio on 35 mm film. Mathematical principles were used to develop lenses that were "anamorphic," which stemmed from the Greek words meaning "formed again." Ultra Panavision used counter-rotated prisms, Technirama used curved mirrors and reflection principles, and CinemaScope used a cylindrical lens. However, the lenses created distortion problems as compared to spherical lenses. In the twenty-first century, mathematics continues to play a role in anamorphic widescreen processes.

CRT Television

While electromechanical televisions such as the Nipkow disk were being developed, an electronic alternative that used a CRT rather than mechanical parts was also being explored. Philo Farnsworth and Vladimir Zworykin, among others, worked independently on this technology in the United States in the late 1920s. The diameter of the round picture tube, which was also the diagonal of the rectangular cover, was the critical parameter. Televisions are still measured on the diagonal in the twenty-first century.

The innovation involved harnessing electrical properties of matter. At the receiving end is a CRT—a glass vacuum tube, which receives the incoming transmitted

3D Television

Stereoscopic effects produced by special televisions add a perceived depth of dimension to standard television that has previously been represented by only height and width, though this technology was still in its infancy at the start of the twenty-first century. Mathematics plays an important role in the evolution of home 3D technology. For example, it is used in determining the proper viewing distance and angle, which depend on the geometry of the display and the location of the viewer in a (often) small space. However, there is a great deal of variability in the process. At the start of the twenty-first century, some people complain of headaches caused by improper parallax, interocular distance in the images or display, or difficulty in interpreting the motion.

signal that represents the picture, known as the "video signal" (audio and visual components are transmitted separately). At one end of the CRT is a cathode, which is heated so that it will radiate electrons (negatively charged particles) that are then attracted along the circuit to the other end of the tube (called the "anode end"), which is at positive electric potential for this purpose. This beam of electrons is focused electrically by charged plates and can be delicately manipulated by interactions with a magnetic field produced by electric current passing through coils.

At this end of the tube is a photosensitive phosphor-coated screen, which has the property of responding to the beam of electrons by emitting light that is proportional in intensity, point for point, to the beam that is moved across it. The video signal is synchronized with the electron beam so that the variations in the beam relay image information. The beam moves line-by-line, lighting the phosphor that illuminates the screen on which the image is viewed. Color images necessitate a more complicated technology than black-and-white images: three signals, one for each of the primary colors (red, green, and blue) and three electron beams are exploited to produce color images.

LCD and Plasma

CRT television was standard through the 1980s but the line-by-line sweeping of the electron beam across the screen takes time and faster technology is available on high-definition television (HDTV), which depends on either an LCD or a plasma screen. The image received via these newer technologies is still comprised of small units, called "pixels" (an abbreviation of "picture elements"), but these operate differently. In an LCD system, each pixel is deployed by an electrically stimulated liquid crystal, which undergoes internal molecular rearrangement in such a way as to polarize (filter) light that is shone from the back. Intensity of light is adjusted by a blocking procedure similar to sunglasses. In a plasma screen, however, each pixel functions like a miniature fluorescent light, since it contains a mixture of gases and mercury that respond to electric charge by radiating energy that in turn causes phosphor on a screen to emit light.

Further Reading

Abramson, Albert. *The History of Television, 1880–1941.* Jefferson, NC: McFarland & Company, 1987.

———. *The History of Television, 1942–2000.* Jefferson, NC: McFarland & Company, 2003.

Noll, A. Michael. *Television Technology: Fundamentals and Future Prospects.* Norwood, MA: Artech House, 1988.

Todorovic, Aleksandar Louis. *Television Technology Demystified: A Non-Technical Guide.* Philadelphia: Elsevier, 2006.

Connie Wilmarth

See Also: Digital Images; Digital Storage; Electricity; Energy; Television, Mathematics in.

Temperature

Category: Space, Time, and Distance.
Fields of Study: Algebra; Measurement.
Summary: Scientists and mathematicians have developed and investigated a variety of principles and scales associated with the measurement and definition of temperature.

Quantification of temperature is necessary for many reasons, including scientific experiments, weather prediction, and many manufacturing processes. Temperature, by its formal definition, measures the movement of molecules in an object. Greater movement results in higher temperatures; conversely, less movement results in lower temperatures. The byproduct is heat, so temperature is often thought to measure the heat of an object. Mathematicians, many of whom are also physicists, have made significant contributions in quantifying heat and developing the temperature scales widely used in the twenty-first century.

History

Joseph Fourier began heat investigations in the early nineteenth century. His work *On the Propagation of Heat in Solid Bodies* was controversial at the time of its publication in 1807. Joseph Lagrange and Pierre-Simon Laplace argued against Fourier's trigonometric series expansions; however, Fourier series are widely used in a variety of theories and applications in the twenty-first century. Jean-Baptiste Biot, Simone Poisson, and

Laplace objected at various times to Fourier's derivation of his heat transfer equations. In 1831, Franz Neumann formulated the notion that molecular heat is the sum of the atomic heats of the components. Studying mixtures of hot and cold water, which did not produce water that was the average of the two temperatures, he concluded that water's specific gravity increases with temperature. This relationship was later shown by other researchers to be true only for a certain range of temperatures. In the late nineteenth century, James Maxwell and Ludwig Boltzmann independently developed what is now known as the "Maxwell–Boltzmann kinetic theory of gases," showing that heat is a function of only molecular movement. Their equations have many applications, including estimating the heat of the sun.

Around the same time, Josef Stefan proposed that the total energy emitted by a hot body was proportional to the fourth power of the temperature, based on empirical observations. In the twentieth and twenty-first centuries, scientists continued to study heat and have developed mathematical and statistical models to estimate heat. These models are used in areas like astronomy, weather prediction, and the global warming debate.

Measuring Tools and Temperature Scales

Heat can be difficult to quantify. Scientists and mathematicians developed many methods and instruments to measure and describe perceived temperature. Some of the earliest were called *thermoscopes*, often attributed to Galileo Galilei. In the early 1700s, Gabriel Fahrenheit created mercury thermometers and marked them with units that became known as "degrees Fahrenheit." He empirically calibrated his thermometer using three values. Icy salt water was assigned temperature zero. Pure ice water was labeled 30. A healthy man would show a reading of 96 degrees Fahrenheit. Later, Fahrenheit would measure the temperature of pure boiling water as 212 degrees Fahrenheit, adjusting the freezing point of water to be 32 degrees Fahrenheit so there was 180 degrees between the freezing and boiling point of water.

Anders Celsius created a different temperature scale in the mid-1700s. The Celsius temperature scale was numerically inverted with respect to Fahrenheit. He used 100 to indicate the freezing point of water and 0 for the boiling point of water. Because there were 100 steps in his temperature scale, he referred to it as a "centigrade" (*centi* means "a hundred" and *grade* means "step"). A few years later, Carolus Linnæus alleg-

edly reversed the scale to make zero the freezing point and 100 the boiling point.

About a century after Celsius created his scale, William Thomson, Lord Kelvin, is given credit for the idea of an absolute zero, a temperature so cold that molecules do not move. The Kelvin scale was precisely defined much later after scientists and mathematicians better understood the concept of conservation of energy. Near-absolute zero conditions produce many interesting problems in mathematics and science. For example, clumping of atoms as they approach an unmoving state can be studied as a classic packing problem, which has extensions in areas like materials science and digital compression. The Kelvin temperature scale uses the same scale as centigrade, with absolute zero about 273 degrees below the freezing point of pure water. Converting from degrees centigrade to Kelvin is as simple as shifting the scale by adding 273.

In the mid-twentieth century, the centigrade scale was replaced with the Celsius scale. The changes were relatively minor, so one estimates the freezing and boiling points of water to be 0 degrees Celsius and 100 degrees Celsius. In actuality, 100 degrees Celsius (the boiling point of water) is now 99.975 degrees Celsius. Converting from degrees Celsius to degrees Fahrenheit, or degrees Fahrenheit to degrees Celsius, involves multiplicative rescaling, not just translation, since 1 degree Celsius is 1.8 times larger than 1 degree Fahrenheit.

Further Reading

Callen, Herbert B. *Thermodynamics and an Introduction to Thermostatistics*. Hoboken, NJ: Wiley, 1985.

Chang, Hasok. *Inventing Temperature: Measurement and Scientific Progress*. New York: Oxford University Press, 2004.

Lipták, Béla G. *Temperature Measurement*. Radnor, PA: Chilton Book Co., 1993.

Pitts, Donald R., and Leighton E. Sissom. *Schaum's Outline of Theory and Problems of Heat Transfer*. New York: McGraw-Hill, 1998.

Quinn, T. J. *Temperature*. San Diego, CA: Academic Press, 1990.

CHAD T. LOWER

See Also: Climate Change; Clouds; Cooking; Geothermal Energy; Measuring Tools; Thermostat; Weather Forecasting.

Textiles

Category: Arts, Music, and Entertainment.
Fields of Study: Geometry; Representations.
Summary: Mathematics is integral to creating both traditional and modern weave patterns in textiles.

Textiles are flexible sheets made out of fibers. Natural textiles are made from plants, animals, or minerals; artificial textiles use human-made fibers, like plastic or synthetic proteins. Woven textiles combine longer fiber threads either by hand or by using looms or knitting machines. In nonwoven textiles, like felt, short or microscopic fibers are bonded by chemical or physical treatments. Nonwovens are often meant to be highly durable or disposable and have many applications in health, construction, and filtration technologies. Mathematical methods are used to design, produce, and analyze textiles. In 1804, Joseph Jacquard invented a weaving system using cards with patterns of holes to control loom threads. These cards were later modified by Charles Babbage into computer punch cards. Weaver and mathematics teacher Ada Dietz wrote *Algebraic Expressions in Handwoven Textiles* in 1949. She outlined a method for using expansions of multivariate polynomials to generate weaving patterns.

Weave Formulas

On a loom, "warp" threads are held parallel and "weft" threads are passed over and under them. A pattern formed in one pass of weft can be either repeated exactly, transposed, or otherwise changed in the next passes. Let *A* stand for warp threads on top and *E* stand for the weft thread on top. In plain fabric, a pattern *AEAE*... indicates that the weave is transposed by one thread in the next row. Basket weave uses *AAEEAAEE*..., so the pattern is repeated for two rows and then transposed by two (or some other whole number) for the next two rows. Satin is *AAAAEAAAAE*..., giving four repeats followed by one transposition. A satin weave results in the majority of the threads being parallel, so light is minimally scattered, producing the characteristic sheen. In contrast, twill has a distinct, textured diagonal pattern formed by using an *EEAEEA*... weaving scheme. Patterns may be added to plain weaves by printing or dying the fabric. The U.S. group Complex Weavers provides a forum for sharing advanced weaving methods and patterns, such as manifold twill.

Patterns and other factors like the thread intersections per area also dictate other properties. For example, plain weave fabrics tear the easiest, because force is applied to the single thread immediately next to the tear. Crimp is how easily the fabric morphs under tension. Plain weaves generally morph the easiest. Wrinkle resistance is the opposite; the more freedom of movement threads have, the easier it is for them to return to smoothness. Satin is an example of a wrinkle-resistant weave. On the other hand, satin silks shrink the most because their weave pattern is loose. Twill has a relatively high resistance to tearing, which makes fabrics such as jean popular for working clothing.

Cultural Textiles

Textiles are a significant cultural art form for many people in Africa. The three most well-known forms are *kente*, *adire*, and *adinkra*. Kente cloth is woven in long narrow strips, traditionally by Asante and Ewe men, and then sewn together into larger pieces of fabric that may be used for clothing or household goods. The cloth was often a sign of wealth and kept for special occasions. There are more than 300 known kente patterns, many of which represent people or historical events. Widely found adira cloth has patterns made by resistance dying. The cloth is tied, stitched, or stenciled, often with geometric patterns, to prevent the dye from adhering to some portions of the cloth. Adinkra cloth is printed, usually by drawing a square grid and stamping symbols into each square. This highly developed symbol language expresses concrete and abstract concepts, such as transformation or unity. Like kente cloth, adinkra often tells stories or proverbs. Tessellations and other repeating patterns are also common. In Ghana, the cloth was originally worn for mourning and some is still reserved for that purpose.

In Scotland, tartans represent families, clans, or regions. A "sett" is a specific plaid pattern, specified by sequences and widths of colored stripes. The pattern is formed by interweaving bands of stripes at right angles. Most are symmetrical, which means the sett is reflected 90 degrees around a pivot or center stripe. Asymmetrical setts have no pivot point. Symmetry has implications in kilt making. A kilt "pleated to the sett" has pleats folded to visually reproduce the tartan pattern across the back of the kilt, often not possible with an asymmetric pattern. Tartan patterns have been investigated with mathematical methods, such as group theory, and they are used in classrooms as examples of symmetry. Artist Andrew Hennessey has proposed "stella tartan" in which tartan setts would be woven radially and overlap in irregular polygon patterns.

High-Technology Textiles

The Industrial Revolution made rapid mass production of textiles feasible and the textile industry has since used many mathematical and computational techniques to continue its evolution. These techniques include differential equations, numerical methods, image processing, pattern recognition, and statistics. Computer-aided design (CAD) and computer-aided looms (CAL) are widespread. Application areas include supply chain management, quality control, and product development. The latter may involve structural modeling and simulation, as well as thermal or biomechanical bioengineering, particularly for specialty textiles. Some competitive swimwear has tiny triangular projections that mimic shark skin to reduce drag. An absorbent, nonwoven textile called *air-laid paper* is used in diapers. Integrating tiny light-emitting diodes into fabric allows clothes to change color or display text or animation. Thermal self-regulation may be achieved with phase-changing microcapsules that become fluid for cooling or solid to release heat, as needed. Weak link theory and bundle theory, as well as research in twisted continuous filaments, helical modeling of yarns, two-dimensional elasticity theory, aerodynamics, and many other investigations have also revolutionized the individual threads that compose fabric, often changing its properties even when using traditional weaves.

Further Reading

Dietz, Ada. *Algebraic Expressions in Handwoven Textiles.* Louisville, KY: The Little Loomhouse, 1949. http://www.cs.arizona.edu/patterns/weaving/monographs/dak_alge.pdf.

Harris, Mary. *Common Thread: Women, Mathematics and Work.* Staffordshire, England: Trentham Books Limited, 1997.

Zeng, Xianyi, Yi Li, Da Ruan, and Ludovic Koehl. *Computational Textile.* Berlin: Springer, 2007.

Maria Droujkova

See Also: Algebra in Society; Crochet and Knitting; Engineering Design; Matrices.

Thermostat

Category: Architecture and Engineering.
Fields of Study: Algebra; Measurement.
Summary: Thermostats are mathematically calibrated according to physical principles to regulate temperature in a variety of settings.

Thermostats and thermometers are related instruments that perform different tasks. A thermometer measures ("meter") heat ("thermo") to determine and display a current temperature. On the other hand, a thermostat is designed to keep the heat ("thermo") stationary ("stat") to help maintain a desired temperature. Inventor and college professor Warren S. Johnson produced the first electronic room thermostats in 1883. He installed them in classrooms to keep students more comfortable in cold weather and to minimize outside interruptions. In the twenty-first century, thermostats are most commonly found inside vehicle engines and as a part of residential, commercial, or industrial heating systems—though they can also be found in appliances, like gas stoves.

Automobiles

In an automobile engine, the thermostat helps regulate temperature so that the engine operates properly and efficiently. The thermostat acts as a control valve for the coolant fluid, which flows within an engine and to a separate radiator that helps to cool the hot coolant. When an engine is first started, the thermostat is closed, and the coolant flowing within the engine cycles through only the engine until it warms up to an ideal temperature. The thermostat measures the temperature change using a special type of wax. Initially, the wax is solid but as the temperature of the surrounding coolant increases, the wax melts and expands to allow hot fluid to flow from the engine to the radiator and cooler fluid to flow from the radiator back in to the hot engine. If the engine gets too hot, the thermostat will open more to allow coolant from the radiator to permeate through the engine. On the other hand, if the engine begins to get too cold, the thermostat will begin to close, allowing less coolant into the radiator and more coolant to cycle through the engine to heat it back up. The thermostat is mathematically calibrated to the engine type and will automatically make the needed corrections as the vehicle is in use.

Buildings

A thermostat used to control temperature in a building similarly does not directly heat (or cool) the rooms. In this situation, it controls a heating (or cooling) unit, which is used to help regulate the temperature. In many systems, a bimetallic strip is used to measure the temperature of a room. Metals expand and contract as they heat and cool. Bimetallic strips work because different metals expand and contract at different rates. A strip of steel and a strip of copper (or brass) will be placed together and the ends secured to each other. If the temperature does not change, the strip remains flat. When the temperature changes, the different rate of expansion or contraction will cause the flat strip to develop a curve toward the metal that has changed less. The amount of curvature can be matched mathematically to a specific degree or range of change in temperature, triggering the system to adjust accordingly.

To increase the sensitivity of the thermostat, most bimetallic strips are long and coiled inside the thermostat. The coil loosens or winds more tightly with a change in room temperature. At a certain point, the bimetallic strip's movements will trigger the heating unit to turn either on or off. Once turned on, the thermostat uses weights or magnets to keep the heating unit from turning off too quickly. Without these devices, the thermostat would create short cycles (turning on and off quickly), which are generally inefficient and could cause a premature failure of the heating unit. Since the bimetallic strip's movement depends directly on the temperature of the immediately surrounding air, the thermostat should not be placed in a location that would cause an inaccurate reading. One common mistake is placing the thermostat by a heat register, where hot air flowing out will trigger the thermostat to turn the heating unit off before the rest of the room has acclimated.

Electronic Variations

More advanced thermostats frequently use electronic rather than electromechanical sensors and may have more than a simple on-off setting. Setpoint staging uses one type of heating process, or *stage*, when the room temperature is within two degrees of the thermostat setting and another when the difference is greater than two degrees from the thermostat setting. Time-based staging activates a secondary stage or unit after the first stage runs for a predetermined amount of time, indicating that the room is colder or hotter

Other Thermostat Applications

The term "thermostat" is also used in statistical thermodynamics, which applies probability theory to systems made up of a large number of particles. This field of study helps relate the large-scale properties of materials observed by people in everyday life to the microscopic properties of the atoms and molecules from which they are made. Here, a thermostat mathematically maintains a constant temperature in computer simulations of molecular dynamics by realistically exchanging the energy of endothermic and exothermic processes that happen during the simulation. For example, the Gaussian thermostat, named for mathematician Carl Friedrich Gauss, maintains system temperature by rescaling the velocities of the simulated atoms at each individual step of the simulation.

than some preset value. Multistage thermostats analyze variables such as the current room temperature, the desired temperature, and the amount of time it takes for a space to warm or cool one degree to determine mathematically when to use a second heating stage.

Further Reading

Automatic and Programmable Thermostats. Merrifield, VA: Energy Efficiency and Renewable Energy Clearinghouse, 1997.

Brumbaugh, James E. *Audel HVAC Fundamentals, Heating Systems, Furnaces and Boilers.* 4th ed. Hoboken, NJ: Wiley, 2007.

Cleveland, Cutler J., et al. *Dictionary of Energy.* Expanded ed. Oxford, England: Elsevier, 2009.

Miles, Victor Chesney. *Thermostatic Control; Principles and Practice.* London: Newnes, 1965.

CHAD T. LOWER

See Also: Auto Racing; Budgeting; Measuring Tools; Temperature.

Tic-Tac-Toe

Category: Games, Sport, and Recreation.
Fields of Study: Geometry; Number and
Operations; Problem Solving.
Summary: Traditional Tic-Tac-Toe has a limited
number of possible games, which can lead players
to quickly discover an unbeatable strategy as long as
they move first.

Tic-tac-toe is a famous game often played by children.
It requires a playing board of a 3-by-3 arrangement of
square cells, usually quickly drawn by making two vertical lines cross two horizontal lines and imagining an
outer border. Two players alternate marking cells with
either an X (usually the first player) or an O (the second player). Each attempts to put three of their marks
in a straight line, while trying to block the attempts of
the other. The winner is the player who first makes the
three-in-a-row line. Unfortunately for the challenge of
the game, the first player can always win by putting an
X in the center cell and playing carefully. Children often
learn this strategy and the game can become mundane
if this strategy is always employed.

Play Possibilities

However, tic-tac-toe is simple enough that it can serve
as a fairly easy example of game analysis, where all possible positions and plays are determined. Most other
games are so complex that such analyses are over-
whelmingly complex.

Ignoring symmetric patterns, there are three possible first plays—a corner, a side, or the center. The
second play patterns are based on these three openings. Again, ignoring symmetries, the corner opening
leads to five possible second moves, the side opening
also allows five possible second moves, but following
a center opening there are only two possible second
plays. Hence, there are a total of 12 noncongruent,
nonsymmetrical second plays. Similar exploration
of the possibilities shows a total of 66 possible third
moves, though 26 are duplications, so there are only
40 noncongruent arrangements after the third play.
Then, it becomes much more complicated because
of overlaps of first- and third-move Xs and second-
and fourth-move Os. This fact demonstrates that even
in such a simple game as tic-tac-toe, the full analysis
becomes quite complex.

Variations

The 3-by-3 magic square (with numbers 1–9 arranged
in the cells so that each row, column, and diagonal sums
to 15) looks like a tic-tac-toe board with numbers. A
game can be played where players take turns choosing
numbers 1–9 (without repeats), trying to reach a sum of
15 with three numbers. Playing this game and placing
the numbers onto the 3-by-3 magic square turns out to
follow the same general games strategies as tic-tac-toe.

Tic-tac-toe can become a much more interesting—and challenging—game by expanding the board
to three dimensions. If the game is played on a stack
of three 3-by-3 boards (a cube of 27 cells), any row of
three is a win. Some have suggested that a 4-by-4-by-4
cube, with a line of four to win, is a smoother game.
Winning lines can lie entirely on a horizontal level,
drop vertically from top to bottom, slant along a vertical plane, or go from one corner to the opposite corner along the body diagonal. New players often have
difficulty even noticing winning lines! For even more
complexity, the game can be played in four dimensions, usually displayed as a two-dimensional array of
two-dimensional boards, assuming the boards can be
stacked in any of the horizontal, vertical, or diagonal
ways, with winning lines in any of the stacks according
to the three-dimensional patterns, a variation that can
be either 3-by-3-by-3-by-3 or 4-by-4-by-4-by-4.

Alternatively, the traditional board can be imaged to
extend infinitely, allowing more possibilities for winning lines. One version keeps the traditional board but
assumes the left column wraps to be next to the right
column, so a line of three can be the upper center, the
right center, and the left bottom corner. Similarly, the
top and bottom rows can be considered as wrapping
around to be next to each other.

Nine-Men's Morris

Many games from around the world pick up on the
ideas of tic-tac-toe, especially the goal of making three
(or more) counters in a row. Probably the most famous
is called Nine-Men's Morris in English (also called
"mill" or, in French, *merelles* or *morelles*); some suggest early versions were even played in ancient Egypt.
The board is three concentric squares connected in the
middles of the sides, with each junction and corner
marked with a dot. Two players each have nine counters, marked to distinguish those of each player. They
take turns playing their counters onto the dots of the

board, trying to get three in a row, which is called a "mill." After players use up the nine counters each, play continues by sliding already-played counters along the lines on the board. Anytime a row of three is made by one player, the player is allowed to remove one of the other player's counters (but they cannot take a counter that is already in a mill). Eventually, one player either has no counters left or cannot move any remaining counters, and the other player wins.

Further Reading

Beck, Jozsef. "Combinatorial Games: Tic-Tac-Toe Theory." In *Mathematical Constants: Encyclopedia of Mathematics and its Applications*. Edited by Steven R. Finch. New York: Cambridge University Press, 2008.

Malumphy, Chris. "3-D Tic-Tac-Toe." http://home .earthlink.net/~cmalumphy/3d.html.

Masters, James. "Nine Mens Morris, Mill—Online Guide." http://www.tradgames.org.uk/games/ Nine-Mens-Morris.htm.

Smit, William. "4-D Tic-Tac-Toe Game." http://www .ugcs.caltech.edu/~willsmit/4d/index.html.

Zaslavsky, Claudia. *Tic Tac Toe: And Other Three-In-A Row Games From Ancient Egypt to the Modern Computer*. Toronto: Crowell, 1982.

LAWRENCE H. SHIRLEY

See Also: Acrostics, Word Squares, and Anagrams; Board Games; Dice Games; Sudoku.

Tides and Waves

Category: Weather, Nature, and Environment.
Fields of Study: Geometry; Number and Operations.
Summary: Mathematicians study and model the forces that cause tides and waves.

Approximately 70% of the Earth's surface is covered with water, most of which is in a constant state of motion. The causes of this motion include the gravitational pull of celestial bodies in space, like the sun and moon; the rotation and shape of the Earth; and the influence of natural phenomena, like wind and earthquakes. Mathematicians have long studied tides and waves, following in the path of ancient scholars and others who sought to understand these phenomena for many spiritual and practical reasons, such as sailing. In the twenty-first century, people still travel both above and below the surface of the oceans for research, commerce, and pleasure, and there are many problems old and new to be explored. Some interesting mathematical investigations related to tides and waves at the start of the twenty-first century include three-dimensional modeling of extreme waves (also called "rogue waves"), such as those observed during the 2004 Indian Ocean tsunami and the Hurricane Katrina storm surges in 2005. Mathematicians, scientists, and engineers have also explored methods and developed technology to harness tide and wave power as an alternative energy source, including methods that actually create waves in addition to using naturally-occurring ones. Some colleges and universities teach courses on tides and waves that involve substantial mathematics. The theme of Mathematics Awareness Month in 2001 was "Mathematics and the Ocean," underscoring the importance and relationship of ocean phenomena and mathematics, as well as the depth and breadth of the topics studied.

Tides

Water in Earth's oceans moves in a variety of ways, including many scales of currents, tides, and waves. Mathematicians and scholars from ancient times up through the Renaissance observed, identified, and quantified tidal patterns. The term "tides" generally refers to the overall cyclic rising and falling of ocean levels with respect to land—though tides have been observed in large lakes, the atmosphere, and Earth's crust, resulting largely from the same forces that produce ocean tides.

The daily tide cycles are caused by the moon's gravity, which makes the oceans bulge in the direction of the moon. A corresponding rise occurs on the opposite side of the Earth at the same time, because the moon is also pulling on the Earth itself. Most regions on Earth have two high tides and two low tides every day, known as "semidiurnal tides," which result from the daily rotation of the Earth relative to the moon. Since the angle of the moon's orbital plane also affects gravitational pull on Earth's curved surface, some regions have only one cycle of high and low, known as "diurnal tides." The height of tides varies according to many variables, including coastline shape; water depth ("bathymetry");

Officers of the National Oceanic & Atmospheric Administration Corps photographed the devastation caused in New Orleans by the 2005 Hurricane Katrina's storm surges.

latitude; and the position of the sun, which also exerts gravitational force. "Spring" tides, not named for the season, are extremely high and low tides that occur during full and new moons when the sun and moon are in a straight line with the Earth, and their gravitational effects are additive. A proxigean spring tide occurs roughly once every 1.5 years when the moon is at its proxigee (closest distance to Earth) and positioned between the sun and the Earth. Neap tides minimize the difference between high and low tides. They occur during the moon's quarter phases when the sun's gravitational pull is acting at right angles to the moon's pull with respect to the Earth.

A few of the many contributors to the theory and mathematical description of tides include Galileo Galilei, René Descartes, Johannes Kepler, Daniel Bernoulli, Leonhard Euler, Pierre Laplace, George Darwin, and Horace Lamb. Some mathematicians, like Colin Maclaurin and George Airy, won scientific prizes for their research. Work by mathematician William Thomson (Lord Kelvin) on harmonic analysis of tides led to the construction of tide-predicting machines.

Waves

There are many mathematical approaches to the study of waves in the twenty-first century, and some mathematicians center their research around this topic. In contrast to tides, a wave is a more localized disturbance of water in the form of a propagating ridge or swell that occurs on the surface of a body of water. Despite the fact that surface waves appear to be moving when observed, they do not move water particles horizontally along the entire path of the wave. Rather, they combine limited longitudinal or horizontal motions with transverse or vertical motions. Water particles in a wave oscillate in localized, circular patterns as the energy propagates through the liquid, with a radius that decreases as the water depth or distance from the crest of the wave increases. Wind is a primary cause of surface waves, because of frictional drag between air and water particles. Larger waves, like tsunamis, result from underwater Earth movements, such as earthquakes and landslides.

The Navier–Stokes equations, named for Claude-Louis Navier and George Stokes, are partial differential equations that describe fluid motion and are widely used in the study of tides and waves. Solutions to these equations are often found and verified using numerical methods. The Coriolis–Stokes force, named for George Stokes and Gustave Coriolis, mathematically describes force in a rotating fluid, such as the small rotations in surface waves. A few examples of individuals with diverse approaches who have won prizes in this area include Joseph Keller, who has researched many forms and properties of waves, including geometrical diffraction and propagation; Michael Lighthill and Thomas Benjamin, who jointly posed the Benjamin–Lighthill conjecture regarding nonlinear steady water waves, which continues to spur research in both theoretical and applied mathematics; and Sijue Wu, who has researched the well-posedness of the fully two- and three-dimensional nonlinear wave problem in various function spaces, using techniques like harmonic analysis. In other theoretical and applied areas, some techniques from dynamical systems theory, statistical analysis, and data assimilation, which combines data and partial differential equations, have been useful for formulating and solving wave problems.

Further Reading

Cartwright, David. *Tides: A Scientific History*. Cambridge, England: Cambridge University Press, 2001.

Johnson, R. I. *A Modern Introduction to the Mathematical Theory of Water Waves*. Cambridge, England: Cambridge University Press, 1997.

Joint Policy Board for Mathematics. "Mathematics Awareness Month April 2001: Mathematics and the Ocean." http://mathaware.org/mam/01.

SARAH J. GREENWALD
JILL E. THOMLEY

See Also: Coral Reefs; Gravity; Mapping Coastlines; Marine Navigation; Moon; Radiation; Swimming.

Time, Measuring

See *Measuring Time*

Time Signatures

Category: Arts, Music, and Entertainment.
Fields of Study: Measurement; Number and Operations.
Summary: Musical time signatures are mathematically defined and are cyclical in nature.

A time signature is a musical notation that defines the meter of a particular composition or a portion of a composition. It establishes a hierarchical, cyclic relationship among beats and among the subdivisions of those beats, which are inherently mathematical in nature. The history of time signatures is somewhat unclear. Some suggest that time signatures first made their appearance around 1000 C.E., though they may not have looked like the ones used in the twenty-first century. Others date the development of the fractional-form time signature closer to the fifteenth century. Nearly all modern Western music uses time signatures or some type of grouped pulses. Along with tempo (rate of beats), musicians use time signatures to gain an understanding of the relation of the elements of a piece of music to one another in time, particularly with regard to a contextual temporal metric.

A time signature normally consists of two integers,

$$n$$
$$b$$

written with one directly above the other. Although it is often notated in prose as a fraction (for example, n/b), it is not a fraction and does not contain a dividing bar or solidus. A time signature appears in the first measure of a composition (in the staff following the clef and key signature), where it defines the default meter for the composition as a whole or until any subsequent time signature occurs that establishes a new default.

Meters and Beat

Time signatures may define various types of meters: simple, compound, complex, additive, or open. In simple meters (those in which the beats have a binary division), the upper integer indicates the number of beats in any one measure. The lower integer is conventionally expressed as a power of two, $b = 2^m$, and specifies what rhythmic value receives the beat. For instance, the time signature

$$\frac{2}{4}$$

indicates a simple meter in which every measure contains two beats and the quarter-note value is the relative duration of each beat. In compound meters (where beats divide into triples), the upper integer n, which is larger than three and divisible by it, designates that each measure contains $n/3$ beats. The lower integer, $b = 2^m$, indicates that the dotted $1/2^{m-1}$-th note receives the beat (the total relative duration of a $1/2^{m-1}$-th note and a $1/2^m$-th note). For example,

$$\frac{6}{8}$$

is the time signature for the compound meter in which each measure has two beats, and the dotted quarter-note duration (a quarter-note value plus an eighth-note value, or equivalently three eighth-notes) represents the beat.

Meters: Complex and Open

Complex meters incorporate beats that normally divide into a mixture of twos and threes. For example, the time signature

$$\frac{5}{8}$$

(each measure has the duration of five eighth-notes) might divide into two unequal beats: one with two subdivisions and one with three. The time signature for a complex meter might also be notated as an additive meter, wherein the upper value is actually an arithmetic expression that agrees with this pattern. For instance, the complex meter

$$\frac{5}{8}$$

could be indicated by the time signature

$$\frac{2+3}{8}$$

An open meter is notated by the symbol $\mathbf{0}$ in place of a more traditional time signature. It indicates that the duration of each measure is defined merely by the rhythmic values or graphic spacing of the notes it contains and does not incorporate a recurring or otherwise specified pattern of beats.

Cyclic Groups

Because of its cyclic nature, meter suggests a modular temporal space, similar to clock time. Algebraically, one might use cyclic groups to model different types of meters. The time signature is useful in determining the order of such a cyclic group, n from above, and what relative duration represents a generating unit, b from above. Then, the first beat of a measure, beginning at time-point zero, would associate with the identity element of the cyclic group, and so on through the nth beat of the measure. Any subsequent measures would represent additional cycles through these sequential group elements.

Interesting Time Signatures

Some time signatures are frequently used, like the lilting rhythm of the following:

$$\text{the waltz } \frac{3}{4} \text{ or the quick Sousa march } \frac{6}{8}.$$

A mathematician might argue that the number of time signatures is limited because the number of beats per measure quickly becomes divisible by a smaller number, making it a multiple of another time signature. However, in music theory, time signatures have a broader meaning in terms of tempo and musical phrasing, not just counts of beats. Interesting compositions have been constructed by considering the mathematical properties of time signatures. Robert Schneider of indie rock band The Apples in Stereo composed a score for a play written by mathematician Andrew Granville and his sister Jennifer Granville in which all the time signatures had only prime numbers of beats per measure. It also included Greek mathematics related to primes in musical form. An entire subgenre of music called *math rock*, which emerged in the 1980s, is typified by uncommon time signatures such as

$$\frac{13}{8} \text{ or } \frac{7}{8}.$$

These complex rhythms can also be found in some mainstream music, such as the song "Anthem" by Rush, which is partially written in

$$\frac{7}{8} \text{ time.}$$

Further Reading

Lewin, David. *Generalized Musical Intervals and Transformations*. New Haven, CT: Yale University Press, 1987.

Mazzola, Guerino. *The Topos of Music: Geometric Logic of Concepts, Theory, and Performance*. Basel, Switzerland: Birkhäuser, 2002.

Rastall, Richard. "Time Signatures." *Grove Music Online*. Edited by L. Macy. http://www.grovemusic.com.

Wright, David. *Mathematics and Music*. Vol. 28 of *Mathematical World*. Providence, RI: American Mathematical Society, 2009.

ROBERT W. PECK

See Also: Ballet; Ballroom Dancing; Composing; Popular Music; Step and Tap Dancing.

Toilets

Category: Architecture and Engineering.
Fields of Study: Algebra; Geometry; Measurement.
Summary: The geometry of modern toilets has been analyzed by engineers using a variety of mathematical and statistical methods.

In all human societies, the disposal of bodily waste has been a primary health concern. It has been estimated that the average human being produces one to two liters of urine and one-quarter to one-half kilogram of feces each day. Fecal matter, in particular, can contribute to the spread of a wide range of diseases, as bacteria and other pathogens can enter food and water when waste is not treated properly. Such problems are especially prevalent in areas of high population density and limited water resources. Over time, a range of toilets and treatment systems have been developed to deal with sewage. Because of the lack of resources and infrastructure, many places in the world in the twenty-first century still contend with waterborne diseases that originate in human waste.

History

Given that many mammals, including most primates, choose to defecate in selected areas in their habitat, it is likely that humans have had specific defecation sites throughout history. Dry toilets, such as pit latrines and outhouses, are ways communities formalized the locations in which humans defecate and are still used in many parts of the world in the twenty-first century. In these systems, waste is concentrated in one place, ideally where it will not infect drinking water. The earliest sitting toilets that used running water to carry waste away date to at least 2500 B.C.E. in the civilizations of the Indus Valley, in what is now India and Pakistan. In 1596, Queen Elizabeth I's godson, Sir John Harrington, invented the first indoor flushing toilet. In 1775, Alexander Cummings, a Scottish watchmaker who studied mathematics, filed a patent for a flush toilet. However, it was not until the late 1700s in Europe and 1800s in America that further modifications and inventions ushered in an age of modern plumbing.

Design and Operation

The geometry of modern toilets is essential to their efficiency and is extensively analyzed by design engineers using a variety of mathematical and statistical methods. The modern home tank toilet consists of a storage tank, a bowl, and an *s*-shaped siphon. Water is stored in the tank. When the toilet is flushed, this water is released into the bowl through rim jets on the underside of the toilet's rim and through a tube called the "siphon jet" that allows most of the water to flow directly into the bowl. The bowl is attached to an "s"-shaped tube, and the influx of water from the tank into the bowl pushes the waste and water over the lip of the "s" and down to an attached waste system. The bowl clears because of the siphon-action created. When the toilet finishes flushing, air enters the siphon tube and stops the siphon. Meanwhile, a flapper valve in the toilet tank closes the connection between the tank and the bowl and allows the tank to refill.

New Developments

The flush toilet takes a large volume of water to operate. In an era of increasingly limited resources, there has been a movement to create low-flush and no-

Coriolis Effect

There is a frequently recurring question of whether the swirl of the water in toilets in the southern hemisphere is opposite that in the northern hemisphere. This notion has been perpetuated in many ways, including popular television shows and scientific programming or textbooks. It is true that large oceanic and atmospheric phenomena, such as hurricanes, will spin in opposite directions in the two hemispheres because of the Coriolis effect. In a small-scale system like a toilet, the geometry of the apparatus, along with water turbulence or temperature, is a much more important factor—a fact that has been verified through systematic experimentation.

flush toilets. For example, toilets manufactured in the United States prior to 1994 used 13 liters of water per flush. The Energy Policy Act of 1992 required that toilets use six liters or less per flush, and as of 2011, high-efficiency toilets used 4.8 liters per flush. In Europe, dual flush toilets are common, providing the user with a choice of how much water to use depending on whether urine or feces is being flushed. Other technologies, including composting toilets that require no water and allow waste to biodegrade for use as fertilizer, have been developed for use by ecologically conscious consumers and people in areas of the world where water or sewage treatment facilities are limited. In addition, a number of toilets have been developed that include warmed seats, water and air jets for cleaning and drying the user, and built-in stool and urine analysis for health assessments.

Modeling Toilet Use

Many modern homes now have multiple toilets and ensuring adequate toilet facilities in public places requires planning and calculation. Two statistical studies of public-restroom use in the late 1980s are still referenced into the twenty-first century. They focused on the amount of time men and women spent in the restroom and they provided some of the first quantitative evidence that women take longer and thus require more toilets. This equity principle is known as "potty parity" and has been enacted into law in many places.

Further Reading

George, Rose. *The Big Necessity: The Unmentionable Subject of Human Waste and Why It Matters.* New York: Henry Holt and Co., 2008.

Raum, Elizabeth. *The Story Behind Toilets.* Chicago: Heinemann Library, 2009.

JEFF GOODMAN

See Also: Energy; Green Design; Water Quality.

Tools, Measuring

See *Measuring Tools*

Tornadoes

See *Hurricanes and Tornadoes*

Tournaments

Category: Games, Sport, and Recreation.
Fields of Study: Algebra; Data Analysis and Probability; Number and Operations; Representations.
Summary: Mathematical methods can be used to seed the bracket for a tournament.

A tournament is any of a variety of competitions in which a relatively large number of players or teams compete at a sport, game, or other competitive activity. While formats differ widely, tournaments generally involve teams or individuals playing a large number of games in a relatively brief period of time. Typically, the ostensible purpose is to determine a single overall winner when the total number of players is (much) larger than the number of players who can participate in a single match. Tournaments of various kinds are held for most competitive activities.

Considerable mathematics goes into the design of tournaments and the choice of format for a particular tournament, often drawing from disciplines such as combinatorics and graph theory. Different choices about the rules of the tournament affect the appeal of the tournament for participants and spectators and, more importantly, can affect which players will be more likely to win. The situation is somewhat analogous to voting systems in which the outcome of a decision can change based on the form of ballot, even when the voters' preferences are unchanged.

Common Types of Tournaments

In a single-elimination knockout tournament, the players compete in pairs. The loser of each game is eliminated from the tournament; the winners go on to the next round. This process continues until only one player is left, who is declared the winner. If the number of competitors is not a power of two, then some competitors sit out one or more initial rounds, automatically advancing to the next round. Which players sit out can be determined randomly or based on some prior rankings. The schedule for which players meet in the first round, the winners of which of these games will meet in the second round, and so on, is called the *bracket* for the tournament. In situations where the competitors are ranked in advance (for example, seeds in a tennis tournament), care must be taken in designing the bracket. It would be undesirable for a player to gain an advantage in a tournament by deliberately underperforming in order to obtain an artificially low prior ranking. The most commonly used brackets involve the highest ranked player meeting the lowest-ranked player in the first round and are used because they are optimized to prevent such manipulation. Double-elimination and triple-elimination tournaments (participants are not eliminated until suffering a second or third loss also exist, though the latter are rather rare. These formats are tolerant of one (or two) lost matches by the player or team that will go on to be champion but the problem of arranging the brackets and scheduling the matches can be more complicated.

In a round-robin tournament, each participant competes against every other participant. Typically, each pairing competes in a single match but variants exist in which more games are played. Such a format gives more information about the relative strength of the players at the expense of requiring more games. Another drawback is that it is generally difficult to identify a canonical choice for first-place champion after a round-robin tournament.

Of course, much more complicated systems exist. Consider, for example, the FIFA World Cup. In the 2010 format, the 32 competing teams are first randomly divided into eight groups. The teams within each group all play against one another. Based on the results of these round-robin matches, a winner and a runner-up emerge from each group. These 16 teams then compete in a single-elimination knockout; the first round of knockout matches involve the group winners each competing against the runner-up from another group.

Graph-Theoretic Tournaments

The term "tournament" is also used with a specialized meaning in the subject of graph theory. A tournament in this sense is a collection of any number of vertices and arrows, where each pair of vertices is connected by a single arrow. Such a picture can represent a round-robin tournament in which each participant competes against every other participant exactly once, and there are no ties. The vertices are the players, and the direction of the arrow indicates who won each game (the arrow points from the winner of the game to the loser). Such configurations were originally studied by H. G. Landau to study the dominance relationships among populations of chickens. Tournaments have gone on to find important applications to social voting theory and public choice.

Further Reading

Froncek, Dalibor. "Scheduling a Tournament." In *Mathematics and Sports*. Edited by Joseph Gallian. Washington, DC: Mathematical Association of America, 2010. http://mathaware.org/mam/2010/essays/FroncekTournament.pdf.

Schwenk, A. J. "What Is the Correct Way to Seed a Knockout Tournament?" *Journal of American Mathmatical Monthly* 107, no. 2 (2000).

MICHAEL "CAP" KHOURY

See Also: Competitions and Contests; Rankings; Sport Handicapping.

Traffic

Category: Travel and Transportation.
Fields of Study: Algebra; Geometry; Problem Solving.
Summary: Mathematical models and statistical analysis of traffic flow suggest solutions.

Traffic flow is studied using mathematical and statistical techniques and computer simulations in order to better understand the movement of vehicles on roads and highways. Americans drive their vehicles almost 3 trillion miles per year on approximately 4 million miles of public roads. Mathematical models have shown that the behavior of even a single driver can have a broad impact on overall traffic flow in this dynamic system. As every driver knows, traffic patterns can often be unpredictable and frustrating, leading to driver stress, accidents, pollution, wasted fuel, and wasted time. Mathematical analysis of traffic congestion can provide transportation engineers with insights leading to improvements in efficiency and safety in the transportation of goods and people. A mathematical understanding of traffic flow patterns can also provide guidance for the design of roadways and provide more accurate calculations of trip itineraries and real-time driving times. These can be disseminated to the public and used in intelligent transportation systems.

The use of mathematics to describe traffic flow patterns slowly originated in the 1930s in order to study road capacity and also to begin to address traffic-related questions, such as how does traffic move through intersections. The mathematical investigations of vehicular traffic increased rapidly in the 1950s, mainly because of the expansion of the highway system after World War II. In the twenty-first century, theoretical models of traffic are utilized by high-performance computers, which can simulate the motions of vehicles on virtual road networks of entire cities and regions.

Traffic engineers distinguish between uninterrupted traffic flow situations (for example, traffic streams on highways and other limited-access roads) and interrupted flow circumstances (for example, where two or more traffic streams meet at a road intersection). The methods suited to analyze a particular traffic scenario depend on whether the flow is interrupted or uninterrupted. When formulating a mathematical description or model of traffic, one must attempt to account for the interplay between the vehicles and the drivers, the layout of the road system, traffic lights, road signs, and other factors.

Queuing theory, which is essentially the mathematical theory of waiting lines, is a probabilistic framework used for analyzing various traffic flow problems, such as optimizing vehicle passage through an intersection or traffic circle, calculating vehicle waiting times at tollbooths, and other similar waiting problems. On the other hand, car-following models and hydrodynamic modeling are deterministic approaches for analyzing traffic flow on long stretches of road.

Car-Following Traffic Models

Car-following models, also known as *microscopic models*, are considered from the point of view of tracking the movements of a line of $n = 1, \ldots, N$ individual cars driving in the same direction down a road in order to try to predict their exact positions $x_n(t)$, velocities $v_n(t)$, and accelerations $a_n(t)$. The starting point for car-following problems is to model how the driver of a car reacts when the vehicle directly in front of it changes speed (it is assumed for simplicity that there no passing is allowed). As a first crude estimation, one could assume a driver adjusts instantaneously according to the relative speed of the driver's car and the vehicle in front:

$$a_n(t) = C\left[v_{n-1}(t) - v_n(t)\right]$$

where C is a constant of proportionality, called the *sensitivity parameter*, which can be measured experimentally. A more realistic assumption would be that a driver adjusts with a lag response time of about one or two seconds, to a maneuver by the vehicle in front of it:

$$a_n(t) = C\left[v_{n-1}(t-T) - v_n(t-T)\right]$$

where T is the time lapse because of the driver's delayed reaction. Equations with delays such as these are then solved to keep track of each vehicle as the traffic moves. Numerous additional assumptions and effects have been incorporated into more sophisticated theories of car-following, such as considering the impact of spacing between cars, the effect of aggressive or cautious driving, and the effect of drivers looking ahead in the road and reacting to the motions of multiple vehicles in front of it.

Hydrodynamic Traffic Models

Hydrodynamic modeling, also called "continuum modeling," considers the flow of a traffic stream to be analogous to the flow of a compressible fluid in a pipe. Continuum traffic models do not keep track of the positions of individual vehicles, like car-following models, but track averaged, macroscopic quantities. For a long stretch of crowded road, such as an interstate highway, three important quantities of interest are flow rate (Q in vehicles per hour), vehicle speed (V in miles per hour), and vehicle density (ρ in number of vehicles per mile). These variables, of course, can vary along the stretch of road in both space and time, and their relationship is described algebraically as $Q = \rho V$. Furthermore, based on observations of traffic patterns over the years, it has been posited that for a given stretch of road, there exists a direct relationship between the flow rate and density. What has essentially been observed is that, on a road having some maximum flow rate, there is a critical vehicle density below which speed is not severely impacted but above which speed reduces. As the density continues to increase, then eventually flow rate reduces, and traffic becomes completely congested. For a concrete example, Greenshield's model postulates a simple linear relation between vehicle speed and density,

$$ V = V_{free}\left(1 - \frac{\rho}{\rho_{jam}}\right) $$

where the parameter V_{free} is the free flow speed of a vehicle that is unencumbered, and ρ_{jam} is the density corresponding to bumper-to-bumper traffic. Then, the flow-density relation would be given by

$$ Q = V_{free}\,\rho\left(1 - \frac{\rho}{\rho_{jam}}\right). $$

This parabolic function begins to capture some of the flow-density behavior that is observed on some real roads, although it is certainly an oversimplification. If the traffic density is zero ($\rho = 0$), then the flow rate must also be zero ($Q = 0$). Additionally, in bumper-to-bumper traffic ($\rho = \rho_{jam}$), the flow rate is zero, or very nearly zero in reality.

In the Lighthill–Whitham–Richards (LWR) theory of traffic, a long stretch of road is considered that has no entries or exits. On such a stretch of road, the number of vehicles must be conserved, and this fact combined with a flow-density relation gives rise to an equation, called a "conservation law," that predicts how vehicle density varies along the stretch of road. When a traffic jam occurs, it manifests as a sudden disturbance, or shock-wave, in the vehicle density along the road. LWR theory and other much more sophisticated continuum models of traffic can predict conditions under which traffic jams will form, propagate, and dissipate. Common reasons for traffic jams are accidents, construction, lane merges, and other changes in road capacity. However (as all drivers have experienced) sometimes "phantom jams" occur on highways for no apparent reason. These phantom jams can also be explained by continuum traffic models.

Further Reading

Daganzo, Carlos F. *Fundamentals of Transportation and Traffic Operations*. Oxford, England: Pergamon-Elsevier, 1997.

Gazis, Denos C. *Traffic Theory*. Norwell, MA: Kluwer Academic Publishers, 2002.

May, Adolf D. *Traffic Flow Fundamentals*. Upper Saddle River, NJ: Prentice Hall, 1990.

ANTHONY HARKIN

See Also: Auto Racing; Highways; Smart Cars; Travel Planning.

Trains

Category: Travel and Transportation.
Fields of Study: Algebra; Geometry; Measurement; Number and Operations.
Summary: Trains and railways present interesting mathematical problems related to force and load, scheduling, and geometry.

Railroads influenced nearly every aspect of nineteenth and early twentieth century U.S. society. Companies building infrastructure for railroads (and railroads themselves) dominated the U.S. economy as more goods and people were transported via rail. Investors clamored to profit from the railway boom, inspiring engineers and mathematicians to improve the technology used in the railway system. As more people traveled by train, punctuality and reliability needed to improve. Time zones in the United States were established primarily because competing rail companies used different standard times for their schedules. In addition, Christophorus Buys-Ballot and others conducted experiments using trains to explore the Doppler effect, named for mathematician and physicist Christian Doppler. At the start of the twenty-first century, wooden and electric railway sets remain popular toys with children of all ages, while railroad enthusiasts design elaborate model train layouts in various scales reflecting the days when towns were centered around train stations.

Locomotives

Locomotives are classified using the Whyte system, named for mechanical engineer Frederick Whyte, which utilizes numbers to describe the wheel arrangement of the engine. For example, a 4-8-4 type locomotive has four wheels in the front, 8 driving wheels in the middle, and 4 wheels in the rear. The capacity of a locomotive depends on the amount of friction the driving wheels have with the track and the weight of the engine over the driving wheels. These quantities are related by the equation $F = MW$, where F represents the maximum pulling force of the train, M represents the coefficient of friction between the wheels and the track, and W is the portion of the weight of the locomotive over the driving wheels. While this relationship indicates that heavier trains can pull larger loads, more power is needed to move the train, leading to higher fuel costs. Increasing the coefficient of friction gives the train better traction and thus more pulling force, so most locomotives have a sandbox on the front from which sand is sprayed onto the track when the rails are slippery. Though friction is needed to get the train started, reducing M increases efficiency once the train is in motion, lowering operating costs.

Modern diesel-electric locomotives use high-tech designs to achieve more horsepower while reducing engine weight significantly. Equipped with a sophisticated array of sensors, onboard computers, and control systems, twenty-first-century trains maintain their hauling capacity while reducing fuel consumption and emissions. The future may see more magnetic levitation (Maglev) trains, which use magnetic fields to suspend the train above the track. The first commercial Maglev train opened in 1984 in Birmingham, United Kingdom, but ceased operations in 1995 in part because of design problems. A Maglev train in Japan recorded a maximum speed of 581 kilometers per hour (361 miles per hour) in 2003, the highest ever speed for a Maglev transport.

Passengers and Timetables

Commercial trains, whether passenger trains or freight trains, follow carefully written schedules. Composing these intricate timetables is a daunting task. Railways must ensure that trains do not collide on the tracks, and that goods and people are transported in a timely and efficient manner. In 2006, the Netherlands introduced a new railway timetable for all trains and mathematical modeling played a key role in developing the timetable. To determine how a set of trains should be routed through a station, researchers listed all feasible routes through the station for every train. Each combination of a train and a feasible route is represented by a node on a graph. Nodes on this large graph are connected if they belong to the same train or if there is a routing conflict between the train/route combinations. Presenting the scheduling problem in graph form enables sophisticated computer programs to generate a usable timetable. Additional modifications improve the efficiency of the timetable in the case of unexpected delays.

Railway passengers expect trains to be on time and to have sufficient space for a comfortable ride. Timetables can be fine-tuned to meet these customer demands using another type of mathematical modeling called "peak load management." Consultants work with railways to determine when trains are the most crowded and when passenger demand is highest. Mathematicians quantify the notion of "attractiveness," a measure of how satisfied a rider on a given train will be as a function of the journey time on the train, the time the passenger would like the train to arrive at its destination, and the actual arrival time. Another constant is added to the equation to determine how much attractiveness is reduced for each minute the actual arrival time differs from the customer's ideal arrival time. More terms can be added to measure the crowding on the train—overcrowding having a significant impact on attractiveness. Using this model, railways can develop timetables that increase the probability that a customer will ride on an "attractive" train. Further refinements to the model attempt to minimize the chance that a passenger will need to stand while riding.

Trains as Teaching Tools

Creative elementary school teachers have devised ways to use the appeal of toy trains to teach addition and subtraction. A colorful cardboard train is taped to a bulletin board and children count the number of cars on the train. Train cars are easily removed or added and the students see addition and subtraction in action by counting the number of cars on the new train. Wooden railway systems with magnetic couplings between cars also allow for easy joining and separating, making these toys excellent mathematical manipulatives when working with small groups of children. Older students may encounter the Two Trains puzzle. Two trains are on the same track traveling toward one other at a constant speed. A fly starts on the front of one train and flies toward the other train at a constant speed, faster than either train. Once the fly reaches the other train, the fly immediately turns around and continues buzzing back toward the first train. How far does the fly travel before being smashed when the two trains collide?

Track Geometry

Freight yards use combinations of switches, sidings, and turnaround loops to sort railway cars, assembling them into trains bound for various destinations. The fact that trains cannot pass each other on a single track leads to many challenges. The optimal arrangement of freight cars in the most efficient manner is another problem for mathematical modeling, but these fascinating switching systems have inspired mathematicians to investigate interesting questions involving train track layouts and railway switching puzzles.

A switch (also known as a "turnout," or "point") is a Y-shaped structure used to split tracks into two lines or to combine two lines into one. The directional nature of a switch makes the dynamics interesting: trains entering at the "top" of the Y will always exit through the bottom branch, but trains entering through the bottom have the option of traveling on the left branch or the right branch. Switches are used to sort cars in freight yards, enable locomotives to move onto a siding to allow a train traveling the opposite direction on the track to pass, and make it possible via a turnaround loop for a train traveling one direction to reverse direction.

How can two trains traveling in opposite directions, say eastbound and westbound, pass one another? If there is a siding long enough to contain one of the trains, the problem is easy. But what if only one car can occupy the siding at a time? Variations on this train-passing puzzle have been around for over a century. The trains can still pass each other through clever use of the siding. The eastbound train leaves its cars behind, moves onto the siding, and waits for the westbound train to pass through. After the eastbound engine emerges from the siding, the westbound train backs through the siding, bringing along one of the eastbound train's cars and leaving that car on the siding. After the westbound train has pulled forward past the siding, the eastbound train can pick up its car, and the process repeats until the entire eastbound train is through.

Imagine a child playing with a toy railroad. Given a set of switches and plenty of track, how many different layouts can the child make? To determine whether two track layouts are different, the structure is transformed into a graph, with nodes representing lengths of track. Nodes are connected if there is a switch allowing a train to travel from one length of track to another. Layouts are said to be different if their graphs are the same. A child with two switches can make five distinct layouts. Using more switches and combinations of other types of switches, like the three-way pitchfork-shaped variety, even more layouts can be made and counted using mathematics.

Further Reading

England, Angela. "Train Math Lesson Plan." http://www .suite101.com/content/train-math-lesson-plan-a45144.

Gent, Tim. "Model Trains." http://plus.maths.org/ content/model-trains.

Hayes, Brian. "Trains of Thought." *American Scientist* 95, no. 2 (2007).

Kroon, Leo. "Mathematics for Railway Timetabling." *ERCIM News* 68 (2007).

Lynch, Roland H. "Locomotives." *Ohio State Engineer* 23, no. 5 (1940).

Peterson, Ivars. "Ivars Peterson's MathTrek: Laying Track." http://www.maa.org/mathland/mathtrek _01_08_07.html.

Mark R. Snavely

See Also: Bus Scheduling; Graphs; Mathematical Modeling.

Trajectories

See *Learning Models and Trajectories*

Transformations

Category: History and Development of Curricular Concepts.
Fields of Study: Communication; Connections; Geometry.
Summary: Numerous mathematicians since antiquity have studied and worked on the concept of transformations.

In mathematics, transformations have a rich history that connects various disciplines, including geometry, algebra, linear algebra, and analysis with applications in statistics, physics, computer science, architecture, art, astronomy, and optics. In general, a transformation changes some aspect while at the same time preserving some type of structure. For example, a dilation of an object will shrink or enlarge it but will preserve the basic shape; while a reflection of the plane will produce a mirror image, which flips figures while preserving distances between points. Mathematicians and geometers often transform an object, equation, or data to something that is easier to investigate, such as trans-forming coordinates to simplify algebraic expressions. The theory of transformations has important implications as well. There are many types of transformations including geometric transformations, conformal transformations, z-score transformations, linear transformations, and Möbius transformations, named for August Möbius. Geometric transformations have long been implicitly used in aesthetically pleasing design patterns in pottery, quilts, architecture, and art, such as tessellations in the Moorish Alhambra Palace. Historians and anthropologists compare and contrast these patterns to track the spread of groups of people. Mathematical transformations can be represented in a variety of ways, such as matrix representations of linear transformations, which are useful in algorithms and computer graphics. In school, young children study geometric transformations and this study continues through high school, where students represent various geometric and algebraic transformations using coordinates, vectors, function notation, and matrices. Students also investigate transformations using computers and calculators.

Early History

The early development of geometric transformations is tied to motions that were useful in modeling the Earth and the stars and in creating artistic works, architectural buildings, and geometric objects. The Pythagoreans thought that points traced lines and lines traced surfaces. Aristotle objected to the use of physical concepts like movement in these abstract mathematical objects. Euclid of Alexandria mostly avoided the concept of motion in his work. However, he used the notion of "superposition," where one object is placed on top of another, in triangle congruence theorems, such as in his proof of side-angle-side congruence. In modern proofs, mathematicians would likely use transformations in order to place these triangles on top of each other. Euclid also defined a sphere as the rotation of a semicircle, and he defined a cylinder as the rotation of a rectangle. Archimedes of Syracuse investigated axial affinity motions in his work on ellipses, and Apollonius of Perga explored inversion. Marcus Vitruvius described the projections that were important in architecture, and he also investigated the concept of stereographic projection, which was useful in astronomy and map making.

Mathematicians around the world generalized these motions and applied them to a variety of fields. Most mathematicians in later times relied on transformations in geometry, although Omar Khayyám criticized Ibn al-Haytham's extensive use of motion by questioning how a line could be defined by a moving point when "it precedes a point by its essence and by its existence." Both Thabit ibn Qurra and his grandson Ibrahim ibn Sinan investigated "affine" transformations of the plane that preserved straight lines, like dilations. Alexis Clairaut and Leonhard Euler defined and explored general affine transformations. Sir Isaac Newton investigated various coordinate systems and the transformations between them, such as what are referred to as "rectangular and polar coordinates." Girard Desargues systematically investigated projective transformations, although many earlier mathematicians had investigated perspective drawing and projection in mathematics, art, and optics. Edward Waring and Gaspard Monge also studied projective transformations. Mobius represented affine and projective transformations analytically in terms of homogeneous coordinates.

Carl Friedrich Gauss linked transformations with linear algebra when he represented linear transformations of quadratic forms as rectangular arrays of numbers. A linear transformation of the plane is a map that preserves addition and scalar multiplication of vectors. Linear transformations of the plane are combinations of rotations, reflections, dilations, shears, and projections, and they are important in modeling movement in computer graphics. In general, a linear transformation is a map between vector spaces that preserves addition and scalar multiplication. Linear transformations of coordinates were important in the development of analytic geometry and some multivariate statistical methods and linear transformations were also linked to projective geometry and Mobius transformations, which are also called "fractional linear transformations." Henri Poincaré connected these transformations to hyperbolic geometry. Gotthold Eisenstein and Charles Hermite tried to extend Gauss's work on forms and in this context they defined the addition and multiplication of linear transformations. Arthur Cayley defined a general notion of matrices and recognized that the composition of linear transformations could be represented using them. James Sylvester explored properties of matrices that were preserved under transformations and defined the nullity of a matrix. Matrices continued to be connected to linear transformations and the theory of linear transformations extended to infinitely many dimensions.

Modern Developments

At the beginning of the twentieth century, Felix Klein revolutionized mathematics and physics with the idea of a transformation group. In his Erlanger Program, the properties of a space were now understood by the transformations that preserved them. Thus the classification, algebraic structure, and invariants of these transformations provided information about the corresponding geometries. His ideas unified Euclidean and non-Euclidean geometry and became the basis for geometry in the twentieth century. Klein's collaboration with Sophus Lie impacted the development of the Erlanger program. Lie also developed the notion of continuous transformation groups and associated these with a differential equation. Physicists and mathematicians continue to study the local structure of a so-called Lie group by the infinitesimal transformations in the Lie algebra. Earlier mathematicians and physicists had already used invariants in a several ways. For instance, Cremona transformations are named for Luigi Cremona, who studied birational transformations. These transformations were important in the study of algebraic functions and integrals. Max Noether investigated the invariant properties of algebraic varieties using birational transformations. In physics, Hermann Minkowski explored Maxwell's equations for electromagnetism, named after James Maxwell. These equations were invariant under Lorentz transformations, named for Hendrik Lorentz, and led to a geometry of space-time and the beginning of relativity theory.

Further Reading

Kastrup, H. A. "On the Advancements of Conformal Transformations and Their Associated Symmetries in Geometry and Theoretical Physics." *Annalen der Physik* 17, no. 9–10 (2008).

Kleiner, Israel. *A History of Abstract Algebra*. Boston: Birkhauser, 2007.

Rosenfeld, B. A. *A History of Non-Euclidean Geometry: Evolution of the Concept of a Geometric Space*. New York: Springer, 1988.

Sarah J. Greenwald
Jill E. Thomley

See Also: Animation and CGI; Composing; Coordinate Geometry; Equations, Polar; Quilting; Symmetry.

Transplantation

Category: Medicine and Health.
Fields of Study: Algebra; Data Analysis and Probability; Number and Operations.
Summary: Locating and allocating available compatible organs is an important task of surgery, as is determining the likelihood of success and survival.

Organ transplantation involves replacing a damaged organ or body part with an organ taken from another body, a location on the patient's own body, or sometimes another source. Relatively common organ transplants include hearts, lungs, livers, corneas, bone marrow, and skin. In the twenty-first century, there are increasing instances of transplantations involving parts that have proven more difficult in the past, including a human face in 2010. Transplantation is one of few medical fields where practice is driven by statistical analysis of large-scale national datasets. Collecting comprehensive data about transplantation in the United States is mandatory, and researchers use statistics to inform clinical practice and national policy. Still, there are too few living and deceased organ donors to meet the need. Optimization tools make the best use of scarce resources, like donated organs. With kidney paired donation, optimization can even increase the supply of available organs. An artificial pancreas employing control theory was under development in 2010.

Statistics

Statistical analyses inform transplant policy and individual decisions. The transplant community seeks equity in allocating organs, so the allocation system is frequently analyzed for gender and racial disparities. Understanding outcomes with and without transplantation helps patients decide if they will benefit from a particular transplant.

Survival analysis is the branch of statistics concerned with the distribution of time to an event. Survival analysis is commonly used in medicine to study time-to-death but can also be used to study time to any event, such as time from joining a transplant waitlist until receiving a transplant. The survival function $S(t) = \Pr(T > t)$ indicates the probability that the random time of an event T is later than a given time t.

Complications

Survival analysis is complicated by censoring; not all patients in a study have reached the event of interest. In a time-to-death analysis, some patients are likely still alive. The first technique for estimating a survival function with censoring was the product-limit estimator of statisticians Edward Kaplan and Paul Meier.

Confounding is another challenge. One could perform a survival analysis of the association between gender and time-to-transplantation to see whether men and women receive transplants at the same rate. However, not all patients are expected to wait the same amount of time. Other factors (such as age and blood type) confound studies of the effect of the factor of interest (gender) on time-to-transplantation. Cox proportional hazards analysis methods, named for statistician David Cox, can account for confounding, using a regression model based on the hazard function $\lambda(t)dt = \Pr(t \leq T < t + dt \mid T \geq t)$, which indicates the

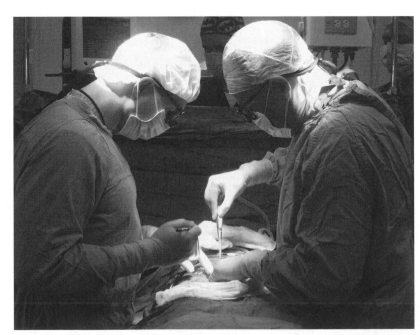

Doctors from Walter Reed Army Medical Center Organ Transplant Service perform Guyana's first kidney transplant operation in 2008.

instantaneous probability of an event at some time (*t*) conditional on having survived to at least that time.

Optimization

Donated organs are scarce and each organ must be allocated to one of many potential recipients. Optimization techniques allocate scarce resources by maximizing an objective function. A person's Lung Allocation Score is largest when the transplant has the largest lifespan benefit, and available lungs are offered to the nearby person with the largest score.

Kidney paired donation in which two living donors who are incompatible with their intended recipients exchange kidneys for compatible transplants requires more complex optimization techniques. More people can obtain better transplants when the paired donations are arranged using either a maximum weight matching in a graph or a maximum weight cycle decomposition (if more than two donors and recipients are involved in each exchange). By optimizing an individual's outcome rather than the overall good, a Markov decision process model, named for mathematician Andrei Markov, can determine whether it is better for a patient to accept a certain organ offered or wait until a possibly better organ is offered later. Another Markov decision process model can establish the best time for a patient to receive a liver transplant from a living donor.

Control

Control theory studies systems where adjustments over time maintain some desired set point, like a thermostat heating or cooling a room to maintain a comfortable temperature. In transplantation, control theory is used in an experimental artificial pancreas. A healthy person's pancreas maintains blood glucose levels over time by regulating insulin in response to eating a meal or exercising. An artificial pancreas uses a blood glucose monitor and a mathematical control system to drive an insulin pump. The control algorithms are tested on mathematical models of blood glucose levels before being tested in human subjects.

Further Reading

Cox, David R. "Regression Models and Life Tables." *Journal of the Royal Statistical Society* 34, no. 2 (1972).

Harvey, R. A., et al. "Quest for the Artificial Pancreas." *IEEE Engineering in Medicine and Biology Magazine* (March/April 2010).

Kaplan, Edward L., and Paul Meier. "Nonparametric Estimation From Incomplete Observations." *Journal of the American Statistical Association* 53 (1958).

Segev, D. L., S. E. Gentry, D. Warren, B. Reeb, and R. A. Montgomery. "Kidney Paired Donation: Optimizing the Use of Live Donor Organs." *Journal of the American Medical Association* 293 (2005).

Sommer Gentry
Dorry Segev

See Also: Cochlear Implants; Disease Survival Rates; Life Expectancy; Surgery.

Travel Planning

Category: Travel and Transportation.
Fields of Study: Geometry; Measurement; Problem Solving; Representations.
Summary: Mathematical models are used to plan and evaluate short- and long-term transportation infrastructure decisions.

Travel planning is a broad field that covers everything from individual journey planning (for example, deciding which form of transportation to use when commuting) to regional transportation planning (for example, deciding on the layout for a new train line or arterial road). Professional transportation planners use equations and computer programs to directly compare different transportation modes and routes. These equations can be very simple with just a handful of terms, or incredibly complex models with hundreds of interacting variables. Regardless of the complexity of the analysis, the ultimate goal is to satisfy the objectives of the planning project and often the most important objective is to minimize journey costs.

Governments aim to maintain effective road networks and public transportation systems. Within certain cities, however, there can be a distinct bias toward either private or public transportation. This bias reflects the fact that governments prioritize certain planning decisions over others. These decisions fall into two broad categories: long-term and short-term planning.

Professional transport planners use equations and computer programs to compare different modes and routes when trying to maintain effective road networks and public transportation systems.

Long-Term Planning

Long-term planning includes decisions such as the use of land and placement of new freeways or bypasses. The objectives for these projects are often manifold: reduce costs, reduce pollution, reduce noise, maintain traffic flow, and maintain priority for public transport and carpool vehicles. It is the challenge of the transport planner to balance these objectives and ensure that the final decision satisfies these criteria.

Long-term planning also incorporates projections of future effects. For example, a wider freeway leads to better accessibility in certain urban areas, which eventually leads to more construction in those areas. Construction, in turn, leads to more traffic on the freeway and a renewed need to widen the road, thus creating a cycle. Long-term projects typically take several years to implement and even longer to monitor their impact.

Short-Term Planning

Short-term planning includes the introduction of bus priority lanes, changing the timing of traffic light signals, using trains with a greater numbers of cars during peak travel times, changing the price of parking in a particular area, changing taxi regulations, introducing new public transport fare systems, and so on. These are changes that can be implemented and evaluated within weeks as opposed to years.

Comparing Alternatives

To make any long-term or short-term planning decisions, it is necessary to compare a range of alternatives, side by side, using as few indices as possible. As a simple example, a set of four time and money measurements can be reduced to a single measurement of cost (C) using the equation

$$C = a(P) + b(t_T) + c(t_W) + d(t_J)$$

where P is the fare price, t_T is the transit time, t_W is the wait time, t_J is the journey time, and the coefficients a, b, c, and d are used to weight the components relative to one another (for example, for a given individual, wait time may be perceived to be twice as costly as journey time). This equation is particularly useful for comparing different forms of public transport. The single cost values will paint a very clear picture of which mode and route has the optimum mix of short times and low costs. Cost is often expressed in minutes, as opposed to dollars, as this measure will remain stable even as prices increase.

Some other measures used by transport planners include traffic density (number of vehicles on a given stretch of road), traffic flow (number of vehicles passing through a given stretch of road every minute), and performance index (an aggregate measure of the delays experienced in a given transport network). Each of these measures must be interpreted in context because acceptable ranges for the values will vary depending on road type, city size, and network connectivity.

Another common method used to estimate the amount of traffic passing between two zones (for example, a neighborhood and a commercial center) is called the "gravity model." It was given this name because the form of the equation is similar to Isaac Newton's equation of gravity. The traffic passing between two zones, A and B, is proportional to the product of the traffic originating in zone A and the traffic arriving in zone B but inversely proportional to a function of the distance between the two zones.

Governmental transport planners use these measures to test for weaknesses in the transport network—places where demand exceeds supply—and to gauge the effects of previous planning decisions. The act of planning is therefore firmly rooted in the interpretation of numerical output from mathematical analyses.

Further Reading

Banister, David, ed. *Transport Planning*. 2nd ed. New York: Taylor & Francis, 2002.

Black, John. *Urban Transport Planning*. London: Croom Helm, 1981.

Button, Kenneth. *Transport Economics*. Cheltenham, England: Edward Elgar Publishing, 2010.

O'Flaherty, Coleman. *Transport Planning and Traffic Engineering*. Oxford, England: Elsevier, 1997.

Wells, Gordon Ronald. *Comprehensive Transport Planning*. London: Griffin, 1975.

EOIN O'CONNELL

See Also: Bus Scheduling; City Planning; Traffic; Traveling Salesman Problem.

Traveling Salesman Problem

Category: Travel and Transportation.
Fields of Study: Geometry; Problem Solving.
Summary: The traveling salesman problem is a notable applied mathematics problem that is simply constructed and may be unsolvable.

Imagine a salesperson that needs to travel to 30 cities. The salesperson wants to begin in his or her hometown, visit every city exactly once, and return to the hometown. In what sequence should the salesperson visit the cities in order to minimize the total amount of traveling time on the road between cities? The significance of the traveling salesman problem (TSP) lies in the fact that many other problems can be translated into a traveling salesman formulation and that a brute force check-all-the-possibilities approach will take prohibitively long—even for moderately sized problems (like the example) and with the use of fast computers.

Many problems can be translated to the TSP. The travel time between cities can be replaced by distance, cost, or other measures. Hence, in essence, this problem captures many sequencing problems where a number of tasks have to be sequenced and the costs can be modeled appropriately. Problems as diverse as optimizing the routes of garbage trucks, planning the sequence of motions performed by a robot, and ordering genetic markers on a chromosome have been modeled by the TSP.

Solving the TSP

Why is solving the TSP hard? If one decides to solve the problem by checking all the possibilities and then choosing the best one, then the sheer number of possibilities will make the problem impossible to solve. For example, with 30 cities and starting at a hometown, initially there are 29 cities to choose as a first destination. Regardless of the first choice, there are 28 cities to choose from next and so on. The total number of possible ways to start from a hometown, traverse each of the 30 cities exactly once and return to the hometown is

$$
\begin{aligned}
29! &= 29 \times 28 \times \ldots \times 3 \times 2 \times 1 \\
&= 8,841,761,993,739,701,954,543,616,000,000 \\
&\approx 8.8 \times 10^{30}
\end{aligned}
$$

possibilities.

Even if a computer checked a million possibilities per second, checking all the possibilities would take more than 200,000,000,000,000,000 years—much longer than the age of the universe. Making the computer twice or 10 times faster still will not be enough to make the problem worth attempting.

Solution Through Algorithms

Could there be clever algorithms that solve the TSP faster? The TSP is among the problems that computer scientists call *NP-hard*. Given any algorithm for solving the TSP, certainly the number of steps needed by the algorithm grows as the size of the problem—namely the number of the cities—grows. If the number of steps in an algorithm as a function of the size of the problem is a polynomial, then it is generally believed that the problem is tractable. In other words, if there is one such polynomial time algorithm, then one can hope to find other more efficient ones and be able to solve even large-sized problems efficiently. At the start of the twenty-first century, it is not known whether the TSP has such a polynomial time algorithm. But it is known that if there is such an algorithm, then there is also efficient algorithms for a host of other problems of interest to computer scientists. For many years, researchers have looked for such algorithms and have not been able to find one, and the strong prevailing opinion is that no such algorithm exists (this is the famous $P \neq NP$ problem).

Even though the TSP is a difficult problem to solve in general, progress has been made in developing algorithms that do much better than the brute force method. In fact, very large instances—for example, one with 85,900 cities—of the TSP have been solved exactly. On another front, many approximation algorithms have been devised. These algorithms do not aim to find the absolute best solution but rather find a solution that is close to the best one. A simple approximation algorithm using minimum spanning trees, for example, can find a solution that is guaranteed to be no worse than twice the optimal solution. More sophisticated algorithms can find a solution within a few percentages of the optimal solution for a problem with the number of cities in the millions.

Further Reading

Applegate, David L., Robert E. Bixby, Vašek Chvátal, and William J. Cook. *The Traveling Salesman Problem: A Computational Study*. Princeton, NJ: Princeton University Press, 2006.

Gutin, Gregory, and Abraham P. Punnen. *The Traveling Salesman Problem and its Variations*. Dordrecht, The Netherlands: Kluwer Academic Publishers, 2002.

SHAHRIAR SHAHRIARI

See Also: Birthday Problem; Bus Scheduling; Cocktail Party Problem; Scheduling; Tournaments.

Trigonometry

Category: History and Development of Curricular Concepts.
Fields of Study: Communication; Connections; Geometry; Measurement.
Summary: Trigonometry is one of the most essential branches of mathematics in engineering and science.

The principal value of trigonometry rests in its numerous and practical applications throughout the world. It has always been viewed as one of the most applied areas of mathematics. When combined with the latest technologies, trigonometry impacts society in immeasurable ways. It is essential in engineering and in all of the sciences. Some of the applications of trigonometry include astronomy, aviation, architecture, engineering, geography, physics, seismology, surveying, oceanography, cartography (mapmaking), navigational systems, space sciences, medical imaging, music, and video games. Clearly, the applications and influences of trigonometry are fully embedded within contemporary society. To appreciate modern applications of trigonometry and the contributions of societies throughout the world, it is important to consider its historical evolution.

Origins of Trigonometry

The beginnings of trigonometry date back to prehistoric cultures and mirror the evolution of civilization itself. The word "trigonometry" comes from the Greek word *trigonon* (meaning "triangle") and *metron* (meaning "measurement"). However, trigonometry did not originate as the study of triangles. It was initially viewed as a combination of geometry and astronomy used in studying the movements and locations of celestial bodies in the sky and in time keeping. The foundations of trigonometry originated in prehistoric cultures with their vigilant observations of the night sky. Some of their discoveries are inherent in the designs of Stonehenge and ancient Egyptian monuments. The early civilizations of Egypt, Babylonia, and India (c. 2000 B.C.E.) contributed significantly to the origins of trigonometry. The Egyptians applied the properties of geometry and astronomy in constructing the pyramids, and the Babylonians (c. 1900 B.C.E.) used angles and ratios to keep track of the motions of celestial bodies in the sky. They followed the paths of planets, the lunar and solar eclipses, and gave coordinates to the stars, all of which required familiarity with angular distances measured on the "celestial sphere." While these early civilizations used trigonometric principles for astronomical measurements, in designing monuments, and in time keeping, they did not fully develop the mathematical system that is now known as "trigonometry."

The invention of trigonometry emerged as a defined body of knowledge from work conducted by Greek astronomers around 350 B.C.E. in the city of Alexandria, Egypt, the intellectual center of the ancient world. The Greek astronomer, Hipparchus (190–120 B.C.E.), is frequently called "the Father of Trigonometry. " He used ratios to determine the distances of the Earth from the sun and the moon and was responsible for tabulating the measures of arcs and their corresponding chord lengths for angles within a circle. The trigonometry of ancient Alexandria would now be called "spherical trigonometry" since it was confined primarily to the study of the properties of great circles on spherical bodies.

The earliest recorded work on spherical trigonometry appears in the book, *Sphaerica*, written by Menelaus, a Greek mathematician working in Alexandria (70–140 C.E.). It describes Menelaus' theorem, which associates the ratios of the lengths of intersecting arcs of great circles on a sphere.

Development of Trigonometric Functions

A fundamental difference between ancient Alexandrian trigonometry and modern trigonometry is that the former used arcs and chords in its trigonometric tables instead of the trigonometric functions that are used in the twenty-first century: sine, cosine, and tangent. The oldest surviving table of arcs and chords was created by Ptolemy of Alexandria (90–168 C.E.). However, when the chords of the circle are rotated to a vertical position and radii are inserted to connect the endpoints and midpoint of the chord to a common center, it is possible to translate half-chord lengths to a sine function. Thus, Ptolemy's famous table of arcs and chords was equivalent to a modern table of sines. His most famous mathematical work was the *Almagest*. It included a table of chords for angles for one-half degree to 180 degrees, in increments of half degrees. His chords were accurate in length to five significant digits. In addition, Ptolemy proved (using chords) the formulas for the sum and difference of two angles that are equivalent to the

current sine functions of two combined angles. These discoveries were applied to astronomy, time keeping, and in locating the direction of Mecca for the daily five prayers required by followers of Islam.

In the first century B.C.E., trigonometric principles were used primarily in navigation and map-making. Fortunately, Ptolemy recorded all of the geographical knowledge collected by the ancient world in his eight volumes, titled *Geographia*. It included the latitudes and longitudes of 8000 places on Earth and was the world's first atlas, similar to those used in the twenty-first century. It took astronomers nearly 400 years to shift from using tables of angles and chords to a reliance on tables of sines. Indian astronomers of the fifth century gave trigonometry its current interpretation of the sine function, which quickly spread to Arab and Islamic astronomers.

It is important to recognize that the six trigonometric functions that are used in the twenty-first century were developed at the end of the tenth century by Arab and Islamic astronomers. The West learned about Arab and Islamic trigonometry at the beginning of the twelfth century through translations of Arabic and Islamic astronomy handbooks. Indeed, the maps created by the Alexandrian Greeks and the trigonometric functions developed by the Arab and Islamic astronomers were employed during the world explorations of the fifteenth and sixteenth centuries. Christopher Columbus (1451–1506) utilized these materials to guide him to discover the "new world" in 1492. There is no denying the meritorious impacts of trigonometry on exploration and navigation during the fifteenth and sixteenth centuries.

Trigonometry of the Renaissance

Calculations using trigonometric functions to determine the sides of right triangles did not become prevalent until the sixteenth century. German mathematician Bartholomew Pitiscus (1561–1613) is recognized for creating the word "trigonometry," meaning triangle measurement. He chose *Trigometria* for the title of his book, which described the applications of trigonometry in surveying.

During the seventeenth century, numerical calculations in trigonometry were simplified by the invention of logarithms by the Scottish mathematician John Napier (1550–1617). Fifty years following Napier's invention of logarithms, English mathematician Isaac Newton (1642–1727) invented the calculus in which he represented trigonometric functions as infinite series in powers of *x*. Specifically, Newton discovered that

$$\sin(x) = x - \frac{1}{6}x^3 + \frac{1}{120}x^5 - \frac{1}{5040}x^7 + \frac{1}{362880}x^9 - \cdots$$

and

$$\cos(x) = 1 - \frac{1}{2}x^2 + \frac{1}{24}x^4 - \frac{1}{720}x^6 + \frac{1}{40320}x^8 - \cdots.$$

These representations of the sine and cosine functions continue to play important roles in applied and pure mathematics in the twenty-first century.

For most of the history of trigonometry, angles were measured in degrees, which were defined as fractions of the circumference of a circle. This practice was not efficient nor consistent because the radius of the circle was not fixed. The creators of tables of sines often chose a radius convenient for their calculations. Ptolemy used a radius of 60 since his fractions were expressed in 60ths. The Austrian mathematician Georg Rheticus (1514–1574) used a radius of 10^{15}, which permitted him to tabulate the six trigonometric functions with 15-digit accuracy without the use of decimals or fraction manipulation.

Further innovations and interpretations in trigonometry were developed in the eighteenth century by the Swiss mathematician Leonhard Euler (1707–1783). He explained that computations of trigonometric functions would be more efficient if the line lengths were measured in the same unit. Consequently, he chose "1" to be the radius of a circle centered at the origin. Thus, the circle's circumference would be 2π; the arc for 45 degrees would be $\pi/4$; the arc for 30 degrees would be $\pi/6$; and so on. This system led to the later development of radian measures for angles and arc lengths.

More significantly, Euler discovered that trigonometric functions could be defined in terms of complex numbers (the union of real and imaginary numbers). Specifically, Euler's famous equation states that for any real number *x*,

$$e^{ix} = \cos(x) + i \sin(x).$$

The relevance of this equation was that trigonometry could then be viewed as only one of the numerous applications of complex numbers. This formula served as a unifying concept for the study of mathematics as a whole.

Contemporary Applications of Trigonometry

Although trigonometry originated in ancient civilizations from studying the movements of astronomical bodies and triangle relationships, its current applications encompass far more. Trigonometry now spans the diverse fields of architecture, engineering, science, music, navigation, medicine, digital imaging, and games of entertainment.

Trigonometry is a perfect partner for architecture and engineering. Contemporary buildings with curved surfaces in glass and steel would be impossible without trigonometry. Although the surfaces are perceived as curved, they are frequently composed of numerous triangles. Furthermore, since the triangle is an ideal shape for evenly distributing the weight of a structure, an understanding of relationships among the parts of triangles is essential in the design and construction of buildings, bridges, and monuments. Specifically, if an engineer knows the lengths of the beams that will be attached to a structure, the angles at which they must be attached can be calculated using trigonometry. Additionally, in the architectural design of an amphitheatre, the engineer's task is to design the structure so that all sounds from the stage are funneled into the audience's ears. Engineers and architects use trigonometry to identify the perfect shape to balance this sound as it reflects off the walls and ceilings.

Since trigonometry facilitates the understanding of space, it has numerous applications in the physical sciences. In optics and statics, trigonometric functions are vital in understanding the behavior of light and sound. It is particularly useful for modeling the periodic processes found in music because of the cyclical and periodic nature of trigonometric functions (sine, cosine, and tangent). Harmonics are determined by the form of their sine waves with respect to their periods and frequencies. The period of a sine wave is the length of the interval of repetition of the sine wave. The frequency of a sine wave is the number of cycles a sine wave goes through in a standard distance or time interval. In music, the frequency is often expressed in units of hertz, (Hz), where 1Hz means one period per second. For example, the pitch of every note in music is determined by the length of its sine wave (period) and by its frequency. Musical notes with wide sine waves are lower in pitch because they have fewer cycles per second, while notes that have narrow sine waves are higher in pitch and have more cycles per second.

Trigonometry plays a vital role in modern technologies through the process of triangulation. It is used by global positioning systems (GPS), in computer graphics, and in gaming. Specifically, computer generation of complex images is made possible by coloring numerous, microscopic squares (called "pixels") that define the precise location and points on the image. The technique of triangulation is used to make the image highly detailed and clearly focused. In GPS, triangulation is used for object location. Similar imaging technologies have also revolutionized the medical fields through the development of Computed Axial Tomography (CAT) scans, ultrasounds, and magnetic resonance imaging (MRI).

Trigonometry Instruction

It has been shown that trigonometry emerged approximately 4000 years ago through careful observations of the movements of celestial objects in the sky. These observations gave rise to the study of the celestial sphere and relationships among arcs, chord lengths, and central angles of great circles. In the twenty-first century, this form of trigonometry is called "spherical trigonometry." It was not until the end of the tenth century that Arab and Islamic astronomers defined the six trigonometric ratios. It took another 500 years before right triangle trigonometry became prominent in surveying, navigation, and architecture throughout the Western world. Therefore, it is interesting that this evolutionary sequence of the development of trigonometry is not reflected in the order in which it is typically introduced to students.

The current order of instruction is to first introduce the six trigonometric functions as ratios between the sides of right triangles. These concepts are then applied in finding the missing parts of right triangles. Finally, the unit circle and the periodic nature of trigonometric functions are introduced. This instructional order is often supported by the fact that right triangle trigonometry is a natural extension of the study of the Pythagorean theorem and ratios between parts of triangles, both of which are covered in the middle grades. Furthermore, trigonometric ratios are not as difficult to comprehend as periodic functions.

Some educators may argue that instruction should follow the evolutionary sequence of a topic. In support of that argument, students of today's technological world (with microwaves, wifi's, electrocardiograms, and electronic music) gain early familiarity with the periodicity of the sine and cosine functions. They are also more likely to model scientific and social phenomena with trigonometric functions than to apply right triangle trigonometry in finding the missing parts of triangles. Thus, for research purposes in identifying "best practices" for instruction, some educators have considered assessing the effects of teaching trigonometry in the same order that it historically evolved.

In conclusion, trigonometry is more valuable to society today than ever before in recorded history. The foundations of trigonometry emerged about 4000 years ago in the Egyptian and Babylonian civilizations. It developed into a well-defined mathematical discipline in the city of Alexandria, Egypt, circa 350 B.C.E. In the centuries that followed, trigonometry continued to evolve through the contributions and insights of diverse cultures and societies throughout the world. Trigonometry, with its advanced measurement facilities and associated technologies, remains one of the most applicable and practical fields of mathematics, vital to the advancement of the sciences, engineering, and technologies.

Further Reading

Baumgart, John K., ed. *Historical Topics for the Mathematics Classroom*. 2nd ed. Reston, VA: National Council of Teachers of Mathematics, 1989.

Bressoud, David M. "Historical Reflections on Teaching Trigonometry." *Mathematics Teacher* 104, no. 2 (September 2010).

Buchberg, Jerrold T., et al. *The Essential Physics of Medical Imaging*. 2nd ed. Philadelphia: Lippincott Williams & Wilkins, 2002.

Joyce, David. "Applications of Trigonometry" (1997). http://www.clarku.edu/~djoyce/trig/apps.html.

Katz, V., ed. *The Mathematics of Egypt, Mesopotamia, China, India, and Islam*. Princeton, NJ: Princeton University Press, 2007.

Van Brummelen, Glen. *The Mathematics of the Heavens and the Earth: The Early History of Trigonometry*. Princeton, NJ: Princeton University Press, 2009.

SHARON WHITTON

See Also: Animation and CGI; Functions; Geometry and Geometry Education; Geometry in Society; GPS; Harmonics; Measurements, Length.

Tunnels

Category: Architecture and Engineering.
Fields of Study: Algebra; Geometry; Measurement; Number and Operations.
Summary: Tunnels have long presented interesting mathematical and engineering problems.

A tunnel is a connecting passageway through materials like rock, earth, or water. Tunnel engineers must take into consideration issues like seepage and weight. Scientists and mathematicians create mathematical models of tunnels to investigate aspects like aquifers and safety issues. Analytic and closed form solutions are useful in engineering. Mathematical fields like graph theory, differential equations, geometry, probability, and trigonometry are important for modeling and measuring tunnels.

Mathematically Challenging Tunnels

Five centuries after it was completed, Hero of Alexandria gave a theoretical explanation that may explain how the Tunnel of Samos was constructed. Mathematical physicist Renfrey Potts had an undergraduate degree in mathematics. He worked as a consultant for General Motors and created car-following models. This work led to experiments on a testing track with just two cars that successfully predicted the optimum speeds for congested traffic in the Holland Tunnel in New York, named for engineer Clifford Holland. The Channel Tunnel between England and France represented a significant engineering and mathematical challenge. At the time of its building and into the twenty-first century, it had the longest undersea length of any tunnel in the world. It presented significant challenges including problems related to the topology and geology of the rock through which it was bored; significant water

If the height and width of a parabolic tunnel are known, one can determine the tunnel's height at different distances from the base center by modeling using coordinate geometry and equations.

pressure; ventilation; communication; and the fact that construction was started at the same time from both ends, requiring exceptional precision to meet in the middle. This tunnel serves as a model for other underwater tunnel projects and many teachers use it to present mathematics concepts. Scientists and mathematicians also experiment with digital and physical wind tunnels as well as quantum tunnels.

Ancient Tunneling

The problem of delivering fresh water to large populations has been an ongoing human endeavor since ancient times. In the sixth century B.C.E., a one-kilometer tunnel was dug through a large hill of solid limestone to bring water from the mountains to the main city on the island of Samos. The Eupalinian aqueduct on Samos was designed by the ancient Greek engineer Eupalinos of Megara. The tunnelers worked from both ends and met in the middle, with an error less than 0.06% of the height. To achieve this remarkable result, Hero of Alexandria theorized that the tunnelers used a method based on similar triangles in order to determine the correct direction for tunneling. Mathematicians and scientists continue to debate the pros and cons of various theories of how this engineering marvel was constructed.

Modeling Tunnels

Tunnels can be modeled using coordinate geometry and equations. For example, knowing the height and width of a parabolic tunnel, one can determine the tunnel's height at different distances from the base center. To solve this problem, one needs to find the equation for the parabola choosing convenient x-y axes.

Frictionless Tunnels

The possibility of mathematical modeling allows for innovative and challenging ideas. What if a frictionless tunnel would be bored through Earth's center? Paul Cooper, a mathematician fond of Jules Verne's books, tried to answer this question in an issue of the *American Journal of Physics*. He set up and solved by computer a set of differential equations for tunnels that would provide minimum gravity-powered travel time between any two cities on Earth.

According to Cooper's differential equations, by freefalling in airless, frictionless, straight-line tunnels, passenger vehicles powered only by the pull of gravity could theoretically travel between any two points on the Earth's surface in a total time of only 42.2 minutes. Accelerated by the force of gravity on the first half of the trip, the vehicle would gain just enough kinetic energy to coast up to the other side of the Earth. However, significant obstacles make such a project impossible in the twenty-first century. Subterranean temperatures reach extremes, even for relatively shallow tunnels of only a few miles deep, requiring huge cooling systems for vehicles. Also, it is almost certainly impossible to create a completely frictionless path without a rail or track of some type, leaving the vehicle with insufficient kinetic energy to complete its trip without a source of additional power. Consequently, such a tunnel is still science fiction more than science.

Further Reading

Apostle, Tom. "The Tunnel of Samos." *Engineering and Science* 1 (2004).

Cooper, P. W. "Through the Earth in Forty Minutes." *American Journal of Physics* 34, no. 1 (1966).

Lunardi, Pietro. *Design and Construction of Tunnels: Analysis of Controlled Deformations in Rock and Soils.* New York: Springer, 2008.

Oxlade, Chris. *Tunnels.* Portsmouth, NH: Heinemann-Raintree, 2005.

FLORENCE MIHAELA SINGER

See Also: Caves and Caverns; Coordinate Geometry; Energy; Traffic; Wind and Wind Power.

U

Ultrasound

Category: Medicine and Health.
Fields of Study: Algebra; Geometry;
Representations.
Summary: Ultrasound uses mathematical principles
to create images of the human body.

Although ultrasound cannot be heard by humans, it has
been produced and used for a vast number of applications in many different fields. In industry, ultrasound has
been used as a technique to assess the structural integrity
of materials. The interaction between ultrasound and
live systems has been studied since the 1920s. During
the 1960s, it was used in medicine, initially as a therapeutic option and then later as a diagnostic resource. In
the twenty-first century, ultrasound is a major medical imaging technology widely used in clinical facilities around the world because it causes no harm to the
human body and results can be achieved in real time,
besides the fact it is considerably cheap and easy to use.
The available technologies using ultrasound are in constant development. Every new application depends on
the advance of computer sciences that work with many
concepts of physics and the solution of mathematical
problems in this field seems inexhaustible.

Sound is a form of energy consisting of the vibration
of molecules of an environment that can be air, water,
solid, or biological tissues (such as bones and muscles).
This kind of energy propagates across the medium in
the form of waves. Sound is a mechanical wave whose
fundamental characteristics are amplitude, which is the
distance between the highest and lowest point of the
wave and frequency, which is the number of cycles that
occur in a second, measured in hertz (Hz). Humans are
able to detect sounds with a frequency of 20–20,000
Hz–the normal limits of the human hearing. The term
"infrasound" refers to sound waves that have a frequency lower as 20 Hz, and sounds with a frequency
higher than 20,000 Hz are called "ultrasound." Unlike
humans, some animals, such as bats, dolphins, whales,
dogs, cats, and mice can hear ultrasound.

Imaging the Human Body

While traversing a material, the properties of ultrasound change in intensity and speed of propagation,
which means that ultrasound waves travel at different speeds depending on the material. Consider two
samples of human bone, one from a 30-year-old person and the other from an 80-year-old person. If ultrasound waves cross these two bony samples, the speed at
which the sound propagates in the bones can be represented algebraically by the following equation:

$$v = \sqrt{\frac{E}{\rho}}$$

where ν is the speed of ultrasound in the bone sample, E is the modulus of elasticity of the bone sample, and ρ is the density of the bone sample.

The speed of sound (ν) can be calculated by measuring the time required for the wave to propagate through the bone and then dividing by the width of the bone. Knowing the density of the bones (ρ), this equation could be used to determine the values of the modulus of elasticity (E) that indicates the elastic properties of the bone. In a 30-year-old person, the speed of the sound through the bone is approximately 4000 m/s. In an 80-year-old person, this rate drops to 3800 m/s. This fact means that the higher the speed of the sound through the bone, the better is the quality of bone. A low speed could reveal a bony fragility and a fracture probability. This principle is used in ultrasonometry, a technique used to estimate the bony fracture or osteoporosis risk in patients. Ultrasound medical imaging is one of the most powerful diagnostic tools in modern medicine. Along with other imaging methods, it is based on advanced mathematical techniques and numerical algorithms that are necessary to analyze the data and produce readable pictures or three-dimensional images of inner body structures without surgery or use of radiation. It has been widely used to identify the sex or to detect malformations in fetuses during gestation.

Further Reading

Ammari, Habib. *An Introduction to Mathematics of Emerging Biomedical Imaging*. Berlin: Springer, 2009.

Gibbs, Vivien, et al. *Ultrasound: Physics and Technology*. Philadelphia: Churchill Livingstone, 2009.

MARIA ELIZETE KUNKEL

See Also: Diagnostic Testing; Digital Images; Harmonics; Medical Imaging.

Unemployment, Estimating

Category: Government, Politics, and History.
Fields of study: Algebra; Data Analysis and Probability.

Summary: Unemployment rates are calculated using intricate statistical models and sampling methods.

An unemployed person is generally defined as an individual who is available for work but who currently does not have a job. Overall unemployment is typically quantified using the unemployment rate, which represents the number unemployed people as a percent of the labor force. The Bureau of Labor Statistics is an independent statistical agency of the U.S. federal government primarily responsible for measuring labor market activity. Many mathematicians and statisticians are involved in data collection, modeling, and estimation of employment activity, including the highest levels of direction and management. For example, Janet Norwood was the first woman commissioner of the U.S. Bureau of Labor Statistics and frequently spoke to the Joint Economic Committee and other congressional Committees. She was also president of the American Statistical Association and chair of the Advisory Council on Unemployment Compensation. Regarding her work, she noted, "These data figure very prominently in most of the political debates, so it is extremely important that they be accurate and of high quality, and that they be released in a manner that is totally objective."

Economist John Maynard Keynes's revolutionary work, *The General Theory of Employment, Interest and Money*, was published in 1935–1936. The Industrial Revolution and shift away from an agrarian economy had significantly changed the way in which researchers in many fields looked at economic measures, including employment, and the Great Depression brought even greater attention and emphasis to these concepts. Because of labor-market volatility in the late 1920s, the 1930 U.S. census attempted the first comprehensive federal measure of unemployment, but data from the decennial census were not timely enough to be useful in assessing the effectiveness of Depression legislation to aid unemployed workers. Statisticians used newly emerging polling methods to develop better measures and mathematical models. Better methods also changed, at times, the definition of unemployment. Overall, it is commonly accepted that unemployment induces negative effects on the financial and economic status of societies and individuals with respect to many variables. As workers become unemployed, the goods and services that they could have produced are lost along with the purchasing power of these workers, thus leading to the unemployment of more workers. In addition, a large unemployment rate can induce significant social changes and has been the foundation of civil unrest and revolutions. Mathematicians and statisticians continue to create explanatory and forecasting models that are used to guide policies and decisions intended to stabilize economies and aid unemployed workers at local, state, and national levels. These models draw from mathematical ideas and techniques in a wide range of areas, including time series analyses, equilibrium modeling, structural component modeling, neural networks, and simulation.

Sample Design and Collection of Unemployment Data

In most countries, the task of collecting and analyzing unemployment-related information is assigned to certain governmental agencies. In the United States, the Current Population Survey (CPS), conducted by the Census Bureau for the Bureau of Labor Statistics since the mid-twentieth century, provides most of the necessary data. Counting every unemployed person each month is impractical in terms of both cost and time, so the Census Bureau conducts a monthly survey of the population using a sample of households that is designed to represent the civilian population of the United States. At the start of the twenty-first century, the (CPS) surveyed about 50,000 households per month. The selection is generally a multistage stratified sample selected from many different sample areas. The sample provides estimates for the nation and serves as part of model-based estimates for individual states and other geographic areas.

In the first stage of sampling, the United States is divided into primary sampling units (PSUs) that usually consist of a metropolitan area, a large county, or a group of smaller counties. PSUs are then grouped into strata based on some factor that divides the population into mutually exclusive homogeneous groups. The homogeneity of the stratum ensures that the within-strata variability is very small compared to the variability between strata. One PSU is then randomly selected from each stratum with a probability of selection proportional to the PSU's population size. The second stage of sampling consists of randomly selecting small groups of housing units from the sample PSUs. Elements from this sample of housing units are called "secondary sampling units" (SSUs). These households are usually selected from the lists of addresses obtained from the last decennial census of the population. Housing units from blocks with similar demographic composition and geographic proximity are grouped together in the list. The final sample is usually described as a two-stage sample but occasionally, a third stage of sampling is necessary when actual SSU size is extremely large. In this situation, a third stage, called "field subsampling," is needed in order to keep the surveyor's workload manageable. This involves selecting a systematic subsample of the SSU to reduce the number of sample housing units to a more convenient number. Once a survey is designed and the sample is drawn, field representatives and computer-assisted telephone interviewers contact and interview a responsible person living in each of the sample units selected to complete the interview.

Seasonal Adjustment of Unemployment Data

The collected data by the CPS are subjected to a series of transformations and adjustments before the analytical tools are applied to fit adequate models to the unemployment rate and explain its behavior in terms of relevant factors. Because some types of employment are seasonal or cyclical over time, such as December holiday retail sales or fall farm harvesting, adjustments must often be made to account for such cycles. In fact, throughout a one-year period, the level of unemployment experiences continuous variations because of such seasonal events as changes in weather, major holidays, agricultural harvesting, and school openings and closings. Since seasonal events follow an almost regular periodic pattern each year, their influence on the overall pattern can be easily estimated and eliminated. There are two popular methods for removing seasonality. The first estimates the seasonal component using a regression model with time series errors. The explanatory variables in the regression equation are 12-period harmonic terms. Once the regression coefficients are estimated, the fitted values are evaluated for each month subtracted from the corresponding actual values leading to seasonally adjusted series. The second method consists of simply taking seasonal differences of the unemployment series. The removal of the anticipated seasonal component makes it easier for data analysts to observe fundamental variations in the unemployment level, such as trends, gains, nonseasonal intrinsic cycles, and effects of external events, especially those related to economic factors.

Rate Estimation and Prediction

Since the unemployment survey is conducted in the same manner on a monthly basis, the type of data collected is called "time series data." Dependence or autocorrelation among the observations in such data is common, which means that most classical mean-variance types of statistical models are not applicable for estimation and prediction with most unemployment data. Mathematical and statistical models that take into account the particularity of time-dependent data are called "time series models." Among the most popular and useful are autoregressive integrated moving average (ARIMA) models and their seasonal extension (SARIMA). Such models can be used to describe the relationship between a current unemployment rate and past ones using differencing operations and linear equations. As a consequence, the model can also be used to predict future realizations of the unemployment rate. The ARIMA models are very flexible in the sense that they allow for the inclusion of external factors, which can help explain the movement of the unemployment rate and lead to estimators and predictors with smaller variability errors.

Further Reading

Downey, Kirstin. *The Woman Behind the New Deal.* New York: Anchor Books, 2010.

Flenberg, Stephen. "A Conversation With Janet L. Norwood." *Statistical Science* 9, no. 4 (1994).

Pissarides, Christopher. *Equilibrium Unemployment Theory.* Cambridge, MA: The MIT Press, 2000.

Zbikowski, Andrew, et al. *The Current Population Survey: Design and Methodology. Technical Paper 40.* Washington, DC: Government Printing Office, 2006.

Mohamed Amezziane

See Also: Census; Forecasting; Gross Domestic Product (GDP).

Units of Area

Category: Space, Time, and Distance.
Fields of Study: Algebra; Geometry; Measurement; Number and Operations; Representations.

Summary: Numerous units of area have been used throughout history for measuring land.

Specific measurements of land area date back to ancient times to define land ownership (for the purposes of taxation, among other reasons). Some of these measurements are still used in the twenty-first century.

Ancient Units of Measurement

In Mesopotamia, land area was divided into a *bur* (an estate), which covered about 64,800 square meters. The bur, in turn, was divided into *iku* (fields), each of which covered about 3600 square meters. Further measurements and sub-divisions are recorded on surviving land documents. The Egyptians also had their own system based on the *kha-ta* (100,000 square cubits), which in turn was divided into 10 *setat*, which consisted of 10 *kha* (1000 square cubits, or 275.65 square meters).

The Romans had a very specific system of measuring land with the basic measure being an *actus quadratus* (acre), which covered about 1260 square meters. Smaller measurements were described as being a *pes quadratus* (square foot) or *scripulum* (or square perch). These measurements were based on the *pes* (foot) being the basic unit of measurement throughout the Roman Empire, a length that was fixed throughout the Empire.

By contrast, the Greeks used a different system of land measurement by which land was divided into a "plethron"—a variable area of land that consisted of the amount of land a yoke of oxen were able to plough in a single day. As a result, the exact measurment varied from some parts of Greece to other parts (and indeed for different parts of a city), although it was thought to approximate to about four English acres. In rocky and hilly areas, the land area was larger than in other parts of the city state. This method of measuring land area—based on what could be done with it—is quite different to the Roman system and largely emerged from a method of equitable taxation by which those with poorer land could be taxed fairly alongside those with more fertile land.

This Greek concept of land measurement was later followed by the Anglo-Saxons in England with their use of the "hide" as a measure of land. This measurement was used in the *Domesday Book* in 1086 and continued until the end of the twelfth century. Traditionally, it was thought that a hide consisted of the land needed to support 10 families, because it is used instead of the term *terra x familiarum* (land of 10 families) in the Anglo-Saxon version of Bede's ecclesiastical history. In Scotland during the same period, the term "groatland" was used to describe the land that could be rented for a particular coin—in this case, a groat. It would represent a larger area for poorer agricultural land than for richer land.

Medieval Era

By medieval times, in Europe and especially England, the terms of measuring land were standardized, and these tended to follow the Roman measurements of a "perch," a "rood," and an "acre." In spite of these measures (although the "hide" was being phased out), there were other measures including the "carucate," which covered the land that an eight-ox team could plough in a year (approximately 120 acres); a "virgate," which covered land that could be ploughed by two oxen in a year (about 30 acres); and a "bovate," which covered the land that could be ploughed by a single ox in a year. There was also an area known as a "knight's fee," which was the land expected to be able to produce a single armed soldier in times of war. Although early in medieval England, an acre was supposed to be the land that could be ploughed in single day, by late medieval times, it had been formalized as 4840 square yards.

Other Systems of Measuring Area

Elsewhere in the world, many other places had their own system of measuring area. The Chinese had a system based on the *li* (7.9 square yards), the *fen* (10 *li*), the *mu* (10 *fen*), the *shi* (10 *mu*), and the *qing* (10 *shi*). The Japanese also had a system of measurement by *tsubo*, which covered the land that was the same size as two tatami mats (about 3.306 square meters). In Korea, there is a similar measure called the *pyeong*, which covers 3.3058 square meters. These measures are generally used to measure the size of rooms and buildings rather than large areas of land. The *tsubo* and the *pyeong* are both still used in the twenty-first century to help describe the size of houses or apartments for sale, in the same way as the term "square" is used by Australian estate agents (approximating to 100 square feet, or 9.29 square meters).

Metric System

The metric system was devised during the 1790s following the French Revolution in an attempt to standardize measurements and it was adopted by the French after Napoleon Bonaparte came to power in 1799. It focuses on the meter as the main measurement of length, and the square meter as the measurement of area. This is used throughout most of the world in the twenty-first century.

Further Reading

Anderton, Pamela. *Changing to the Metric System.* London: Her Majesty's Stationery Office, 1965.

Balchin, Paul N., and Jeffrey L. Kieve. *Urban Land Economics.* New York: Palgrave Macmillan, 1985.

Boyd, Thomas D. "Urban and Rural Land Division in Ancient Greece." *Hesperia* 50, no. 4 (October–December 1981).

Kanda, James. "Methods of Land Transfer in Medieval Japan." *Monumenta Nipponica* 33, no. 4 (Winter 1978).

Snooks, Graeme D., and John McDonald. *Domesday Economy: A New Approach to Anglo-Norman History.* Oxford, England: Clarendon Press, 1986.

Walthew, C. V. "Possible Standard Units of Measurement in Roman Military Planning." *Britannia* 12 (1981).

JUSTIN CORFIELD

See Also: Measurements, Area; Roman Mathematics; Units of Length; Units of Mass; Units of Volume.

Units of Length

Category: Space, Time, and Distance.
Fields of Study: Algebra; Geometry; Measurement; Number and Operations; Representations.
Summary: Numerous units of length exist and are used according to the distance measured.

Measuring length or distance has been necessary as far back as the oldest hunter-gatherer peoples in order to perform necessary tasks, such as traveling and finding or hunting food. Many of the first units of length were derived from bodily measurements. Modern units of length can broadly be divided into two categories: the U.S. customary system and the international system. The U.S. customary system is more commonly known as the "American system." The International system is more commonly known as the "metric system." The basic unit of length in the American system is the foot, while the basic unit of length in the metric system is the meter. The American system is used more often in the United States, while the metric system is more common in other parts of the world. Scientific journals almost always report measurements in metric units. The exact values of length measurements depend on the units chosen but certain constants (like π) that are fundamental to related measurements (like circumference) are unitless.

American System

The American system of lengths is similar to the British imperial system from which the American system takes its historical roots. The basic unit of measurement is the foot (ft), which originally was set to be the length of an adult man's foot. Each foot is approximately 0.3048 meters. Smaller distances in the American system are typically measured in inches (in) or less commonly in mils. There are 12 inches in a foot and 1000 mils in an inch. Rather than mils, it is much more common to use fractions of an inch to obtain additional accuracy in the American system. Longer distances in the American system are usually measured in yards (yd) or miles (mi). There are three feet in a yard and 1760 yards in a mile (5280 feet in a mile).

Metric System

The meter was originally established in France as one ten-millionth of the distance from the Earth's equator to the North Pole along the meridian passing through Paris. However, in 1983 it was defined as 1/299,792,458 of the distance traveled by light in a second in a vacuum. Smaller units of length in the metric system are often measured in centimeters (cm), millimeters (mm), micrometers (μm, also known as the *micron*), and nanometers (nm). There are 100 centimeters in a meter and 1000 millimeters in a meter. Similarly, there are 1 million micrometers in a meter and 1 billion nanometers in a meter. Longer distances are usually measured in kilometers (km). There are 1000 meters in a kilometer, which are sometimes referred to as "klicks" in the military. The fermi and the angstrom are also units of length in the metric system, though they are not officially part of the international system. There are 10^{15} fermis in a meter and 10 trillion angstroms in a meter. Because of their small length, the fermi and the angstrom are best

suited for very small distances. Less common units of length in the metric system include the decimeter (one-tenth of a meter), picometer (10^{-12} meters), decameter (10 meters), megameter (1 million meters), gigameter (1 billion meters), and petameter (10^{15} meters).

Atomic and Astronomic Measurements

Atomic measurements are also given in terms of either Planck length or the Bohr radius. The Planck length is defined in terms of Planck's constant, the gravitational constant, and the speed of light in a vacuum. The result is that the Planck length is based entirely on universal constants rather than human constructs, such as the second. A Planck length is approximately 1.61625×10^{-35} meters. The Bohr radius is defined as the expected distance between the nucleus of a hydrogen atom and its electron in the Bohr model of the atom. The Bohr radius is approximately 5.29177×10^{-11} meters.

Astronomical distances are typically given in terms of light-years, astronomical units, or parsecs. The light-year is defined as the distance light travels in a vacuum in a Julian year (365.25 days). The light year is approximately 9,460,730,472,581 kilometers or 5,878,630,000,000 miles. Distances such as the light-second, the light-minute, and the light-month are defined analogously to the light-year. The astronomical unit is defined as the average distance between the Earth and the sun, approximately 149,597,871 kilometers or 92,955,807 miles. The parsec is defined in terms of the astronomical unit and an angle with measure one arc second. The imaginary right triangle that defines the parsec has one angle with measure one arc second. The opposite side of the triangle from this angle has length equal to one astronomical unit. The length of the adjacent side to this angle is defined as a parsec and can be derived using basic trigonometry. There are approximately 3.26 light-years in a parsec.

Other Measurements

There are a number of units of length that are based on the American system and still in use in certain professions in the twenty-first century. A furlong is often used in horse racing and is defined as one-eighth of a mile (220 yds). The hand is a unit of length used to describe the height of a horse and is equivalent to four inches. Rods (5.5 yds) and chains (66 ft) are often used in surveying. A fathom is often used to measure the depth of water and is equal to six feet. A nautical mile is approximately equal to one minute of latitude. Thus, there are

1872 meters (approximately 6076 feet) in a nautical mile. Fathoms and nautical miles are often used by mariners.

There are also a number of archaic units of length that may be familiar to the reader, most significantly the cubit (1.5 ft) and the league. The dimensions of Noah's Ark as well as other Biblical artifacts are given in cubits. The league has several different values, however the most common is the distance that a person can walk in an hour (approximately three miles). The league was featured in the title of Jules Verne's *Twenty Thousand Leagues Under the Sea*.

Further Reading

Glover, Thomas G. *Measure for Measure*. Littleton, CO: Sequoia Publishers, 1996.

Hopkins, Robert A., *The International (SI) System and How It Works*. Tarzana, AK: AMJ Publishing, 1975.

Liflander, Pamela. *Measurements & Conversions*. Philadelphia: Running Press, 2003.

Wildi, Theodore. *Metric Units and Conversion Charts*. Hoboken, NJ: Wiley-IEEE Press, 1995.

Young, Hugh D., et al. *University Physics With Modern Physics*. 12th ed. Boston: Addison-Wesley, 2007.

Robert A. Beeler

See Also: Measurements, Length; Units of Area; Units of Volume.

Units of Mass

Category: Space, Time, and Distance.
Fields of Study: Algebra; Geometry; Measurement; Number and Operations; Representations.
Summary: A variety of measurement systems have been used throughout history to measure weight and mass.

Throughout history, there have been many ways of measuring mass. Until modern times, these methods were those used to measure what was known as "weight." A number of ways of assessing weight existed in prehistoric times. The Sumerians used a system similar to that later used throughout the ancient Middle East, with 180 grains making a shekel (or *gin*), and 60 of these forming a pound (or *ma-na*), and 600 of these making a load (or *gun*). A wall painting from ancient Egypt, dating from 1285 B.C.E., shows the god Anubis weighing the heart of Hunefer using scales, indicating that the Egyptians had a system of using weights and measures. There were, however, slight differences between the Middle Kingdom and the New Kingdom in Egypt.

Greeks and Romans

The Greeks, with the extensive use of coinage, used a scale that was based on the barley corn but it was actually more fixed on the weight of individual coins. The Romans adapted the Greek system for their own use, with the basic measure of an *uncia* (or ounce). Twelve of these made up one *as*, with different names were given to parts of an *as*: *quadrans* were a quarter of an *as* and *semis* were half an *as*.

Middle Ages

During the Middle Ages in Europe, there were a number of measures that were used for a variety of purposes. For apothecaries, jewelers, and the making of coins, there were "grains," "scruples," and "drams." Two systems were heavily used in Western Europe. The Troy weights, named after the French city of Troyes, were based on the troy ounce (the name "ounce" coming from the Roman "uncia"). By contrast in England, until 1526, there was the Tower ounce, which was slightly lighter than its continental measure (18.75 dwt/pennyweight, rather than the Troy ounce which was 20 dwt). For both measures, 12 ounces made up a pound. In England, eight pounds equaled a "butcher's stone," and 12 pounds a "mercantile stone." The larger measurements were in tons, which consisted of 2240 pounds—now known as a "long ton." The United States later adopted a measure in which 2000 pounds equals a "short ton."

Throughout Europe, there were regional varieties and customary names. Scotland was divided between using the "Troy" measures, and the "Tron" measures, the latter being used in Edinburgh—the system was standardized in 1661. The Portuguese used a system maintained at a national level and was based on the *onca* (ounce), with 16 of these making an *arratel* (pound), 128 *arrateis* making a *quintal*, and 1728 making a *tonelada*. These Portuguese measures, also used in Brazil, were abandoned when both countries adopted the metric system: Portugal and its colonies (or overseas provinces) in 1852, and Brazil 10 years later. The

Standardization

An attempt to standardize the measurement of mass started in France, which, on December 10, 1799, passed a new law establishing a kilogram that consisted of 18,827.15 grains, although the kilogram had already beem used for the previous four years. It was defined as being a 1 cubic decimeter of distilled water at 4 degrees centigrade, its maximum density. This standardization led to the metric system, and in turn it led to the introduction of what became known as the International System of Units (SI units).

The SI units define mass in kilograms as the base unit. It is almost exactly the same as one liter of water, although the exact measure is the same as a piece of platinum-iridium alloy, which is called the International Prototype Kilogram and is stored in a vault in France. An anomaly meant that it was the only base unit with an SI prefix "kilo-" (meaning thousand). There are a number of multiples and submultiples, but only some are commonly used. A one-thousandth part of a gram is called a "milligram," a millionth part called a "microgram," and 10^{-9} g called a "nanogram." Going the other way, although there are terms such as a "zettagram" (10^{21} g) and a "yottagram" (10^{24} g), these are rarely used. One curiosity is that instead of using the term "megagram," for 1000 kg, the term "tonne" is used; its spelling denoting its difference from the pre-decimal "ton." It has been relatively easy to convert from the old "imperial" system of pounds to the metric SI system, with one kilogram essentially being 2.2 pounds. Most of the world uses kilograms in the twenty-first century, the United States being a prominent exception.

Russians also had their own system, which had emerged from that used by the Mongols—although Peter the Great (r. 1682–1725) overhauled the system and used one based on the English system.

Asia

Elsewhere in the world, there were many other systems of measuring mass. The Chinese used a system with 1000 *cash* making a *tael*, and ten *taels* equaling a *catty*, and 100 of those making up a *picul*. The Japanese system relied on the *momme* (about 3.75 g), with 100 of these forming a *hyakume*, 160 of them making one *kin*, and 1000 of them equaling one *kan*. The *momme* is still used as a measure of mass in the pearling industry, which is still dominated by Japan.

Further Reading

Anderton, Pamela. *Changing to the Metric System*. London: Her Majesty's Stationery Office, 1965.

Fenna, Donald. *A Dictionary of Weights, Measures, and Units*. New York: Oxford University Press, 2002.

Grayson, Michael A. *Measuring Mass: From Positive Rays to Proteins*. Darby, PA: Chemical Heritage Foundation, 2005.

Moody, Ernest A. *The Medieval Science of Weights*. Madison: University of Wisconsin Press, 1960.

Zupko, Ronald Edward. "Medieval Apothecary Weights and Measures: The Principal Units of England and France." *Pharmacy in History* 32 (1990).

Justin Corfield

See Also: Roman Mathematics; Units of Area; Units of Length; Units of Volume.

Units of Volume

Category: Space, Time, and Distance.
Fields of Study: Algebra; Geometry; Measurement; Number and Operations; Representations.
Summary: A variety of units are used to measure volume throughout the world.

Measuring of volume intrigued many scientists in the ancient world. For the most part, crops, stones, and other items were measured by weight rather than volume because of the relative ease of doing so—especially given the irregular shapes of many items. For solid items with irregular shapes, it seemed far too complicated to work out their volume, even if this could be done with any degree of accuracy. This notion changed dramatically with the ideas that have been attributed to the famous Greek mathematician and inventor Archimedes of Syracuse (c. 287–212 B.C.E.). The tale of Archimedes is that he was given the task of determining the purity of the gold used to create the crown of King Hiero II—the king was worried that silver or base metals might have been used in its manufacture and were cleverly disguised. Pondering the problem while getting into a bathtub, Archimedes, according to the story, noticed that the water rose and the amount it rose was equal to the size of the parts of his body that were submerged. This led Archimedes to deduce that water could be used to measure the volume of a particular item, such as the king's crown. It could then be weighed against a block of pure gold of the same volume. It is said that when he realized that this could be done, Archimedes shouted "Eureka!" ("I have found it!") and ran through the streets to tell everybody of his discovery, forgetting that he had not put his clothes on.

Whether or not the story of Archimedes is actually true—and some historians doubt its veracity, although Galileo stated that he believed that it might well be true—the story does illustrate the use of fluid displacement, which can be used to easily measure the volume of irregularly shaped objects. This method does not seem to have been known before the Greeks. Certainly, the ancient Egyptians had major problems working out volume and there are complicated equations and formulae on the Rhind Mathematical Papyrus, which dates to about 1700 B.C.E., illustrating that the Egyptians were already grappling with the subject.

The Romans had two systems for recording the measurement of volume. The first and most often used was for liquid measures and was based on a sextarius (from *sester*), which is roughly 0.54 liters. Six *sesters* made up one *congius*; four of these made up one *urn*; and two *urns* make one *amphora*. For dry measures, although a *sester* was still used and was of equivalent size, eight *sesters* equaled a gallon; two gallons made up one modius (also called "peck"); and three of these one quadrantal (also called "bushel").

English

In use from the Middle Ages, the English ended up with an extremely complicated system of measuring volume, which was formalized as Imperial Measurements. The smallest measure was a "mouthful," with two of those making a "pony," two ponies making a "jack," two jacks making one "gill," two gills making one "cup," and two of those making a "pint." The system continued with two pints making a "quart" (from "quarter gallon"), with two quarts equaling one "pottle," and two pottles making a "gallon." The next levels of measurements were "pecks," "kennings," "bushels," "strikes," "coombs," "hogsheads," and "butts" (also called "pipes"). A slightly different scale was used to measure wine and beer. Even when British adopted the metric system in 1965, some of the old terminology (and measures) were still used, especially pints (for milk) and gallons (for gasoline). A bushel is also the standard measurement for wheat and some other items in agriculture.

Metric System

Although the English had numerous terms, medieval and early modern France had a vast range of measures of volume, which varied from one part of France to another, most arising for customary reasons. After the French Revolution, the new government sought to standardize all systems of measurement, including volume. This process saw the introduction, under Napoleon Bonaparte, of metrication, and in turn it led to the International System of Units (SI units).

The original metric system had liters (or litres) as the measure of volume, and from 1901 until 1964, it was defined as being the volume of one kilogram of pure water heated to 4 degrees Centigrade and measured under a pressure of 760 millimeters of mercury. When it came to devising the SI units, the liter was dropped as a measure, and the official measurement was in cubic meters. The difference is not significant in all but scientific terms, although liters continued to be used by many people throughout the world.

Gas Volume

While it was possible to measure the volumes of liquids easily (and also of solid objects by measuring the displacement of a water of a similar quantity), the measuring of the volume of gas has long posed a problem. The problem was solved by the British civil engineer Samuel Clegg (1781–1861), who had worked on natural gas flues and was able to design a dry meter and then a water meter, which were able to measure the amount of gas used by consumers. This invertion helped the gas industry in Britain—and later in other countries—measure gas and thereby charge customers based on usage.

Further Reading

Falkus, M. E. "The British Gas Industry Before 1850." *The Economic History Review New Series* 20, no. 3 (December 1967).

Hirshfeld, Alan. *Eureka Man: The Life and Legacy of Archimedes*. New York: Walker, 2009.

Jaeger, Mary. *Archimedes and the Roman Imagination*. Ann Arbor: University of Michigan Press, 2008.

Lawn, Richard E., and Elizabeth Prichard. *Measurement of Volume*. Cambridge, England: Royal Society of Chemistry, 2003.

JUSTIN CORFIELD

See Also: Archimedes; Cubes and Cube Roots; Measurements, Volume; Roman Mathematics; Units of Length; Units of Mass.

Universal Constants

Category: Space, Time, and Distance.
Fields of Study: Number and Operations; Measurement.

Summary: Universal constants help describe the universe and are believed to be fixed for all times and places in the universe.

A universal constant is a physical quantity whose value remains fixed throughout the universe for all time. However, most constants are known only approximately; humans started measuring them relatively recently and it is an assumption that they are—and have always been—fixed. There may be other assumptions that scientists and mathematicians have implicitly made that turn out to be false and undermine the universality of these constants. For example, the ratio of the circumference of a circle to its diameter in Euclidean space is π, but with Albert Einstein's conceptualization that the universe could have non-Euclidean geometry, this circumference-to-diameter ratio in the real world may be some value not equal π.

The international Committee on Data for Science and Technology defines and modifies physical constants and quantifies their levels of certainty. Three constants in particular are fundamental to the current understanding of the physical world. Together, they underlie the mathematics of gravity, relativity, and quantum physics. They are G (the gravitational constant), c_0 (the velocity of electromagnetic radiation in a vacuum (in other words, the speed of light), and h (Planck's constant).

Universal Constant: "G"

G first appeared in Isaac Newton's famous equation $F = Gm_1m_2/r^2$, which quantifies the force (F) of gravitation between two masses (m_1 and m_2), where r is the distance between their centers of mass. G is approximately $6.67 \times 10^{-11} \mathrm{m^3kg^{-1}s^{-2}}$ (meters-cubed per kilogram per second-squared), which is a very small number. Gravity is thus a very weak force. Although every mass is attracted to every other mass, the effects of gravity are obvious only when the masses involved are very large (such as with planets).

Using another of Newton's equations, $F = ma$, it follows that the acceleration due to gravity on Earth is the same for all masses. This acceleration is known as g and its value is around 9.81 ms^{-2} at sea level. This value varies with distance from the Earth's center of mass (r in the equation above), so acceleration due to gravity decreases to around 9.78 ms^{-2} at the top of Mount Everest. Knowing g to be about 9.81 ms^{-2} and the radius of the Earth to be roughly 6,378,000 meters, one can use G to show that the mass of the Earth is about 5.98×10^{24} kg. One can also estimate the mass of the Sun and other celestial bodies, such is the applicability of G.

Universal Constant: "c_0"

The velocity of light in a vacuum, c_0, is probably the most widely known universal constant. Since the length of a meter is defined by it, c_0 is fixed at exactly 299,792,458 ms^{-1}. The constancy (or invariance) of c_0 is a principle that was made famous by Albert Einstein in his theory of special relativity. Einstein's principle states that no matter how fast you or the light source are travelling, you will always measure c_0 to be 299,792,458 ms^{-1}. This principle is counterintuitive, but both the constancy of c_0 and related predictions of relativity theory have been verified empirically. From relativity theory, it is known that as velocity increases, measurements of time and space change because duration and displacement are relative—they depend on how fast one is moving. The amounts by which they change are determined by c_0.

What is actually traveling at c_0 in electromagnetic radiation are massless particles called "photons."

As carriers of the electromagnetic force, all light, electricity, and magnetism are the result of photon motion. The relationship between the photon energies and the frequency of their electromagnetic radiation is the basis of quantum physics and the third constant, h.

Universal Constant: "h"

Named after Max Planck, h has an approximate value of 6.63×10^{-34} kgm^{2}s^{-1}. The units of h can be understood as joule-seconds, also known as "action." This unit is distinct from *power*, which is joules *per* second; for example, 10 joules expended every second for 10 seconds is 100 joule-seconds.

The first appearance of h was in the Planck's relation $E = hv$. Planck discovered that photons only had certain discrete energy values, the $E = hv$ equation relates the energy (E) of the photon to the frequency (v) of its electromagnetic radiation. The fact that h exists implies that energy comes in discrete lumps, not in a continuous stream. The unit of h appears in a number of important and fundamental relations, such as Werner Heisenberg's uncertainty principle and Niels Bohr's model of the atom.

Further Reading

Carnap, Rudolf. *An Introduction to the Philosophy of Physics*. New York: Dover Publications, 1995.

Feynman, Richard. *Six Easy Pieces: Essentials of Physics Explained by Its Most Brilliant Teacher*. Jackson, TN: Perseus Books, 1995.

———. *The Character of Physical Law*. Cambridge, MA: MIT Press, 2001.

Finch, Steven. *Mathematical Constants*. Cambridge, England: Cambridge University Press, 2003.

Fritzsch Harald. *The Fundamental Constants: A Mystery of Physics*. Translated by Gregory Stodolsky. Singapore: World Scientific Publishing, 2009.

Magueijo, Joao. *Faster Than the Speed of Light: The Story of a Scientific Speculation*. Jackson, TN: Perseus Books, 2002.

Eoin O'Connell

See Also: Einstein, Albert; Elementary Particles; Gravity; Pi; Relativity.

Universal Language

Category: Space, Time, and Distance.
Fields of Study: Communication; Connections Representations.
Summary: Mathematics has been proposed as a universal language; attempts have been made at a mathematics notation that would be recognizable on any planet.

From the beginnings of humanity, people needed to establish connections. Along with speaking, counting developed from the early stages of human evolution. Numbers and counting were necessary in the first civilizations to describe ownership, for trade, or for calculating taxes. Shapes and measures were needed to make furniture, buildings, and ritual places, as well as in landscaping, time-keeping, sky-charts, and calendars. Mathematics is present everywhere in the real world: in science, art, entertainment, business, and leisure. People use mathematics to describe the universe, and mathematics is commonly referred to as the "language of science or the universe." Albert Einstein questioned:

At this point an enigma presents itself, which in all ages has agitated inquiring minds. How can it be that mathematics, being after all a product of human thought which is independent of experience, is so admirably appropriate to the objects of reality? Is human reason, then, without experience, merely by taking thought, able to fathom the properties of real things?

Some take this idea a step further and view mathematics as a universal or interstellar language or explore the creation of a universal language.

Debate

Those who consider that mathematics is a universal language reason that because mathematics arises naturally and humans possess the ability to be literate in the shared language of mathematics then it must be universal. Others criticize this viewpoint and note that learning mathematics is challenging for many people. Some scientists and mathematicians point to the fact that despite differences between cultures and natural languages, the discoveries in mathematics are the same all over the world because mathematics is so well-suited to describe reality. Discoveries that were simultaneous, like the formulations of calculus by physicist Sir Isaac Newton and mathematician and philosopher Gottfried Leibniz, appear to give even more credence to this viewpoint. However, Newton and Leibniz were able to share ideas and build upon the contributions of the same earlier mathematicians and they developed different mathematical approaches and terminology. In some examples of simultaneous discoveries, like for mathematicians in the Soviet Union and the United States, the researchers were quite separated. Other philosophers and mathematicians assert that humanity invents mathematics and distorts reality in accepting its postulates.

Physicist Werner Heisenberg's uncertainty principles seem to give rise to questions about whether anyone can objectively measure or quantify reality. Attempts to model the universe on a quantum and grand scale have led to both calls for and rejection of a theory of everything.

Creating a Universal Language

Scientists, mathematicians, philosophers, and linguists have long contemplated a language that is universal. Linguists explore languages for commonalities, and Search for Extraterrestrial Intelligence (SETI) researchers analyze signals for mathematical patterns. Some visual or graphical representations are also viewed as universal. In *De Arte Combinatoria*, Leibniz imagined

> . . . a general method in which all truths of the reason would be reduced to a kind of calculation. At the same time this would be a sort of universal language or script . . . for the symbols and even the words in it would direct the reason . . . It would be very difficult to form or invent this language or characteristic, but very easy to understand it without any dictionaries.

Leibniz cited earlier attempts at universal languages, such as correspondances that converted words into numbers by physician Johann Becher or scholar Athanasius Kircher. George Dalgarno had invented a system for translating numbers into words. In 1678, Leibniz also developed this type of system: 81,374 would be written and pronounced as *mubodilefa*. For Leibniz, the digits 0–9 became the first nine consonants of the alphabet and powers of 10 were represented using vowels. Leibniz also planned to explore the logical foundations of geometry via a universal language but he did not continue this work.

Philosopher Sundar Sarukkai noted that: "The search for 'universal' language or 'pure' language is part of human history in all civilizations. In part, this reflects an enormous distrust of ambiguity in meaning." However, he also asserts that, "it is semantic ambiguity that allows individuals and societies to develop and flourish."

Further Reading

Ballesteros, Fernando. *E.T. Talk: How Will We Communicate With Intelligent Life on Other Worlds?* New York: Springer, 2010.

Jeru. "Does a Mathematical/Scientific World-View Lead to a Clearer or More Distorted View of Reality?" *Humanistic Mathematics Network Journal* 26 (June 2002).

Rutherford, Donald. "The Logic of Leibniz by Louis Couturat, Chapter 3 Translation." http://philosophyfaculty.ucsd.edu/faculty/rutherford/Leibniz/ch3.htm.

Sarukkai, Sundar. "Universality, Emotion and Communication in Mathematics." *Leonardo Electronic Almanac* 11, no. 4 (2003).

Yench, John. *A Universal Language for Mankind.* New York: Writers Club Press, 2003.

SIMONE GYORFI

See Also: Calculus and Calculus Education; Mathematics: Discovery or Invention; Mathematics, Utility of; Universal Constants; Visualization.

V

Vectors

Category: History and Development of Curricular Concepts.
Fields of Study: Communication; Connections; Measurement; Number and Operations.
Summary: Vectors express magnitude and direction, and have applications in physics and many other areas.

There are some quantities, like time and work, that have only a magnitude (also called "scalars"). If one says the time is 6 A.M., it is adequate. When discussing velocity or force, however, then magnitude is not enough. If a particle has a velocity of five meters per second, this is not sufficient information because the direction of movement is unknown. Quantities that require both a magnitude and a sense of direction for their complete specifying are called "vectors." Pilots use vectors to compensate for wind to navigate airplanes, sport analysts use vectors to model dynamics, and physicists use vectors to model the world.

History and Development of Vectors
The term "vector" originates from *vectus*, a Latin word meaning "to carry." However, astronomy and physical applications motivated the concept of a vector as a magnitude and direction. Aristotle recognized force as a vector. Some historians question whether the parallel law for the vector addition of forces was also known to Aristotle, although they agree that Galileo Galilei stated it explicitly and it appears in the 1687 work *Principia Mathematica* by Isaac Newton. Aside from the physical applications, vectors were useful in planar and spherical trigonometry and geometry. Vector properties and sums continue to be taught in high schools in the twenty-first century.

The rigorous development of vectors into the field of vector calculus in the nineteenth century resulted in a debate over methods and approaches. The algebra of vectors was created by Hermann Grassmann and William Hamilton. Grassmann expanded the concept of a vector to an arbitrary number of dimensions in his book *The Calculus of Extension*, while Hamilton applied vector methods to problems in mechanics and geometry using the concept of a "quaternion." Hamilton spent the rest of his life advocating for quaternions. James Maxwell published his *Treatise on Electricity and Magnetism* in which he emphasized the importance of quaternions as mathematical methods of thinking, while at the same time critiquing them and discouraging scientists from using them. Extending Grassman's ideas, Josiah Gibbs laid the foundations of vector analysis and created a system that was more easily applied to physics than Hamilton's quaternions. Oliver Heaviside independently created a vector analysis and advocated for vector methods and vector calculus. Mathematicians such as Peter Tait, who preferred quaternions, rejected the methods of Gibbs and Heaviside. However, their methods were eventually accepted and they are taught as part of the field of linear algebra. The quaternionic method of Hamilton remains extremely useful in the twenty-first century. Vector calculus is fundamental in understanding fluid dynamics, solid mechanics, electromagnetism, and in many other applications.

During the nineteenth century, mathematicians and physicists also developed the three fundamental theorems of vector calculus, often referred to in the twenty-first century as the "divergence theorem," "Green's theorem," and "Stokes' theorem." Mathematicians with diverse motivations all contributed to the development of the divergence theorem. Michael Ostrogradsky studied the theory of heat, Simeon Poisson studied elastic bodies, Frederic Sarrus studied floating bodies, George Green studied electricity and magnetism, and Carl Friedrich Gauss studied magnetic attraction. The theorem is sometimes referred to as "Gauss's theorem." George Green, Augustin Cauchy, and Bernhard Riemann all contributed to Green's theorem, and Peter Tait and James Maxwell created vector versions of Stokes theorem, which was originally explored by George Stokes, Lord Kelvin, and Hermann Hankel. Undergraduate college students often explore these theorems in a multivariable calculus class.

The concept of a space consisting of a collection of vectors, called a "vector space," became important in the twentieth century. The notion was axiomatized earlier by Jean-Gaston Darboux and defined by Giuseppe Peano, but their work was not appreciated at the time. However, the concept was rediscovered and became important in functional analysis because of the work by Stefan Banach, Hans Hahn, and Norbert Wiener, as well as in ring theory because of the work of Emmy Noether. Vector spaces and their algebraic properties are regularly taught as a part of undergraduate linear algebra.

Mathematics

A vector is defined as a quantity with magnitude and direction. It is represented as a directed line segment with the length proportional to the magnitude and the direction being that of the vector. If represented as an array, it is often represented as a row or column matrix. Vectors are usually represented as boldface capital letters, like A or with an arrow overhead: $\vec{A}$.

The Triangle Law states that while adding, "if two vectors can be represented as the two sides of a triangle taken in order then the resultant is represented as the closing side of the triangle taken in the opposite order" (see Figure 1).

Any vector can be split up into components, meaning to divide it into parts having directions along the coordinate axes. When added, these components return the original vector. This process is called "resolution into components" (see Figure 2). Clearly, this resolution cannot be unique as it depends on the choice of coordinate axes. However, for a given vector and specified coordinate axes, the resolution is unique. When two vectors are added or subtracted, these components along a specific axis simply "add up" (like $2 + 2 = 4$ or $7 - 2 = 5$) but the original vectors do not, which follow the rule of vector addition that can be obtained by the Parallelogram Law of Vector Addition. Vector addition is commutative and associative in nature.

Multiplication for vectors can be of a few types:

1. For scalar multiplication (multiplication by a quantity that is not a vector), each component is multiplied by that scalar. Vector multiplication by a scalar is commutative, associative, and distributive in nature.
2. For the multiplication of two vectors, one can obtain both a scalar (dot product) or a vector (cross product). For a cross product the resultant lies in a plane perpendicular to the plane containing the two original vectors. Dot product is both commutative and distributive. But cross product is neither commutative nor associative in nature because the result is a vector and depends on the direction.

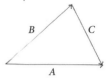

Figure 1. **B** and **C** add up to **A**.

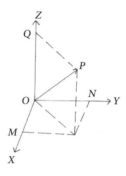

Figure 2. **OP** can be split into mutually perpendicular components **OM**, **ON**, and **OQ**.

Applications

Theoretical sciences have a wide spread of applications of vectors in nearly all fields:

- *Obtaining components*: Occasionally, one needs a part (or component) of a vector for a given purpose. For example, suppose a rower intends to cross over to a point on the other side of a river that has a great current. The rower would be interested to know if any part of that current could help in any way to move in the desired direction. To find the component of the current's vector along any specified direction, take the dot product of that vector with a unit vector (vector of unit magnitude) along the specified direction. This method is of particular importance in studying of particle dynamics and force equilibria.
- *Evaluating volume, surface, and line integrals*: In many problems of physics, it is often necessary to shift from either closed surface integral (over a closed surface that surrounds a volume) to volume integral (over the whole enclosed volume), or from closed line integral (over a loop) to surface integrals (over a surface). To accomplish these shifts, it is often very useful to apply two fundamental theorems of vector calculus, namely Gauss's divergence theorem and Stokes's theorem, respectively.
- *Particle mechanics*: In the study of particle mechanics, vectors are used extensively. Velocity, acceleration, force, momentum, and torque all being vectors, a proper study of mechanics invariably involves extensive applications of vectors.
- *Vector fields*: A field is a region over which the effect or influence of a force or system is felt. In physics, it is very common to study electric and magnetic fields, which apply vectors and vectorial techniques in their description.

Further Reading

Katz, Victor. "The History of Stokes' Theorem." *Mathematics Magazine* 52, no. 3 (1979).

Matthews, Paul. *Vector Calculus*. Berlin: Springer, 1998.

Stroud, K. A. , and Dexter Booth. *Vector Analysis*. New York: Industrial Press, 2005.

ABHIJIT SEN

See Also: Function Rate of Change; Gravity; Matrices; Numbers, Complex.

Vedic Mathematics

Category: Government, Politics, and History.
Fields of Study: Number and Operations; Problem Solving.
Summary: Vedic mathematics involves challenging mental calculations and was transmitted orally.

Vedic mathematics is a system of mathematics associated with India's Upper Indus Valley prior to 1000 B.C.E. Originally transmitted orally, the Vedic mathematics known in the twenty-first century was abstracted from ancient Sanskrit texts, known as "Vedas." Sri Bharati Krsna Tirthaji rediscovered the Vedas in the early 1900s, but his scholarly results were not published until 1965.

The Vedas covered all areas of knowledge, with the mathematics created to support this knowledge. Since recording mechanisms were not available, Vedic mathematics involves creative mental calculations, often at very challenging levels. Through Arab and Islamic writers in the 770s C.E., some Vedic mathematics was transmitted and became part of European mathematics, including elements such as the Arabic numerals, the multiplication sign, and a symbol for zero. However, the mental aspects of Vedic mathematics were not known until 1965, and these "secrets" have provided scholars, mathematicians, and students interesting explorations into multiple areas, including basic arithmetic computations, factoring, exponents, algebra in the form of linear through cubic equations, elementary number theory, analytic geometry involving the conic sections, the Pythagorean theorem, and differential calculus.

The Sutras

Sixteen formulas (or *Sutras*, which means "thread") form the foundation of Vedic mathematics, along with fourteen "sub-Sutra" corollaries. Expressed as word phrases, each formula acts as a "thread" woven throughout the Vedic mathematics system, assuming the role of a unifying element.

For example, Sutra #2 states: "All from 9 and the Last from 10." Sutra #3 states: "Vertically and Crosswise." The combined importance of both Sutras is best explained within the context of mental multiplication, such as finding the "sum" 88×98. Both numbers are close to the "base" 100, involving "deficiencies of 12 and 2," respectively. The desired product is obtained using these deficiencies (Sutra #2), then represented either mentally or symbolically (by Sutra #3):

$$
\begin{array}{r}
88 \text{ -- } 12 \\
98 \text{ -- } 2 \\
\hline
86/24
\end{array}
$$

In these operations, the deficiencies 12 and 2 are placed to the right of the original numbers, 88 and 98. The 86 is found by subtracting a deficiency from the other number in the product ($98 - 12 = 86 = 88 - 2$), while the 24 is the product of the deficiencies. Finally, the desired result is found: $88 \times 98 = 8624$, as the 86 actually represented 8600. Though this process involves a sense of magic, it is much easier than the modern computational algorithm commonly used in the twenty-first century.

It is not only important to investigate why this Sutra-based technique works, but also determine possible constraints or exceptions. For example, applying the Sutra to the product 25×57, the process becomes:

$$
\begin{array}{r}
25 \text{ -- } 75 \\
57 \text{ -- } 43 \\
\hline
-18/3225
\end{array}
$$

Because the desired product can be obtained via $-1800 + 3225 = 1425$, the power and the limitations of the Sutra become more evident, especially the emphasis on the numbers 9 and 10. The technique is not useful in this example because the large internal products and need for a negative quantity become obtrusive. However, the method does work, and it can be proven true algebraically. Suppose the desired product is $a \times b$, where a and b are whole numbers less than 100. Using the respective deficiencies ($100 - a$ and $100 - b$), the Sutra's process leads to the algebraic identity $ab = 100\left[b - (100 - a)\right] + (100 - a)(100 - b)$. Thus,

the numbers a and b could be any numerical values—positive, negative, fractions, irrational, or even complex numbers.

Finding Decimals

As another example, Sutra #1 states: "By One More than the One Before." This Sutra is used in the construction of the number system, as each whole number is one greater than its predecessor (akin to the Peano postulates formulated in the nineteenth century). However, the Sutra's power is its application in other situations as well. Suppose the problem was to find the repeating decimal equivalent to the "vulgar" fraction 1/19, usually obtained by laboriously dividing 19 into 1. The Sutra suggests a focus on "one more than the number before" the 9, or the number 2, which is one more than the 1 which appears before the 9. The 2 (called *Ekadhika* for "one more") becomes the new divisor in lieu of the troublesome 19. The "strange" decimal resulting from this division of 2 into 1 is

$$
0._{1}05_{1}263_{1}1_{1}5_{1}7_{1}89_{1}47_{1}3_{1}68421 \ldots.
$$

To explain this strange expression, start with a 0 and a decimal point. Then, 1 divided by 2 is 0 remainder 1, represented by placing a 0 in the decimal expression, preceded by a subscripted 1 as the remainder. The process is repeated, where 10 (or the visual of the subscripted 1 and adjacent 0) is divided by 2, resulting in 5 with remainder 0. Thus, the 5 in the decimal expression now is not preceded by a subscripted number. Next, 5 divided by 2 results in 2 remainder 1, which are represented as before with the remainder becoming the preceding subscript. And, in subsequent divisions, 12 divided by 2 is 6 remainder 0, 6 divided by 2 is 3 remainder 0, 3 divided by 2 is 1 remainder 1, and so on. Finally, to get the final value of the decimal expression for 1/19, the subscripted values are removed: $1/19 = 0.052631578947368421$, as they are needed only as "mental" reminders of the division process by 2. The mathematical explanation underlying this process is quite complex, but can be found in Chapter 26 of Tirthaji's *Vedic Mathematics*.

These two examples illustrate the enjoyment of investigating Vedic mathematics. On one level, the 16 Sutra and their corollaries provide efficient mental algorithms that become very powerful and efficient in special instances. On a second level, the careful examination of the Sutra and its application provides a rich opportunity to understand the role of generalization and algebraic identities.

Further Reading

Bathia, Dhaval. *Vedic Mathematics Made Easy.* Mumbai, India: Jaico Publishing, 2006.

Howse, Joseph. *Maths or Magic? Simple Vedic Arithmetic Methods.* London: Watkins Publishing, 1976.

Tirthaji, B. K. *Vedic Mathematics.* Delhi: Motilal Banarsidass, 1965.

Williams, Kenneth, and Mark Gaskell. *The Cosmic Calculator: A Vedic Mathematics Course for Schools.* (Books 1, 2, and 3). Delhi: Motilal Banarsidass, 2002.

JERRY JOHNSON

See Also: Arabic/Islamic Mathematics; Asia, Southern; Multiplication and Division.

Vending Machines

Category: Architecture and Engineering.
Fields of Study: Algebra; Geometry.
Summary: Ubiquitous vending machines use algebra and Boolean logic to function.

Vending machines are finite state machines, also known as "automata," that transition between states based on customer input data, such as product selection. Vending machine designers use mathematical models and Boolean algebra to determine the states the machine should transition into based on input data variables, with the outcome often expressed as a table. The control unit reads the data as either "true," meaning the machine recognizes the input language, or "false," meaning that it does not.

The first documented vending machine, invented by the Egyptian mathematician Hero of Alexandria, appeared c. 215 B.C.E. By the twentieth century, vending had developed into a billion dollar industry, and vending machines dispensed a variety of products. Older vending machines relied on the mechanical activity of knobs or levers activated by the customer to dispense the desired product. Vending machine operators utilize mathematics to determine potential and actual expenses and profits, as well as to process sales and stock data. For example, net income can be determined through the simple formula: Net Income = Income − Expenses.

Modern vending machines, however, utilize basic computing system processors to analyze customer input data, such as a letter and number, that corresponds to the desired product, which is then electronically dispensed. Modern advances in vending machine technology include card validators for debit and credit

A woman buying a beverage on a Tokyo street. Vending machines are extremely popular in Japan and there are machines that sell ramen noodles, alchoholic beverages, fruit and vegetables, batteries, and even clothing.

cards; voice activation; electronc message displays for insufficient funds, lack of change, or sold out products; and remote wireless diagnostics and data collecting to alert venders of the need for restocking or repair.

Vending machine control units are part of a class of abstract machines known as "finite state machines" or "automata"; in particular, they are deterministic or discrete finite state automata (DFA). Finite state machines are always in a position known as a "state," transitioning between these states based on input data. Designers use mathematical models in the design of finite state machines, such as vending machines. The machines are designed to recognize a regular language, converting computation into language recognition. Each state is labeled either "true" (accept the data) or "false" (reject the data) based on whether the machine recognizes the language of the input data.

Vending machine design utilizes Boolean logic or algebra, or algebra based on two logical values, in this case the values of "true" and "false." The general Boolean function is expressed through the formula

$$y = \sum (x, \ldots)$$

where $(x, \ldots)$ is equal to a set of Boolean variables with the values "true" or "false." Diagrams of the various states of the vending machine and the possible transitions between them can be converted into Boolean operations.

The control unit reads each string of input data, generally input from the vending machine customer, such as the diameter, thickness, or number of ridges of coins followed by product selection codes. Transition functions tell the machine which state it should enter based on input data. Transition functions are often represented in tabular form. The control unit changes its state with each data string entered until the final input, after which it outputs either "true" or "false" based on its final state. Vending machines also use the algebraic relationship between range and domain, where the range is the machine's output and domain is the customer's input. For example, a customer must input an equal or greater amount of money than the cost of the desired product.

Further Reading

Hopcroft, John E., and Jeffrey D. Ullman. *Introduction to Automata Theory, Languages, and Computation.* Reading, MA: Addison-Wesley, 1979.

Salomaa, Arto. *Computation and Automata.* New York: Cambridge University Press, 1985.

Salyers, Christopher D. *Vending Machines: Coined Consumerism.* Brooklyn, NY: Mark Batty Publisher, 2010.

Segrave, Kerry. *Vending Machines: An American Social History.* Jefferson, NC: McFarland & Company, 2002.

Marcella Bush Trevino

See Also: Algebra in Society; Closed-Box Collecting; Functions.

Video Games

Category: Games, Sport, and Recreation.
Fields of Study: Algebra; Data Analysis and Probability; Geometry.
Summary: Video games use the mathematical concepts of algorithms, matrices, and random numbers as part of their programming.

Video games are pervasive in modern society, from computers to television-based systems to applications that can be downloaded easily onto cell phones. There is an ongoing debate over what should be called the first video game. The narrower definition is a game generated by a computer and displayed on a video device. Others consider it to be any electronically based game displayed with video output. The most likely candidate is a 1940s invention by physicists Thomas Goldsmith and Estle Ray Mann. Their "Cathode-Ray Tube Amusement Device" was inspired by World War II radar displays and allowed the player to shoot virtual missiles at targets. Though patented, at the time it was too costly to produce commercially and only a few prototypes were ever made. Much of the mathematics used to design and operate computers also applies to video games and the various fields and professions are closely connected. Video game design programs offered by many colleges emphasize physics and mathematics education along with computer programming, as these skills are necessary to represent the real world in increasingly realistic ways. The new generation of body-sensing game controllers uses optics to detect a player's motion in three axes and translate it to corresponding movements within the game environment. While most people think of video games as entertainment, they are increasingly being incorporated into the classroom and other learning applications. In 2009, U.S. President Barack Obama initiated a campaign called Educate to Innovate, which seeks to use interactive games, among its other strategies, to improve the mathematical and scientific abilities of American students.

Simple Modeling Using Polygons

Any video game that has graphics needs to have a way of drawing a picture on the screen. A very basic program can take a turtle (or curser) on a screen and move it forward and rotate its direction clockwise. Many geometrical shapes are easy to draw using a turtle. For example, to tell the turtle to draw a rectangle, a simple program might tell the turtle to move 100 steps (which could be measured by pixels on the screen), turn 90 degrees, move 50 steps, turn 90 degrees, move 100 steps, turn 90 degrees, and move 50 steps. At this point, a 100×50 rectangle has been drawn, and the turtle is perpendicular to the position where it started.

A circle (or any object with a curve) would be much more difficult to draw using these commands because of the thickness of a pixel and the fact that the turtle cannot move half degrees. A user could try to tell the turtle to move one step then turn one degree. After repeating those commands 360 times, the turtle will be back where it began, and will have drawn a circle that is slightly less than 115 steps across. Technically, it did not draw a circle, but rather a polygon with 360 sides. A slight modification may be to tell the turtle to move two steps then turn one degree. After 360 repetitions, the turtle will appear where it started and the shape appears to be a circle that is twice as wide as the first shape drawn, about 229 steps wide.

There is a big gap between 115 steps and 229 steps wide. If a programmer needs a circle between those dimensions (or beyond those dimensions), the programmer can use mathematics to adjust the step length to get a circle of the desired size. The length across a circle is called the "diameter" and the distance around a circle is called the "circumference." The relationship between these two measurements is $C = \pi d$, where C is the circumference, and d is the diameter.

Since the turtle will be tracing the outside of the circle, it will travel the length of the circumference. The turtle will also be making 360 turns during its travel. Since each step should be the same length, one can find the length of each step by taking the circumference and dividing by 360. Since π is approximately 3.14, one can estimate the length of the step by multiplying 3.14 and the desired diameter and then dividing by 360.

Depending on the video game being created, a programmer will probably desire to draw more than circles and polygons. Using the above steps for a circle but only repeating the steps 180 times will yield a half circle, which could approximate the shape of a setting sun, the top of a silo, or the ice cream in a cone. More complex shapes, like drawing a long-haired cat, could be made by the turtle but the programmer now has a time concern. The programmer creating the directions to draw the cat and the fur on the cat would require a long time to type in the programming for the cat—and even more if the cat is supposed to move—since the repeat step would be used sparingly, if at all. On the users end, a large program with a lot of steps would take a long time to draw, depending on the speed of the computer or gaming system on which it is to be played.

Although video games are displayed on a two-dimensional screen, programmers now commonly create elements of the game in three dimensions. To mimic the body of an object, programmers create the outer shell of the object using a mesh of triangles or quadrilaterals. Depending on the detail desired, more meshes could be created. Once the object is created, it needs to be displayed on the screen. This process involves using a "point-of-view camera," which will change how the object is drawn based on where the camera is and how far away it is from the object. The triangle mesh of the object is adjusted accordingly. For example, as the object approaches the camera (gets closer to the screen), the triangles will elongate and become larger. A programmer that wants the object to get closer to the camera and rotate will use vectors and matrices (linear algebra) to adjust the size and the dimensions of each triangle in the meshes. Once the computer does the mathematical calculations to modify the triangle mesh, the point-of-view camera creates a two-dimensional image of the three-dimensional mesh in the orientation it has been set to. This two-dimensional image then gets projected to the viewing screen.

Interesting geometry is also found in the movement of objects through the game. In some cases, like the games Portal and the older PacMan, players can exit

A game programmer working with multiple monitors. The screen on the right shows a portion of the large amount of source code used.

combines them to make a six-digit color number by placing the intensities in order for red, then green, and finally blue. For example, pure red would be FF0000 (intensity 255 for red and intensities 0 for both green and blue). Similarly, 00FF00 would be pure yellow and 0000FF would be pure blue. The color white would be represented FFFFFF (a combination of all three colors), whereas black would be 000000 (no light whatsoever).

Random Number Algorithm

Many video games that have been created offer a storyline or, at least, a progression to get from one stage or level to the next. Moving to the next level often requires a certain level of skill or collecting certain objects. On the other hand, there are video games that are created, like video poker or Tetris, where skill alone is not enough to do well. There is a certain random element that will determine the outcome. However, computers are not capable of creating random numbers. Instead, the video game console is pre-programmed with a list of pseudo random numbers. For example, every TI-84 calculator that has its memory reset will create the number 0.94359740249213 as its first "random" number. Obviously, if everyone obtains the same result, it cannot be random.

Using the TI example, every random number it produces will be a decimal between zero and one. If the game requires a number higher than one, the programmer merely multiplies the random number times the highest number they desire. For example, in Tetris, there are seven tetrominoes that could be selected for the next drop. A programmer may want a random number generated to determine the shape of the next piece. As a result, the programmer would create a random number, and then multiply it by seven to get a number between zero and seven; however, this number is still a decimal. The computer programmer can then tell the console to truncate the number, which would ignore everything beyond the decimal point giving an integer between zero and six. A final (optional) step would be to add one to this truncated integer resulting in a number between one and seven. Each block would

the playing field on one side of the screen and return from another side or in a different orientation. This property involves concepts like a torus and higher-dimensional analogs.

Color

When programming colors (assuming the screen is not monochrome), a programmer needs to remember that the primary colors for light are different than the primary colors for pigment. When drawing on paper, the three colors magenta (red), cyan (blue), and yellow can be combined in such a way as to create almost any other color. For example, many color printers only use three colors to print. Since most screens work based on a projection of light (whether a computer monitor or a television screen), the primary colors of light must be used. For light, the colors red, blue, and green are the primary colors; with these colors, any other color can be created. All three together make white, and no light at all makes black.

When coding colors, each of the three primary light colors is given an intensity value $0-255$. This value is then converted to a two-digit hexadecimal number, where 00 is the decimal number zero and FF is the hexadecimal number 255. The hexadecimal number 12 would be an intensity level of 18. The hexadecimal number A0 would be an intensity level of 160. The programmer then takes these three intensity numbers and get assigned a number, and the pseudo-random number that resulted would select the next block.

Further Reading

Dunn, Fletcher, and Ian Parberry. *3D Math Primer for Graphics and Game Development*. Plano, TX: Worldware, 2002.

Egan, Jill. *How Video Game Designers Use Math*. New York: Chelsea Clubhouse, 2010.

Flynt, John P., and Boris Meltreager. *Beginning Math Concepts for Game Developers*. Boston: Thomas Course Technology, 2007.

CHAD T. LOWER

See Also: Animation and CGI; Polygons.

Vietnam War

Category: Government, Politics, and History.
Fields of Study: All.
Summary: Because of the importance of cryptography in World War II and the emergence of game theory in the 1950s, mathematics was heavily involved in the Vietnam War.

The Vietnam War, a conflict transpiring in Vietnam, Cambodia, and Laos from 1955 to 1975, involved the Communist forces of North Vietnam, the Viet Cong, the Khmer Rouge, the Pathet Lao, the People's Republic of China, the Soviet Union, and North Korea, against the anti-Communist forces of South Vietnam, the United States, South Korea, Australia, the Philippines, New Zealand, Thailand, the Khmer Republic, Laos, and the Republic of China. Most American involvement was concentrated from 1963 to 1973, with the last U.S. troops leaving with the fall of Saigon in 1975. It eventually resulted in a Communist victory, with U.S. forces and their allies withdrawing, Communist parties taking control of Laos and Cambodia, and South Vietnam unified with the North under Communist rule.

Mathematicians and the War

Mathematicians fell on both sides of the disagreement regarding the Vietnam War. Some served in the war effort, such as William Corson, an economist with an undergraduate degree in mathematics who later wrote the book *The Betrayal*. Grace Murray Hopper returned to active duty in 1967 because of an increased demand for naval computer systems. Others engaged in war-related research. Warren Henry helped develop the hovercraft for nighttime fighting during the 1960s while working at Lockheed Space and Missile Company and this was used in the war.

In 1966 and 1970, mathematicians at the International Congress of Mathematicians appealed to their colleagues to avoid war-related work. Mathematicians around the world organized or participated in protests, including Alexander Grothendieck in France and Steven Smale in the United States. Mathematicians in Japan at the University of Kyushu in South Japan organized "demonstrations of the 10" against the war on the 10th, 20th, and 30th of the month. Funding originally designated for teacher development during the New Math movement was instead directed to the war. Some have asserted that this diversion of funds was one of the main reasons that the educational movement failed. Mathematics played a role in the war in a number of ways, including war strategy, precision weapons, airplane computers, cryptography, and a statistically flawed 1969 draft drawing. Statisticians and others have used statistical techniques to study the long-term effects of Agent Orange on soldiers. Decision theory has been used to model the war. Systems analysis and game theory may have contributed to U.S. involvement and defeat, such as in the decisions of Secretary of Defense Robert McNamara.

Game Theory

One of the key political leaders of the American forces during the Vietnam War was Robert McNamara, a student of game theory, who served as the secretary of defense from 1961 to 1968—the period corresponding with the nation's first serious engagement with the war and its major expansions and escalations. McNamara was also responsible for the policy of Mutually Assured Destruction (MAD), a nuclear policy grounded in game theory. It said that the best deterrent to full-scale use of nuclear weapons was for opposing sides to each possess sufficient firepower to completely destroy the other so that neither side dares attack, knowing it cannot survive the counterattack. A chilling take on foreign policy, history may be on McNamara's side with the Cold War. The escalating war in Vietnam is another story. From a game theory perspective, those escalations make perfect sense. Consider that fact that North Vietnam had the options to escalate or to negotiate a peace. The United States also had those options as well as the option to pull out. The only way for the United States to gain a military advantage—and potential victory—was to escalate, with the worst possible outcome of such escalation being a stalemate. Despite increased desertion and plummeting morale, as well as growing anti-war sentiment at home, McNamara continued to escalate the engagement because it was the most promising option he was trained to see.

This was later used as an example of "escalation of commitment," a phenomenon identified in Barry Straw's 1976 paper "Knee Deep in the Big Muddy: A Study of Escalating Commitment to a Chosen Course of Action," wherein cumulative prior investment becomes the motive to continue to escalate one's investment even when rational thought says it is the wrong choice. That initial error of judgment becomes the motive to continue, to stay committed to the course of action, in order to justify it. The more one continues, the greater error one must admit to if one disengages, which is why psychologists sometimes refer to this phenomenon as the "commitment bias," a natural tendency to want to believe that one has been making the right choices and to ignore evidence to the contrary.

Further Reading

Batterson, Steve. *Stephen Smale: The Mathematician Who Broke the Dimension Barrier*. Providence, RI: American Mathematical Society, 2000.

Bosse, Michael. "The NCTM Standards in Light of the New Math Movement: A Warning!" *The Journal of Mathematical Behavior* 14, no. 2 (1995).

Bunge, Mario. "A Decision Theoretic Model of the American War in Vietnam." *Theory and Decision* 3, no. 4 (1973).

Starr, Norton. "Nonrandom Risk: The 1970 Draft Lottery." *Journal of Statistics Education* 5, no. 2 (1997).

BILL KTE'PI

See Also: Asia, Southeastern; Cold War; Game Theory; Infantry; Military Draft; Predicting Attacks; Vietnam War.

Viruses

Category: Medicine and Health.
Fields of Study: Algebra; Geometry.
Summary: The spread of viruses in a population—and the internal structure of viruses themselves—can be analyzed mathematically to help epidemiologists study viral infections.

A virus is a parasite. It cannot reproduce on its own. Instead, it must invade a cell of another organism and use the host cell's machinery to make copies of itself. The newly replicated viruses then leave the host cell and infect other cells. In the process, the virus often damages the host. For example, different viruses cause measles, polio, and influenza in people; hoof-and-mouth disease in cattle; and leaf curl in many vegetables. Mathematics provides a language to describe viral structures. Furthermore, mathematical models of the spread of a virus in a population are powerful tools in public health policy.

Capsid Geometry

A virus consists of genetic material (either DNA or RNA) surrounded by a protein coat called a "capsid." Viruses have much less genetic material and are much smaller than single-celled organisms like bacteria. With limited genetic material, a virus can encode only a few proteins of its own, and so must use them efficiently. Often, the entire capsid is assembled from many copies of a single protein, which means the capsid should be highly symmetric.

One of the first virus structures to be determined was that of the Tobacco Mosaic Virus (TMV). Copies of the TMV capsid protein are arranged in a helix around the viral RNA. Many other viruses have helical capsids as well. In contrast, poliovirus, the Hepatitis B virus, tomato bushy stunt virus, and other viruses have icosahedral capsids. Figure 1 shows a computer-generated image of the poliovirus capsid with protein subunits colored to highlight the icosahedral symmetry. Other, more complicated capsid shapes are possible.

While the capsids do not have flat triangular faces, they have axes of five-fold rotational symmetry, like those through the vertices of the icosahedron; axes of three-fold rotational symmetry, like those through the centers of the triangular faces of the icosahedron; and axes of two-fold rotational symmetry, like those through the centers of the edges of the icosahedron.

Modeling the Spread of Viruses

Models of virus transmission in a population help researchers understand which interventions might slow the spread of a virus. The SIR model, first proposed by W. O. Kermack and A. G. McKendrick in 1927, is one of the simplest and is suitable for viruses such as measles and influenza. Each person in a population is in one of three categories: (1) susceptible to the virus, (2) infected and infectious, or (3) recovered and immune.

Let S, I, and R be the proportion of the population that is susceptible, infected, and recovered, respectively. The SIR model is given by the following system of differential equations:

$$\frac{dS}{dt} = -\beta SI, \; \frac{dI}{dt} = \beta SI - \gamma I, \text{ and } \frac{dR}{dt} = \gamma I$$

where the constant β depends on the probability that an infected person transmits the virus to a susceptible person, and the constant γ depends on how long it takes an infected person to recover. This model does not lead to simple expressions for S, I, and R as functions of time but it can be explored computationally. One simple way to do so is to treat time discretely and approximate

$$\frac{dS}{dt}$$

by $(S_{t+1} - S_t)$, where S_t is the value of S at time step t. This method yields the difference equations

$$S_{t+1} = S_t - \beta S_t I_t$$

$$I_{t+1} = I_t + \beta S_t I_t - \gamma I_t$$

$$\text{and } R_{t+1} = R_t + \gamma I_t.$$

The basic SIR model can be modified to fit other scenarios. For example, immunity might wear off over time, or some part of the population might be at higher risk of infection, or a vaccination campaign might begin.

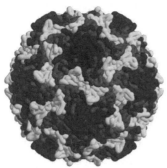

Figure 1: Poliovirus capsid.

The SIR model assumes that all possible contacts between infected people and susceptible people are equally likely (hence the factor of SI in dS/dt). Modifying the model to reflect the social structure of the population allows researchers to ask crucial questions. If the supply of influenza vaccine is limited, is it more effective to vaccinate school children, who spread the disease, or the elderly, who may suffer more complications from infection? Will closing airports slow an epidemic enough to justify the costs to travelers? In such situations, mathematical models allow public health officials to test the effects of different interventions before choosing a course of action.

Further Reading

Carrillo-Tripp, Mauricio, et al. "VIPERdb2: An Enhanced and Web API Enabled Relational Database for Structural Virology." *Nucleic Acids Research* 37 (2009).

Keeling, Matt J., and Pejman Rohani. *Modeling Infectious Diseases in Humans and Animals*. Princeton, NJ: Princeton University Press, 2008.

Levine, Arnold J. *Viruses*. New York: W. H. Freeman and Company, 1992.

CATHERINE STENSON

See Also: Diseases, Tracking Infectious; HIV/AIDS; Mathematical Modeling; Polyhedra; Symmetry.

Vision Correction

Category: Medicine and Health.
Fields of Study: Geometry; Measurement; Problem Solving; Representations.
Summary: Modern optometry depends on precise measurements to construct corrective lenses.

Human vision is subject to a variety of ailments and disorders. Some are congenital; others are age-related. Faulty vision results in blurriness, coupled with headaches and ocular tiredness. However, for many years, humans have been perfecting the art of using external implements to aid vision. Technologies exist in the twenty-first century that can restore perfect vision to people suffering from common vision-related problems, such as myopia or astigmatism. The methods used to diagnose vision issues and to construct corrective lenses rely on precise mathematical measurements and understanding of the geometric principles behind light refraction. Vision may also be modeled in various ways, including using a concept called "orthonormal polynomials," such as the Fourier series and optic wavefronts. This has many applications, including laser vision correction. In stereoscopic vision, two-dimensional projections of the world onto the retina of each eye are combined and compared to form a three-dimensional image. It was once thought of as virtually impossible to cure stereoblindness, but in the early twenty-first century, vision therapists use a variety of techniques to help patients perceive stereoscopic depth in three spatial dimensions.

Lens Power

The optical power of a lens, also known as "dioptic power," "refractive power," or "focusing power," is a measure of the curvature of the lens and the degree to which a lens converges or diverges light. It is equal to the reciprocal of the focal length of the lens in meters. Its unit is "diopter." Prescriptions for eyeglasses specify the optical power of the lenses. The human eye has a refractive power of 60 diopters. Stacking lenses helps to combine their optical power.

Eyeglasses and Bifocals

A simple pair of eyeglasses contains nothing more than two pieces of glass shaped in such a way that they act like a pair of lenses. Lenses exploit the physical property of light called "refraction." Refraction occurs when light travels between mediums of different densities, such as air and glass. The change in the medium causes light to bend in a certain calculable way. This property of lenses is suitable to refocus the image back onto the retina in people suffering from long-sightedness and short-sightedness.

The focal length of a lens in air can be calculated using the lensmaker's equation, given by

$$\frac{1}{f} = (n-1)\left[\frac{1}{R_1} - \frac{1}{R_2} + \frac{(n-1)d}{nR_1R_2}\right]$$

where f is the focal length of the lens, n is the refractive index of the material, R_1 is the radius of curvature of the lens surface closest to the light source, R_2 is the radius of curvature of the lens surface farthest from the light source, and d is the thickness of the lens.

To address people suffering from vision problems such as myopia, hyperopia, and astigmatism, bifocal lenses were invented. These lenses have a section of magnification at the lower portion of the frames to allow the wearer to read small print. Benjamin Franklin is generally associated with the invention of the first pair of bifocals.

Contact Lenses

Contact lenses are corrective or cosmetic lenses placed on the cornea of the eye. Their performance is similar to that of eyeglasses but they can be shaped somewhat differently. Spherical lenses are the typical shape of contact lenses on both the inside and the outside surfaces, whereas toric contact lenses, often used for people with astigmatism, are created with curvatures at different angles and cannot move on the eye. Contact lenses are extremely lightweight and are virtually invisible when compared to eyeglasses. However, they are also not held in place by a rigid framework like glasses. Mathematical models are useful for understanding the various movements of lenses within the eye, especially hard contact lenses.

In the twenty-first century, technology has advanced to a level where it is possible to imprint electronics onto the contact lenses themselves, resulting in the ability to project a virtual display onto the eye directly. While this technology by itself does not directly correct any vision problems, it could be used to assist people in their everyday activities, such as locating objects, or reading street signs by magnifying letters.

LASIK

Laser-assisted in situ keratomileusis (LASIK) is becoming an increasingly popular alternative to contact lenses and eyeglasses. LASIK is a type of refractive surgery performed using a laser. A "laser" (Light Amplification by Stimulated Emission of Radiation) is a highly concentrated beam of light capable of focusing high energy in a small area.

The technology was invented by a Colombia-based Spanish ophthalmologist Jose Barraquer. His technique involved cutting thin flaps in the cornea and altering

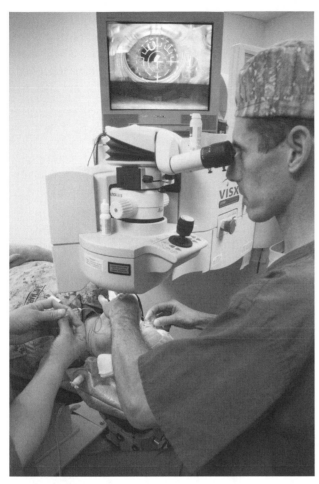

LASIK VISX surgery being performed by a U.S. Navy surgeon. A close-up of the eye is seen on the monitor.

its shape. After the laser was invented, Dr. Bhaumik, in 1973, announced the breakthrough in using lasers to treat vision problems.

LASIK involves creating a flap of corneal tissue, remodeling the cornea underneath the flap with the help of a laser, and then repositioning the flap. Mathematical computations are used to determine the depth of the cuts used in the surgery, and these are often a function of the average cornea thickness of 550 micrometers. One alternative is to leave some fixed tissue depth.

Further Reading

Barry, Susan. *Fixing My Gaze: A Scientist's Journey Into Seeing in Three Dimensions.* New York: Basic Books, 2009.

Dai, Guang-ming. "Wavefront Optics for Vision Correction." *SPIE Press Monograph* PM 179 (2008).

Hecht, Eugene. *Optics.* 4th ed. Addison Wesley, 2002.

ASHWIN MUDIGONDA

See Also: Light; Surgery.

Visualization

Category: History and Development of Curricular Concepts.
Fields of Study: Algebra; Communication; Connections; Data Analysis and Probability; Geometry; Representations.
Summary: Visualization is a useful practice when doing or learning mathematics and computers can help create visualizations of difficult concepts.

The ability to form a mental image is a fundamental process and has been incorporated in many theories about knowledge acquisition. The advent of the printing press and perspective drawings allowed for an unprecedented sharing of realistic pictures, graphs, and inventions, and this led in part to the Industrial Revolution.

The development of coordinate geometry gave rise to graphical representations of data and algebraic concepts. With the popularity of computers and computer graphics, mathematicians, artists, and programmers have created visualizations of mathematical objects and huge amounts of data. Mathematicians also found new ways to visualize and share abstract ideas such as the fourth dimension. Dynamic image manipulation features, such as rotation or zooming, further increased the accessibility of visualized objects by facilitating new perspectives and comprehension of hard-to-see surfaces. Mathematical visuals have been fundamental in both research and entertainment contexts like for computer-generated imagery (CGI) used in modeling, computational geometry, or movies. Various types of visualization, including spatial visualization and visuals of data and graphs, are important components of all levels of mathematics and statistics classrooms in the twenty-first century. Visualization is an interdisciplinary topic and researchers from a diverse range of fields contribute, including mathematicians, computer scientists, psychologists, engineers, and neuroscientists. Educators and researchers create visualizations, study visualization ability, and design new ways to help students visualize.

Early History

Visualization has been as important in mathematics and statistics research as in education and mathematicians in many fields throughout history created visual representations. Representations of maps are as ancient as the earliest societies from which there exists evidence of stone tablets and animal skins. Another important historical research area related to visualization and mathematics was the field of optics. For example, ancient people created lenses. Euclid of Alexandria investigated geometry and perspective in his book on optics. Many mathematicians and scientists worked to understand vision, including mathematician Abu Ali al-Hasan ibn al-Haytham, who wrote a seven volume work on optics and visual perception, which is noted by some as the first work to correctly demonstrate understanding that light is reflected from an object to the eye. Self-taught mathematician and scientist Tobias Mayer was one of many to formulate a theory for color perception and he also modeled the limits of vision, noting, "there is a certain visual angle below which an object presented to the eye appears either not distinct enough or not even distinct at all, but only confused and as though it had vanished from sight.... We shall call this angle the limit of vision, and we shall investigate its angle by experiment."

In the seventeenth century, René Descartes made significant progress in coordinate geometry. The Cartesian plane that is named for him allowed for new representations of data and algebraic equations. Mathematicians, statisticians, social scientists, and others began to investigate ways to visually present graphs and data to facilitate analysis, interpretation, and understanding. Social issues motivated many researchers in the nineteenth century. For example, William Playfair created color-coded graphical representations of the English national debt and the trade balances between England and other countries. Adolphe Quetelet graphed the distributions of anthropometric data to show both the center and variability, leading in part to the measure now known as Body Mass Index. Florence Nightingale developed the polar area chart as part of her campaign for improved sanitation in medical facilities. John Snow

used graphical mapping techniques to trace the source of a London cholera outbreak. Graphs of mortality statistics and many other naturally occurring phenomena also proliferated. Philosopher and logician John Venn developed Venn diagrams in 1881, which are also used in many mathematics classrooms.

Recent Developments

The rise of computers in the twentieth century led to mind-bending visualizations and new fields of research in mathematics as well as beautiful artistic forms. Mathematician Benoit Mandelbrot popularized the field of fractals. The computer visualization of some objects helped clarify their mathematical properties. One example is Enneper's surface, which had been introduced by Alfred Enneper in the nineteenth century. In the mid-twentieth century, Steven Smale proved that it was possible to turn a sphere inside out in three dimensions without creating any creases. This idea stretched the imagination and mathematicians tried to visualize it. For instance, mathematicians at the Geometry Center for the Computation and Visualization of Geometric Structures produced a video called *Outside In*, which visualized William Thurston's sphere eversion method. Geometer Thomas Banchoff pioneered visualizations of four-dimensional objects. Mathematicians in the twenty-first century attempted to visually model the Internet using hyperbolic geometry in order to reduce the load on routers. Researchers from interdisciplinary fields have participated in conferences on topics like visualization algorithms or data visualization. Mathematicians have designed visualization software and techniques for many areas in mathematics, including linear algebra, group theory, and complex analysis. Some of these visualizations are used in classrooms, while others are the focus of research investigations or artistic exhibitions.

Visualization Ability

The connections between visualization ability and mathematical success also have a long and varied history. In the nineteenth century, scientist and mathematician Sir Francis Galton conducted studies to examine the relationship between visual imagery and abstract thought. Some have noted that nineteenth-century mathematician Henri Poincaré had poor eyesight as a student and scored a zero on an entrance exam for the École Polytechnique; however, he had a great memory because he was able to mentally translate concepts he heard aurally into visual representations of the same concepts. Poincaré later wrote about the ability to form retina images and what he referred to as "pure visual space." The Poincaré disc model of hyperbolic geometry is named for him, and twenty-first-century students explore this in interactive computer models that are designed to help visualize and explore mathematical topics, including the variation in the sum of the angles for differently sized triangles.

Other visual challenges, like "stereoblindness" and "subitizing" difficulties, have also been tied to mathematics. Stereoblindness, the inability to properly combine images in the mind to see in three dimensions, was once thought of as impossible to cure. Subitizing is the ability to rapidly perceive and differentiate the number of distinct items in a small group of objects, like dots on a cube. Some researchers in the first part of the twentieth century investigated the importance of subitizing to the understanding of numbers, counting, and abstract thinking and educational psychologists in the second half of the twentieth century continued this work and developed a variety of theories. While the specific mechanisms are still the topic of debate, in the twenty-first century, vision and subitizing therapies have been successfully implemented in the optometry profession and are thought to help mathematics students. Some proponents of left-brain versus right-brain dominance theories assert that visualization is focused in the right brain, while other mathematical skills, like logic and analysis, are focused in the left side of the brain. Psychobiologist Roger Sperry was awarded the Nobel Prize in 1981 in part for his split-brain experiments. However, medical imaging scans of people performing mathematical tasks has shown regions from both sides of the brain highlighted and researchers continue to investigate this issue.

Gender

In the latter half of the twentieth century, researchers investigated gender differences in spatial visualization ability. In 1978, geneticists Steven Vandenburg and Allan Kuse developed a mental rotation test that has been used in part to quantify spatial visualization ability. In 1980, Camilla Benbow and Julian Stanley, referred to as psychologists and educators, asserted that gender differences in mathematics might result from "greater male ability in spatial tasks." Their state-

ments were widely publicized in the media. Later researchers found that visual training by video games or certain changes in testing conditions, like removing "I don't know" as an answer or eliminating time constraints, could reduce these observed gender differences. Research on stereotype vulnerability, where the effort to counter societal perceptions about a whisper of inferiority can negatively impact performance, has further complicated visualization research efforts.

Education

Various educational learning models and theories stress the importance of visualization. In Piaget's theory, named for epistemologist Jean Piaget, spatial skills develop at various age levels or stages and according to experience. For instance, he proposed that young children could understand two-dimensional space, while the mental manipulation of three-dimensional objects in space comes later on. Mathematician Walter Whiteley has proposed research questions related to visualization and suggested a variety of ways in which teachers might intentionally train students to "see like a mathematician." He noted:

> Curriculum suggests that 2-D is easier than 3-D, although it is cognitively less natural for many modes of reasoning, and 3-D skills are the needed goal for later work. The domination of analytic over synthetic reasoning encourages the pattern that 2-D is the starting point, and the disconnection between early childhood reasoning, and latter problem solving both of which engage 3-D reasoning.

The van Hiele model of geometric thought, developed by educators Dina van Hiele-Geldof and Pierre van Hiele, listed visualization as its first level. Additional learning models presented by mathematicians and educators have also stressed the importance of interweaving visualization training with other skills.

Further Reading

Barry, Susan. *Fixing My Gaze: A Scientist's Journey Into Seeing in Three Dimensions*. New York: Basic Books, 2009.

Clements, Douglas. "Subitizing: What Is It? Why Teach It?" *Teaching Children Mathematics* 5, no. 7 (1999).

Friendly, Michael. "Milestones in the History of Data Visualization: A Case Study in Statistical

Historiography." In *Classification: The Ubiquitous Challenge*. Edited by Claus Weihs and Wolfgang Gaul. New York: Springer, 2005.

Friendly, Michael, and Daniel Denis. "Milestones in the History of Thematic Cartography, Statistical Graphics, and Data Visualization." http://www.datavis .ca/milestones/.

Gallagher, Ann, and James Kaufman. *Gender Differences in Mathematics: An Integrative Psychological Approach*. New York: Cambridge University Press, 2005.

Hege, Hans-Christian, and Konrad Polthier. *Visualization and Mathematics: Experiments, Simulations and Environments*. Berlin: Springer, 2003.

Malcom, Grant. *Multidisciplinary Approaches to Visual Representations and Interpretations*. Amsterdam, Netherlands: Elsevier, 2004.

Nelson, Roger. *Proofs Without Words: Exercises in Visual Thinking*. Washington, DC: Mathematical Association of America, 1997.

Whiteley, Walter. "Visualization in Mathematics: Claims and Questions Towards a Research Program." *The 10th International Congress on Mathematical Education* (2004). http://www.math.yorku.ca/Who/ Faculty/Whiteley/Visualization.pdf.

Zimmerman, Walter, and Steve Cunningham. *Visualization in Teaching and Learning Mathematics*. Washington, DC: Mathematical Association of America, 1991.

SARAH J. GREENWALD
JILL E. THOMLEY

See Also: Animation and CGI; Coordinate Geometry; Graphs; Maps; Optical Illusions; Painting; Sculpture; Telescopes.

Volcanoes

Category: Weather, Nature, and Environment.
Fields of Study: Data Analysis and Probability; Geometry.
Summary: Mathematical models and data analysis can help geologists better understand the activity of volcanoes and the fluid dynamics of their eruptions.

Volcanoes are openings of channels connecting the molten interior of a planet with its surface. Active volcanoes emit magma, ash, and gasses, and inactive volcanoes are reminders of past eruptions, consisting of solidified lava and ash. The science of studying volcanoes is known as "volcanology." Many scientists and philosophers throughout history, including mathematicians Johannes Kepler and René Descartes, theorized about their nature and formation. Mathematics continues to play a role in modern volcanology through both the coursework and degrees that are required and in the mathematical research prevalent in the exploration of various volcanic phenomena. Computer-based numerical simulations and digital imagery, often from satellite observation, combined with mathematical and statistical methods, such as neural networks and data mining, are increasingly used to model, describe, and visualize the complex mathematical representations of volcanic processes. Predicting eruptions is also a challenge, which is necessary not only for safety and response at the time of the eruption but also for larger issues such as global climate change. Benjamin Santer of Lawrence Livermore National Laboratory, who specializes in mathematical and statistical analyses of climate data, has used volcanoes as one variable in explaining climate change. Scientists at the Yellowstone Volcano Observatory also collect data to monitor and mathematically study the enormous Yellowstone caldera, sometimes known as the Yellowstone supervolcano.

Measuring Volcanoes

The most destructive volcanic effect comes from pyroclastic flow, which is a mixture of solid to semi-solid fragments of rock, ash, and hot gases that flows down the sides of the volcano. It is a type of gravity current, similar to an avalanche, that can be modeled with theories and equations from fluid dynamics. A useful metric for comparing eruptions is the volume of volcanic ejecta. For example, the 1980 eruption of Mount St. Helens produced about 1.3 cubic kilometers of ash, but the ancient eruption of the Toba volcano on Sumatra around 75,000 years ago produced more than two thousand times more ash. It is possible to measure the fragmentation of the airborne volcanic matter, called "tephra," even for ancient eruptions. Fragmentation is associated with the strength of the volcanic explosion. The dispersion of tephra over an area has been found to be related to the height of the eruption column. Finding and analyzing dispersion allows estimation of heights for ancient eruptions and an additional way to measure heights for modern eruptions. Volcanologists have created the Volcanic Explosivity Index (VEI), which takes into account the volume of ash and the height and duration of the eruption. There are nine types of volcanoes according to VEI, scaled 0–8. For example, the low-strength, low-height Type 0 is called "Hawaiian," and the high-strength, low-fragmentation Type 6 through Type 8 are called "Plinian eruptions," named for Roman historian Pliny the Younger, who described in detail the first century eruption of Mount Vesuvius that destroyed Pompeii. Plinian eruptions can have global environmental effects. Similar to the Richter scale, VEI is logarithmic: each level type is about 10 times greater in magnitude than the previous level.

Geometry of Volcanoes

Shapes of volcanoes depend on their explositivy, viscosity of magma, the composition of the surrounding crust, and other geological factors. The familiar, iconic cone shape such as Mount Fuji defines a "stratovolcano," so named because of its many layers (or "strata") of ash and hardened lava. Eruptions of these volcanoes have high explosivity and low-viscosity lava, making lava and tephra deposit near the opening in layers of diminishing thickness, thus forming the cone.

In contrast, broad, very fluid lava fields produce shield volcanoes that resemble a rather flat warrior shield. Lava domes, as the name suggests, are proportionally higher than shield volcanoes and more rounded than cone volcanoes, resembling semispheres. Lava domes are formed by high viscosity lava combined with low explosivity, where lava either accumulates under the crust and pushes it up, or flows over the crust and solidifies in the dome shape.

Eruption Forecast

Because volcanic eruptions depend on many variables, eruption forecasting relates to such areas of science and mathematics as chaos theory and systems science. Overall, prediction means collecting multi-variate data in volcano observatories and matching variable patterns to those that occurred before eruptions of similar types of volcanoes in the past. For example, the pattern of earthquakes becoming stronger and shallower with time, called "earthquake swarm," can be used to forecast the eruption time. Mathematical models of volcanoes are

After the 1980 eruption of Mount St. Helens, 24 square miles were filled by a debris avalanche.

based on equations from thermodynamics, fluid dynamics, and solid mechanics. The systems science principles of prediction describe qualitative trends in variables. For example, the principle of coinciding change says that unrelated, co-evolving trends in several parameters are more significant than changes in any one parameter.

Further Reading

Marti, Joan, and Gerald Ernst. *Volcanoes and the Environment.* Cambridge, England: Cambridge University Press, 2005.

Zeilinga de Boer, Jelle, and Donalt Sanders. *Volcanoes in Human History.* Princeton, NJ: Princeton University Press, 2002.

MARIA DROUJKOVA

See Also: Earthquakes; Geothermal Energy; Measurement, Systems of; Measuring Tools; Plate Tectonics; Prehistory; Probability.

Volleyball

Category: Games, Sport, and Recreation.
Fields of Study: Data Analysis and Probability;
Geometry; Measurement.
Summary: Mathematics is fundamental to player
motion, strategy, and scoring in volleyball.

Volleyball, which began in the late nineteenth century
as a non-contact recreational sport, quickly developed
into a globally popular competitive sport. Two teams,
typically with two to six players, face one another on
opposite sides of a rectangular court divided by a net.
Beach volleyball is played on sand courts rather than
a hard surface. Game strategy uses mathematical concepts such as angles, rotation, and parabolic motion
in an effort to impart optimal trajectories, speeds, and
spins on the ball to prevent the other team from successfully returning it. The receiving team must understand three-dimensional motion and vectors in order to
intercept the ball and change its direction, often using
a sequence of hits coordinated among several players.
The strategies of beach volleyballers often differ from
those of hard court volleyballers because of differences
in the ability to jump or dive for an incoming ball.
Mathematics is also used to analyze and model body
kinetics, such as the motions of a player's shoulders and
arms while serving. Statistics are used to analyze and
describe both team and individual proficiencies and
success. These include measures like number of attacks,
kills, and assists; hitting percentages; and kill average
and efficiency as a function of total attempts.

General Game Play and Scoring

Volleyball teams work together to hit the ball over the net
in such a way as to prevent the other team from returning it. A match consists of three or five games. The third
game of a three-game match or the fifth game of a five-
game match is the deciding game. A single sequence of
back and forth hitting is known as a "rally," which begins
with one side serving the ball and ends when one team
or the other fails to legally return it. Each side gets three
attempts and the same player may not touch the ball
twice in a row. At the end of the rally, the winning side
may earn a point, the right to serve the ball, or both.

There are two different scoring systems used in volleyball. In side-out scoring, only the serving team may
earn a point. In rally scoring, either side earns a point.

Winners always get the serve. Deciding games are played
to 15 points; nondeciding games are played to 25. However, the winning team must be ahead by at least two
points or play continues. Sometimes, a scoring cap is
used, which nullifies this requirement. Statistical analyses show that rally point scoring makes matches shorter
and match lengths more predictable versus side-out
scoring. However, there appears to be no significant
effect on scoring margins between teams; on average,
after an even number of serve changes, points awarded
to non-serving teams balance. In addition to statistics,
Markov chains are useful for analyzing volleyball games
in terms of the proportion of points won and the probabilities of winning a point, game, and match.

Player Roles and Strategy

Hard court teams typically consist of six players with
specialized roles, with the left, center, and right forwards in a row along the frontcourt and the left, center,
and right backs in a row along the backcourt. However,
players usually rotate through positions during play,
requiring analysis of permutations and the timing of
substitutions. Beach volleyball teams typically consist
of two players each, generally front and back. Players
seek to control the ball through the angle, force, and
timing with which the ball is struck and by choosing
whether or not to impart spin on the ball. The volleyball typically travels along a parabolic path, modified by its spin and additionally influenced by player
efforts and external factors, such as air resistance. The
basic skills used in volleyball include the serve, pass, set,
spike, block, and dig. A variety of serves can be used as
the server hits the ball into the opponent's court. Different types of serves affect the ball's direction, speed,
and acceleration with the goal of increasing the difficulty of handling the ball for the opposing team. Serves
that have flatter parabolic paths tend to preserve more
of the initial force and velocity and are usually more
difficult to return.

The opposing team's first reception of the ball is
known as the "pass," the second contact is known as
the "set," and the third contact is known as the "attack"
(also called "spike"), though a team may not opt to
use all three contacts in every play. A block is a team's
attempt to prevent the opposite team from spiking the
ball into their court, and a dig is an attempt to prevent
a ball from hitting the court. Shots include the hard
angle, deep angle, seam shot, line shot, angled line shot,

swiping shot, high and hard, and the save. Achieving different shots relies on affecting the ball's speed, spin, and angle of trajectory through shoulder and hip positions, aiming at gaps between opposing players, and the amount of force applied. Spin tends to make the ball more difficult to return successfully, since the appropriate counterforce to control the ball and change its directional vector is more difficult to determine and apply quickly.

Further Reading

Calhoun, William, G. R. Dargahi-Noubary, and Yixun Shi. "Volleyball Scoring Systems." *Mathematics and Computer Education* (Winter 2002).

Kiernan, Denise. *Sports Math*. New York: Scholastic Professional, 1999.

USA Volleyball. *Volleyball Systems and Strategies*. Champaign, IL: Human Kinetics, 2009.

MARCELLA BUSH TREVINO

See Also: Curves; Kicking a Field Goal; Mathematical Modeling.

Voting

See *Elections*

Voting Methods

Category: Government, Politics, and History.
Fields of Study: Algebra; Number and Operations; Problem Solving.
Summary: Social choice theory concerns itself with the mechanics of group decisions such as elections and the impact methodology can have.

Voting theory (also known as "social choice theory") is concerned with how group decisions are made when there are a number of alternatives from which to choose (for example, finding the winner of an election). When there are only two options, voting is straightforward—

the winning alternative (also called the "social choice") should be the one that receives the most votes. However, when the choice is among three or more alternatives, determining the social choice is significantly more complex. There are many reasonable methods for selecting a winner and the methods can produce different winners even when given the same sets of votes. All voting methods have inherent flaws and, regardless of the method used, strange and paradoxical situations can occur. For example, in the 2000 U.S. presidential election, George W. Bush and Al Gore were major party candidates, while Ralph Nader, representing the Green Party, had much less support. Although Bush won the election, exit polls at the time indicate that had Nader not been on the ballot in some states, Gore almost surely would have won the election. In other words, in the U.S. electoral system, the presence (or lack thereof) of an "also-ran" candidate can have a profound outcome on the winner. This disturbing property is one of many that interests mathematicians, economists, and political scientists who study voting theory.

Preference Ballots

Preference ballots, where voters rank the alternatives in order of preference, are among the most useful ways of gathering information from voters. A voting method aggregates these preferences in some way and determines a social choice (or choices, in the case of ties). In this way, a voting method can be thought of as a function whose typical input is a set of individual ballots and whose output is the winning alternative, or—in the case of a social welfare function—a ranking of the alternatives, perhaps with ties. Many such functions are possible:

- *Plurality method:* A procedure that returns as the social choice the alternative that is the top preference on the most ballots (the candidate with the most first place votes).
- *Weighted voting method:* Also called the "positional method," this process assigns points to an alternative based on its position on a ballot, with higher placings on a ballot earning more points. The winning alternative is the one having the most points.
- *Borda count:* A special positional method whereupon a voter's lowest-ranked alternative earns zero points, the voter's second lowest-

ranked alternative earns one point, and so on, with the voter's top choice earning $n-1$ points, assuming n candidates.

- *Hare system:* Also called "instant runoff voting" or "plurality with elimination," this method arrives at the social choice by successively eliminating less desirable outcomes. In this procedure, ballot-counting proceeds in rounds, with the candidate having the fewest first-place votes eliminated at the end of each round. A ballot on which an eliminated candidate was the top choice has its vote transferred to the highest ranking remaining candidate on the ballot. The process of elimination continues until one candidate has more than half the first place votes (a "majority"), in which case that candidate is declared the winner.
- *Dictatorship:* In a dictatorship, one voter is specially designated so that the social choice is always the alternative that this voter has at the top of his or her ballot.

For example, suppose that there are 100 voters in an election, and three candidates (A, B, and C). Suppose that the voters express their votes as shown in the following table:

Number of Voters	40	35	25
1st Choice	A	B	C
2nd Choice	C	C	B
3rd Choice	B	A	A

Note that 40 of the voters prefer A as their top choice, 35 prefer B as their top choice, and 25 prefer C as their top choice. If this election were decided using the plurality method, then candidate A would win with 40 first place votes (with B and C earning 35 and 25 first place votes, respectively). Using the Borda count, A would tally $40 \times 2 = 80$ points, B would earn $\left(35 \times 2\right) + 25 = 95$ points, and C would win with $\left(25 \times 2\right) + 75 = 125$ points. Using the Hare system, candidate C would be eliminated in Round 1, and C's votes would transfer to candidate B, because B is second on all 25 ballots. In Round 2, B has 60 first place votes to A's 40, so B is the winner.

This example demonstrates that different methods can yield different results. As Donald Saari writes, "Rather than reflecting the voters' preferences, the outcome may more accurately reflect which election procedure was used."

It should be noted that there are other methods of voting that do not require preference ballots. In a system called "approval voting," a voter may vote for as many candidates as desired. The winner is the candidate receiving the most votes. No distinction is made among the candidates of which the voter "approves," and the voter can vote for any combination of the candidates.

Fairness

By aggregating voters' preferences and producing a social choice, an election method should reflect, in some way, the will of the people. Given the vast library of possible election methods, it is natural to ask whether there is a method that captures this will in an ideal way. Social choice experts have developed different ways of assessing the quality of voting methods, and the notion of "fairness" has emerged as a prime consideration. When there are two alternatives, it can be expected that any reasonable voting method will be anonymous (all voters are treated equally), neutral (the two candidates are treated equally), and monotonic (if a voter changes his vote from candidate A to candidate B, then that should not hurt candidate B). Mathematician Kenneth May proved in 1952 that if the number of voters is odd and ties are not allowed, then only one voting method is anonymous, neutral, and monotonic: "majority rule," the procedure where the candidate with more than half the first place votes is declared the winner.

When there are three or more alternatives, there are many desirable properties for voting methods. The following list is far from exhaustive:

Majority criterion: This method requires that when some alternative is the first choice on more than half the ballots, that alternative should be the social choice. The plurality method satisfies this criterion, for if a candidate has a majority, no other candidate can have as many first place votes. On the other hand, the Borda count violates the majority criterion, for there are elections where a candidate can have a majority but still lose.

Condorcet winner criterion: This is a slightly weaker condition: if an alternative is preferred head-to-head over every other alternative in a one-on-one matchup that ignores the other alternatives, then that candidate should win the election. The example above shows that the plurality method violates this criterion. While candidate C is preferred over A on 60 of the ballots, and C is preferred over B on 65 of the ballots, C loses the plurality election to A. The Hare system and the Borda count fail the Condorcet winner criterion as well.

Pareto condition: This method asserts that for every pair x and y of candidates, if all voters prefer x to y, then y should not be a social choice. This is a relatively weak criterion, and all of the methods described above satisfy it.

Monotonicity criterion: According to this method, if x is a social choice and someone changes a ballot in such a way that x is moved up one spot (in other words, x exchanged with the alternative immediately above x on the ballot), then x should still be a social choice. In other words, making a change to a ballot that is favorable only to a winning candidate should not hurt the candidate. The plurality method and the positional voting methods satisfy monotonicity, but the Hare system does not.

Independence of Irrelevant Alternatives: Also called "binary independence," this method states that if x is a social choice while y is not, and if a voter changes a ballot in a way that does not change the relative positions of x and y on the ballot, then y should still not be a social choice. In other words, changing the positions of other "irrelevant" candidates on a ballot should not affect the relative position of x over y or y over x in the outcome. This is precisely the difficulty that occurred in the 2000 U.S. presidential election, where Nader's presence in the election affected the relative rankings of Bush and Gore.

Although each of these criteria is, in turn, a reasonable expectation of a voting method, Kenneth Arrow, in 1952, proved the mutual exclusivity of them. In his "impossibility theorem," Arrow showed that if there are at least three alternatives and a finite number of voters, then the only social welfare function that satisfies both the Pareto condition and "independence of irrelevant alternatives" is a dictatorship. This profound result, which earned Arrow the Nobel Prize in Economics in 1972, argues against the possibility of a theoretically perfect democracy. Nevertheless, Arrow himself encourages continuing to search for voting methods that work well most of the time. He writes:

My theorem is not a completely destructive or negative feature any more than the second law of thermodynamics means that people don't work on improving the efficiency of engines. We're told that you'll never get 100% efficient engines . . . It doesn't mean you wouldn't like to go from 40% to 50%.

Sincere and Strategic Voting

Strategic voting is the practice of voting against one's true preferences in order to achieve a better outcome in an election. This contrasts with sincere voting, where one votes according to one's true preferences. Strategic voting most often occurs in situations where a voter's preferred candidate has little chance of winning, or where the voter's top candidate is most threatened by his second or third candidate. While strategic voting can affect the outcome of an election, its effects can be disastrous. Election results should reflect the aggregate will of the people, and if voters do not express their individual preferences truthfully, then the voting method has little hope of determining the socially desired outcome. Therefore, voting methods that tend to encourage strategic voting are unattractive. It should be noted that for strategic voting to be at all effective, there must be at least three candidates in the election, and the voters need a thorough understanding of both the voting method being used and the preferences of other voters.

For example, in the 2000 election, exit polls in Florida indicated that Nader voters widely supported Gore as their second choice, far beyond both the margin of error of the polls and Bush's margin of victory. Had these voters instead voted strategically for Gore, Gore would likely have carried Florida and its 25 electoral votes, thereby winning the presidency. The U.S. electoral college notwithstanding, this shows how powerfully the plurality method encourages strategic voting. Had the 2000 election been decided by the Borda count, one could imagine that a conservative voter might have Bush as the top choice, Gore as the second choice, and Nader last, but might insincerely rank the candidates in the sequence Bush, Nader, Gore in an attempt to maximize the point differential between Bush and Gore.

Some voting methods prove resistant to strategic voting. One of the major advantages of the Hare system is that it tends to encourage sincere voting. In the 2000 election, for example, a Nader supporter would have less reason to vote strategically for Gore if it is known that the vote will transfer to Gore should Nader be eliminated. Nevertheless, there are situations where even with the Hare system, strategic voting can prove beneficial to a voter.

In the 1970s, Allan Gibbard and Mark Satterthwaite proved that no voting method is completely immune to strategic voting. Any non-dictatorial system that uses preference ballots and allows at least the possibility of any candidate winning will necessarily lead to situations, however hypothetical, where strategic voting can be beneficial. This proof serves as a result analogous to Arrow's, but in the realm of strategic voting. As with Arrow's result in fairness, it is important to note that the degree to which a voting method encourages sincerity still serves as an important criterion for selection.

Further Reading

Brams, Steven. *Mathematics and Democracy: Designing Better Voting and Fair-Division Procedures*. Princeton, NJ: Princeton University Press, 2007.

Saari, Donald. *Basic Geometry of Voting*. New York: Springer, 2003.

Taylor, Alan, and Allison Pacelli. *Mathematics and Politics: Strategy, Voting, Power and Proof*. New York: Springer, 2008.

STEPHEN SZYDLIK

See Also: Elections; Government and State Legislation; Rankings.

Water Distribution

Category: Architecture and Engineering.
Fields of Study: Algebra; Data Analysis and Probability; Geometry; Measurement; Number and Operations; Problem Solving.
Summary: Mathematicians have long studied issues related to optimizing water distribution.

Water distribution has two separate but interrelated meanings: the natural physical distribution of water in the world and the way in which people choose to distribute available water. In some regions, accessing and distributing fresh water for human needs, like drinking and irrigation, can be a significant challenge. Roughly 70% of the Earth's surface is covered with water but most is saline (salty). Much of Earth's fresh water is in glaciers or underground. Some is polluted from human activities. In the early twenty-first century, approximately 20% of Earth's population lived in areas with insufficient fresh water because of climate or geography. About the same number lived in areas in which water existed but where technological or economic barriers limited effective distribution. Many systems have been devised throughout history and in different societies to access and distribute water. It is so valuable a resource that armed conflicts been fought over water. Mathematicians, scientists, and others who work on water distribution problems use mathematical techniques to design, build, optimize, and monitor water distribution and associated wastewater systems. For example, graph theory is used to model water distribution networks. Graph edges may represent pipes and nodes represent intersections, junctions, and access points. Statistical and topological methods can be used to compare networks in terms of capacity and reliability against failure.

Irrigation

Irrigation is an ancient practice that allows food to be grown where it might otherwise not thrive. Evidence shows that it was used as early as the sixth millennium B.C.E. in Mesopotamia, Egypt, and Persia, and the fifth millennium B.C.E. in South America. In the early twenty-first century, agriculture is still globally the greatest consumer of fresh water, though it varies widely by location. For example, the United Kingdom's abundant rainfall means that it requires almost no irrigation. Mexico and India, on the other hand, use it extensively. The green revolution of the twentieth century, which greatly increased the agricultural yield of many developing countries, relied in part on irrigation. One criticism was that the increased food production in these areas resulted in accelerated population growth that placed further burdens on scarce water resources. This criticism is supported by some statistics and mathematical models, which show that the demand for water grew at rates that exceeded population increases, raising per-capita water requirements.

Mathematicians and others who study ancient systems of irrigation in order to better understand them (and perhaps improve modern methods) have noted that some societies appear to have created and implemented complex and efficient water distribution methods without using mathematical methods for planning. Others have sought to build mathematical models of irrigation systems. The paddy field system used for growing rice generally requires the creation of intricate structures of terraces, canals, and reservoirs in order to ensure that all fields receive adequate water. It is believed to have been used as early as 4000–3500 B.C.E. in China and Korea. Researchers who have investigated mathematical models to describe a paddy field system have noted that it may not be possible to create a reliable model by including only variables based on physical measures such as amount of water available and rate of evaporation. A variable describing an ethic of cooperation among owners of the various fields, a factor that is difficult to quantify, was also required to ensure that water would be used fairly. For example, if owners on the upstream end of a water source took more than their fair shares, the owners farther downstream would not receive sufficient water for their crop, regardless of the values of some other variables.

Industry

Industry is the second largest category of global water use. Most industrial processes need water in some way, though some are more readily visible, such as hydropower generation of electricity and water extraction of minerals in mining. At the start of the twenty-first century, per capita water use is typically higher in industrialized nations than in developing countries, though this gap is closing. Some economists use the term "virtual water" to refer to the water that is used in the entire chain of manufacturing a product or growing an agriculture commodity. Similar to a carbon footprint, which is often used to quantify the quantity of greenhouse gasses emitted by a process, a water footprint represents the total amount of water used to create a good or service. Calculating water footprints provides an additional metric for assessing and comparing the environmental impact of competing products and services. For example, in 2010, the Water Footprint Network estimated that production of 1 kilogram of beef required about 16,000 liters of water, while one kilogram of rice required 3000 liters of water, and one liter of milk required 1000 liters of water.

Sanitation

The creation of sanitary systems of water supply and wastewater disposal or treatment is a major factor in the general improvement of public health from about the mid-nineteenth century onwards. Large cities, such as London, New York, and Boston, were among the first to establish municipal water supply systems. They were motivated in part by data collected by statisticians and others such as physician John Snow, who demonstrated via statistical methods that an 1854 cholera outbreak in London could be traced to the local water pump.

Mathematical methods may be used to model different aspects of supply systems. The fluid pressure necessary for water to flow through a system is affected by variables like gravity. Water stored in a rooftop tank will deliver water at a higher pressure to lower floors versus higher ones. Mathematical calculations show that a vertical foot of water exerts a pressure of 0.433 pounds per square inch (psi) at its bottom surface. The flow of water through the system is a function of the cross-sectional area of the pipe: $Q = A \times V$, where Q is the flow of water through the system, A is the cross-sectional area of the pipe, and V is the velocity of the water.

Municipal water systems tend to be quite complex, involving massive networks of storage tanks, pipes, pumps, and valves. Mathematical models are used to describe and manage these systems. Navier–Stokes equations, named for mathematicians Claude-Louis Navier and George Gabriel Stokes, are partial differential equations that describe fluid flow and velocity, while the Reynolds number, named for mathematician Osborne Reynolds, quantifies "laminar" (smooth) and turbulent fluid flow through a pipe. Contamination is an ever-present risk because of the natural physical deterioration of system components over time (such as corroded pipes) as well as the possibility of accidental or deliberate introduction of contaminants. Researchers are developing systems that can sense when a contaminant has been introduced into the water distribution system, allowing for rapid identification of the time and location of its introduction. For example, experiments done by the U.S. Environmental Protection Agency showed statistically that chlorine and total organic carbon, which are routinely monitored in municipal water systems, were sensitive and reliable predictors of contamination.

Further Reading

Cohen, Y. Koby. *Problems in Water Distribution: Solved, Explained, and Applied.* Boca Raton, FL: CRC Press, 2002.

Gates, James. *Applied Math for Water Distribution, Treatment, and Wastewater Operators.* Dubuque, IA: Kendall Hunt Publishing, 2010.

Sarah Boslaugh

See Also: Canals; Carbon Footprint; Farming; Floods; Tides and Waves; Water Quality.

Water Quality

Category: Weather, Nature, and Environment.
Fields of Study: Algebra; Data Analysis and Probability; Measurement.
Summary: Water quality standards and data are mathematically modeled and analyzed to help keep drinking water safe.

Water is fundamental for human life. Approximately 70% of the Earth's surface is covered by water but only a very small fraction is consumable fresh water, and much of that has chemical or biological contaminants. Drinking water comes from a variety of sources. Underground water, such as aquifers or springs, may be tapped by wells; surface water, such as rivers and streams, are diverted for use; precipitation may be collected or allowed to flow into other sources; and plants may be processed for moisture. Desalinization (the process of removing salt from water) makes seawater drinkable. Waterborne diseases in open water sources like rivers are endemic to many parts of the developing world. Natural disasters may spread contamination via flooding. Some global warming researchers predict that increased rainfall, flooding, and warmer weather will result in more waterborne disease worldwide. In developed countries, water is commonly piped to end users and may be recycled via sewage treatment. The standards for potable water in many countries are set by government agencies, though the regulation of bottled water differs from piped and well water. Even in nations with extensive closed water distribution systems and sewage treatment, contamination occurs in a number of ways, including agricultural runoff, dumping of manufacturing byproducts into streams and rivers, and degradation of systems that may contain outdated materials such as lead. One of the Millennium Development Goals adopted by the United Nations and other international organizations is to cut in half the proportion of people that do not have reliable access to safe drinking water by 2015. Mathematicians and mathematical methods contribute significantly to the discovery, testing, and delivery of potable water.

How Safe is Your Drinking Water?

The Environmental Protection Agency (EPA) sets the standards for drinking water in the United States. For each potentially harmful substance, the EPA identifies the maximum contaminant level (MCL) allowed and the maximum contaminant level goal (MCLG). The MCLG is the level below which there is zero expected risk to human health. While it would be best to have levels of a substance like arsenic at or below the MCLG, the EPA sets MLC requirements at concentrations that can be higher. U.S. citizens who receive water from a community water system should receive a Water Quality Report each year.

Providing access to safe drinking water around the world is one of the goals of the United Nations.

Those curious about water quality at work may request a copy from the building owner. Each report includes the source of the water (such as a river or lake); a list of all detected regulated contaminants and their levels; potential health effects of contaminants detected that violate the standards; information for people with weakened immune systems; and contact information for the company or agency that supplies the water. The report will alert the public to violations of the EPA safe drinking water standards and, equally important, will list information about potentially harmful substances that are below the legal limit. For example, a report may list arsenic, describe that it is measured in parts per billion (ppb), give the highest level measured, and list the range measured in the water. The report will also provide the MCLG (0.0 ppb for Arsenic) and the MCL (10.0 ppb). If the report states that the water ranges from 0.5 to 2 ppb for arsenic, water consumers will know that it is safe to drink according to EPA standards. However, upon comparing the MCL and MCLG, consumers may consider drinking water from other sources or request additional information from the water company since 0.5 ppb is higher than the 0.0 ppb MCLG.

Mathematical Analysis and Modeling

The management of water resources is increasingly reliant on mathematical modeling and analysis. For example, the dynamics and kinetics of surface water, along with distributions and dispersal over time of contaminants, have been extensively modeled and simulated. Reactive transport (RT) models use coupled equations to examine particle transportation through porous surfaces, which are widely used to model infiltration of contaminants into ground water. They may utilize mathematical and statistical concepts such as stochastic differential equations, which can be traced in part to physicist Paul Langevin's work on the mathematical theories of dynamic molecular systems. Animal behavioral responses to variables like water quality have been successfully modeled using the Eulerian–Lagrangian–Agent Method (ELAM). The Eulerian framework, named for mathematician Leonhard Euler, mathematically models environment factors affecting the animal agents, while the Lagrangian framework, named for mathematician Joseph Lagrange, governs the perception and movement of individual agents.

Near-continuous water quality monitoring provides a wealth of data and facilitates time series analyses and other statistical models of water quality as functions of variables like land use and precipitation patterns, as well as other measurable human behaviors and natural occurrences. Model calibration, verification, and sensitivity analysis often require comparing mathematical equations and simulation results with observed data. Mathematicians, engineers, and scientists have improved systems for remote water quality monitoring and assessment using data, mathematical methods, and theories from many sciences. Some applications include remote automated stations with the ability to wirelessly network and transmit data, artificial intelligence algorithms that can adaptively sample in response to problems or concerns, and satellite or aircraft observation and analysis of large areas.

These analyses also influence public policy and legislation, such as the U.S. Safe Drinking Water Act and the Clean Water Act. Scientists in many fields continue to seek methods to provide easily accessible clean water for everyone.

Further Reading

Chapra, Steven. *Surface Water-Quality Modeling*. Long Grove, IL: Waveland Press, 2008

U.S. Environmental Protection Agency (EPA). *Drinking Water and Health: What You Need to Know!* Washington, DC: EPA, 1999. http://www.epa.gov/safewater/dwh/dw-health.pdf.

Christine Klein

See Also: Farming; Floods; Water Distribution.

Waves

See *Tides and Waves*

Weather Forecasting

Category: Weather, Nature, and Environment.
Fields of Study: Connections; Data Analysis and Probability; Problem Solving; Representations.

Summary: Accurate weather forecasting requires the use of advanced mathematical models and powerful supercomputers to handle the vast number of calculations.

Weather prediction, or forecasting, is the application of science and technology to predict the future state of the atmosphere at a given location using available past and present data from the surrounding area. The word "weather" describes the state of the atmosphere at a particular time, or short time period, while the word "climate" is an average of these conditions over long time periods—often months or years. The weather is typically described in terms of temperature, wind speed, wind direction, air pressure, density, and atmospheric composition (for example, water vapor, liquid water, or carbon dioxide content). The intensity of solar and terrestrially emitted radiations is also a fundamental determining factor. A forecast typically includes the prediction of these meteorological variables and helps people make more informed daily decisions that may be affected by the weather. Moreover, it helps predict dangerous weather phenomena, such as hurricanes, which might endanger human life.

History

People have tried to forecast the weather for thousands of years and throughout history, farmers, hunters, warriors, shepherds, and sailors understood the importance of accurate weather predictions for planning daily activities. Ancient civilizations appealed to the gods of the sky: the Egyptians looked to Ra, the sun god; the Greeks sought out Zeus; and in the ancient Nordic culture, Thor was believed to govern the air with its thunder, lightning, wind, rain, and fair weather. The Aztecs used human sacrifice to satisfy the rain god, Tlaloc, while Native American and Australian aborigines performed rain dances.

The Babylonians were predicting the weather from cloud patterns as well as astrology by 650 B.C.E., but the earliest scientific approach to weather prediction occurred circa 340 B.C.E. when Aristotle described his theories about the earth sciences and weather patterns in *Meteorologica*. The ancient Greeks invented the term "meteorology," which derives from the Greek word *meteoron* which refers to any phenomenon in the sky. The Greek philosopher Theophrastus, one of Aristotle's successors, compiled the ultimate weather text *The Book of Signs*, which contained a collection of weather lore and forecast signs and served as the definitive weather book for over 2000 years.

Weather forecasting advanced little from these ancient times to the Renaissance. Beginning in the fifteenth century, Leonardo da Vinci designed an instrument for measuring humidity, Galileo Galilei invented the thermometer, and his student Evangelista Torricelli came up with the barometer. With these tools, people could objectively monitor the atmosphere. In 1687, Sir Isaac Newton published the physics and mathematics that govern the motion of all bodies and can be used to accurately describe the atmosphere. To this day, his principles are the foundation for modern mathematical analysis and computer prediction of weather.

However, scientifically accurate weather forecasting was not feasible until the early twentieth century, when meteorologists were able to collect and organize data about current weather conditions from observation stations in a timely fashion. Vilhelm and Jacob Bjerknes developed a weather station network in the 1920s that allowed for the collection of regional weather data. The data collected by the network could be transmitted nearly instantaneously by use of the telegraph, invented in the 1830s by Samuel F. B. Morse. This system allowed knowledge of the weather conditions upwind to be incorporated into downwind forecasts, improving their quality.

Great progress was made in the science of meteorology during the twentieth century. The possibility of numerical weather prediction was proposed by Lewis Fry Richardson in 1922, although computers did not yet exist. It was consequently impossible to perform the vast number of calculations required to produce a forecast before the predicted events actually occurred. Practical use of numerical weather prediction began in 1955, spurred by the development of programmable electronic computers.

Numerical Weather Prediction

Numerical weather prediction is the science of forecasting weather using computer simulations built from mathematical models. In this process, the atmosphere is divided into a three-dimensional lattice of grid points, and at each point the various atmospheric variables of interest are represented. These values are initialized with a state determined through analysis of past and present conditions. This state is then evolved

Weather forecasters prepare their forecasts at PC workstations with weather analysis software.

forward into the future by solving, at each grid point, the classical laws of (fluid) mechanics and thermodynamics, which are known to accurately approximate the behavior of the atmosphere. The output from the model provides the basis of the weather forecast.

The equations that govern how the state of a fluid changes with time contain many variables and require a great deal of computer processing resources to solve. Weather prediction centers have access to supercomputers containing thousands of processors on which to run a forecasting model. The required calculations are shared among the processors and computed simultaneously to produce a complete forecast in a fraction of the time possible with a single computer. This system is essential to ensure that an accurate prediction can be made within a useful time frame.

Good weather forecasts depend upon an accurate knowledge of the current state of the weather system, also called the "starting point" or "initial condition." The initial conditions are determined from global measurements of the state of the atmosphere. Surface weather observations of atmospheric pressure, temperature, wind speed, wind direction, humidity, and precipitation are made near the Earth's surface by trained observers, automatic weather stations, or buoys. The initial state has a degree of uncertainty since there are an insufficient number of measurements to initialize all meteorological variables at every grid point. Furthermore, the locations of the measurements do not usually coincide with the numerical grid points and there is also a degree of error in the actual measure-

ment. The problem of determining the initial conditions for a forecast model is very important, highly complex, and has become a science in itself (known as "data assimilation").

The atmosphere is an incredibly complex dynamical system and the approximation of its behavior is only compounded by the inability to measure its state at each and every grid point in the model. The limit on useful weather forecasts using present technology is typically one week. The forecast errors are initially localized, leading to incorrect predictions in small regions, but are generally accurate enough to be useful in most of the forecast area. The longer the simulation is run, the more the measurement and model approximation errors begin to dominate the calculation. However, steady improvements in computer power and prediction models in the twenty-first century have led to a three-day forecast being as accurate as a two-day forecast from the 1990s. Weather forecasting centers are constantly reviewing the accuracy of their forecasts and set themselves annual targets for accuracy improvements.

The raw output from the simulation is often modified before being presented as a forecast. Modifications include either the use of statistical techniques to remove known biases in the model or adjustments to take into account consensus among other numerical weather predictions. Accurate forecasts of precipitation for a specific location are particularly challenging because of the chance that the rainfall may fall in a slightly different place (such as several kilometers away) or at a slightly different time than the model forecasts, even if the overall quantity of precipitation is correct. Therefore, daily forecasts give fairly precise temperatures but put probabilistic values on quantities such as rain, based on knowledge of the uncertainty factors in the forecast.

Probability of Precipitation

A Probability of Precipitation (PoP) is a formal measure of the likelihood of precipitation that is often published from weather forecasting models, although its definition varies. In U.S. weather forecasting, PoP is the probability that greater than 1/100th of an inch of precipitation will fall in a single spot, averaged over the forecast area. For instance, if there is a 100% probability of rain covering one side of a city and a 0% probability of rain on the other side of the city, the PoP would be 50%. A 50% chance of a rainstorm covering the entire city would also lead to a PoP of 50%. The mathematical definition of

PoP is defined as $PoP = C \times A \times 100$, where C is the confidence that precipitation will occur somewhere in the forecast area, and A is the percent of the area that will receive measurable precipitation, if it occurs at all.

For example, a forecaster may be 40% confident that precipitation will occur and that, should rain happen to occur, it will happen over 80% of the area. This results in a PoP of 32%: $0.4 \times 0.8 \times 100 = 32\%$.

The Future
Over the years, the quality of the models and methods for integrating atmospheric observations has improved continuously, resulting in major forecasting improvements. The power of supercomputers has increased dramatically, allowing for the use of much more detailed numerical grids and fewer approximations in the operational atmospheric models. Small-scale physical processes (such as clouds, precipitation, turbulent transfers of heat, moisture, momentum, and radiation) have been more accurately represented within the model. Finally, the use of increasingly accurate methods of data assimilation and the integration of satellite and aircraft observations has resulted in improved initial conditions for the models, which ultimately lead to a better forecast.

Further Reading
Kalnay, Eugenia. *Atmospheric Modeling, Data Assimilation and Predictability*. Cambridge, England: Cambridge University Press, 2003.
Pasini, Antonello. *From Observations to Simulations: A Conceptual Introduction to Weather and Climate Modeling*. Singapore: World Scientific Publishing, 2005.

Silvia Liverani

See Also: Climate Change; Data Analysis and Probability in Society; Forecasting; Parallel Processing; Statistics Education; Temperature; Weather Scales.

Weather Scales

Category: Weather, Nature, and Environment.
Fields of Study: Algebra; Connections; Measurement.

Summary: Weather scales and tools are used to help measure and classify atmospheric conditions.

Weather affects virtually every aspect of human life, including afternoon showers that might inconvenience commuters; tremendously destructive episodes, like hurricanes; and long-term occurrences, like drought, which impact agriculture and increase the likelihood of other events like wildfires. Meteorology is an interdisciplinary science that focuses on weather and short-term forecasts, typically up to a few weeks. Climatology is a science that looks at long-term average weather. In fact, many define the word "climate" in terms of the average of weather over time, both locally and globally. Mathematics plays a critical role in weather science, enabling people to quantify, compare, model, and predict weather. Valid and reliable comparisons are facilitated by the development of scales and standard systems of quantification, along with mathematical and statistical models that use those measures.

It is thought that some ancient peoples had methods for predicting the weather, though historical evidence is mixed. In the early twentieth century, mathematician Vilhelm Bjerknes and colleagues examined several measurable variables of weather and derived equations to connect them to one another. Mathematician Lewis Richardson, who contributed significantly to mathematical weather prediction and pioneered the use of finite differences in the field, reformulated the Bjerknes equations. However, they remained impractical for rapid forecasting until the introduction of computers. Another product of his work, the Richardson number, is a function of density and velocity gradients that helps predict fluid turbulence in weather and other applications. Mathematicians continue to contribute and modern forecasting involves a wide variety of mathematical techniques and models, drawing in depth from such areas as chaos theory, data assimilation, statistical analyses, scale cascades of error (related to the so-called butterfly effect), numerical analysis, vectors, fluid dynamics, and entropy. Climatologists, scientists, and mathematicians also research related phenomena like geomagnetic and solar storms.

Temperature, Pressure, and Humidity
One of the most pervasive and intuitively obvious variables used to characterize the weather is air temperature—along with air pressure and humidity in

most modern reports and forecasts. Strictly speaking, air temperature is a measure of the average kinetic energy of the air molecules, measured by a variety of types of thermometers. The most common scales used to quantify temperature are the Celsius (or centigrade) scale used throughout most of the world and the Fahrenheit scale used primarily in the United States. Atmospheric pressure is measured by a barometer, whose invention is attributed to various sources including Galileo Galilei and mathematicians Gasparo Berti and Evangelista Torricelli.

There are many common units for pressure, including inches of mercury, pounds per square inch, pascals, named for mathematician Blaise Pacsal, and atmospheres. One atmosphere is defined as the mean atmospheric pressure at mean sea level, originally measured with respect to the latitude of Paris, France. Millibars are often used in weather reports and forecasts. A hygrometer measures the amount of water vapor in the air. How much water vapor the air can hold is a function of temperature and relative humidity expresses the quantity of water vapor as a unitless fraction or percentage of the possible amount of water for a given temperature.

Humidity can be used in probability models to predict precipitation, dew, and fog. Further, high humidity changes the subjective feeling of the air temperature for people because high humidity reduces the evaporation of sweat. This effect is quantified as a heat index, with assumptions about many variables such as wind speed, body mass, clothing, physical activity, and exposure to sunlight. A similar concept is wind chill, which relates the subjective perception of cold. Scientist Robert Steadman has researched and mathematically modeled both of these effects and they have become a common part of weather forecasts.

Wind

Another weather variable is wind speed. In 1805, Sir Francis Beaufort, an Irish hydrographer, developed what is now called the Beaufort scale to describe and categorize the strength of the wind. The scale has 13 points ranging from zero (calm air) to 12 (hurricane-force winds). On the scale, the Beaufort number two is identified as a "light breeze," with wind speed 6–11 kilometers per hour (km/hr) producing wind that is felt on the face, leaves that rustle, movement of a wind vane, and on the water, small, short wavelets that do not break. Further along the scale is Beaufort number five, a "fresh breeze," with wind speeds between 29 and 38 km/hr. At this point, small, leafy trees will sway, moderate waves become longer, and there are many whitecaps and some spray. Wind speeds between 62 and 74 km/hr are classified as a "gale," Beaufort number eight. Twigs and small branches break off trees. At sea, there are moderately high waves of greater length. Beaufort number 10 is used when wind speeds are between 89 and 102 km/hr and are "storm-force" winds. Trees are broken and uprooted and structural damage occurs. At sea, there are very high waves with overhanging crests and visibility is reduced.

Table 1: Fujita scale of tornado strength.

Scale	Wind Speed (km/hr)	Damage
F-0	65–118	Light
F-1	119–181	Moderate
F-2	182–253	Considerable
F-3	254–332	Severe
F-4	333–419	Devastating
F-5	420–513	Incredible

Table 2. Saffir-Simpson scale of hurricane strength.

Scale Number	Wind Speed (km/hr)	Storm Surge (meters)	Central Pressure (millibars)	Damage
1	121–154	1–2	≥ 980	Minimal
2	155–178	2–3	965–979	Moderate
3	179–210	3–4	945–964	Extensive
4	211–250	4–6	920–944	Extreme
5	>250	>6	<920	Catastrophic

The terms and descriptions make it clear that as wind speed rises so does its destructive power. In fact, the force exerted by wind increases as the square of the velocity such that a doubling of the wind's velocity leads to a quadrupling of the force: $F \sim V^2$. Some of the most powerful winds experienced on Earth are found in hurricanes and tornadoes. Their destructive power can be astounding and has been the subject of much study and research. The Fujita scale, presented in Table 1, is used to categorize tornado strength in terms of rotational wind speed (given in km/hr) and damage inflicted by the wind. While tornadoes are generally associated with severe thunderstorms and are seldom more than 1.5 km in diameter, hurricanes can involve whole systems of thunderstorms and may be several hundred kilometers in diameter. The Saffir–Simpson scale, used to categorize hurricanes, is presented in Table 2.

Further Reading

Ahrens, Donald C. *Meteorology Today.* Belmont, CA: Thompson Brooks/Cole, 2007.

Lynch, Peter. *The Emergence of Numerical Weather Prediction: Richardson's Dream.* Cambridge, England: Cambridge University Press, 2006.

Moran, Joseph P., and Lewis W. Morgan. *Meteorology: The Atmosphere and the Science of Weather.* Edina, MN: Burgess Publishing, 1986.

Mark Roddy
Sarah J. Greenwald
Jill E. Thomley

See Also: Clouds; Doppler Radar; Hurricanes and Tornados; Temperature; Weather Forecasting.

Weightless Flight

Category: Travel and Transportation.
Fields of Study: Algebra; Geometry.
Summary: The forces to experience the sensation of weightlessness, or zero-*G*, can be calculated and achieved in a variety ways.

Gravity is the mutual attraction of two masses. Important aspects of the mathematics and the theory of gravity were described centuries ago by Galileo Galilei and Isaac Newton. Albert Einstein's work was critical to the modern understanding of gravity and weightlessness. Mass is the measure of the amount of matter in an object. For living beings, weight can be thought of as the subjective experience of muscles resisting the pull of the much larger Earth on their smaller masses. On the Earth's surface, gravitational acceleration is about 9.8 meters per second (one gravity or *g*). Other planets have different gravity. For example, an Earth person would feel about 2.5 times heavier on Jupiter. Infants learn to accommodate gravity's pull when performing the activities of daily life until the force feels natural and largely unnoticed. However, sometimes people experience other forces acting on their bodies that counter the pull of gravity and change their perceptions of weight. For example, the quick start or stop of an elevator can make a person feel heavier or lighter. Roller coasters purposely induce similar effects for amusement. Parabolic drops, turns, and loops exert temporary linear or angular forces on a moving body, some of which act along a different directional vector than gravity and combine mathematically to alter the body's perception of weight. Mathematicians, scientists, and engineers precisely calculate the net effect of gravity and other forces on objects for a wide range of applications, such as banked curves on racetracks and highways, the movement of subatomic particles, launching spacecraft to the moon, and of course, ever more thrilling amusement park rides.

Zero-*G*

The planet's mass exerts a strong gravitational pull even on objects in space. This force is what keeps satellites in position. However, many people have seen video images of astronauts who are floating around as if they are weightless. This effect is known as *zero-G* or, more accurately, "microgravity" (about 1×10^{-6} *g*). Like roller coasters, this effect results from a combination of forces acting on the body. At any given instant in time, the astronauts are accelerating freely toward the Earth inside an object that is accelerating freely at the same rate. They can be visualized in that instant as falling on a straight line drawn from the spaceship to the Earth, perpendicular to a tangent line drawn at the ship's current position in its curved orbit. However, the ship's directional vector is constantly changing because of its curved orbit, so it perpetually "falls" in a

Astronauts aboard an aircraft that flies a parabolic pattern to provide weightlessness training.

astronauts, and the weightless effects seen in the 1995 movie *Apollo 13* were produced using parabolic flight. Several commercial companies also offer the experience to the general public. A privately funded experimental "spaceplane" called SpaceShipOne achieved suborbital flight in 2004. A revised commercial version called VSS Enterprise flew for the first time in 2010 and is taking reservations for future commercial flights that will launch passengers into suborbital space.

Further Reading

Clement, Giles, and Angeli Bukley. *Artificial Gravity*. New York: Springer, 2007.

Erickson, Lance. *Space Flight: History, Technology, and Operations*. Lanham, MA: Government Institutes, 2010.

Sparrow, Giles. *Spaceflight: The Complete Story From Sputnik to Shuttle—and Beyond*. New York: Dorling Kindersley Publishers, 2007.

JULIAN PALMORE

See Also: Airplanes/Flight; Gravity; Interplanetary Travel; Planetary Orbits; Ride, Sally; Spaceships.

new direction—around the Earth, instead of toward it. The spacecraft's precisely calculated inertial trajectory effectively counters the astronauts' constant "falling." As a result, the astronauts do not move with respect to their immediate surroundings, so they look and feel as if they are floating weightlessly. A spacecraft lands by altering its curved orbit so that the gravity is no longer sufficiently opposed.

Free-fall or zero-G can be achieved in several ways without leaving Earth's atmosphere. NASA's Neutral Buoyancy Simulator uses the world's largest indoor pool, containing over six million gallons of water, to simulate weightlessness without flying or falling, while their Zero Gravity Research Facility can achieve just over five seconds of free fall in a 467-foot long steel vacuum chamber, which is used to test microgravity effects on phenomena such as combustion and fluid physics. As part of a series of experiments in the 1960s, Air Force Captain Joseph Kittinger parachuted from a gondola at an altitude of almost 103,000 feet. He achieved a speed of over 600 miles per hour on his descent but he reported having no real subjective sensation of the incredible speeds. Standard aircraft can be used to create brief periods of weightlessness, about 30 seconds, by flying in a parabolic pattern or "Kepler curve," named for Johannes Kepler. NASA uses this method to train

Wheel

Category: Travel and Transportation.
Fields of Study: Algebra; Geometry.
Summary: Wheels help humans perform work and travel by providing a mechanical advantage.

Circles are present in many places in nature and mathematicians studied them long before the common use of the wheel. A wheel is traditionally a cylinder rotating around an axle. Together, a wheel and an axle form a simple machine that can change direction and magnitude of forces. Wheels are widely used in transportation as gears, as handles and knobs, and for converting the energy of water, animals, or people into work. The notion of curvature is of interest to many mathematicians, scientists, engineers, and others. In geometry, wheels are often modeled as circles or as concentric circles. In addition to standard circles or cylinders, mathematicians have explored the properties of wheels of other shapes along with varying surfaces. Aristotle's Wheel paradox, named for Aristotle of Stagira, is an interesting mathematical problem involving the paths traced by a wheel made of two concentric circles. It seems to imply that the circumferences of different sized circles are equal. This is one of many mathematical questions that arise from rotating concentric circles or exploring the curves generated by wheels.

History and Mechanical Advantage

Wheeled vehicles were invented about 6500 years ago, but they were not used widely until the rise of large, organized, road-building societies. This discrepancy between the discovery and its wide adoption, because of the lack of infrastructure, is frequent in science. Using wheels as levers to change the magnitude of force for applications like grinding grain was more widespread in many societies. The force advantage that a wheel provides is equal to the radius of the wheel divided by the radius of the axle. For example, a ship's capstan with the radius of eight feet and the axle radius of one foot multiplies the force of sailors using it by eight. This relationship is the reason that water wheels on small, weak streams that do not provide much force have to be larger than on fast-moving streams—a weak stream will not provide enough force to turn a small wheel. Rotating handles or knobs, grinders, drills, and old-fashioned water wells all use the wheel's mechanical advantage.

Geometry and Physics of Rolling: Work Smart, Not Hard

Rolling vehicles on wheels save work compared to dragging the same weight along the ground. Friction between the ground and a dragged object occurs along the length of the path. The work needed to overcome this friction is proportional to the friction coefficient, which depends on the surfaces of the object and the path. On smooth surfaces, such as ice, the friction coefficient is lower than on rough surfaces, such as rock. Work is also proportional to the weight of the object and the length of the path. When an object is rolled, its weight presses the axles to the wheels. Instead of the object-road friction, the force to overcome is now the axle-wheel friction, which is also proportional to the weight. When a wheel turns around, the vehicle travels the distance equal to the wheel's circumference. If the radius of the axle is one-tenth of the radius of the wheel, then the distance the axle slides within the

wheel is one-tenth of the distance the vehicle travels and the required work is divided by 10. It is relatively easy to reduce axle-wheel friction many times by using smooth surfaces, oil, and ball bearings. Vehicles for heavier loads usually have more wheels to distribute the force of the load.

Reinventing the Wheel

Since wheels are essential to most human endeavors, there are many wheel-related sayings. "Reinventing the wheel" means "needlessly duplicating a well-known method." Ironically, wheels themselves are being constantly reinvented. For example, roller bearings first appeared in Leonardo da Vinci's drawings in the sixteenth century but were patented and used widely only in the nineteenth century. Magnetic bearings reduce axle-wheel friction to essentially zero and, therefore, promise huge increases in machine efficiency; their development started in 1980s. In the 1990s, mathematics and science museums began to feature bikes with square wheels that move smoothly over special surfaces consisting of "catenaries," which are hyperbolic shapes resembling hanging lengths of chains.

Further Reading

Farris, Frank. "Wheels on Wheels on Wheels—Surprising Symmetry." *Mathematics Magazine* 69, no. 3 (1996).

Goodstein, Madeline P. *Wheels! Science Projects With Bicycles, Skateboards, and Skates*. Berkeley Heights, NJ: Enslow Publishers, 2009.

Helfand, Jessica. *Reinventing the Wheel*. New York: Princeton Architectural Press, 2002.

MARIA DROUJKOVA

See Also: Bicycles; Curves; Pi; Street Maintenance; Windmills.

Wiles, Andrew

Category: Mathematics Culture and Identity.
Fields of Study: Algebra; Connections; Geometry.
Summary: Over 350 years after its conjecture in a marginal comment, Fermat's Last Theorem was finally proven by British mathematician Andrew Wiles.

Andrew Wiles is most well-known for solving Fermat's Last Theorem, and he has received many awards, including the prestigious MacArthur Fellowship. For seven years, Wiles worked in unprecedented secrecy, struggling to solve Fermat's Last Theorem, a problem that had perplexed and motivated mathematicians for three centuries. Wiles's solution of Fermat's Last Theorem brought him both fame and personal satisfaction. He said of his accomplishment, "I had this very rare privilege of being able to pursue in my adult life what had been my childhood dream." This work also brought him pain when a subtle but fundamental error was discovered in his proof. Wiles eventually fixed the mistake, solidifying his magnificent achievement and permanent place in history.

Fermat's Last Theorem

Fermat's Last Theorem states that the equation $x^n + y^n = z^n$ has no positive whole number solutions for $n > 2$. In other words, while the Pythagorean Theorem $x^2 + y^2 = z^2$ has whole number solutions (such as $x = 3$, $y = 4$, and $z = 5$), similar equations with larger exponents, like $x^3 + y^3 = z^3$ and $x^4 + y^4 = z^4$, have no positive whole number solutions. French mathematician Pierre Fermat (1601–1665) wrote in the margin of a book that he had discovered a remarkable proof for this theorem, but that the margin was too small to contain it. For the next three centuries, the best mathematicians in the world sought a solution to this problem, and these attempts inspired many new mathematical ideas and theories.

Wiles's Proof

As a 10-year-old, Andrew Wiles already loved solving mathematical problems. He read about the history of Fermat's Last Theorem in a library book about mathematics. Despite its long history, this problem was simple enough for him to understand, and it fascinated and motivated him. As his mathematical knowledge became more advanced, he realized that there were no new techniques available to solve Fermat's Last Theorem. When Fermat's Last Theorem became linked to modern mathematical methods in algebraic geometry, he resumed his work. The quest to find a proof of Fermat's Last Theorem finally came to an end when Wiles announced his results in 1993. Wiles had worked in isolation on the problem for many years while on the faculty at Princeton University, and his announcement came as a surprise to the mathematics community. Wiles's work combined two fields of mathematics, elliptical functions and modular forms, to solve the elusive problem.

Wiles directly proved what is known as the Taniyama–Shimura Conjecture. Goro Shimura and Yutaka Taniyama were two Japanese mathematicians who, in the 1950s, conjectured that there was a relationship between elliptical equations and modular forms. Later, thanks to the earlier work of mathematicians Gerhard Frey, Ken Ribet, and Barry Mazur, it was shown that if the Taniyama–Shimura Conjecture were true, then so was Fermat's Last Theorem. His results were presented in a dramatic series of lectures at a conference in Cambridge, England.

However, not long after Wiles announced his discovery, an error was found in one section of the long and difficult proof. With the help of one of his former students, Richard Taylor, Wiles was able to make the necessary changes. However, these corrections took over a year to complete, illustrating the complexity of the proof that Wiles had constructed.

Methods

Many people may wonder how Andrew Wiles was able to solve a problem that had eluded so many others skilled mathematicians. Wiles himself has said that he does not always know exactly where his new techniques come from, but he defines a good mathematical problem by the mathematics it generates, not by the problem itself. He never uses a computer in his work, preferring to doodle, scribble, or find patterns via calculations. As do most scholars, he also reads previous research for methods that he can adapt to his work. When he gets stuck working on a problem, he reportedly tries to change it into a new version that he can solve or steps away from it entirely to relax and allow his subconscious to work. He has described his personal process by the following analogy:

Perhaps I could best describe my experience of doing mathematics in terms of entering a dark mansion. One goes into the first room, and it's dark, completely dark. One stumbles around bumping into the furniture, and gradually, you learn where each piece of furniture is, and finally, after six months or so, you find the light switch. You turn it on, and suddenly, it's all illuminated. You can see exactly where you were. Then you move into the next room and spend another six months in the dark. So each of these breakthroughs, while sometimes they're momentary, sometimes over a period of a day or two, they are the culmination of—and couldn't exist without—the many months of stumbling around in the dark that proceed them.

Further Reading

Aczel, Amir D. *Fermat's Last Theorem: Unlocking the Secret of an Ancient Mathematical Problem*. New York: Basic Books, 2007.

Hardy, G. H., Edward M. Wright, Andrew Wiles, and Roger Heath-Brown. *An Introduction to the Theory of Numbers*. New York: Oxford University Press, 2008.

Singh, Simon. *Fermat's Enigma*. New York: Doubleday Press, 1997.

WGBH Science Unit. "NOVA: Transcripts: The Proof." http://www.pbs.org/wgbh/nova/transcripts/ 2414proof.html.

Todd Timmons

See Also: Cubes and Cube Roots; Mathematics, Theoretical; Proof; Pythagorean Theorem.

Wind and Wind Power

Category: Weather, Nature, and Environment.
Fields of Study: Data Analysis and Probability; Geometry; Measurement.
Summary: Wind and wind power have been mathematically studied for centuries as an energy source and promise to be increasingly important energy sources.

Wind is omnipresent. There are few parts of the world that are not affected by the wind, from the pleasant breezes off a lake to the terrifying destruction of hurricanes and tornados. Historically, wind was one of the most important sources of energy; it drove sailing ships and was key to driving some pre-industrial revolution machines, such as windmills. Being able to master the wind was a key component in the fate of empires. For example, in 1588, it is said that the Spanish Armada of Catholic King Philip II was defeated by a "strong Protestant wind" that forced his fleet off course and prevented a vulnerable England under the reign of Queen Elizabeth I from being invaded. In the wake of the steam engine, developed by James Watt in the 1760s, and the emergence of coal-powered machines during the Industrial Revolution, the age of wind and sail began to decline for much of the industrialized world. Many cite this shift to fossil fuel sources as a cause of the rise in carbon dioxide, other greenhouse gasses (GHGs), and the global warming phenomenon, and there is a movement toward returning to wind as one source of clean energy.

Mathematicians and scientists have long been involved in the study of wind and wind energy. Posidonius of Rhodes (c. 135–51 B.C.E.) theorized about clouds, mist, wind, and rain. Francis Beaufort (1774–1857) developed a mathematical scale to describe wind speed. Twenty-first-century engineer Michael Klemen has explored mathematical issues of wind data acquisition as a function of time and estimated wind resource availability for power generation. Mathematicians continue to contribute to these fields and to the exploration of related phenomena like solar winds, which are believed to have first been observed by astronomer John Herschel during his observations of Halley's comet in 1835.

History

Seventeenth-century mathematician Evangelista Torricelli was reputed to be skilled in making instruments and he is often credited with inventing the barometer. He also conducted research about weather and is believed to have given the first correct explanation of wind when he said, "winds are produced by differences of air temperature, and hence density, between two regions of the earth." In the seventeenth and eighteenth centuries, mathematician Philippe de La Hire studied instruments to measure climate, including temperature, pressure, and wind speed. He went on to collect data using these instruments at the Paris Observatory. In the nineteenth century, William Ferrel proposed a model for wind circulation, which was the first recorded theory to explain the westerly winds in the middle latitudes of both the northern and southern hemispheres. Ferrel cells are phenomena where air flows eastward and towards the pole near the Earth's surface, but westward and toward the equator at higher altitudes. The Beaufort wind scale was also named in the nineteenth century after Francis Beaufort, a British Rear Admiral who reportedly extended the work of many individuals in trying to standardize wind measurement and description. The invention of the cup anemometer by astronomer and physicist John Robinson in the middle of the same century aided in measuring winds and reputedly helped popularize the measure. The Beaufort wind scale was later revised by meteorologist George Simpson in the early twentieth century. Mathematician Lewis Richardson is widely considered a pioneer of mathematical weather prediction. He applied the method of finite differences and other mathematical methods in his Weather Prediction by Numerical Process in 1922. Wind is often mathematically modeled as a fluid, and some of Richardson's work was an extension of studies regarding water flow in peat. The Richardson number is a function involving gradients of temperature and wind velocity. Edward Milne, his contemporary, studied wind and sound, helping to refine huge binaural listening trumpets used to detected aircraft at night during World War I. In the twenty-first century, mathematicians often model various aspects of wind and wind power, including the wind movement through plant canopies using first and second order closure techniques; the probability of bird collisions with wind turbine rotors using statistical methods and calculus; descriptions and predictions of surface wind in mountainous terrain using statistical methods, geometry, vectors, and other mathematical functions; and the wind flow or turbulence over many types of surfaces, including turbine blades, ocean waves, automobiles, and structures.

U.S. Wind Research and Applications

The first wind system to generate electricity in the United States was built by Charles Brush in the late nineteenth century. However, there was relatively little development in that area until the energy crises of the 1970s, which motivated people to seek alternative sources of electricity, such as wind. The 1990s and the 2000s saw technological advances, decreasing turbine costs, and the emergence of popular and political support for wind energy. At the start of the twenty-first century, the U.S. government aimed to have 20% of all electricity generated by wind by 2030. Moreover, statistical studies and other data suggest that wind should be able to compete on a cost-effective basis with traditional fossil fuel sources. Some reports even estimate that wind will account for 26% of the increase in renewable energy production by 2035, though this extrapolation may not be reliable. Wind has shown a number of advantages compared to other forms of electricity production: it does not emit greenhouse gasses while in operation, it is freely available, it is not subject to energy security concerns, there are no waste products, and the maintenance costs are relatively low compared to traditional or nuclear generating facilities. For energy sources such as wind and nuclear, the emissions occur during the construction phase and tend to be associated with the amount of concrete and steel used in the facilities. Wind energy also faces technological problems with intermittency, as electricity can only be produced while the wind is blowing and this problem had been studied by mathematicians.

For example, the Weibull correlation model, based on the Weibull distribution named for mathemati-

Wind Tunnels

Wind tunnels allow scientists and mathematicians to create wind under controlled conditions to test theories and applications. Mathematicians Benjamin Robins and George Cayley constructed simple spinning devices to model drag and other aerodynamic forces in the nineteenth century but the flow is difficult to control under such conditions. Engineer Francis Wenham is credited with the invention of the first enclosed wind tunnel, in 1871, with colleague John Browning. Wind tunnels were used by Orville and Wilbur Wright in developing their airplane prototypes as well as by German scientists at the famous World War II Peenemünde research facility. With advances in computer technology, the properties of wind are often modeled using computational fluid dynamics rather than physical data collection in wind tunnels, or the two methods are used to compare and cross-validate results. The foundations of these methods are the Navier–Stokes equations, which are systems of nonlinear partial differential equations developed by mathematicians Claude-Louis Navier and George Stokes.

cian (Ernst) Waloddi Weibull, estimates energy outputs with reduced uncertainty versus previous models, which is potentially useful for preventative operation and maintenance strategies. The National Renewable Energy Laboratory offers both wind data sets and has developed many mathematical models to explore wind energy grids, economic impact of wind energy, and even a model called Village Power Optimization Model for Renewables (ViPOR), which is a computational tool that facilitates the design of a village electrification system using the lowest cost combination of centralized and isolated power generation. Beyond land-based power generation, scientists and engineers like Maximillian Platzer and Nesrin Sarigul-Klijn are exploring the potential benefits of a return to wind energy as a supplement for large, ocean-going ships.

Further Reading

Huler, Scott. *Defining the Wind: The Beaufort Scale and How a Nineteenth-Century Admiral Turned Science Into Poetry*. New York: Three Rivers Press, 2004.

Shepherd, William, and Li Zhang. *Electricity Generation Using Wind Power*. Singapore: World Scientific Publishing Company, 2010.

Walker, Gabrielle. *An Ocean of Air: Why the Wind Blows and Other Mysteries of the Atmosphere*. London: Bloomsbury, 2007.

JASON L. CHURCHILL

See Also: Hurricanes and Tornadoes; Tides and Waves; Weather Forecasting; Weather Scales.

Wind Instruments

Category: Arts, Music, and Entertainment.
Fields of Study: Geometry; Number and Operations; Representations.
Summary: The frequency and pitch of wind instruments are determined by their shape, length, and other factors.

Wind instruments convert the energy of moving air into sound energy—vibrations that are perceptible to the human ear. Under this definition, wind instru-

ments include the human voice; pipe organs; woodwind instruments, such as the clarinet, oboe, and flute; and brass instruments, like the trumpet. The nature of this vibration and the associated resonator tube are responsible for the unique timbre of each type of wind instrument.

Sources of Vibrations

In the human voice, the flow of air from the lungs causes the vocal cords (also called "vocal folds") in the larynx to open and close in rapid vibration. This periodic stopping of the air stream creates oscillatory pulses of air pressure, or sound. The frequency of this vibration and the pitch of the resulting sound are determined by the length and tension of the cords. A singer or speaker controls these factors using the musculature of the larynx.

The rapid open-close vibration of the vocal cords is present in many wind instruments. In brass instruments, such as the trumpet, trombone, French horn, and tuba, the lips of the musician form a small aperture that opens and closes in response to air pressure. Brass instruments are sometimes called "lip-reed" instruments. In single-reed instruments, like the clarinet and saxophone, a thin cane reed vibrates in oscillatory contact with a specially shaped structure (the mouthpiece) to bring about the open-close effect. The oboe and bassoon utilize two cane reeds held closely together with a small space between them that opens and closes in response to flowing air, controlled by the muscles of the lips.

A third important mechanism for converting the energy of moving air into vibration is utilized in the flute and the so-called flue pipes of the pipe organ. In these instruments, vibration occurs when flowing air passes over an object with a distinct edge that splits the airstream. The resulting turbulence gives rise to oscillatory vibration. With the modern flute, the flutist's lip muscles actively control the interaction between the airstream and the edge. With the recorder and other whistle-type instruments, as well as flue pipes of the organ, the interaction is controlled by the mechanical design of the instrument alone.

Tube Resonators and Overtones

With the exception of the human voice, all wind instruments are constructed with a tube resonator enclosing a column of air that functions in much the same way as the vibrating string. Oscillations in air pressure inside the tube reflect from the ends, resulting in significant feedback with the primary vibrating medium. The relationship between the vibration frequency and length of a string fixed at both ends is explained by the concept of "harmonics." In idealized settings, changing the string length by small integer factors (for example, 1/2, 1/3, or 1/4) results in frequency changes that are recognizable as musical intervals (for example, an octave, an octave plus a fifth, or two octaves). The resonating air column in wind instruments behaves similarly to a vibrating string.

An important performance practice on most wind instruments is overblowing. Not to be confused with simply playing overly loudly, the term "overblowing" refers to the fact that changes in the airflow can cause the resonating air column to vibrate at an overtone above its fundamental frequency. Overblowing allows performers on modern instruments to achieve a large range of pitches (often two octaves or more) from a relatively compact resonating tube. Instruments with cylindrical tubes open at both ends, such as in some flutes, overblow at the octave, as do conical instruments that are closed at one end, such as the oboe and saxophone. On the other hand, cylindrical tubes closed at one end, such as the clarinet, overblow at the twelfth—an octave plus a fifth. The relative weakness of the overtone at the octave and other even-numbered overtones account for the particular timbre of the clarinet.

Altering the Tube Length in Performance

Just as the length of a vibrating string determines the frequency or pitch of the vibration, the length of the resonating air column accounts for the pitch of notes played by a wind instrument. In reed instruments, the resonating tube is perforated along its length with holes. By systematically covering some of the holes but not others, the musician effectively changes the length of the resonating column. This change, in turn, causes the vibrating reed assembly to assume the frequency of the air column. Most brass instruments have secondary lengths of tubing that are brought into play by mechanical valves by which the performer alters the length and the fundamental frequency of the vibrating air column. The exception to this is the slide trombone, which features a concentric tube arrangement by which the outer tube can move to lengthen the air column resonator.

Further Reading

da Silva, Andrey Ricardo. *Aeroacoustics of Wind Instruments: Investigations and Numerical Methods.* Saarbrücken, Germany: VDM Verlag, 2009.

Miller, Dayton Clarence. *The Science of Musical Sounds.* Charleston, SC: Nabu Press, 2010.

Sundberg, Johan. *The Science of Musical Sounds: (Cognition and Perception).* San Diego, CA: Academic Press, 1991.

Wood, Alexander. *The Physics of Music.* 7th ed. London: Chapman and Hall, 1975.

ERIC BARTH

See Also: Geometry of Music; Harmonics; Percussion Instruments; Pythagorean and Fibonacci Tuning; Scales; String Instruments.

Windmills

Category: Architecture and Engineering.
Fields of Study: Algebra; Geometry; Measurement.
Summary: The amount of power that a windmill can harness can be determined mathematically according to its size and design.

For centuries, windmills have captured peoples' imaginations through their form, function, and romantic appeal. Immortalized by Miguel de Cervantes in his book *Don Quixote*, windmills have transformed over the years from broad, short structures with an even number of sails to tall, sleek, three-sailed structures equipped with turbines for capturing energy from the wind. Windmills utilize natural power sources to perform a variety of functions, including energy production and food processing. Wind-driven prayer wheels have been used since the fourth century in Tibet and China. Historians believe that people in ancient Persia built the first practical windmills for both grinding grain and pumping water. From there, they spread through the Middle East and parts of Asia, as well as to India. They can be documented in Europe by the twelfth century. Wind turbines developed primarily in the twentieth century. Mathematics has been important for both the design of windmills and in calculating and modeling their output. Interestingly, English mathematician and physicist George Green was also a miller, and he is believed to have done much of his mathematics work in his windmill.

Designs

Windmills have had a wide variety of designs and appearances. Some of the earliest windmills rotated along a vertical axis, with the main rotor placed vertically in relation to the ground and giving a look similar to a helicopter. Some modern wind turbines have retained this engineering design in areas where wind direction is variable. This design is advantageous because vertical-axis windmills have an axis of rotation perpendicular to the ground, so the sails react similarly to all wind directions. On the other hand, horizontal-axis windmills have an axis of rotation that is parallel to the ground, resembling the more common image of a windmill such as that found in *Don Quixote*. The structure of horizontal-axis windmills gives the advantage of allowing their potential work to be maximized with respect to a specific wind direction. It is important to place a horizontal-axis windmill in line with the prevailing wind.

Windmills have traditionally been designed symmetrically, including an even number of sails. Historically, workers would place food and other substances in special locations inside the windmill to be ground by stones or other clashing materials. The grinding materials were sometimes connected to a system of gears and pulleys to increase the power beyond the mere rotation of the sails. Most modern wind turbines continue to have a sleek, symmetric design but have three sails. The insides of these turbines are devoted mostly to the attainment of electric power.

Number of Blades

The number of blades on a windmill is in direct correlation to the power generated, although the coefficient is quite small. The amount of power generated increases nearly linearly with each additional blade but the increase in power beyond just two or three blades is quite small for modern wind turbines. Physicists have determined that the power generated by a wind turbine is proportional to the cube of the wind speed and can be found algebraically by

$$P = EA\frac{1}{2}dv^3$$

where E is the power efficiency of the rotor, A is the swept area, d is air density, and v is wind speed. The swept area relates to the circle created by a rotation of a sail, calculated by

$$A = \frac{1}{2}\pi l^2$$

where l is the length of the sail. The theoretical maximum of E, known as the "Betz limit," is 0.59. The Betz limit is named for Albert Betz, a German physicist who was also interested in wind power. However, this theoretical value is reduced significantly when common physical constraints, including friction and drag on the rotors, are considered. One can calculate the maximum power produced by a windmill algebraically as

$$P_{max} = \frac{8}{27}A\frac{1}{2}dv^3.$$

It is difficult to put tight parameters on the variables that determine the amount of power produced by a wind turbine. However, a good estimate of the production of power for a 10-foot diameter sail in 12 miles per hour average winds is 2300 kilowatts of power. In a wind farm, several turbines are interconnected by a power collection system and communications network to pool their output and connect to a power grid. Probabilistic mathematical models are used to estimate and describe the output of networks of wind turbines.

Further Reading

Betz, A. *Introduction to the Theory of Flow Machines.* Translated by D. G. Randall. Oxford, England: Pergamon Press, 1966.

Brooks, L. *Windmills.* New York: Metro Books, 1989.

Gipe, Paul. *Wind Power, Revised Edition: Renewable Energy for Home, Farm, and Business.* New York: Chelsea Green Publishing, 2004.

Gorban, A. N., A. M. Gorlov, and V. M. Silantyev. "Limits of the Turbine Efficiency for Free Fluid Flow." *Journal of Energy Resources Technology,* 123, no. 4 (2001).

DAVID SLAVIT
GISELA ERNST-SLAVIT

See Also: Carbon Footprint; Electricity; Energy; Wind and Wind Power.

Wireless Communication

Wireless communication has become ubiquitous in the twenty-first century. Consider all of the aspects of one's life that are impacted by wireless communications, including text messaging and voice calls over a cellular network, and e-mail and Web surfing over a wireless Internet connection. Wireless communication consists of encoding information onto radio waves and passing them through the atmosphere—not unlike how an amplitude modulation (AM) or frequency modulation (FM) radio signal is sent and received. Wireless communication would not be possible without mathematics, and mathematicians contribute in many ways to creating, sustaining, and studying wireless processes and technologies.

Information theory plays a central role in wireless communications; its origins are attributed to mathematician Claude Shannon in the mid-twentieth century. Sergio Verdu, who is cited as a world-renown researcher in wireless communications noted, "Claude Shannon was the archetypical seamless combination of mathematician and engineer. . . . Shannon's theory has been instrumental in anything that has to do with modems, wireless communications, multi-antenna and so on."

Many other theoretical and applied mathematical methods have also been fundamental in wireless communication. For example, methods like stochastic calculus, stochastic modeling, control theory, graph theory, game theory, signal processing, wavelets, simulation and optimization, and multivariate statistical analysis have been used to develop communication networks, quantify or predict performance characteristics like network traffic, and to create protocols for signal transmission, encryption, and compression. Some mathematical models have been used by developers to quantify and compare wired versus wireless communication systems.

Mathematicians and engineers working in wireless communications must consider the properties of the waves and how the information is encoded. Information, whether an e-mail, telephone, video, or other data, is encoded onto the sinusoidal waveform by combining changes in frequency, amplitude, and phase. This encoding is accomplished by modifying various properties of

Figure 1. Amplitude modulation (AM).

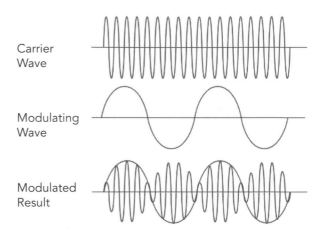

Source: M. Qaissaunee.

a periodic sinusoidal function—the carrier wave—to embed information or message wave on the carrier. Figure 1 shows a simple example for the case of AM. The height or amplitude of the carrier wave is modified to represent or information or modulating wave.

Researchers also consider the variety of factors that can affect the strength and quality of the signal. A communications engineer or technician is most often concerned with behaviors that will affect the propagation of the radio wave through the air. These include absorption, attenuation, diffraction, free space path loss, gain, reflection, refraction, and scattering. A combination of these factors will impact the signal quality and determine the likelihood of a successful transmission.

One common number associated with a wireless signal is the frequency. Frequency is a measure of how many cycles occur for a given time period. A signal cycle occurs every time a waveform repeats. Frequency is measured in cycles per second, which are also called "hertz" (Hz) after German physicist Heinrich Hertz. A waveform that repeats once every second has a frequency of 1 hertz. Waves used in communications are at much higher frequencies, so some prefixes must be used to measure radio frequencies. The wireless networks used for laptops and smartphones at the beginning of the twenty-first century often operate at the 2.4 GHz and 5 GHz frequencies of the spectrum. AM and

FM radio are in the kHz or MHz frequencies, while satellites operate at very high frequencies—often in the hundreds of GHz.

MICHAEL QAISSAUNEE

See Also: Cell Phone Networks; Satellites; Telephones.

Further Reading

Agrawal, Prathima, Daniel Andrews, Philip Fleming, George Yin, and Lisa Zhang. "Wireless Communications." In *IMA Volumes in Mathematics and its Applications Series.* Vol. 143. New York: Springer, 2007.

Boche, Holger, and Andreas Eisenblatter. "Mathematics in Wireless Communications." In *Production Factor Mathematics.* Edited by Martin Grotschel, Klaus Lucas, and Volker Mehrmann. Berlin: Springer, 2010.

Leong, Y. K. "Mathematical Conversations—Sergio Verdu: Wireless Communciations, at the Shannon Limit." *National University of Singapore Newsletter of the Institute for Mathematical Sciences* 11 (September 2007). http://www.princeton.edu /~verdu/singapore.pdf.

Women

Category: Mathematics Culture and Identity.
Fields of Study: Communication; Connections.
Summary: Historically, women have been underrepresented in mathematics careers and professions.

Questions are raised periodically about women's participation or lack thereof in mathematics. This issue has been investigated from the perspectives of various disciplines, among them, history, psychology, neuroscience, economics, and statistics. Each of these perspectives has strengths and weaknesses and sheds light on different aspects of the issue.

Differences of era, place, and culture can affect findings; thus, results for one population do not always extend to others and findings from one decade may not hold for the next. In all cases, various forms of bias may affect the selection and interpretation of the information presented—on the part of newspapers, journals, researchers, and writers, as well as their audiences.

Pre-College and College Participation

Historical research has documented how the proportions of women in mathematics and other fields have waxed and waned with changes in societal norms, institutional policies, and mathematical practices.

In the antebellum, nineteenth-century United States, schooling was not compulsory, and most adolescents did not attend school. Mathematics, other than arithmetic, was not a college prerequisite and the adolescent girls enrolled in school did often not study the Greek and Latin required of college-bound boys. By the 1890s, about 7% of 14–17-year-olds attended high school. Girls outnumbered boys in mathematics courses at public high schools, sometimes outperforming them.

The proportion of adolescents attending high school increased rapidly. By 1940, almost three-quarters of 14–17-year-olds attended high school. However, many high schools de-emphasized or eliminated mathematics requirements and smaller proportions of students enrolled in advanced courses. The percentages of girls in these courses declined to parity in the early 1900s and decreased further until the 1950s. By the 1970s, their proportions had increased and 2005 statistics showed them at or above parity.

In every epoch on record, girls have predominated in high school, but before 1900 and between 1930 and 1980, women were a minority of undergraduates. Women's share of mathematics and statistics baccalaureates was similar to their share of all baccalaureates in 1950 but later lagged, remaining at 40% to 50%, although their overall share has since risen.

Recent Research on College and Pre-College Populations

Cognitive factors such as spatial abilities have been analyzed independently and with respect to mathematical performance. A 1985 meta-analysis by Marcia Linn and Anne Petersen grouped spatial abilities into three categories: spatial perception, spatial visualization, and mental rotation. They found little evidence of gender differences for the first two categories but found large gender differences on mental rotation tasks for which scores depend on speed and accuracy. Subsequent research reports that these differences have diminished and training studies conducted by Nora Newcombe, Sheryl Sorby, and others show this ability can be improved. Mental rotation appears more important for careers such as engineering and fashion design than mathematics.

Another line of research has focused on mathematical aptitude, often as measured by the mathematics section of the Scholastic Aptitude Test (SAT or SAT-M). One finding, frequently cited as evidence for innate gender differences in mathematical aptitude, concerns the SAT-M scores from "talent searches" among middle school student volunteers. Between 1980 and 1982, the ratio of boys to girls scoring 700 or above was 13:1. Later, larger samples have yielded different, smaller ratios; a 2005 ratio is 2.8:1.

Although the first finding received extensive media coverage and is widely cited, the drop has received little publicity and few citations. Underlying causes may be related to those of the file-drawer effect—the tendency for findings that fail to reject a null hypothesis to remain unpublished.

Use of the SAT-M as a measure of mathematical aptitude or ability has been criticized on the grounds of construct validity and predictive validity. Studies of the latter find that the SAT-M underpredicts women's undergraduate mathematics course grades and overall grade point averages relative to those of men.

Possible reasons for gender gaps in SAT-M scores include differences in strategies (documented by Ann Gallagher and her collaborators) and the phenomenon of stereotype threat identified by Claude Steele and Joshua Aronson. An individual may be vulnerable to stereotype threat in a particular context if the individual is a member of a group that is stereotyped as performing poorly in such contexts. For example, reminding a woman of such stereotypes can hamper her mathematical performance, particularly when she cares about doing well in mathematics.

Using imaging techniques, researchers have found gender differences in brain areas used for processing when subjects were asked to calculate or solve mathematics problems. These have been popularly interpreted as "hard-wired" gender differences. However, the subjects of these studies are adults. Thus, these differences may result from differences in experience. Moreover, the studies are small in scale, and their findings are not always consistent.

International assessments for primary and secondary education are administered by the Trends in Mathematics and Science Survey and the Programme for International Student Assessment. Scores on these assessments, representing 493,495 students, were analyzed in 2010 by Nicole Else-Quest and her colleagues. They concluded that, on average, males and females differ little in mathematical achievement, despite more positive attitudes toward mathematics among males and substantial variability across nations. The most powerful predictors of cross-national variability in gender gaps were gender equity in school enrollment, women's share of research jobs, and women's parliamentary representation.

Graduate and Faculty Participation

Since the nineteenth century, a standard credential for professors at four-year academic institutions has been a Doctor of Philosophy degree (Ph.D.). In the United States, a Ph.D. is a terminal degree—the highest degree given in scientific fields. Thus, for modern times, Ph.D. attainment is a frequently used measure of women's participation in mathematics.

The first American woman to be awarded a Ph.D. in mathematics was Winifred Edgerton Merrill, in 1886. (A decade earlier, Christine Ladd-Franklin had completed a dissertation in mathematics at Johns Hopkins University. However, her Ph.D. was not awarded until 1926.) Before 1890, most Ph.D. programs in the United States did not allow women to enroll, making them less likely to frequent many mathematics departments. Other obstacles were quotas and professors who refused to have women as Ph.D. students. Reflecting societal norms, qualified women were sometimes not considered for academic positions, paid less, promoted more slowly or not at all, or expected to quit their positions if they married. University anti-nepotism rules were often used to exclude wives from paid employment at a spouse's institution (except during extreme circumstances, such as World War II). Such policies were likely to have affected women in mathematics more than women in many other disciplines. Then, as now, husbands and partners of female mathematicians and scientists tended to also be mathematicians and scientists. Unlike experimental scientists, female mathematicians had few opportuni-

Winifred Edgerton Merrill

ties for professional employment outside academia or as laboratory researchers within academia.

Despite these factors, the numbers and percentages of women earning Ph.D.s in mathematics increased until the 1940s. Between 1950 and 1970, women's numbers stalled while the numbers of men earning Ph.D.s in mathematics and science increased. Part of this increase was because of the influx of World War II veterans whose college and graduate tuition was supported by the Servicemen's Readjustment Act of 1944, known as the GI Bill. The lack of any corresponding increase in women's numbers may have been because of neglect of the female veterans who were nominally beneficiaries of the GI Bill together with changes in social norms and science policy. Consistent with these factors, women who were called to teach at colleges and universities during the war were displaced by men returning from war projects.

Changes in science policy and views of science may have had an especially damping effect on women's participation in mathematics, intensifying what was often seen as a dichotomy between teaching (associated with women) and research (associated with men). Margaret Murray writes that the "myth of the mathematical life course" became the prevailing model of how a mathematical career should unfold—a trajectory more compatible with societal expectations of men than women. In this view, mathematical talent emerges in childhood—creative achievements begin early and are quickly recognized. The mathematician focuses on research, ignoring distraction or shielded by a spouse or relative. Accomplishments continue, without interruption, until the mathematician's early 40s.

Faculty Participation After 1970

With the women's movement of the 1970s, percentages of women in mathematics and other fields increased. In 1971, the American Association of University Professors and the Association of American Colleges issued official policy statements urging that anti-nepotism rules be rescinded. However, the absence of anti-nepotism policies does not always solve a "two-body problem"—finding appropriate professional employment in the same geographical area for two Ph.D.s.

Another important event was the passing of the Educational Amendments Act of 1972. Its Title IX prohibits discrimination against women at educational institutions that receive federal funding and mandates periodic reviews of these grantees by federal agencies.

Elimination of anti-nepotism policies and prohibition of sex discrimination were major changes. However, for two decades, proportions of women had been very small in many mathematics departments and elsewhere in academia. Changes in institutional policies and federal regulations were no guarantee of change in individual expectations and departmental policies.

Individual expectations may be affected by evaluation bias. One example is a study conducted by Linda Fidell in 1970. Sets of 10 fictitious "résumés" of psychologists were sent to psychology department heads with the request to indicate the appropriate professorial rank at which each person described should be hired. Six of the résumés carried a male's name and the others female names. These were rotated so that the same résumé would sometimes carry a female name and sometimes a male name. The department heads assigned different ranks to identical qualifications, depending on the names they carried. Those with female names received lower ranks than those with male names. Later research suggests that this phenomenon is more complex than originally hypothesized because ratings are affected by social context. An explanatory mechanism identified by Virginia Valian is the notion of "gender schemas"—implicit hypotheses, usually unarticulated, that affect expectations and evaluations of women and men.

Although the percentage of women earning Ph.D.s in mathematics has continued to increase by at least 5% every decade since the 1970s, the presence or absence of departmental policies, such as family leave, may weigh more heavily on women. Moreover, sociological research suggests that women in science have fewer professional interactions within their departments or workplaces and are thus less likely to be aware of expectations conveyed informally. For example, a study of science departments found that departments with written guidelines for graduate students about courses of study, exams, and other expectations tended to have a larger percentage of women who earned Ph.D.s.

A variety of empirical findings suggest that, since the 1970s, the cumulative effects of individual actions, departmental practices, and institutional policies have changed to filter out fewer women. One factor may have been individual and class-action lawsuits brought on the grounds of Title IX violation. In contrast, a 2004 Government Accountability Office study found that federal agencies that fund scientific and mathematical research had not conducted the compliance reviews of their grantees mandated by Title IX.

In 2006, a National Academies report recommended that Title IX and other federal antidiscrimination laws be enforced and that federal agencies work with scientific societies to host mandatory workshops on gender bias. In 2007, the Gender Bias Elimination Act was introduced in Congress, which would have authorized such workshops and directed funding agencies to better enforce federal antidiscrimination laws. This bill did not pass, and similar bills were introduced in 2008 and 2009.

Recent Survey Findings

Every five years, the Conference Board of the Mathematical Sciences surveys a representative sample of two- and four-year academic institutions. The 2005 survey found that women were 50% of the full-time permanent mathematics faculty at two-year colleges (up from 34% in 1990 and 40% in 1995). At four-year institutions, the percentages of women in tenure-track (entry-level) and tenured (permanent) positions also increased, with the exception of tenure-track positions at B.A.-granting institutions (see Table 1).

Table 1. Percentages of Women on Mathematics Faculties of Four-Year Institutions

	1995	2000	2005
Tenured women (% of tenured faculty)			
Ph.D.-granting departments	317 (7%)	346 (7%)	427 (9%)
M.A.-granting departments	501 (15%)	608 (19%)	532 (21%)
B.A.-granting departments	994 (20%)	972 (20%)	1373 (24%)
Tenure-track women (% of tenure-track faculty)			
Ph.D.-granting departments	158 (20%)	177 (22%)	220 (24%)
M.A.-granting departments	235 (29%)	276 (32%)	337 (33%)
B.A.-granting departments	748 (43%)	517 (32%)	693 (28%)

Source: Conference Board of the Mathematical Sciences 2000 and 2005 Surveys.

The Survey of Doctorate Recipients has collected longitudinal data about 40,000 science and engineering Ph.D. recipients who earned their degrees from institutions within the United States. Recent analysis of data from this survey and the National Survey of Postsecondary Faculty—together with results from surveys of research-intensive departments and faculty members conducted in 2005—found few gender differences on key measures such as grant funding and salary for faculty members. In mathematics, women published fewer articles than men, and the proportions of women applying for jobs were slightly smaller than the proportion earning Ph.D.s. Overall, for the six scientific fields surveyed, the likelihood that a position would have female applicants was affected by institutional characteristics, the presence of family-friendly policies, the proportions of women on search committees, and the gender of the search committee chair. On average, men and women in research-intensive departments reported similar allocations of time on research and teaching but differences in professional interactions. Women were more likely to have mentors; men were more likely to engage with their colleagues on a wide range of topics from research to salary. Women were less satisfied with their jobs, and indirect evidence suggests that women were more likely than men to leave before tenure consideration.

Organizations

Organizations such as the Association for Women in Mathematics (AWM), European Women in Mathematics, and Korean Women in the Mathematical Sciences are dedicated to supporting and promoting women and girls in the mathematical sciences. Student organizations at colleges and universities include AWM chapters and Noetherian Ring groups, the latter named for mathematician Emmy Noether, who is well-known for her pioneering work in abstract algebra.

These and other organizations document women's participation in mathematics. Biographies of past and present women in mathematics are available online at the MacTutor History of Mathematics Archive, Biographies of Women Mathematicians at Agnes Scott College, and Mathematicians of the African Diaspora. The biographies of 228 women who earned Ph.D.s in mathematics at U.S. institutions before 1940 are maintained at the Web site for the book *Pioneering Women in American Mathematics*.

Further Reading

Case, Bettye Anne, and Anne Leggett, eds. *Complexities: Women in Mathematics.* Princeton, NJ: Princeton University Press, 2005.

Committee on Women in Science, Engineering, and Medicine, et al. *Gender Differences at Critical Transitions in the Careers of Science, Engineering, and Mathematics Faculty.* Washington, DC: National Academies Press, 2010.

Eliot, Lise. *Pink Brain, Blue Brain: How Small Differences Grow Into Troublesome Gaps—And What We Can Do About It.* New York: Houghton Mifflin, 2009.

Green, Judy, and Jeanne LaDuke. *Pioneering Women in American Mathematics: The Pre-1940s Ph.D.s.* Providence, RI: American Mathematical Society and London Mathematical Society, 2009.

Kenschaft, Patricia. *Change Is Possible: Stories of Women and Minorities in Mathematics.* Providence, RI: American Mathematical Society, 2005.

Rossiter, Margaret. *Women Scientists in America: Before Affirmative Action 1940–1972.* Baltimore, MD: Johns Hopkins University Press, 1995.

CATHY KESSEL

See Also: Educational Testing; Measurement in Society; Minorities.

World War I

Category: Government, Politics, and History.
Fields of Study: All.
Summary: World War I saw an increased emphasis on applied mathematics but ultimately disrupted mathematics research.

Although mathematicians were not as heavily involved with the conduct of World War I as they would be with World War II, the four years of conflict impacted the field of mathematics in two main ways: they severed international ties among researchers, thus slowing collaborative research efforts; and the war provided the circumstances for applied mathematics to develop more fully through military research. Many mathematicians contributed their knowledge and abilities to the war effort. At the same time, others published papers unrelated to the military, worked to encourage reconciliation among mathematicians of warring nations, or strove to end the war outright. World War I, which was fought from 1914 to 1918, was precipitated by the assassination of Archduke Franz Ferdinand of Austria. After the initial declaration of war on Serbia by Austria, countries with various political alliances joined the fighting, with the result that more than 30 countries on five continents were ultimately named as combatants. The massive scope of this first truly global war led U.S. President Woodrow Wilson to refer to it as the "war to end all wars."

Mathematics Applied to Military Research

Some mathematicians turned their attention to more practical and applied uses of the field. World War I saw extensive use of both trench warfare, which the United States had already experienced somewhat during the U.S. Civil War; and potent chemical weapons, like mustard gas. In the United States and in Europe, mathematicians researched ballistics and aeronautics as the warring countries sought advantages in firepower on land and began to realize the potential of air power. Mathematician John Littlewood performed research on ballistics and improved tables for the British Royal Garrison Artillery. In the United States, important figures such as Gilbert Bliss, Oswald Veblen, Norbert Wiener, and Forest Ray Moulton worked at the U.S. Army's Aberdeen Proving Ground, Maryland, in ordnance and improvement in ballistics calculations. The American Mathematical Society published, in 1919, a list of over 175 mathematicians working in some capacity to support the war effort. The National Advisory Committee on Aeronautics also began construction of the Langley Laboratory in 1917, although research did not fully get underway until a few years later.

Similarly, Europeans conducted research with the aim of improving military operations. The British mathematician Frederick William Lanchester devised a formula to calculate the likely outcome of a battle between opponents of different strengths. He also published a series of articles on the military potential of aeronautics, which in 1916 were collected into a book. At Göttingen, Germany, Felix Klein and others instituted the Aerodynamic Proving Ground in 1917. In Italy, Mauro Picone investigated new methods for calculating ballistics tables, and Vito Volterra proposed using helium in airships.

As was the case in many wars dating back into antiquity, codes and cyphers played an important role. For example, "trench codes" consisting of three number or letter groups were used for rapid communications of tactical situations but they were fairly easily cracked and were quickly supplanted by more complex structures. The Germans widely employed the ADFGVX cypher, so named because only those six letters were used in coded messages. They had been chosen to minimize operator error because when those letters are sent by Morse code, they sound very different from one other. The code was a fractionating transposition cipher using a modified Polybius square, named for second-century B.C.E. historian Polybius of Megalopolis, with a single columnar transposition.

The cypher keys were typically changed every few days and the code was broken in only a few isolated cases during the war. A general solution was found in the 1930s by William Friedman, who is often referred to as the "father of modern U.S. cryptography." The Germans also used some double transposition cyphers, which applied the same transposition key horizontally and vertically to the same matrix. In addition, they proved to be skillful in deciphering the codes of others, and the U.S. Army began to experiment with using Native American languages as military code. Several Choctaw soldiers served in the U.S. Army in Europe during World War I and are credited with helping to win some major battles.

The goal of war-related mathematicians was to improve the efficiency of military action. In the United States, this goal also applied to the home front. Allyn A. Young, the president of the American Statistical Association, proposed in a December 1917 address that a central statistical office or commission be established to aid the coordination of various boards and agencies then gathering statistics related to the war.

A greater division between mathematics research and teaching concerns also occurred around the time of World War I, as evidenced by the founding and branching off of the Mathematician Association of America in 1915 and the National Council of Teachers of Mathematics in 1920 from the more research-focused American Mathematical Society.

Suspension of International Cooperation

The war ended or made much more difficult the international relations among mathematicians that had developed in previous decades. National organizations of mathematicians publicly condemned their colleagues in enemy countries. International meetings were abandoned. Even after the war, an international congress did not fully accept German members again until 1928. A mathematician of one nation working in or visiting a hostile country might run the risk of being stranded, or worse, face arrest and imprisonment. As a whole, there were few mathematicians who made efforts during the war to maintain relations with their counterparts and such efforts were sometimes limited to individual statements of protest against a severing of ties among nations. The division of researchers slowed the development of some fields, like topology and set theory.

Non-Military Research During the War Years

Although much mathematical work from 1914 to 1918 related to improving military capability, there were many other notable advances that did not have immediate effects on war power. For instance, Albert Einstein published his general theory of relativity in 1915. David Hilbert also published field equations about that time. While a prisoner of war in Russia, the Polish mathematician Waclaw Sierpinski published a paper on his fractal triangle. Together with Godfrey Harold Hardy, after arriving at Cambridge University on Hardy's invitation, Srinivasa Iyengar Ramanujan published a series of papers on number theory during the war.

Efforts for Peace and Reconciliation

At the same time, some mathematicians focused not on improving the conduct of war or other research, but instead on ending the conflict and reconciling with their colleagues in the peace that would follow. Perhaps the most famous case is that of the British mathematician Bertrand Russell, who soon after the turn of the century had identified a paradox that challenged assumptions of set theory and in the years immediately before the war had co-authored *Principia Mathematica* with Alfred

North Whitehead. Repulsed by the battlefield slaughters and the general support of his countrymen for the war, Russell became an increasingly active pacifist, eventually taking part in public demonstrations and spending six months in prison for his antiwar writings.

Less dramatically, but still forcefully, the German David Hilbert made a point of recognizing the accomplishment of colleagues in enemy countries. The Dutch geometer Luitzen Egbertus Jan "L. E. J." Brouwer worked after the war to bring German mathematicians back into recognition. Gosta Mittag-Leffler, a Swedish mathematician, deliberately published English, French, and German papers in his journal *Acta Mathematica*. After the war, he and Godfrey Harold Hardy worked to encourage reconciliation with German researchers.

Approaching the cause of peace from another angle, the Quaker mathematician Lewis Fry Richardson, who had served in an ambulance unit in France during the war, worked to understand the causes of wars so as to better prevent them. A limited printing of his first paper on the subject, "The Mathematical Psychology of War," appeared in 1919. In later decades, as World War II loomed, Richardson would return to the subject.

Conclusion

The death of possible future contributors to the field of mathematics during World War I as a whole was, of course, an incalculable loss. By disrupting the continuity of research and discovery, the war also delayed advances in areas of mathematics such as topology and set theory. At the same time, however, the possible applied uses of mathematics began to receive more attention and appreciation. In addition, national governments became more aware of the military value of mathematicians—a value that they would exploit much more thoroughly and effectively in World War II.

Further Reading

Dauben, Joseph W. "Mathematicians and World War I: The International Diplomacy of G. H. Hardy and Gosta Mittag-Leffler as Reflected in Their Personal Correspondence." *Historia Mathematica* 7 (1980).

Newman, James R. "Commentary on a Distinguished Quaker and War." In *The World of Mathematics*. Vol. 2. Edited by James R. Newman. New York: Simon & Schuster, 1956.

Price, G. Baley. "American Mathematicians in World War I." In *AMS History of Mathematics, Volume I:*

A Century of Mathematics in America, Part I. Providence, RI: American Mathematical Society, 1988.

Siegmund-Schultze, Reinhard. "Military Work in Mathematics 1914–1945: An Attempt at an International Perspective." In *Mathematics and War.* Edited by Bernhelm Booss-Bavnbek and Jens Hoyrup. Basel, Switzerland: Birkhauser Verlag, 2003.

United States Cryptologic Museum. *The Friedman Legacy: A Tribute to William and Elizabeth Friedman.* 3rd ed. Fort Meade, MD: National Security Agency, 2006. http://www.nsa.gov/about/_files/cryptologic _heritage/publications/prewii/friedman_legacy.pdf.

CHRISTOPHER J. WEINMANN

See Also: Airplanes/Flight; Artillery; Mathematics, Applied; Predicting Attacks; Professional Associations; World War II.

World War II

Category: Government, Politics, and History.
Fields of Study: All.
Summary: World War II saw significant mathematical advances in cryptography, operations research, and navigation.

World War II was fought between two major alliances of countries, the "Axis" and the "Allies." The beginning might be traced to pacts signed in 1936 and 1937 by the three primary Axis powers: Germany, which came to control much of the European continent; Italy, which influenced the Mediterranean; and Japan, which governed much of East Asia and the Pacific. The ultimately victorious Allies coalition, led by Great Britain, the United States, and the Soviet Union, gained the surrender of Italy in 1943 and Germany and Japan in 1945. Well over 50 countries participated in the war, and there were millions of military and civilian deaths, some of the most controversial being those that resulted from the United States' use of the atomic bomb in Japan. Mathematics played a critical role in many aspects of the war effort, notably in coding and encryption, which achieved levels unseen in previous wars and led to additional developments in the subsequent cold war era, such as mathematician Claude Shannon's ideas on information theory. New areas of applied mathematics, such as operations research, also emerged from technologies and problems created during or inspired by the war. Many mathematicians served in the military or worked for military agencies, such as the U.S. Aberdeen Proving Grounds. An Applied Mathematics Panel was formed in 1942 to solve war-related mathematical problems. Mathematicians were involved in the Manhattan Project to develop the atomic bomb, a matter that is widely discussed even in the twenty-first century with regard to the ethics of mathematics research and social obligations of mathematicians as citizens of the world. The immediate prewar era and wartime would also result in a flood of mathematicians and scientists emigrating to the United States and many other Allied countries, fleeing religious or political persecution, particularly in Nazi-controlled Europe. It also likely accelerated the growth of participation of women in mathematical and scientific careers. These individuals would shape both research and teaching for decades to come.

Codes and Cyphers

Through World War I, most encrypted messages either used a paper-and-pencil cipher or a "book code" in which the enciphered version of each word was looked up in a codebook. Between the world wars, two new types of cryptography emerged: superencypherment and rotor machines.

With superencypherment, the text to be enciphered was converted into a string of digits. Then, a string of random digits (known as "additives") was added with non-carrying addition. If the additives were never used again, the result was the "one-time pad" cipher. However, if the string of additives digits is reused, it is possible for code-breakers to break the cipher. In the 1930s, American cryptographer William Friedman developed the "kappa test," a statistical test to determine when a superencypherment string was being reused.

The Japanese Navy used a codebook to convert plain text into numeric code groups, which were then superencyphered using a book of 50,000 random digits. During wartime, the number of encrypted messages sent was such that any string of these digits was reused, and the U.S. Navy was able to break the Japanese code.

The main technique was to search for so-called double hits. Suppose two encrypted messages read:

… 77899 45616 <u>27249</u> 31464 68461 …
… <u>77899</u> 81957 <u>27249</u> 81279 59138 …

The double hit is underlined. It could be because of chance but the cryptographer assumes that it is because of the same code words being enciphered by the same stretch of additive. With enough double hits, the cryptographer can recover portions of the additive and start decoding the underlying code words, as well as locating the so-called indicator (numbers hidden in the message to tell the recipient where in the book of additives the sender started). It took months of traffic for enough double hits to appear to break the Japanese naval code, which was changed several times a year. The kappa test could also be used to locate re-used stretches of additive. In 1943, in a project later code-named VENONA, the U.S. Army spotted seven double hits in 10,000 Soviet diplomatic messages. The Soviets, who used the unbreakable one-time pad system, had blundered by re-issuing some 30,000 pages of random additive, and VENONA succeeded in breaking some 2900 Soviet messages.

The Germans and Italians used the "Engima" cipher machine, which consisted of three rotors plus a steckerboard (a plugboard), which added a monalphabetic substitution to the polyalphabetic generated by the rotors. A rotor was a disc with 26 electrical contacts (for the Roman alphabet) on each side. Wiring inside the rotor connected the contacts. Such a rotor creates a monalphabetic cipher—each letter would always be replaced with the same letter. If the rotor is allowed to rotate one contract between letters, it generates a polyalphabetic cipher with a period of 26. If two rotors are connected together, so that the second one advances one space after the first one completes a rotation (in the same way as the rotating numbers in a mechanical car odometer), then the two rotors generate a polyalphabetic cipher with a period of 26×26 (sometimes 26×25, depending on how the two rotors were geared together). Three rotors generate a period of $26 \times 26 \times 26$, and so on. The operator had up to eight rotors available, giving up to

$$\frac{8!}{5!} = 336$$

possibilities for the rotors. For each day, there was a prearranged rotor selection and steckerboard setup and the operator would choose at random an initial rotation for each of the three rotors of the day. An "indicator" giving this random initial position had to be inserted into the message.

In the 1930s, three mathematicians, Marian Rejewski, Zerzy Rozycki, and Henyrk Zygalski of the Polish *Biuro Szyfrow* (Cypher Bureau) had figured out the wiring of the rotors in the Enigma, had worked out techniques for deciphering this indicator; which had been enciphered using the same Enigma, and had invented a machine called a "bomby," which automated much of the work. With these tools and techniques, they were able to read German Enigma messages until the Germans introduced changes in 1938 that defeated the Polish techniques.

The Poles then turned over their work to the British and French. The British took over an estate north of London called Bletchley Park and brought in mathematicians to work on the Enigma and other ciphers. The first four mathematicians were Alan Turing (whose Turing Machine, of 1936 formed the theoretical basis of later computers), Gordon Welchman, John Jeffreys, and Peter Twinn. Bletchley Park's main method for breaking Enigma was to find a crib (a word or words that were highly likely to be in a particular place in the message). Despite the features of Enigma that were supposed to hide any evidence of the plain text, there were certain relationships among the letters of the cyphertext that had to occur when the crib was enciphered. A machine called a "Bombe" then ran through all 26^3 positions of the three rotors, finding the very few that would produce these relationships. Multiple runs would be required for different choices of rotors but Bletchley also developed a statistical technique that—with luck—would eliminate numerous rotor choices.

Searching for a code that would be difficult to break using mathematically based cryptography methods, the U.S. government recruited native Navajo speakers. The Navajo language is very complex with unique phonetics, grammar, and syntax and no written or symbolic alphabet, making it nearly impossible for someone without substantial exposure to understand (no Axis linguists had such exposure) and providing no written cypher that could be analyzed. Several hundred Navajo code talkers served with the U.S. Marines, most in the Pacific theater.

Computers

While general-purpose electronic computers did not exist until after World War II, work during the war helped lead to their development. By 1940, analog computers of considerable sophistication existed. However, there were only a handful of digital computers, all of them electromechanical and not differing much in concept from Babbage's analytical machine of the nineteenth century. At that time, the only design for an electronic computer was from John V. Atanasoff of Iowa State College (now Iowa State University), who with Clifford Berry designed the Atanasoff–Berry Computer (ABC). It was not a general-purpose computer, limited to the solution of sets of linear equations.

In Germany, Konrad Zuse began working on computers in 1936. In 1941, he constructed the electromechanical Z3, which was the first general-purpose programmable computer. It was used for calculations for aircraft design and was destroyed by Allied bombing in 1943. After the war, Zuse built computers commercially and also developed the first programming language, Plankalkül.

In 1941, the Germans invented a new type of cypher for high-level communications. Instead of replacing or scrambling letters, a machine was developed that worked on the bits of the five-bit teletype (Baudot–Murray) code. In principle, this process was a superencypherment in which the bits of the teletype code were superenciphered by a string of binary additives. The additives were not random but were produced by a set of 10 wheels that rotated with different periods.

To solve this cipher, Bletchley Park constructed an electronic device called the "Colossus." Ten were built, each having from 1500 to 2500 vacuum tubes apiece. It was not a general-purpose computer since it could solve only one particular problem but the experience with electronic circuits and the knowledge that a device with thousands of vacuum tubes would work inspired, after the war, three successful British efforts (Turing's ACE, Cambridge University's EDSAC, and Manchester University's Mark I) to build general-purpose electronic computers. This kept the United Kingdom competitive in computer design with the United States through the beginning of the 1960s.

The Ordnance Department of the U.S. Army had the task of computing large numbers of range tables for artillery. Its Ballistic Research Laboratory, in cooperation with the Moore School of Engineering at the University of Pennsylvania, had the foresight—and ambition—to contract for an electronic computer, to be known as Electrical Numerical Integrator and

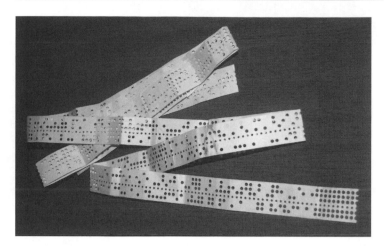

A paper tape with holes from the five-bit teletype cyphered in the early 1940s German "Baudot Code."

Computer (ENIAC). The principal designers of the ENIAC were John Mauchly and John Presper Eckert (later developers of the UNIVAC line of computers), although many of the ideas of the design came from Atanasoff's ABC. The ENIAC did not become operational until 1945. One of its first uses was in designing the hydrogen bomb.

By 1944, the shortcomings of this pioneering design had been realized. It could not handle the workload required for numerical solution of partial differential equations and plans were started for a more advanced computer to be known as EDVAC. In 1945, John von Neumann combined his own ideas, those of Alan Turing, and those of the ENIAC developers into the paper, "First Draft of a Report on the EDVAC," which laid out the principles of the modern computer. This paper led to the "Von Neumann machine" model, still used in the twenty-first century, although most of the ideas came from Turing.

Operations Research

In June 1941, Coastal Command (that portion of the Royal Air Force that operated over the seas from land bases) brought in physicist Patrick M. S. Blackett as an advisor. Blackett decided that instead of designing new weapons, his duty was to analyze how Coastal Command performed its operations and see what he could recommend to improve them. Hence, his work became known to the British as "operational research" (also called "operations research").

Blackett and his colleagues investigated a wide variety of submarine and anti-submarine operations. In one such project, the group figured out that a submarine attacked by an aircraft would not have time to dive very deep (indeed, it might still be on the surface), and that a setting of 25 feet for the depth charges the aircraft dropped had the best chance of lethality to the submarine. Another project was to figure out the optimum size of a convoy. It turned out that the larger the convoy was, the better. A convoy, even a large one, had almost the same chance of avoiding being seen by a submarine as a single ship did. What mattered was not the area of sea the convoy covered but its perimeter, where the escorts were stationed. The perimeter increased much slower than did the number of ships, so if both the number of ships and the number of escorts were doubled, each escort had a smaller length of the perimeter to cover, which gave it a better chance to catch enemy submarines trying to penetrate its portion of the perimeter.

The success of Blackett's original group led to operational research's extension to many other parts of the British forces. In April 1942, the U.S. Navy founded its own Anti-Submarine Warfare Operations Research Group, originally for antisubmarine warfare and later for work throughout the Navy. As Admiral King reported:

The knowledge . . . made it possible to work out improvements in tactics which sometimes increased the effectiveness of weapons by factor or three or five, to detect changes in the enemy's tactics in time to counter them before they became dangerous, and to calculate force requirements for future operations.

Navigation

World War II presented navigation problems not seen in prewar flying, such as how to find a target at night from the air. In the Battle of Britain, the Germans first used the "Knickebein" system for target location at night. Knickebein and it successor "X-gerät" used narrow radio beams that crossed over the target. Later, the Germans introduced "Y-gerät," which used a single ground station, with the aircraft transmitting a return signal from which the distance from the aircraft to the transmitter could be determined by the ground station.

The Allies also developed targeting systems. One was the British "OBOE" in which two stations broadcast signals to which the aircraft responded, allowing each station to determine the distance to the aircraft. The aircraft flew a fixed distance in a circular arc from the first station until it was at a specified distance from the second station. The intersection of these two arcs was the target location. This Y-gerät/OBOE technique, except with the aircraft transmitting and the ground station responding, is still used in the twenty-first century in the Distance Measuring Equipment (DME) system widely used by both military and civilian aircraft for navigating over land.

The British also developed the "GEE" system, which used a different mathematical technique. There was no transmitter on the aircraft. Instead, there was a "primary" or "master" transmitter and at least two "secondary" or "slave" transmitters on the ground. The primary would broadcast a signal, and each secondary would broadcast its own signal as soon as it received the signal from the primary. Any given difference between the arrival times of the signal from a primary and secondary defined one branch of a hyperbola (since a hyperbola is the locus of all points the difference of whose distance from two foci is constant and whether the primary or secondary signal arrived first tells which branch of the hyperbola). The second primary-secondary pair defined one branch of a second hyperbola, and these two branches intersect in exactly two points. Either dead reckoning or a third pair could then be used to determine which of these two intersection points was the aircraft's position.

GEE was soon developed into the Long Range Navigation (LORAN) system, which is still used worldwide for navigation at sea within approximately 1000 kilometers of the LORAN stations. Beyond that distance, the ionospheric bounce of the signals interferes with the ground wave.

The Mathematics Community in World War II

Mathematicians participated in both military service and multiple civilian roles during World War II. Some enlisted voluntarily or were drafted, such as Herman Goldstine, who worked as the army liaison to the ENIAC project. Many stayed in their academic positions, continuing to prepare students and working on war-related

training programs in mathematics. Others left their colleges and universities to work for government programs related to the war effort, including the growing area of operations research, such as G. Baley Price, who worked on applications like bomber accuracy and Philip Morse, who is sometimes referred to as the "father of U.S. operations research" and is credited with organizing the U.S. Anti-Submarine Warfare Operations Research Group. Companies like the Radio Corporation of America (RCA), Westinghouse Electric Corporation, Bell Laboratories, Bell Aircraft Corporation, Grumman Aircraft Engineering Corporation, and Lockheed Corporation recruited mathematicians to help fulfill war contracts. The government also widely recruited nonmilitary mathematicians for groups like the Office of Scientific Research and Development, which had branches conducting medical research, fuse research, and a multi-application area looking at problems like submarine warfare, radar, and rocketry. This body came to include the Applied Mathematics Panel in 1942.

Mathematician and scientist Warren Weaver, a pioneer in the field of machine translation, headed the panel. Some of the problems investigated included gas dynamics and compressible fluids, underwater ballistics and explosions, shock waves in air and water, mechanics and damage in air-to-air combat and anti-aircraft fire, ballistics and firing tables, torpedo spread angles, land mine clearance techniques, and statistical methods. In this time period, women also experienced increasing opportunities to pursue and contribute to a diverse range of careers, including science and mathematics. Hunter College professor Mina Rees took a leave of absence during World War II to contribute to the war effort, working with the Applied Mathematics Panel. Following the war, she became head of the mathematics branch of the Office of Naval Research. The American Mathematical Society said

. . . the whole postwar development of mathematical research in the United States owes an immeasurable debt to the pioneer work of the Office of Naval Research and to the alert, vigorous and farsighted policy conducted by Miss Rees.

Further Reading

Budiansky, Stephen. *Battle of Wits: The Complete Story of Codebreaking in World War II*. New York: The Free Press, 2000.

Goldstine, H. *The Computer From Pascal to von Neumann.* Princeton, NJ: Princeton Univeristy Press, 1972.

Haufler, Hervie. *Codebreakers' Victory: How the Allied Cryptogaphers Won World War II.* New York: New American Library, 2003.

Hodges, Andrew. *Alan Turing: The Enigma.* New York: Simon & Schuster, 1983.

Rees, Mina. "Mathematical Sciences and World War II." *The American Mathematical Monthly* 87, no. 8 (1980).

JAMES A. LANDAU

See Also: Atomic Bomb (Manhattan Project); Coding and Encryption; Intelligence and Counterintelligence; Pearl Harbor, Attack on; Radio; Strategy and Tactics; World War I.

Wright, Frank Lloyd

Category: Architecture and Engineering
Fields of Study: Geometry; Measurement; Representations.
Summary: Frank Lloyd Wright is one of the world's most renowned architects and he revolutionized architecture and design.

Considered one of the greatest American architects of all time, Frank Lloyd Wright was also an interior designer, writer, and educator. Born in Richland Center, Wisconsin, in 1867, he died in Taliesin West, Arizona, in 1959. His mother, who had always expected her son to become an architect, gave him a set of Froebel gifts after visiting the 1876 Centennial Exhibition in Philadelphia. Developed by Friedrich Fröbel in the 1830s, the kindergarten maplewood building blocks allow children to learn the elements of geometric form, mathematics, and creative design while playing. In his autobiography, Wright attests to their influence on his professional career.

Career

After taking engineering courses at the University of Wisconsin, he started working as a draftsman for architect J. Lyman Sielbee and, later, for Louis Sullivan, one of the most prominent members of the Chicago School who coined the famous modernist slogan "form ever follows function." In 1893, Wright established his own practice and in the early 1900s he initiated the series of the Prairie Houses. Rejecting the traditional vocabulary and ornaments of classical architectural styles, he revolutionized the U.S. home by focusing on geometry and the design of volumetric spaces, allowing a free spatial flow between the main living areas. The Robie House, with its low horizontal lines, nearly flat roof, overhanging eaves, central hearth, clerestory windows with delicate geometrical patterns, and open interior spaces is one Wright's finest examples of Prairie architecture. Convinced of the critical role played

Taliesin West was Frank Lloyd Wright's winter home and school in Scottsdale, Arizona. It currently houses the Frank Lloyd Wright School of Architecture, the Frank Lloyd Wright Foundation, and hosts tours year-round.

by architecture in promoting democracy, Wright used similar design principles to develop affordable homes he called "Usonian" during the Great Depression. Simultaneously, he proposed the utopian planning concept of Broadacre City, a low-density, automobile-based, surburban community where each U.S. household would live in a Usonian house on one acre of land.

Fascinated by the integration of the natural world, Wright argued that "form and function are one" and he promoted organic architecture as the modern ideal. He strived to reinterpret the patterns and principles of nature into an architectural language respecting the properties of building materials and the harmonious relationship between the form and function of the structure. Organic architecture is the outcome of an inclusive design process that aims at integrating the various spaces into a coherent aesthetic and functional whole. Wright believed that a building is a unified organism that has an intrinsic relationship not only with people but also with both its site and its time. With such concerns in mind, he designed architectural projects down to their smallest external and internal details including custom-made furniture, stained glass, rugs, light fixtures, and other decorative elements. Fallingwater, the Kaufman house outside Pittsburgh, Pennsylvania, cantilevered over a waterfall, and Taliesin West, which was built with the sand, gravel, and native boulders from the magnificent Arizona desert and mountain setting, exemplify Wright's theories of organic architecture. Another structure reflecting Wright's increased sensitivity to building materials and methods was the 14-story tall Johnson Wax headquarters, whose dendriform columns echoed inside the edifice and clerestories transformed the modern office building into a cathedral of the future.

Later Life

Until the end of his life, Wright increased his range of geometrical and structural themes and after World War II his nonresidential projects gained more significance. To the rectangular forms characteristic of the earlier decades, he added more complex geometries of the plan based on 30 degree and 60 degree angles, polygons, circles, hemicycles, and spirals that he developed in three dimensions. The Guggenheim Museum, Wright's last major work, is also one of the twentieth century's most important architectural landmarks. Its continuous upward spatial helix with sloping walls capped by a glass dome dramatically contrasts with the urban grid of the city of New York and offers a unique spatial experience to the visitor.

Wright left a rich legacy of truly American modern architectural projects unifying art and geometry and an architectural tradition of respect for the natural environment.

Further Reading

Eaton, Leonard K. "Mathematics and Music in the Art Glass Windows of Frank Lloyd Wright." In *Nexus III: Architecture and Mathematics*. Edited by Kim Williams. Pisa, Italy: Pacini Editore, 2000.

———. "Fractal Geometry in the Late Work of Frank Lloyd Wright." In *Nexus II: Architecture and Mathematics*. Edited by Kim Williams. Fucecchio, Italy: Edizioni Dell'Erba, 1998.

Pfeiffer, Bruce Brooks, and Peter Gossel, eds. *Frank Lloyd Wright Complete Works*. 3 vols. Los Angeles, CA: Taschen America, 2009–2010.

Wright, Frank Lloyd. *Frank Lloyd Wright: An Autobiography*. Petaluma, CA: Pomegranate, 2005.

———. *An Organic Architecture: The Architecture of Democracy*. Cambridge, MA, MIT Press, 1970.

Catherine C. Galley
Carl R. Seaquist

See Also: City Planning; Educational Manipulatives; Green Design; Interior Design; Skyscrapers.

Writers, Producers, and Actors

Category: Arts, Music, and Entertainment.
Fields of Study: Communication; Connections.
Summary: Some actors, screenwriters, and producers are also mathematicians or consult with them.

Mathematical scenes can be found in many scripted and unscripted productions. Some of these references are created by mathematically educated people, including writers, producers, or mathematical consultants. Mathematical references can shape society's views of mathematics and some writers or producers have noted that they have this goal in mind during the creation process. Other times, mathematics and mathematicians serve purely as entertainment value and so stereotypes, such as the nerd or mad scientist, proliferate. Actors and actresses may also have mathematical training and some use their popularity to encourage students to succeed in mathematics. Mathematicians and educators showcase these people and their mathematical references or accomplishments in order to interest and motivate students and to highlight the importance, beauty, and usefulness of mathematics, as well as the diverse career options that are available to mathematically talented individuals. Mathematicians also work with writers, producers, and actors in order to increase the realism of the representations.

Similarities Between Production and Mathematics

Numerous writers and producers have likened their work to mathematical processes. As theatrical producer Oscar Hammerstein described:

> A producer is a rare, paradoxical genius: hard-headed, soft-hearted, cautious, reckless, a hopeful innocent in fair weather, a stern pilot in stormy weather, a mathematician who prefers to ignore the laws of mathematics and trust intuition, an idealist, a realist, a practical dreamer, a sophisticated gambler, a stage-struck child. That's a producer.

A producer oversees the script, the hiring process, the budget, editing, music, and advertising. Ronald Bean is a hip-hop producer who uses the name Allah Mathematics. Jeff Westbrook has a bachelor's degree in physics and the history of science from Harvard University and a Ph.D. in computer science from Princeton University. He was an associate professor at Yale University and also worked at AT&T Labs before becoming a television writer and producer for the shows *Futurama* and *The Simpsons*. He noted the similarity between working with a team of people on computer science and mathematics problems and writing:

> Solving story problems is very similar in some ways. Given a problem, how can you fit all the pieces together to make it work? There are a lot of analytical parts to writing and analytical ability is as useful in that as in any field. That's the plus

about mathematics. Nothing trains you better and gives you more analytical skills than mathematics. That skill is useful in the craziest places you might imagine: writing a TV show, writing a cartoon, and lawyering perhaps.

Actress Danica McKellar

Actress Danica McKellar is well-known for her role on the television show *The Wonder Years* (1988–1993) and her other acting projects since then. She obtained her bachelor's degree in mathematics from the University of California, Los Angeles, in 1998. She continues to be interested in mathematics and mathematics education, saying:

> I'd like to show girls that math is accessible and relevant, and even a little glamorous! Math is a fabulous mind strengthener—it's like going to the gym, for your brain. . . . I want them to feel empowered; if they can do math, they can do anything! Math is the only place where truth and beauty mean the same thing.

With that goal in mind, she has written three mathematical books as of 2010: *Math Doesn't Suck: How to Survive Middle School Math Without Losing Your Mind or Breaking a Nail*, *Hot X: Algebra Exposed*, and *Kiss My Math: Showing Pre-Algebra Who's Boss*. Her books have achieved a wide readership and appeared on best-seller lists like the *New York Times*' children's books category.

Other Mathematician Writers, Producers, and Actors

In addition to Danica McKellar and Jeff Westbrook, there have been numerous other mathematically trained writers, producers, and actors. Stewart Burns obtained a master's degree in mathematics and has worked for *The Simpsons*. Shane Carruth was an engineer with a degree in mathematics who wrote, produced, directed, and acted in the movie *Primer*, which won numerous awards including an Alfred P. Sloan Prize, which is awarded for science, technology, or mathematical content. David X. Cohen received a bachelor's degree in physics and a master's degree in theoretical computer science, and he published an article on pancake sorting before working for *The Simpsons* and co-developing *Futurama*.

Gioia De Cari is an actress and playwright who has a master's degree in mathematics. She wrote and performed the autobiographical play *Truth Values: One Girl's Romp Through M.I.T.'S Male Math Maze*. Jane Espenson double-majored in computer science and linguistics as an undergraduate student and was a graduate student at Berkeley in linguistics. She has worked as a writer and producer for shows such as *Buffy the Vampire Slayer*, *Battlestar Galactica*, and *Caprica*. Al Jean earned an undergraduate degree in mathematics and he has been the head writer for *The Simpsons*. Mike Judge was a graduate student in mathematics before developing shows such as *Beavis and Butt-Head* and *King of the Hill*. He has also performed as a voice actor in *King of the Hill* and as an actor in the *Spy Kids* movie franchise. Ken Keeler has a Ph.D. in applied mathematics. He worked for Bell Labs and published an article with Jeff Westbook. He wrote for David Letterman, *The Simpsons*, and *Futurama*. Writer Guillermo Martínez has a Ph.D. in mathematics and was in a postdoctoral position at Oxford University. His novel *The Oxford Murders* was a 2008 movie. There has also been a grant program designed to train mathematicians and scientists to become screenwriters. Robert J. Barker of the U.S. Air Force, who is noted as having approved the grant, justified the program by explaining that: "a crisis is looming, unless careers in science and engineering suddenly become hugely popular."

Goals and Impact

Some writers, producers, and directors state as their motivation the desire to positively impact people's responses to mathematics. Many people learn about mathematicians and scientists from representations in popular culture, and the importance of role models has been well-documented. *Flatland the Movie* film producer Seth Caplan noted, "Our goal is to create a movie that not only entertains, but also inspires. Flatland will help create the next generation of innovative mathematicians and scientists by demonstrating the wonders hidden throughout our universe." Nick Falacci and Cheryl Heuton, writers, producers, and creators of *NUMB3RS* explained: "Our goal first and foremost is to intrigue and tantalize the non-math people out there in TV land. We want people who have never given mathematics a second thought to stop and consider the role that math plays in society and day-to-day life." David X. Cohen hoped that those that appreciated the mathematical references would become die-hard fans of *Futurama*. He also has expressed concern that some of the popular culture portrayals of genius mathematicians with floating numbers

that make it look like a magic power could discourage children who need to see that it takes hard work to become good at mathematics. Research has shown that stereotypical representations of mathematicians can discourage students from pursuing more mathematics. Cohen also apologized for inaccurate references:

One thing I worry about is that when we purposely present inaccurate science in *Futurama* in the name of entertainment, that viewers may hold it against us. We do have genuine respect for science, and we're trying, when we can, to raise the level of discussion of science on television. If we fail sometimes, I hope people still appreciate the frequent attempts to bring real science into the show. I apologize in advance for any failures in the future, because I'm sure there will be many more, hopefully entertaining, failures.

Consultants

Writers or producers sometimes elicit help from mathematical consultants on mathematical references in a script or a blackboard scene. Some consultants are credited as such or acknowledged in interviews or DVD commentaries, while others remain anonymous. Some consultants provide feedback for just one line or scene while others work with a producer or writer for years. Producers, directors, and writers have used consultants in a wide variety of movies, plays, and television shows with mathematical content, including the following examples:

Antonia's Line. In the 1985 movie, the main character's granddaughter was a mathematics professor who lectured about mathematics and homology theory. Wim Pudshoorn was listed as a mathematical consultant.

Arcadia. Teenage mathematics genius Thomasina Coverly worked on Fermat's Last Theorem, named for Pierre de Fermat, Fourier's heat equation, named for Joseph Fourier, and chaos theory in this 1993 play by Tom Stoppard. Mathematician Manil Suri was listed as the production mathematics consultant.

A Beautiful Mind. The 2001 movie explored the life and work of Nobel-Prize–winning mathematician John Nash. Mathematician Dave Bayer was a consultant and his hand appeared in the movie for written blackboard scenes.

Big Bang Theory. The television series debuted in 2007. Young physicists and engineers often discuss their work as well as mathematics. Physicist David Saltzberg has been acknowledged as a consultant.

Bones. The television series first aired in 2005 and the forensic team sometimes engages in mathematical discussions. In addition, the main character was listed as belonging to both a chemistry club and mathematics club in high school. Donna Cline has been acknowledged as a forensic consultant.

Caprica. The television series debuted in 2009 as a spinoff of *Battlestar Galactica.* Among other references on both shows, Dr. Philomon obtained a bachelor's in applied mathematics in addition to other degrees. Physicist Kevin Grazier was a consultant on the original show, and engineer Malcom MacIver has been a consultant on the spinoff.

Cube. The 1997 movie explored the escape attempts by those trapped in interconnected cubes, and some of the plot twists in the movie were also mathematical. Mathematician David Pravica consulted.

Contact. In this 1997 movie based on the novel by Carl Sagan, the main character explained how prime numbers could be used to communicate with aliens. Mathematician Linda Wald and physicist Tom Kuiper were consultants.

Donald in Mathmagic Land. In the 1959 short film, Donald Duck entered a mathematical world filled with references to numbers, geometric objects, and the connections between mathematics and music, architecture, and nature. Physicist Heinz Haber was the chief scientific consultant to Walt Disney productions.

Eureka. The television series began airing in 2006 and focused on scientists in a town where almost everyone worked at a research facility. There have been numerous mathematical references, including mention of a Nobel Prize by scientist and mathematician Nathan Stark, and work by his mathematical savant stepson. Physicist Kevin Grazier consulted.

Futurama. This animated science fiction television series aired 1999–2003 and was brought back to life beginning in 2007. There have been hundreds of references to science and mathematics, written mostly by the scientific writing staff. Astrophysicist David Schiminovich and mathematician Sarah Greenwald consulted on some scenes.

Flatland the Movie. This 2007 movie was based on the well-known work on dimensions by Edwin Abbott. Mathematicians Tom Banchoff, Jonathan Farley, and Sarah Greenwald and mathematics educators L. Charles Biehl and Jon Benson consulted.

Fringe. The television series first aired in 2008. The team sometimes discusses mathematics such as in the episode titled "The Equation." Neuroscientist Ricardo Gil da Costa has consulted.

Good Will Hunting. The main character in this 1997 movie was gifted in mathematics and worked as a janitor at MIT. Physicist Patrick O'Donnell and mathematician Daniel Kleitman were consultants.

Hard Problems: The Road to the World's Toughest Math Competition is a 2008 documentary about the 2006 United States International Mathematical Olympiad Team. The idea for the video was credited to mathematician Joseph Gallian, who also served as an executive producer.

House. Although the television show debuted in 2004, intern Martha Masters, who also had a Ph.D. in applied mathematics, joined the medical team in 2010. Internist Harley Liker has been a consultant.

It's My Turn. In the 1980 romantic comedy, the main character was a mathematician and she proved what is known as the "snake lemma" in the movie. Mathematician Benedict Gross was a consultant.

Madame Curie. Physicist Rudolph Langer consulted in this 1943 movie about physicist Marie Curie.

Medium. This television series aired from 2005 to 2011. The husband of the main character was an applied mathematician. Mathematician Jonathan Farley consulted.

The Mirror has Two Faces. One of the main characters in the 1996 movie is a mathematics professor. Mathematician Henry Pinkham was a consultant.

N is a Number: A Portrait of Paul Erdos. This 1993 documentary listed Donald J. Albers, Gerald L. Alexanderson, Ronald Graham, Reuben Hersh, Charles L. Silver, and Joel Spence as mathematics consultants.

NUMB3RS. This television show aired from 2005 to 2010. Charlie Eppes was a mathematics professor who consulted for the FBI. Each episode featured mathematics as a significant part of the plotline. The mathematics helped with the crime solving. The producers used many mathematical consultants but the most well-publicized were mathematician Gary Lorden and a team from Wolfram Research: Michael Trott, Eric Weisstein, Ed Pegg, Jr, and Amy Young.

The Price Is Right. The television game show aired from 1956 to 1965 and again starting in 1972. Some of the games involved mathematics and mathematicians Bill Butterworth and Paul Coe consulted.

Proof. The 2005 movie was based on David Auburn's Pulitzer Prize winning play. The lead character and her father were both talented mathematicians who also wrestled with the notion of mental illness. Mathematician Timothy Gowers was a consultant.

The Simpsons. This long-running animated television series debuted in 1989. The show's many mathematically talented writers and producers created most of the mathematical references, which have often connected to astrophysics, number theory, geometry, innumeracy, or women in mathematics. Physicist David Schiminovich consulted on some blackboard scenes.

Square One. The mathematics educational television series aired from 1987 to 1994 and featured popular culture parodies. Edward T. Esty was a mathematical consultant.

Sneakers. In this 1992 movie, a mathematician lectures on cryptography. Computer scientist Leonard Adleman consulted.

Team Umizoomi. In this mathematics educational television program, which premiered in 2010, Christine Ricci is listed as an educational consultant.

Watchmen. In this 2009 movie, Dr. Manhattan discusses mathematics. Physicist James Kakalios consulted and is also noted for his *Science of Watchman* video, which also contains mathematical elements.

Some consultants have remarked that the producers and writers were very responsive to their efforts to make the mathematics more realistic. Others have commented that advice was ignored at times in order to focus on entertainment value. Mathematicians and scientists are also members of a Hollywood Math and Science Film Consulting firm and a program run by the National Academy of Sciences called the Science and Entertainment Exchange, which matches scientists with entertainment professionals. In addition to consulting, mathematicians Thomas Banchoff, Sarah Greenwald, and Gary Lorden appeared on mathematical featurettes on movie and television DVDs. In 2003, Scott Frank estimated that approximately 20% of the highest money-making films had scientific or technical consultants.

Connections to Education

Producers of *NUMB3RS* and *Fringe* worked with mathematicians and educators to create worksheet programs based on references in the show. The CBS Network, Texas Instruments, and the National Council of Teachers of Mathematics co-sponsored an educational Web site for *NUMB3RS*. Worksheet authors received a summary of all or part of an episode and designed lesson plans to complement them. Some critiqued the blurred line between entertainment and curricula and questioned the appropriateness of violent representations for middle-grade students or the relationship between the character of Amita and her thesis advisor Charlie. The Fox network partnered with the Science Olympiad organization to create a Science of Fringe Web site of lesson plans.

Actor Portrayals

Actors that portray mathematically talented individuals are sometimes asked about their portrayals in interviews and they have expressed a wide variety of viewpoints regarding mathematics. *Flatland: The Movie* actress Kirsten Bell, who played Hex, noted: "I really enjoyed math when I was growing up. . . . When you actually figure out the solution to a problem it's very rewarding." Martin Sheen acted as Arthur Square in the same movie and stated: "Nothing can happen without math. You can't do anything. You can't build anything. You can't go anywhere without math." *NUMB3RS* actor David Krumholtz, who played the main mathematician Charlie noted: "What's great is that because math is such a universal language, really, our fans come in all shapes and sizes, all ages and genders and races and backgrounds and cultures. . . . I've been more than thrilled to meet a lot of younger people, even as young as 6 years old, who tell me they're inspired by the math and they just think it's a really cool concept." Judd Hirsch, who played his father, stated: "I don't think anybody has to understand all the mathematics in this in order to be interested in it." Navi Rawat, who played a graduate student of Charlie and his eventual wife noted, "Having the chance to help to educate people about the importance of math through the character of Amita makes my job even more rewarding." Lindsay Lohan, who portrayed a mathematically talented high school student in *Mean Girls* stated, "I'm not bad at math. It just wasn't my favorite subject. I just did it just to do it."

Professional Organizations

The professional mathematical community has interacted with writers, producers, actors, and mathematical consultants in a number of ways. They have invited them to speak at conferences or showcase their mathematical work. For example, there have been sessions on mathematics and Hollywood, on using mathematical references in the classroom, and some mathematical films like *Flatland: The Movie* and *Hard Problems: The Road to the World's Toughest Math Competition* have held premiers for the mathematical community at conferences. Mathematicians have also written reviews, columns, articles, and books about the references.

Further Reading

Frank, Scott. "Reel Reality: Science Consultants in Hollywood." *Science as Culture* 12, no. 4 (2003).

Greenwald, Sarah J. "Klein's Beer: Futurama Comedy and Writers in the Classroom." *PRIMUS (Problems, Resources, and Issues in Mathematics Undergraduate Studies)* 17, no. 1 (2007).

Halbfinger, David. "Pentagons' New Goal: Put Science into Scripts." *New York Times* (August 4, 2005). http://www.nytimes.com/2005/08/04/movies/04flyb.html.

Kirby, David. "Science Advisors, Representation, and Hollywood Films." *Molecular Interventions* 3, no. 2 (2003).

McKellar, Danica. *Math Doesn't Suck: How to Survive Middle School Math Without Losing Your Mind or Breaking a Nail*. New York: Plume, 2008.

National Academy of Sciences. "The Science and Entertainment Exchange." http://www.scienceandentertainmentexchange.org/.

Polster, Burkard, and Marty Ross. "Mathematics Goes to the Movies." http://www.qedcat.com/moviemath/index.html.

Silverberg, Alice. "Alice in NUMB3Rland." *MAA FOCUS: The Newsmagazine of the Mathematical Association of America* 26, no. 8 (2006).

SARAH J. GREENWALD
JILL E. THOMLEY

See Also: Movies, Mathematics in; Musical Theater; Plays; Popular Music; Television, Mathematics in; Women.

Z

Zero

Category: History and Development of Curricular Concepts.
Fields of Study: Communication; Connections; Number and Operations.
Summary: The concept of zero took time to be accepted and was explicitly rejected when first introduced to Greek and Roman culture.

Numbers initially served to count property, such as livestock. The numbers needed to count 1, 2, 3, 4, . . . became known as "counting" or "natural" numbers. The number zero is not found among these because one cannot count zero objects. Early civilizations existing over millennia used numbers only to count and so had no need for zero. The word "zero" has various linguistic origins: the French *zéro* and Venetian *zero*, which likely evolved from the Italian *zefiro*. This word came in turn from Arabic *sifr*, meaning "zero or nothing," derived from word *safira*, meaning "it was empty."

Early Development

The ancient Babylonians first introduced zero. With a base-60 system and initially two symbols (a wedge to represent "1" and a double wedge to represent 10), the Babylonians left empty spaces between groups of symbols. The fact that the spaces were not standard-ized in length made it difficult at times to distinguish between numbers because place value could not always be determined. To remedy this situation, the Babylonians developed zero but the zero was not a number in and of itself. It was rather a placeholder used to denote place values that had been skipped.

Independently and across the ocean, the Mayans developed a base-20 number system that included zero. Here, zero was used as a number to mean the absence of something. Zero also appeared in the Mayans' calendar. There was a year zero, and each month had a day zero in it as well. Because of the vast distance between the Mayans and the old world, Mayans' use and understanding of zero did not spread to these other areas.

Rejection by the Greeks and Romans

Despite the Babylonians use of zero, the Greeks and Romans initially rejected its use. Zero was considered dangerous spiritually as it represented the opposite of god and unity. It was associated with the void and chaos. Mathematically, zero presented many dilemmas. While any of the natural numbers (1, 2, 3, 4, . . .) when added to itself yields a larger number, zero added to itself does not. This characteristic violated Archimedes's principal that repeatedly adding a number to itself tends to a sum that is infinitely large. Additionally, a natural number plus any other natural number yields a sum larger than the initial natural number but again zero added to a natural number does not yield a number larger than the original natural number. Finally, multiplication of any number by zero yields zero and division by zero was outside the acceptable norms for these civilizations. The Greeks, known for geometry, often associated geometric figures to the natural numbers but zero could be associated with no figure. They preferred to reject zero as a number altogether.

Zero in India

Indian mathematicians in the fifth century c.e. took ideas from the Babylonians, including the concept of zero. They treated zero as a number that was found in the number line between −1 and 1. They also introduced negative numbers and, in 700, Brahmagupta introduced the idea that $1/0 = \infty$. Thus, infinity and unity depend upon the void and chaos. This idea was troubling to many civilizations, and the Hindu-Arabic numerals commonly used through the twenty-first century were not fully accepted until Leonardo de Pisa (also known as Fibonacci) introduced them to the Western world in his 1202 work *Liber Abaci*. One of the earliest recorded references to the mathematical impossibility of assigning a value to 1/0 occurred in George Berkeley's 1734 work *The Analyst*, which criticizes the foundations of calculus.

Calendars

Zero also caused confusion with the calendar system. Dionysius' calendar, created in 525 c.e., introduced the notation of BC and AD. However, it did not include a year zero. Thus, 1 BC is followed by 1 AD. This omission of zero causes confusion into the twenty-first century. Consider a person born in 1 AD. This person would have to go through 1, 2, 3, 4, 5, 6, 7, 8, 9, and 10 to have lived 10 years, and a new decade would begin at the end of this first decade (10 years). That is, it would begin in 11. Thus, the next decade would begin in 21. The first century would end in 100, and the new one would begin in 101. Thus, the twenty-first century technically began in 2001, not in 2000 when most everyone celebrated it. This confusion rears its head at the start of every decade and century all a result of the omission of a year zero.

Division by Zero

One way in which mathematicians interpret division by zero is to reframe division in terms of other arith-

metic operations. Using standard rules for arithmetic, division by zero is undefined, since division is defined to be the inverse operation of multiplication. While division by zero cannot reasonably be resolved with real numbers and integers, it can be defined using other algebraic structures or analytical extensions.

Zero in the Physical Sciences

Zero is an important value for many physical quantities or measurements. In some cases, zero means "nothing" or an absence of the characteristic, such as in most units of length and mass. However, in some cases, zero represents an arbitrarily chosen starting point for counting or measuring, such as in the Fahrenheit and Celsius temperature scales (though on the Kelvin scale, zero is the coldest possible temperature that matter can reach).

Other more advanced examples can be found in chemistry and physics. Zero-point energy is the lowest possible energy that a quantum mechanical physical system may possess. This energy level is called the "ground state" of the system and is important for investigating concepts such as entropy and perfect crystal lattices. Professor Andreas von Antropoff introduced the term "neutronium" for theoretical matter made solely of neutrons. As early as 1926, he redefined the periodic table with the atomic number zero, rather than the standard hydrogen (Atomic Number 1) in the initial position. More recent investigations suggest that the hypothesized element tetraneutron, a stable cluster of four neutrons with no protons or electrons, could have this atomic number zero.

Zero and Computers

In 1997, the naval vessel USS *Yorktown*'s propulsion system was brought to a dead stop by a computer network failure resulting from an attempt to divide by zero. Mathematical operations like these are problematic for computers, leading to various methods to avoid errors. The floating-point standard used in most modern computer processors has two distinct zeroes: a +1 (positive zero) and a −0 (negative zero). They are considered equal in numerical comparisons but some mathematical operations will have different results depending on which zero is used. For example, $1/-0$ yields negative infinity, while $1/+1$ gives positive infinity, though a "divide by zero" warning is usually issued in either case. Integer division by zero is usually handled differently from floating point, as there is no integer representation for the answer. Some processors generate an exception for integer division by zero, although others will simply generate an incorrect result for the division.

Further Reading

Ifrah, Georges, and Lowell Bair. *From One to Zero: A Universal History of Numbers*. New York: Penguin, 1987.

Kaplan, Robert. *The Nothing That Is: A Natural History of Zero*. Oxford, England: Oxford University Press, 2000.

Seife, Charles. *Zero: The Biography of a Dangerous Idea*. New York: Penguin Books, 2000.

Lidia Gonzalez

See Also: Babylonian Mathematics; Infinity; Number and Operations; Number Theory.

Chronology
of Mathematics

30,000 B.C.E.: System of tallying by groups; an impressive example is a notched wolf shinbone of uncertain date found in Czechoslovakia in 1937. In addition to bone, stones and wood marked with notches have been used for tallying. There is archaeological evidence of counting as early as 50,000 B.C.E. and of primitive geometric art as early as 25,000 B.C.E.

17,500 B.C.E.: The notched Ishango bone, dating from this period, was found at Ishango along the shore of Lake Edward, one of the headwater sources of the Nile River.

2200 B.C.E.: Mythical date of the Chinese *lo-shu* magic square, a square array of numbers in which any row, column, or main diagonal have the same sum.

1850 B.C.E.: Moscow (or Golenischev) papyrus, an Egyptian mathematical text containing 25 numerical problems, dates from this period.

1750 B.C.E. (± 150 years): Plimpton 322, a Babylonian clay tablet containing Pythagorean triples (actually the smallest and largest of the three numbers of each triple) and a column of squares of ratios of the numbers not appearing in the table over the largest number of the triple (leg over hypotenuse), is from this period.

1650 B.C.E.: The Rhind papyrus, an Egyptian mathematical text containing 85 numerical problems copied by the scribe Ahmes from an earlier work, dates from this period.

1600 B.C.E.: Approximate date of the "oracle bones," which is the source of our knowledge of early Chinese number systems.

600 B.C.E.: The Greek mathematician Thales of Miletus is traditionally credited with the beginnings of demonstrative geometry.

540 B.C.E.: Pythagoras of Samos (b. ca. 572 B.C.E.) and the Pythagorean school did considerable work in arithmetic (i.e., number theory) and geometry. Among the accomplishments of the Pythagoreans were several discoveries related to the properties of numbers, work on the Pythagorean theorem, discovery that irrational numbers exist, solution of algebraic equations geometrically, and work with some of the regular solids.

450 B.C.E.: Zeno's paradoxes of motion is attributed to this date.

440 B.C.E.: Hippocrates of Chios made progress in the duplication of the cube problem.

440 b.c.e.: Anaxagoras of Clazomenae (ca. 500–ca. 488 b.c.e.) was the first Greek known to be connected with the quadrature of the circle problem.

430 b.c.e.: Antiphon the Sophist made early important contributions to the problem of squaring the circle with a method that contained the germ of the Greek method of exhaustion.

425 b.c.e.: Hippias of Elis (b. ca. 460 b.c.e.) invented a curve (the quadratrix) that solves the trisection and quadrature problems.

425 b.c.e.: Theodorus of Cyrene (b. ca. 470 b.c.e.) showed the irrationality of several numbers after $\sqrt{2}$ was shown to be irrational.

410 b.c.e.: Democritus of Abdera's work was a forerunner of Bonaventura Cavalieri's method of indivisibles.

400 b.c.e.: Archytas of Tarentum (428–347 b.c.e.) gave a higher geometry solution to the duplication of the cube problem and applied mathematics to mechanics.

380 b.c.e.: Plato (429–347 b.c.e.) founded Plato's Academy around 385 b.c.e that drew scholars from all over the Greek world. Advances toward solving the problems of duplicating the cube and squaring the circle and toward dealing with incommensurability and its impact on the theory of proportion were achieved partly because of Plato's Academy. Much of the important mathematical work of the fourth century b.c.e. was done by friends or pupils of Plato. Plato studied philosophy under Socrates and mathematics under Theodorus of Cyrene.

375 b.c.e.: Theaetetus of Athens (ca. 415–ca. 369 b.c.e.) contributed to the study of incommensurables and the regular solids. Some of his work later became a part of Euclid of Alexandria's *Elements*.

370 b.c.e.: Eudoxus of Cnidus (408–ca. 355 b.c.e.) contributed to incommensurables, duplication of the cube, the method of exhaustion, and the theory of proportion.

350 b.c.e.: Menaechmus did early work on conics. His brother, Dinostratus, also worked in geometry.

340 b.c.e.: Aristotle (384–322 b.c.e.) did important work in systematizing deductive logic. He was the author of *Metaphysics*. Aristotle studied at Plato's Academy.

335 b.c.e.: Eudemus of Rhodes wrote a history of early Greek mathematics that is lost but was referenced by later writers; the *Eudemian Summary* of Proclus is a brief outline of Greek geometry from the earliest times to Euclid.

320 b.c.e.: Aristaeus the Elder did early work on conics and regular solids.

306 b.c.e.: Ptolemy I Soter (d. 283 b.c.e.) of Egypt and his successor Ptolemy II Philadelphus founded the museum and library at Alexandria.

300 b.c.e.: Euclid wrote a number of mathematical works with the most important mathematical text of Greek times, and probably of all times, being his *Elements*. The *Elements* is comprised of 13 books devoted to geometry, number theory, and elementary (geometric) algebra.

280 b.c.e.: Aristarchus of Samos (ca. 310–230 b.c.e) applied mathematics to astronomy. He put forward the heliocentric hypothesis of the solar system.

240 b.c.e.: Nicomedes invented a higher plane curve that will solve the trisection problem.

230 b.c.e.: Eratosthenes of Cyrune served as chief librarian at the University of Alexandria. His most scientific work was a measurement of the earth. He developed a device known as the sieve for finding all prime numbers less than a given number.

225 b.c.e.: Apollonius of Perga (ca. 262–ca. 190 b.c.e.) is most famous for his *Conic Sections*, an extraordinary work that thoroughly examines these curves.

225 b.c.e.: Archimedes of Syracuse (287–212 b.c.e.) is recognized as the greatest mathematician of the ancient world. He worked in numerous areas including measurement of the circle and the sphere, computation of π, area of a parabolic segment, the spiral of Archimedes, infinite series, method of equilibrium, mechanics, and hydrostatics.

140 B.C.E.: Hipparchus of Rhodes (ca. 180–ca. 125 B.C.E.) was an eminent astronomer who played an important part in the development of trigonometry.

75 C.E.: Heron of Alexandria developed a formula for finding the area of a triangle in terms of the sides, now known as Heron's formula. His many works include a detailed work on indirect measurement, a book on mechanics, a handbook of practical mensuration, extraction of roots, and formulas for calculating the volumes of many solids.

100: Nicomachus of Gerasa's *Introduction to Arithmetic*, one of his two works to survive, is devoted to the classification of integers and their relations.

100: Menelaus of Alexandria's *Sphaerica* sheds considerable light on the development of Greek trigonometry.

100: *Nine Chapters on the Mathematical Art*, the most important of the ancient Chinese mathematical texts, was compiled during the Han period of 206 B.C.E.–221 C.E. Our knowledge of very early Chinese mathematics is limited and uncertain. Legend holds that the emperor Qin Shi Huangdi in 213 B.C.E. ordered the burning of all books to suppress dissent, but there is some reason to doubt that this was carried out. Very little work of a primary nature is known to us from the early Chinese civilizations.

150: Claudius Ptolemy (ca. 85–ca. 165) is especially known for his work in trigonometry and astronomy. His definitive Greek work on astronomy is the *Syntaxis mathematica*, better known by its later title the *Almagest*. In the *Almagest*, he gives the value of π as 377/120, or 3.1417.

250: Diophantus of Alexandria played a major role in the development of algebra and exerted influence on later European number theorists.

300: Pappus of Alexandria wrote commentaries on Greek mathematics and did original work in mathematics. Probably his greatest work is *Mathematical Collection*, a combined commentary and guidebook of the existing geometrical works of his time, with propositions, improvements, extensions, and comments.

390: The Greek commentator Theon of Alexandria edited Euclid's *Elements*, the revision that is the basis for modern editions of the work. After Pappus, Greek mathematics ceased to be creative, and its memory was perpetuated by writers and commentators with Theon being one of the earliest.

410: Hypatia of Alexandria (d. 415), Theon's daughter, is the first woman mentioned in the history of mathematics. She wrote commentaries on Diophantus' *Arithmetica* and Apollonius's *Conic Sections*.

460: Proclus Diadochus (410–485) wrote one of our principal sources of information on the early history of elementary geometry, *Commentary on Euclid, Book I*. Proclus had access to historical works now lost to present-day mathematicians.

476: Aryabhata the Elder (b. 476) is the earliest identifiable Indian mathematician. His main work concentrated mainly on astronomy but also contained a wide range of mathematical topics, including, for example, the methods of calculating square and cube roots and what amounts to a special case of the quadratic formula.

480: The mathematician and astronomer Tsu Ch'ung Chih found π to be between 3.1415926 and 3.1415927 and gave the rational approximation 355/113, which is correct to six decimal places.

500: Metrodorus assembled one of the best sources of ancient Greek algebra problems in a collection known as the "Greek Anthology."

505: Varahamihira made contributions to Indian trigonometry and astronomy.

510: The writings of Anicius Manlius Severinus Boethius (ca. 475–524) on geometry and arithmetic became standard texts in the monastic schools.

530: Simplicius wrote commentaries on Aristotle, the first book of Euclid's *Elements*, accounts of Antiphon's attempt to square the circle, of the lunes of Hippocrates, and of a system of concentric spheres invented by Eudoxus to explain the apparent motions of the members of the solar system.

560: Eutocius of Ascalon wrote commentaries on Archimedes's *On the Sphere and Cylinder, Measurement of a Circle, On Plane Equilibriums*, and *On Apollonius' Conic Sections*.

Seventh century: The Bakhshali manuscript, a mathematical manuscript discovered in 1881 in northwestern India, has numbers written using the place value system and with a dot to represent zero. The date of the manuscript is uncertain, but the best evidence available is that the manuscript dates from the seventh century.

625: A work by Wang Xiaotong contained cubic equations without a method of solution given except for a reference to solve according to the rule of cube root extractions.

628: Brahmagupta developed theorems dealing with cyclic quadrilaterals, gave us the well-known dissection proof of the Pythagorean theorem as well as at least one other proof, and did some early work in algebra.

775: Many Indian works had been brought to the Arabian world and they were translated into Arabic, from which they were translated into Latin and other languages.

820: The earliest extant Arabic algebra text was written by Muhammad ibn Musa al Khwarizmi (ca. 780–850). Al Khwarizmi's algebra was ultimately even more influential than his important arithmetical work. The title of al Khwarizmi's algebra work has the word *al-jabr* in it; the word algebra is a corrupted form of *al-jabr*. The earliest extant Arabic geometry is a separate section of al Khwarizmi's algebra text. The work on geometry was not influenced by theoretical Greek mathematics; the geometry work has no axioms or proofs.

850: Mahavira worked in arithmetic and algebra, including giving an explicit algorithm for calculating the number of combinations. Several problems from Mahavira are similar to "word" problems in elementary algebra today.

870: Thabit ibn Qurra (836–901) translated some Greek works, including the first really satisfactory Arabic translation of the *Elements* and especially important versions of some of Apollonius's *Conics*. He also wrote on astronomy, the conics, elementary algebra, magic squares, and amicable numbers.

900: Egyptian mathematician Abu Kamil ibn Aslam (ca. 850–930) wrote an algebra text and wrote a commentary on al-Khwarizmi's algebra that was later drawn upon by Leonardo Fibonacci.

920: Abu Abdallah Mohammad ibn Jabir Al-Battani (ca. 855–929) was an astronomer who also contributed to trigonometry.

980: Abu al-Wafa' (940–998) is known for his translation of *Diophantus*, his introduction of the tangent function into trigonometry, his computation of a table of sines and tangents for 15' intervals, and geometric constructions with compasses of fixed opening.

1000: Gerbert d'Aurillac (945–1003), who became Pope Sylvester II in 999, started a revival of interest in mathematics toward the end of Europe's Dark Ages of about 476–1000. Gerbert's work has the first appearance in the Christian West of the Hindu-Arabic numerals, although the absence of the zero and the lack of suitable algorithms for calculating showed that he did not understand the full significance of the Hindu-Arabic system. Gerbert wrote on astrology, arithmetic, and geometry.

1000: Abu Bakr al-Karaji (d. 1019) was one of the Arabian mathematicians who was instrumental in showing that the techniques of arithmetic could be fruitfully applied in algebra and, reciprocally, that ideas originally developed in algebra could be important in dealing with numbers. Little is known of his life other than that he worked in Baghdad around the year 1000.

Twelfth Century: Many of the major works of Greek mathematics and a few Islamic works were translated from the Arabic into Latin. Some of the translators and a sampling of their translations were Adelard of Bath (fl. 1116–1142; first translation from the Arabic of Euclid's *Elements*), Plato of Tivoli (fl. 1134–1145; Archimedes's *Measurement of a Circle* and Theodosius' *Spherica*), John of Seville and Domingo Gundisalvo (fl. 1135–1153; a work that was an elaboration of al-Khwarizmi's *Arithmetic*), Robert of Chester (fl. 1141–1150; al-Khwarizmi's *Algebra*), Gerard of Cremona (fl.

1150–1185; Euclid's *Elements*, Archimedes' *Measurement of a Circle*, Ptolemy's *Almagest*, and al-Khwarizmi's *Algebra*).

1100: Omar Khayyam (1050–1123), who is best known in the West for his collection of poems known as the "Rubaiyat," is noted in mathematics for systematically classifying and solving cubic equations. He also headed a group that worked to reform the calendar.

1115: An important edition of *Nine Chapters of the Mathematical Art* was printed.

1130: Jabir ibn Aflah did early Islamic work on spherical trigonometry.

1150: Bhaskara II (1114–ca. 1185; called Bhaskara II to distinguish him from an earlier prominent mathematician of the same name) is most noted for his *Lilavati* and *Vijaganita*, which deal with arithmetic and algebra, respectively. Much of our knowledge of Indian arithmetic stems from the *Lilavati*. Among other things in algebra, Bhaskara dealt with indeterminate equations and affirmed the existence and validity of negative as well as positive roots. He gave several approximations for π. The proof of the Pythagorean theorem known as Bhaskara's dissection proof actually appeared much earlier in China.

1202: Leonardo of Pisa, also known as Fibonacci (ca. 1170–1240), wrote several works dealing with arithmetic, algebra, geometry, and statistics. He was one of the earliest European writers on algebra. A trivial problem (the rabbit problem) in his most famous work, the *Liber Abaci*, gives rise to the sequence 1, 1, 2, 3, 5, 8, 13, 21, 34, 55, 89, 144, 233, 377 . . . that Leonardo lists in the margin and notes can be continued indefinitely; this sequence, calculated recursively, is known today as a Fibonacci sequence. The *Liber Abaci* is devoted to arithmetic and elementary algebra and did much to aid the introduction of Hindu-Arabic numerals into Europe. The book contains problems in such practical topics as calculation of profits, currency conversions, and measurement, supplemented by the now standard topics of current algebra texts such as mixture problems, motion problems, container problems, the Chinese remainder problem, and problems solvable by quadratic equations.

1225: Jordanus de Nemore wrote on arithmetic, geometry, astronomy, mechanics, and algebra and was one of the first mathematicians to make some advances over the work of Leonardo.

1250: Nasîr ed-din wrote the first work on plane and spherical trigonometry considered independently of astronomy.

1250: Ch'in Chu-shao (ca. 1202–1261) published his *Mathematical Treatise in Nine Sections* in 1247. *Nine Sections* is the oldest extant Chinese mathematical text to contain a round symbol for zero and is the first in which numerical equations of degree higher than three occur. Ch'in began the custom of printing negative numbers in black type and positive ones in red.

1250: Li Ye (1192–1279) made an original contribution to Chinese mathematical notation by indicating negative quantities by drawing a diagonal stroke through the last digit of the number in question, an improvement over the earlier use of red and black colors, which became the accepted notation in printed works.

1260: Johannes Campanus (d. 1296) made a Latin translation of the *Elements* from the Arabic that became the basis for the first printed edition of the *Elements* in 1482.

1260: Yang Hui gave the earliest extant presentation of Pascal's arithmetic triangle and worked with decimal fractions by essentially our present methods.

1303: Chu Shih-chieh (fl. 1280–1303) wrote works that gave the most accomplished presentation of Chinese arithmetic-algebraic methods that has come down to us and employed familiar matrix methods of today. Chu speaks of what is now known as Pascal's arithmetic triangle as being ancient in his time, so the binomial theorem would appear to have been known in China for a long time.

1325: Thomas Bradwardine (1290–1349) wrote four mathematical tracts on arithmetic and geometry and developed some of the properties of star polygons.

1360: Probably the greatest mathematician of the fourteenth century was Nicole Oresme (ca. 1323–1382), who was associated with the University of Paris. He wrote five mathematical works and translated some of Aristotle. In one of his tracts, he has the first use of fractional exponents (not in modern notation) and in another tract he locates points by coordinates.

1435: Persian astronomer Ulugh Beg (1393–1449) calculated sine and tangent tables for every minute of arc correct to eight or more decimal places.

1450: Nicholas Cusa (1401–1465) was a minor German mathematician who is known primarily for his work on calendar reform and his attempts to square the circle and trisect the general angle.

1460: Georg von Peurbach (1423–1461) wrote an arithmetic and some works on astronomy, and compiled a table of sines. His main work was in Vienna and he made the university there the mathematical center of his generation.

1470: Johann Müller (1436–1476) is more generally known from the Latinized form of his birthplace of Königsberg as Regiomontanus. He wrote *De triangulis omnimodis*, which was the first European exposition of plane and spherical trigonometry considered independently of astronomy.

1478: First printed arithmetic, in Treviso, Italy.

1482: First printed edition of Euclid's *Elements*.

1484: Nicolas Chuquet (d. 1487) wrote an arithmetic known as *Triparty en la science des nombres* in 1484, a work on arithmetic and algebra in three parts. The *Triparty* was the first detailed algebra in fifteenth-century France. Chuquet recognized positive and negative integral exponents and syncopated some of his algebra.

1489: Johann Widman (ca. 1462–1498) wrote an influential German arithmetic that was published in 1489. Here appears for the first time our present + and – signs but not as symbols of operation; they were used to indicate excess and deficiency.

1491: Italian Filippo Calandri wrote one of the less important arithmetics, but it does contain the first printed example of today's modern process of long division.

1494: Italian Luca Pacioli (1445–1509) compiled from many sources the most comprehensive mathematics text of the time. His *Suma de arithmetica, geometrica, proportioni et proportionalita* contained little that was original, but its comprehensiveness and the fact that it was the first such work to be printed made it quite influential. The 600-page book contained practical arithmetic, algebra, and geometry. The *Suma* also contained the first published treatment of double entry bookkeeping.

1510: The German artist-mathematician Albrecht Dürer (1471–1528) wrote the earliest geometric text in German, published in 1525. Dürer felt that German artists needed to know elementary geometrical ideas before they could approach perspective in drawing.

1515: Scipione del Ferro (1465–1526), a professor of mathematics at the University of Bologna, solved algebraically the equation $x^3 + mx = n$. Antonio Maria Fiore (ca. 1506) was del Ferro's pupil who famously challenged Tartaglia to a contest of solving cubic equations.

1518: Adam Riese (ca. 1489–1559) wrote an especially influential German commercial arithmetic, published in 1522. The phrase *nach Adam Riese* (according to Adam Riese) is used even today in Germany.

1525: Christoff Rudolff (ca. 1500–ca. 1545) wrote his *Die Coss*, the first comprehensive German algebra, in the early 1520s.

1530: Nicolaus Copernicus (1473–1543) of Poland was a prominent astronomer who stimulated mathematics; his work necessitated the improvement of trigonometry, and Copernicus himself contributed an arithmetic treatise on the subject. Copernicus's theory of the universe in *De revolutionibus* was completed by about 1530 but was not published until 1543.

1544: Michael Stifel (1487–1567) is perhaps the greatest German algebraist of the 16th century. His best-known mathematical work is *Arithmetica integra*, published in 1544. The book was divided into three parts devoted to rational numbers, irrational numbers, and algebra. He foreshadowed the invention of logarithms by pointing out the advantages of associating an arithmetic progression with a geometric one. He gives the binomial coefficients up to the seventeenth order. Like most of his contemporaries, Stifel did not accept negative roots of an equation. The signs $+$, $-$, and $\sqrt{}$ are used, and often the unknown is represented by a letter.

1545: Cubic and quartic equations were solved by Italian mathematicians in the 16th century. Ludovico Ferrari (1522–1565) solved quartic equations by reducing the complete quartic to a form that could be reduced to a cubic that could then be solved by methods already known. Nicolo Fontana of Brescia (ca. 1499–1557), commonly known as Tartaglia (the stammerer) because of a childhood injury that affected his speech, discovered an algebraic solution to $x^3 + px^2 = n$ and also found an algebraic solution for cubics lacking a quadratic term. Girolamo Cardano, a brilliant but unprincipled mathematician, published his *Ars Magna*, a great Latin treatise on algebra, and in it appeared Tartaglia's solution of the cubic, despite an apparent promise of secrecy when Cardano wheedled the key to the cubic from Tartaglia.

1550: The Teutonic mathematical astronomer George Joachim Rheticus (1514–1574) was the first to define the trigonometric functions as ratios of the sides of a right triangle. He also formed tables of trigonometric functions.

1550: Johannes Scheubel (1494–1570) was one of the German authors to use Pascal's triangle to find roots.

1550: Italian geometer Federigo Commandino (1509–1575) prepared Latin translations of almost all of the known works of many Greek mathematicians.

1556: The first work on mathematics was printed in the New World.

1557: Robert Recorde (ca. 1510-1558) was the most influential English textbook writer of the 16th century. His first book was on arithmetic and he also wrote on astronomy, geometry, and algebra.

1570: The first complete English translation of Euclid's *Elements* was by Henry Billingsley (d. 1606), with a

remarkable preface by English scientist and mystic John Dee (1527–1608) that gave detailed descriptions of some 30 different fields that need mathematics and the relationships among them.

1572: Italian mathematician Rafael Bombelli (ca. 1526–1572) wrote an algebra text that began with elementary material and gradually worked up to the solving of cubic and quartic equations. In his *Algebra*, Bombelli introduced a different kind of cube root that comes in cubic equations of the form $x^3 + mx = n$ when $(n/2)^2 + (m/3)^2$ is negative. Bombelli was the first mathematician to accept the existence of imaginary numbers and presented laws of multiplication for these new numbers.

1575: William Holzmann (1532–1576), also known as Xylander, translated Diophantus's *Arithmetica* into Latin and translated major portions of *Elements* into German.

1580: French mathematician François Viète (1540–1603) wrote a number of works on trigonometry, algebra, and geometry. In his trigonometry book, he developed systematic methods for solving plane and spherical triangles with the aid of all six trigonometric functions. Viète's most famous work is his *In artem analyticam* that did much to aid the development of symbolic algebra. In another work, he gave a systematic process for successively approximating to a root of an equation, and in general contributed to the theory of equations. Viète showed that the trisection and duplication problems both depend upon the solution of cubic equations.

1583: Christopher Clavius (1537–1612) was a German scholar who added little of his own to mathematics, but wrote highly esteemed textbooks on arithmetic and algebra. He also wrote on trigonometry and astronomy and played an important part in the Gregorian reform of the calendar.

1590: Italian mathematician Pietro Antonio Cataldi (1548–1626) wrote a number of mathematical works and is credited with taking the first steps in the theory of continued fractions.

1590: Simon Stevin (1548–1620) is best known in mathematics for his contribution to the theory of decimal fractions. He was born in Belgium, but spent much of his adult life in Holland.

1595: German clergyman Bartholomaus Pitiscus (1561–1613) invented the term "trigonometry" in his treatise on the subject.

1600: Thomas Harriot (1560–1621) is usually considered the founder of the English school of algebraists. His great work in the field, *Artis analyticae praxis*, deals largely with the theory of equations.

1600: Swiss instrument maker Jobst Bürgi (1552–1632) conceived and constructed a table of logarithms independently of Napier, but published after Napier.

1600: Italian astronomer Galileo Galilei (1564–1643) contributed notably to mathematics. Among other contributions, Galileo founded the mechanics of freely falling bodies and laid the foundation of mechanics in general, realized the parabolic nature of the path of a projectile in a vacuum and speculated on laws involving momentum, invented the first modern-type microscope, and made several excellent telescopes (the telescope was invented about 1608 in Holland).

1610: Johann Kepler (1571–1630) discovered the laws of planetary motion, used a crude form of integral calculus to find volumes, and made contributions to the subject of polyhedral and other areas of mathematics.

1612: The Frenchman Bachet de Méziriac (1581–1638) translated Diophantus's *Arithmetica* into Latin; many of Pierre de Fermat's contributions to number theory occur in the margins of his copy of Bachet's work.

1614: Logarithms were invented by Scottish mathematician John Napier (1550–1617). Other contributions by Napier were the "rule of circular parts," a mnemonic for reproducing the formulas used in solving right spherical triangles; "Napier's anologies," useful in solving oblique spherical triangles; and "Napier's rods" or "bones," used for mechanically multiplying, dividing, and taking square roots of numbers.

1619: Savilian professorships in geometry and astronomy were established at Oxford University by mathematician Henry Savile.

1624: Englishman Henry Briggs (1561–1631) constructed a large table of logarithms with base 10, published in his *Arithmetica Logarithmica*, after he and Napier had agreed that logarithms would be more useful with a base of 10. Briggs was the first person to hold the Savilian Chair in astronomy at Oxford University.

1630: French number theorist Marin Mersenne (1588–1648) is especially known in mathematics for what are now called Mersenne primes, or prime numbers of the form $2^p - 1$, which he discussed in his *Cogitata physico-mathematica* of 1644.

1630: William Oughtred (1574–1660) was one of the most influential of the seventeenth-century English writers on mathematics. His *Clavis mathematicae* on arithmetic and algebra helped spread mathematical knowledge in England. Oughtred placed emphasis on mathematical symbols, but only a few of them are still in use. He and another Englishman, Richard Delamain (ca. 1630), independently created a physical version of a logarithm table in the form of a circular (later rectilinear) slide rule.

1630: Albert Girard (1595–1632), who spent much of his life in Holland, gave the first explicit statement of the fundamental theorem of algebra.

1635: Frenchman Pierre de Fermat (1601–1665) made important contributions to analytic geometry and probability, but of his varied contributions to mathematics, the most outstanding is the founding of the modern theory of numbers.

1635: Italian Bonaventura Cavalieri (1598–1647) developed a complete theory of indivisibles, an important pre-calculus development.

1637: Frenchman René Descartes (1596–1650) shares with Fermat early work on analytic geometry that was important in the beginnings of the subject. The work of the two was different in that, to oversimplify a bit, Descartes, in his *La géométrie*, began with a locus and then found its equation whereas Fermat did the reverse. Also in *La géométrie*, Descartes stated without proof the result known today as Descartes's rule of signs, a rule for determining limits to the number of positive and the number of negative roots possessed by a polynomial.

1640: Frenchman Gérard Desargues (1591–ca.1662) did original work on conic sections that was important in the early development of synthetic projective geometry.

1640: Italian Evnagelista Torricelli (1608–1647) is best known for his work in physics and is probably most famous for his discovery of the principle of the barometer in 1643. In mathematics, he did some work with pre-calculus indivisibles and showed that an infinite area, when revolved about an axis in its plane, can sometimes yield a finite volume for the solid of revolution; he used a method similar to the cylindrical shell method of calculus but expressed in terms of indivisibles.

1640: Frenchman Gilles Persone de Roberval (1602–1675) and Torricelli were both accomplished geometers and physicists. Roberval did work in mathematics similar to Torricelli's with questions of priority difficult to settle. Roberval successfully employed the method of indivisibles to find a number of areas, volumes, and centroids.

1650: Frenchman Blaise Pascal (1623–1662) had significant accomplishments in his short life, among them the invention of a calculating machine and the investigation of the action of fluids under the pressure of air. Pascal's triangle appeared in his *Traité du triangle arithmétique*, but he was not the first to exhibit the arithmetic triangle as Chinese writers had anticipated such a triangle several centuries earlier; the work also is famous for its explicit statement of the principle of mathematical induction. The problem of the points, stated by Pacioli in his *Suma* of 1494 and considered by several mathematicians, was important in the origin of probability theory; there was a remarkable correspondence between Pascal and Fermat that largely laid the foundation of this theory.

1650: John Wallis (1616–1703) was appointed Savilian professor of geometry at Oxford in 1649, and occupied this position for 54 years. While at Oxford, Wallis wrote his mathematical works including tracts on algebra, conic sections, mechanics, and of special interest his *Arithmetica infinitorum* that systematized and extended the methods of Descartes and Cavalieri. The *Arithmetica infinitorum* was important for its early calculus work, especially integration; Isaac Newton read Wallis's work and expanded upon what Wallis had done. Wallis was the first to fully explain the significance of zero, negative, and fractional exponents and he introduced the symbol ∞ for infinity.

1650: Dutchman Frans van Schooten the Younger (1615–1660) edited Descartes and Viète.

1650: Belgian mathematician Grégoire de St. Vincent (1584–1667) applied pre-calculus methods to various quadrature problems.

1650: Nicolaus Mercator (1620–1687) lived most of his life in England. He edited Euclid's *Elements* and wrote on trigonometry, astronomy, the computation of logarithms, and cosmography.

1650: Englishman John Pell (1611–1685) extended the factor tables of J. H. Rahn (1622–1676), which had numbers up to 24,000, to 100,000. Pell is incorrectly credited with the Pell equation, actually due to his countryman Lord William Brouncker (1620–1684), the first president of the Royal Society of London.

1650: Belgian René François Walter de Sluze (1622–1685) wrote numerous tracts on mathematics in which he discussed spirals, points of inflection, and the finding of geometric means.

1650: Italian mathematician Vincenzo Viviani (1622–1703) had a number of geometric accomplishments, but is especially noteworthy for setting forth a challenge problem that led to the beginnings of the subject of double integrals in Leibniz's solution to the problem.

1662: The Royal Society was founded in London, followed by the French Academy in Paris in 1666. These were centers where scholarly papers could be presented and discussed.

1663: Lucasian professorship in mathematics was established at Cambridge University, named for donor Henry Lucas.

1670: Englishman Isaac Barrow (1630–1677) gave a near approach to the modern process of differentiation in his *Lectiones opticae et geometricae*. Barrow was probably the first to realize in full generality the fundamental theorem of calculus, that differentiation and integration are inverse operations, which he stated and proved in his *Lectiones*. Barrow was the first occupier of the Lucasian chair at Cambridge, a position he held from 1664 to 1669.

1670: Scottish mathematician James Gregory (1638–1675) was one of the first to distinguish between convergent and divergent series. He expanded functions into series and a series for $\arctan(x)$ that played a part in calculations of π that is know by his name. Gregory is also known for his work in astronomy and optics.

1670: Dutchman Christiaan Huygens (1629–1695) wrote the first formal treatise on probability in 1657, basing his work on the Pascal-Fermat correspondence. He introduced the concept that is now called mathematical expectation.

1670: Sir Christopher Wren (1632–1723) was a famous architect who might have been remembered as a mathematician had it not been for the Great Fire of London in 1666. Wren was Savilian professor of astronomy at Oxford and taught geometry there from 1661 to 1673.

1672: Danish mathematician Georg Mohr (1640–1697) showed that all the constructions of Euclid's *Elements* can be done with a straightedge and a compass of fixed opening.

1680: Englishman Isaac Newton (1642–1727) made numerous contributions to mathematics and physics and is especially noted in mathematics for inventing the calculus.

1680: Dutchman Johann Hudde (1633–1704) gave a rule for finding multiple roots of an equation.

1682: German mathematician Gottfried Wilhelm Leibniz (1646–1716) made numerous contributions to mathematics and shares with Newton credit for the invention of the calculus; the two men worked independently of each other. Leibniz's notation was superior to Newton's and is still in use today.

1690: French nobleman the Marquis de l'Hospital (1661–1704) wrote the first calculus textbook, based on the lectures of his teacher, Johann Bernoulli. The so-called l'Hospital's rule appears in the text.

1690: Edmund Halley (1656-1742), successor of Wallis as Savilian professor of geometry, made major original contributions in astronomy. In mathematics, he restored the lost Book VIII of Apollonius's *Conic Sections* by inference, edited various works of the ancient Greeks with translations of some of them from the Arabic, and compiled a set of mortality tables of the kind now basic in life insurance.

1690: Swiss mathematicians and brothers Jakob (Jacques, or James) Bernoulli (1654–1705) and Johann (John, or Jean) Bernoulli (1667–1748) were among the first in Europe to understand the new techniques of Leibniz and to apply them to solve new problems. They made numerous contributions to mathematics and are part of the famous Bernoulli family of mathematicians.

1691: Frenchman Michel Rolle (1652–1719) is known for the theorem in beginning calculus that bears his name.

1700: Antoine Parent (1666–1716) first systematically developed solid analytic geometry in a paper presented to the French Academy.

1706: Englishman William Jones (1675–1749) first used the symbol π for the ratio of the circumference to the diameter.

1715: Englishman Brook Taylor (1685–1731) and Scotsman Colin Maclaurin (1698–1746) made important contributions to mathematics. They are best known for Taylor's well-known expansion theorem $f(a+h) = f(a) + hf'(a) + h^2f''(a)/2! + \ldots$ with Maclaurin's later expansion being the special case with $a = 0$.

1720: Frenchman Abraham De Moivre (1667–1754) is especially known for his work *Annuities upon Lives*, which played an important role in actuarial mathematics; his *Doctrine of Chances*, which contained much new material in probability; and his *Miscellanea analytica*, which contributed to recurrent series, probability, and analytic trigonometry.

1731: Frenchman Alexis Claude Clairaut (1713–1765) did important work on differential equations. He made a systematic attempt to calculate volumes of certain regions as well as the areas of their bounding surfaces. His definitive work was his *Théorie de la figure de la Terre*, published in 1743.

1733: Italian Girolamo Saccheri (1667–1733) wrote *Euclid Freed of Every Flaw* in which he purported to prove the parallel postulate (Euclid's fifth postulate) by the method of *reductio ad absurdum*.

1734: Irish philosopher Bishop George Berkeley (1685–1753) made one of the ablest criticisms of the faulty foundation of early calculus in his tract *The Analyst*.

1740: Gabrielle Émilie Le Tonnelier de Breteuil, Marquisse du Châtelet (1706–1749) translated Newton's *Principia* into French.

1748: Maria Gaetana Agnesi (1718–1799) contributed to mathematics education by writing a two-volume work, *Instituzioni Analitiche*, in her native Italian instead of the customary Latin. The first volume deals with arithmetic, algebra, trigonometry, analytic geometry, and mainly calculus. The second volume deals with infinite series and differential equations. Included in her work was a cubic curve, $y\left(x^2 + a^2\right) = a^2$ that had been studied by others and is now known, due to a mistranslation, as the "witch of Agnesi."

1750: Swiss mathematician Leonhard Euler (1707–1783) was the most prolific writer ever in mathematics with contributions too numerous to mention in detail here. He made original contributions to almost every branch of elementary and advanced mathematics. Just in elementary mathematics, he conventionalized much of our notation, gave us the formula $e^{ix} = \cos(x) + i\sin(x)$, contributed the method for solving quartic equations that is known as Euler's method, and made significant contributions in elementary number theory.

1770: Johann Heinrich Lambert (1728–1777) was born in Alsace and moved to Switzerland in 1748. Lambert attempted to improve upon Saccheri's work on the parallel postulate in his *Die Theorie der Parallellinien*, a work that places him among the fore-runners of non-Euclidean geometry. Like Saccheri, Lambert used an indirect approach but considered a quadrilateral with three right angles and made three hypotheses as to the nature of the fourth angle (right, acute, or obtuse) whereas Saccheri had considered a quadrilateral ABCE in which angles A and B are right angles with sides AD and BC equal; the hypotheses concerning the other two angles are then the same, as were Lambert's for the one angle. Among Lambert's other accomplishments were his rigorous proof that π is irrational and his systematic development of the theory of hyperbolic functions.

1777: Georges Louis Leclerc, Comte de Buffon (1707–1788) devised his needle problem by which π may be approximated by probability methods.

1788: Italian-born Joseph Louis Lagrange (1736–1813) spent his later years in France. His most important work was *Mécanique analytique*, in which Lagrange extended the mechanics of Newton, the Bernoullis, and Euler and emphasized the fact that problems in mechanics can generally be solved by reducing them to the theory of ordinary and partial differential equations.

1794: French mathematician Gaspard Monge (1746–1818) created descriptive geometry and is considered the father of differential geometry. His work entitled *Application de l'analyse à la géométrie* was one of the most important of the early treatments of the differential geometry of surfaces.

1794: The French *Journal de l'École Polytechnique* was launched. The journal is perhaps the oldest of the current journals devoted chiefly or entirely to advanced mathematics. The nineteenth century saw the rise of a number of mathematical societies and journals devoted to current mathematical research.

1797: Italian Lorenzo Mascheroni (1750–1800) discovered that all Euclidean constructions, insofar as the given and required elements are points, can be made with compasses alone.

1797: Norwegian surveyor Caspar Wessel (1745–1818) presented for the first time the association of the complex numbers with the real points of a plane.

1799: France adopted the metric system of weights and measures.

1800: German mathematician Carl Friedrich Gauss (1777–1855) gave the first wholly satisfactory proof of the fundamental theorem of algebra. His *Disquisitiones arithmeticae* was a work of fundamental importance in the modern theory of numbers. Gauss made the first systematic investigation of the convergence of a series. Gauss was the first to suspect that the parallel postulate is independent of the other axioms and worked with the Playfair form of the parallel postulate by considering the three possibilities: through a given point can be drawn more than one, or just one, or no line parallel to a given line; he shares with Bolyai and Lobachevsky the honor of discovering the geometry that results from having no line parallel to a given line. These are but a few of the ground-breaking results because of Gauss.

1803: French geometer Lazare Nicolas Marguerite Carnot (1753–1823) first systematically employed sensed magnitudes in synthetic geometry.

1805: Frenchman Pierre-Simon Laplace (1749–1827) did his most outstanding work in the fields of celestial mechanics, probability, differential equations, and geodesy. Adrien-Marie Legendre (1752–1833) is known in elementary mathematics for his *Éléments de géométrie,* which attempted to improve pedagogically on Euclid's *Elements* by rearranging and simplifying many of the propositions. Both Laplace and Legendre contributed significantly to advanced mathematics.

1806: Swiss bookkeeper Jean-Robert Argand (1768–1822) published a geometric interpretation of the complex numbers that was similar to the one that had been put forth earlier by Caspar Wessel. The delay in general recognition of Wessel's accomplishment is why the complex number plane came to be called the Argand plane.

1816: Frenchwoman Sophie Germain (1776–1831) was awarded a prize by the French Academy for a paper on the mathematics of elasticity. She later proved that for each odd prime $p < 100$, the Fermat equation $x^p + y^p = z^p$ has no solution in integers not divisible by

p. She introduced into differential geometry the idea of the mean curvature of a surface at a point of the surface in 1831.

1819: Englishman William George Horner (1786–1837) is known for the numerical method of solving algebraic equations that goes by his name, although a similar method had been used by the Chinese much earlier.

1822: French mathematician Jean Baptiste Joseph Fourier (1768–1830) is known for his mathematical theory of heat and especially for Fourier series. Fourier believed that any function can be resolved into a sum of sine and cosine functions. While it is not true that any function can be represented by trigonometric series, the class of functions so representable is very broad and Fourier series are useful in the study of many functions.

1824: Scotsman Thomas Carlyle (1795–1881) made an especially important English translation of Legendre's *Géométrie.*

1826: The principle of duality, important in the development of projective geometry, was enunciated by French mathematician Jean-Victor Poncelet.

1826: The theory of elliptic functions was independently and simultaneously established by German mathematician Carl Gustav Jacobi (1804–1851) and Norwegian mathematician Niels Henrik Abel (1802–1829). In abstract algebra, commutative groups are now called Abelian groups. Both Jacobi and Abel made many other contributions to mathematics.

1827: French mathematician Augustin-Louis Cauchy (1789–1857) strengthened the rigorization of analysis that got underway with the work of Lagrange and Gauss. Cauchy's numerous contributions include researches in convergence and divergence of infinite series, real and complex function theory, differential equations, determinants, and probability. In a paper of 1846, Cauchy introduced the concept of a line integral in n-dimensional space (with the incidental notion of a space higher than three included) and of a theorem today generally known as Green's theorem (George Green, 1793–1841).

1829: Russian mathematician Nicolai Ivanovitch Lobachevsky (1793–1856) published findings on non-Euclidean geometry similar to those of Gauss published later and Hungarian Janos Bolyai (1802–1860) published in 1832. Lobachevsky's publication was first, but all three of these mathematicians share credit for the creation of the geometry that comes from accepting the hypothesis of the acute angle, now known as Lobachevskian or hyperbolic geometry.

1830: French mathematician Siméon-Denis Poisson (1781–1840) had numerous mathematical publications. He applied probabilities to social areas where significant statistical information was available to him.

1830: George Peacock (1791–1858) worked on reforming mathematical study in England. In his *Treatise on Algebra,* he attempted to give algebra a logical treatment comparable to that of Euclid's *Elements.*

1830: English mathematician Charles Babbage (1792–1871) was one of the early mathematicians to work on machines to automatically do a series of arithmetic operations.

1831: German Julius Plücker (1801–1868) developed a coordinate system for the projective plane to deal with points at infinity with his introduction of homogeneous coordinates.

1831: Scotswoman Mary Fairfax Somerville (1780–1872) wrote a popular exposition of Laplace's *Traité de mécanique céleste.*

1832: Frenchman Évariste Galois (1811–1832) essentially created the study of groups that was carried out by his successors. In 1830, he was the first to use the term "group" in its technical sense. He also made contributions to theory of equations. Galois died in a duel at age 21.

1834: Swiss geometer Jacob Steiner (1796–1863) made numerous original contributions to higher synthetic geometry.

1837: Trisection of an angle and duplication of a cube were proved impossible.

1841: *Archiv der Mathematik und Physik* was founded and *Nouvelles annales de mathématiques* was founded a year later, the earliest permanent periodicals devoted to teachers' interests rather than mathematical research.

1843: Czechoslovakian Bernhard Bolzano (1781–1848) produced a function continuous in an interval that has no derivative at any point of the interval, although Karl T. W. Weierstrass (1815–1897) was credited with the first example of this kind. Both men were proponents for rigorization in analysis. Weierstrass is known for being an outstanding teacher of advanced mathematics.

1843: Irish mathematician William Rowan Hamilton (1788–1856) invented an algebra in which the commutative law of multiplication does not hold, his quaternions.

1844: German Herman Günther Grassman (1809–1877) was the first mathematician to present a detailed theory of spaces of dimension greater than three.

1847: German geometer Karl Georg Christian von Staudt (1798–1867) freed projective geometry of any metrical basis in his *Geometrie der Lage.*

1847: English mathematician George Boole (1815–1864) published a pamphlet entitled *The Mathematical Analysis of Logic* in which he maintained that the essential character of mathematics lies in its form rather than in its content; mathematics is not merely the science of measurement and number, but is any study consisting of symbols along with precise rules of operation upon those symbols, the rules being subject only to the requirement of inner consistency.

1849: German mathematician Peter Gustav Lejeune Dirichlet (1805–1859) analyzed the convergence of Fourier series, which led him to generalize the function concept. He also facilitated the comprehension of some of Gauss's more abstruse methods and contributed notably to number theory.

1850: Frenchman Amédée Mannheim (1831–1906) standardized the modern slide rule.

1852: French mathematician Michel Chasles (1793–1880) contributed notably to synthetic geometry.

1854: German mathematician Georg Friedrich Bernhard Riemann (1826–1866) contributed notably to analysis and non-Euclidean geometry. Riemann showed that a consistent geometry can be developed from the hypothesis of the obtuse angle; this geometry is known as Riemannian or elliptic geometry today.

1854: English mathematician George Boole expanded and clarified an earlier pamphlet of 1847 into a book entitled *Investigation of the Laws of Thought*, in which he established both formal logic and a new algebra, the algebra of sets known today as Boolean algebra.

1857: English mathematician Arthur Cayley (1821–1895) devised a noncommutative algebra, the algebra of matrices, which is not commutative under multiplication.

1865: The London Mathematical Society was founded and published the *Proceedings of the London Mathematical Society*. It was the earliest of a number of large mathematical societies that were formed in the second half of the nineteenth century that had regular official periodicals. These became important because they provided forums in which mathematicians could congregate, publish, and set policies.

1872: German mathematician Felix Klein (1849–1925) set forth a definition of "a geometry" that served to codify essentially all the existing geometries of the time and pointed the way to promising geometrical research. The program is known as the Erlanger Programm.

1872: German mathematician Richard Dedekind (1813–1916) published his idea of "Dedekind cuts" as a way of providing an arithmetic definition of the real numbers (he had come up with the idea in 1858). Dedekind, along with Georg Cantor, showed how to construct the real numbers from the rational numbers, and Dedekind completed the process of arithmetizing analysis by characterizing the natural numbers, and hence rational numbers, in terms of sets in a work published in 1888. Dedekind gave a useful definition of an infinite set as one that is equivalent to some proper subset of itself.

1873: French mathematician Charles Hermite (1822–1901) proved that *e* is transcendental.

1874: The Birth of Set Theory. German mathematician Georg Cantor (1845–1918) published a paper in *Crelle's Journal* in which he showed, among other things, that the set of algebraic numbers can be placed in one-to-one correspondence with the natural numbers (countable in later terminology) but that the set of real numbers is not countable. This established for the first time the fact that there are different orders of infinity. Cantor proceeded during the latter quarter of the 19th century to develop naïve (non-axiomatic) set theory.

1877: In 1850, English mathematician James Joseph Sylvester (1814–1897) coined the term "matrix" in the sense that it is used today. Sylvester made important contributions to modern algebra. He came to America in 1877 to chair the mathematics department at the newly opened Johns Hopkins University in Baltimore and helped develop a tradition of graduate education in mathematics in the United States.

1878: Sylvester founded the *American Journal of Mathematics*. It is the oldest mathematics journal in the western hemisphere that has been in continuous publication.

1881: Josiah Willard Gibbs (1839–1903) in America and Oliver Heaviside (1850–1925) in England independently realized that the full algebra of quaternions was not necessary for discussing physical concepts. Gibbs published his version of vector analysis in 1881 and 1883 and Heaviside published his methods in papers on electricity in 1882 and 1883.

1882: German mathematician Ferdinand Lindemann (1852–1939) proved that π is transcendental. From this fact, the impossibility of squaring the circle with Euclidean tools easily follows.

1888: The American Mathematical Society (AMS) was founded (under the name of the New York Mathematical Society) and the *Bulletin of the American Mathematical Society* was begun. In 1900, the society added its *Transactions* and in 1950 its *Proceedings*.

1888: Russian mathematician Sonja Kovalevsky (1850–1891) was awarded the prestigious Prix Bordin for her memoir *On the Problem of the Rotation of a Solid Body about a Fixed Point*.

1889: Italian mathematician Giuseppe Peano 1858–1932 attempted to deduce the truths of mathematics from pure logic in a small tract that contains his famous postulates for the natural numbers.

1892: *Jahresbericht*, the professional journal of Deutsche Mathematiker-Vereinigung (organized in 1890), was founded. The *Jahresbericht* contained a number of extensive reports on modern developments in different fields of mathematics; these reports may be regarded as forerunners of the later large encyclopedias of mathematics.

1895: French mathematician Jules Henri Poincaré (1854–1912) contributed to virtually every area of mathematics. His "Analysis situs" (1895) is the first significant paper devoted wholly to topology.

1896: The French and Belgian mathematicians J. Hadamard (1865–1963) and C. J. de la Vallée Poussin (1866–1962) independently proved the prime number theorem: Let A_n denote the number of primes less than n. Then $\left(A_n / \ln(n) \right)/n$ approaches 1 as n becomes larger and larger.

1899: German mathematician David Hilbert (1862–1943) made highly important contributions in many areas of mathematics. In his *Grundlagen der Geometrie* (1899), Hilbert sharpened the mathematical method from the material axiomatics of Euclid to the formal axiomatics of today. Hilbert founded the formalist school of mathematics.

1904: Henri Lebesgue (1875–1941) generalized the Hiene–Borel theorem to arbitrary infinite collections.

1906: English mathematicians Grace Chisholm Young (1868–1944) and her husband William Henry Young wrote the first comprehensive textbook on set theory and its applications to function theory, *The Theory of Sets of Points*. In 1895, Grace Chisholm became the first woman to receive a German doctorate through the regular examination process (women were not admitted to graduate schools in England at that time).

1906: French mathematician Maurice Fréchet (1878–1973) inaugurated the study of abstract spaces with his introduction of the concept of a metric space.

1908: Dutch mathematician L. E. J. Brouwer (1881–1966) originated the intuitionist school about this time, although some of the intuitionist ideas had been enunciated earlier. The intuitionist thesis is that mathematics is to be built solely by finite constructive methods on the intuitively given sequence of natural numbers.

1910: English mathematicians Bertrand Russell (1872–1970) and Alfred North Whitehead (1861–1947) wrote *Principia Mathematica*. The basic idea of the *Principia* is the identification of much of mathematics with logic by the deduction of the natural number system, and hence of the great bulk of existing mathematics, from a set of premises or postulates for logic itself.

1915: The Mathematical Association of America (MAA) was founded. Although the AMS and MAA are both concerned with university mathematics, the AMS leans more toward research and the MAA more toward teaching. The two organizations together sponsor the Joint Mathematics Meetings every January.

1916: Albert Einstein (1879–1955) introduced his general theory of relativity.

1917: British number theorist G. H. Hardy (1877–1947) and Indian mathematician Srinivasa Ramanujan (1887–1920) reached penetrating results in number theory. Ramanujan had an uncanny ability to see quickly and deeply into intricate number relations. Hardy's efforts brought Rananujan to England to study at Cambridge University, and a remarkable mathematical association resulted between the two of them.

1920: The International Mathematical Union was founded, which was to become a prominent society in the twentieth century and beyond.

1922: German mathematician Amalie Emmy Noether (1882–1935) became extraordinary professor at Göttingen and kept the position until 1933 when she left Germany to accept a professorship at Bryn Mawr College in Pennsylvania and to become a member of the Institute for Advanced Study at Princeton. Her studies on abstract rings and ideal theory have been important in the development of modern algebra. Her father, Max Noether (1844–1921), was also an algebraist.

1923: Polish mathematician (1892–1945) Stefan Banach introduced the notion of what is now called a Banach space, a vector space possessing a norm under which all Cauchy sequences converge.

1931: Austrian logician Kurt Gödel (1906–1978) showed that it is impossible for a sufficiently rich formalized deductive system to prove consistency of the system by methods belonging to the system (Incompleteness theorem). Later, about 1940, Gödel showed that the continuum hypothesis (that the cardinal number of the reals is the next cardinal number after the cardinal number of the natural numbers) is consistent with a famous postulate set of set theory (Zermelo–Fraenkel), provided these postulates themselves are consistent, and conjectured that the denial of the continuum hypothesis is also consistent with the postulates of set theory.

1934: Bourbaki's works started. Nicolas Bourbaki is the collective pseudonym employed by a group of French mathematicians who met in a Paris café to discuss writing a new calculus textbook for French university students. From that beginning, the project grew into the more ambitious undertaking of developing with rigor the essentials of modern French mathematics. The membership has varied over the years. Members must leave the group at age 50.

1940s: IBM's automatic sequence controlled calculator (ASCC) was debuted in 1944, and may be cited as the beginning of the computer age. The electronic numerical integrator and computer (ENIAC), which debuted in 1945, was the first general purpose, completely electrical computer. It used vacuum tubes rather than electromechanical switches. Many mathematicians played a role in computer development, and computers would come to play an important role in many areas of mathematics research and education.

1940: *Mathematical Reviews*, containing abstracts and reviews of the current mathematical literature in the world, was organized by mathematical groups both in the United States and abroad to help researchers keep abreast of mathematical work in their fields. The increase of mathematical specialization in the twentieth century led to the formation of many journals focused on specific subfields of mathematics.

1957: The Soviet Union launched the first satellite, *Sputnik*, into space. The shock in the United States of the unexpected venture into space caused Congress to establish the National Aeronautics and Space Administration (NASA) in 1958, which led to men on the moon and huge breakthroughs in computers. There was a renewed emphasis on mathematics and science education with many government-sponsored programs to support graduate work in these fields.

1960s: The era of so-called new math that emphasized understanding over rote memorization was ushered in at the urging of the National Council of Teachers of Mathematics and other groups concerned with teaching mathematics. The impact was perhaps larger at the elementary school level than at the high school level. In high school, there was consideration of what to do after the traditional two years of algebra and year of geometry. For many years, the courses above these three years of high-school mathematics were a semester each of solid geometry and trigonometry. Experimental courses such as functions and matrices were added as the final year of high school. Many of the reforms fell out of favor after a decade or so with criticism of such things as the emphasis on vocabulary in elementary school mathematics and lack of emphasis on memorizing addition and multiplication tables. Taking algebra I before high school became common resulting in more courses needed at the upper levels of high school. Today, advanced placement calculus is common at the senior year; the study of calculus for decades before had been strictly a college-level course.

1963: Paul J. Cohen (1934–2007) followed up on a conjecture of Gödel some 25 years earlier that a denial of the continuum hypothesis in Zermelo–Fraenkel set theory would not lead to contradictions in the theory. Cohen was able to show that both the continuum hypothesis and the axiom of choice (given a collection of mutually disjoint, nonempty sets, there exists a set which has as its elements exactly one element from each set in the given collection of sets) are independent of Zermelo–Fraenkel set theory (named for mathematicians Ernst Zermelo and Abraham Fraenkel) without the axiom of choice; this makes the situation analogous to that of the parallel postulate in Euclidean geometry.

1969: The National Association of Mathematicians (NAM) was founded to address the needs of the minority mathematical community.

1971: The first pocket calculator was offered for sale in the consumer market. Pocket calculators quickly became cheaper and more sophisticated. Hungarian mathematician John von Neumann (1903–1957) was the person most responsible for initiating the first fully electronic calculator and for the concept of a stored program digital computer. Von Neumann migrated to America in 1930 and became a permanent member of the Institute for Advanced Study at Princeton in 1933. Computers were initially designed to solve military problems, but are now pervasive in the form of personal computers for use in education and business.

1971: The Association for Women in Mathematics was founded.

1976: The four-color theorem, first conjectured in 1852, was established by Kenneth Appel (b. 1932) and Wolfgang Haken (b. 1928) of the University of Illinois. The four-color theorem of topology states that any map on a plane or sphere needs at most four colors to color it so that no two countries sharing a common boundary will have the same color. The Appel–Haken solution of the four-color problem depended on intri-cate computer-based analysis and raised philosophical questions of just what should be allowed to constitute a proof of a proposition in mathematics.

1985: Supercomputers came into general use.

1994: Princeton mathematician Andrew Wiles (b. 1953) completed the proof of Fermat's Last Theorem after correcting a flaw in his 1993 work that had taken seven years to complete. Fermat's Last Theorem states that there do not exist positive integers x, y, z, n such that $x^n + y^n = z^n$ when $n > 2$.

2002: Russian mathematician Grigori Perelman (b. 1966) posted a proof of the long-standing Poincaré conjecture in three installments on the Internet. The Poincaré conjecture states essentially that any closed three-dimensional manifold in which every closed curve can be shrunk to a point is homeomorhic to the three-dimensional sphere. In 2006, the International Mathematical Union awarded Perelman its prestigious Fields Medal. Perelman declined to accept the medal, which also included a million-dollar prize, saying that "everyone understood that if the proof is correct then no other recognition is needed." Subsequently, Perelman decided to drop out of mathematics entirely.

Phillip Johnson
Appalachian State University

Resource Guide

Books

Aaboe, Asger. *Episodes From the Early History of Mathematics*. Washington, DC: Mathematical Association of America, 1975.

Adrian, Yeo. *The Pleasures of Pi and Other Interesting Numbers*. Singapore: World Scientific Publishing, 2006.

Agresti, A. *Categorical Data Analysis*. Hoboken, NJ: Wiley, 2002.

Aho, A. V., J. E. Hopcrotf, and J. D. Ullman. *The Design and Analysis of Computer Algorithms*. Reading, MA: Addison-Wesley, 1976.

Albert, Jim, and Jay Bennett. *Curve Ball: Baseball, Statistics, and the Role of Chance in the Game*. New York: Springer-Verlag, 2001.

Ascher, Marcia. *Mathematics Is Everywhere: An Exploration of Ideas Across Cultures*. Princeton, NJ: Princeton University Press, 2002.

Ball, W. W. Rouse. *A Short Account of the History of Mathematics*. New York: Sterling Publishing Company, 2001.

Barnett, Raymond, Michael Ziegler, and Karl Byleen. *Calculus for Business, Economics, Life Science, and Social Science*. Upper Saddle River, NJ: Prentice-Hall, 2005.

Baumohl, Bernard. *The Secrets of Economic Indicators: Hidden Clues to Future Economic Trends and Investment Opportunities*. 2nd ed. Upper Saddle River, NJ: Pearson Education, 2008.

Beckmann, Petr. *A History of π (Pi)*. New York: Barnes & Noble, 1971.

Behrends, Ehrhard. *Five-Minute Mathematics*. Providence, RI: American Mathematical Society, 2008.

Bell, Eric Temple. *Men of Mathematics*. New York: Simon & Schuster, 1937.

Bennett, Jay, and James Cochran. *Anthology of Statistics in Sports*. Philadelphia, PA: Society for Industrial and Applied Mathematics, 2005.

Berggren, Lennart, Jon Borwein, and Peter Borwein. *Pi: A Source Book*. New York: Springer-Verlag, 1997.

Berlekamp, Elwyn R., John H. Conway, and Richard K. Guy. *Winning Ways for Your Mathematical Plays*. Natick, MA: AK Peters, 2001.

Blackwell, William. *Geometry in Architecture*. Hoboken, NJ: Wiley, 1984.

Blatner, David. *The Joy of π*. New York: Walker & Co., 1997.

Blue, Ron, and Jeremy White. *The New Master Your Money: A Step-by-Step Plan for Gaining and Enjoying Financial Freedom*. Chicago: Moody, 2004.

Blum, Raymond. *Mathemagic*. New York: Sterling Publishing, 1992.

Bodie, Zvi, Alex Kane, and Alan Marcus. *Investments.* Chicago, IL: McGraw-Hill/Irwin, 2008.

Borwein, Jonathan, and Peter Borwein. *A Dictionary of Real Numbers.* Pacific Grove, CA: Brooks/Cole Publishing Co., 1990.

Boyer, C. B. *A History of Mathematics.* Hoboken, NJ: Wiley, 1968.

Boyer, C. B. *The History of the Calculus and Its Conceptual Development.* New York: Dover Publications, 1949.

Brealey, Richard A., Stewart C. Myers, and Franklin Allen. *Principles of Corporate Finance.* 9th ed. New York: McGraw-Hill, 2008.

Bressoud, David. *The Queen of the Sciences: A History of Mathematics.* Chantilly, VA: The Teaching Company, 2008.

Broverman, Samuel A. *Mathematics of Investment and Credit.* Winsted, CT: ACTEX Publications, 2008.

Burkett, Larry, and Brenda Armstrong. *Making Ends Meet: Budgeting Made Easy.* Gainesville, GA: Crown Financial Ministries, 2004.

Burton, David M. *The History of Mathematics: An Introduction.* New York: McGraw-Hill, 2005.

Calinger, Ronald. *A Contextual History of Mathematics.* Upper Saddle River, NJ: Prentice-Hall, 1999.

Clagett, Marshall. *Archimedes in the Middle Ages.* Madison: University of Wisconsin Press, 1964.

Closs, Michael. *A Survey of Mathematics Development in the New World.* Ottawa: University of Ottawa, 1977.

Closs, Michael, ed. *Native American Mathematics.* Austin: University of Texas Press, 1986.

Coe, Michael D. *Breaking the Maya Code.* New York: Thames and Hudson, 1992.

Copeland, Thomas E., J. Fred Weston, and Kuldeep Shastri. *Financial Theory and Corporate Policy.* 4th ed. Upper Saddle River, NJ: Pearson Education, 2005.

Cullen, Christopher. *Astronomy and Mathematics in Ancient China: The Zhou Bi Suan Jing.* Cambridge, England: Cambridge University Press, 1996.

Cuomo, Serafina. *Ancient Mathematics.* London: Routledge, 2001.

Davenport, Harold. *The Higher Arithmetic: An Introduction to the Theory of Numbers.* Cambridge, England: Cambridge University Press, 1999.

Davis, Morton D. *The Math of Money: Making Mathematical Sense of Your Personal Finances.* New York: Copernicus, 2001.

De Mestre, Neville. *The Mathematics of Projectiles in Sport.* Cambridge, England: Cambridge University Press, 1990.

Devlin, Keith. *The Math Gene: How Mathematical Thinking Evolved and Why Numbers Are Like Gossip.* New York: Basic Books, 2001.

———. *The Unfinished Game: Pascal, Fermat, and the Seventeenth-Century Letter That Made the World Modern.* New York: Basic Books, 2008.

Drobat, Stefan. *Real Numbers.* Upper Saddle River, NJ: Prentice-Hall, 1964.

Dudley, Underwood. *Numerology or What Pythagoras Wrought.* Washington, DC: Mathematical Association of America, 1997.

Eastway, Rob, and John Haigh. *Beating the Odds: The Hidden Mathematics of Sport.* London: Robson Books, 2007.

Eglash, Ron. *African Fractals: Modern Computing and Indigenous Design.* New Brunswick, NJ: Rutgers University Press, 1999.

Eves, Howard. *An Introduction to the History of Mathematics.* New York: Saunders College Publishing, 1990.

Flegg, G. *Numbers: Their History and Meaning.* New York: Schocken Books, 1983.

Friberg, Jöran. *Unexpected Links Between Egyptian and Babylonian Mathematics.* Singapore: World Scientific Publishing Co., 2005.

Friedman, Arthur. *World of Sports Statistics: How the Fans and Professionals Record, Compile, and Use Information.* New York: Athenaeum, 1978.

Fries, Christian. *Mathematical Finance: Theory, Modeling, Implementation.* Hoboken, NJ: Wiley, 2007.

Frumkin, Norman. *Guide to Economic Indicators.* Armonk, NY: M. E Sharpe, 2000.

Gamow, George. *One, Two, Three... Infinity.* New York: Viking Press, 1947.

Gardner, David, and Tom Gardner. *The Motley Fool Personal Finance Workbook: A Foolproof Guide to Organizing Your Cash and Building Wealth.* New York: Fireside Books, 2003.

Gardner, Martin. *Mathematics, Magic and Mystery.* New York: Dover, 1956.

Gay, Timothy. *The Physics of Football.* New York: HarperCollins, 2005.

Gerdes, Paulus. *Geometry From Africa: Mathematical and Educational Explorations.* Washington, DC: Mathematical Association of America, 1999.

Gillings, R. J. *Mathematics in the Time of the Pharaohs*. New York: Dover Publications, 1982.

Gutstein, Eric, and Bob Peterson, eds. *Rethinking Mathematics: Teaching Social Justice by the Numbers*. Milwaukee, WI: Rethinking Schools, 2005.

Hadamard, Jacques. *A Mathematician's Mind*. Princeton, NJ: Princeton University Press, 1996.

Hardy, G. H. *A Mathematician's Apology*. Cambridge, England: Cambridge University Press, 1941.

Henry, Granville C. *Logos: Mathematics and Christian Theology*. Lewisburg, PA: Bucknell University Press, 1976.

Hersh, Rueben. *What Is Mathematics, Really?* New York: Oxford University Press, 1997.

Hoyle, Joe Ben, Thomas F. Schaefer, and Timothy S. Doupnik. *Fundamentals of Advanced Accounting*. New York: McGraw-Hill, 2010.

Kalbfleisch, John D., and Ross L. Prentice. *The Statistical Analysis of Failure Time Data*. Hoboken, NJ: Wiley, 2002.

Katz, Victor J., ed. *Mathematics of Egypt, Mesopotamia, China, India, and Islam: A Sourcebook*. Princeton, NJ: Princeton University Press, 2007.

Kellison, Stephen G. *Theory of Interest*. New York: McGraw-Hill, 2009.

Kimmel, Paul D., Jerry J. Weygandt, and Donald E. Keiso. *Financial Accounting: Tools for Business Decision Making*. Hoboken, NJ: Wiley, 2009.

King, Jerry. *The Art of Mathematics*. New York: Plenum Press, 1992.

Klein, John P., and Melvin L. Moeschberger. *Survival Analysis: Techniques for Censored and Truncated Data*. New York: Springer-Verlag, 1997.

Kline, M., *Mathematical Thought From Ancient to Modern Times*. New York: Oxford University Press, 1972.

Koetsier, T., and L. Bergmans, eds. *Mathematics and the Divine: A Historical Study*. Amsterdam: Elsevier, 2005.

Longe, Bob. *The Magical Math Book*. New York: Sterling Publishing, 1997.

Martzloff, Jean-Claude. *A History of Chinese Mathematics*. New York: Springer-Verlag, 1987.

Moses, Robert P., and Charles E. Cobb, Jr. *Radical Equations: Civil Rights From Mississippi to the Algebra Project*. Boston: Beacon Press, 2001.

Mullis, Darrell, and Judith Handler Orloff. *The Accounting Game: Basic Accounting Fresh From the Lemonade Stand*. Naperville, IL: Sourcebooks, 2008.

Nahin, Paul J. *Dr. Euler's Fabulous Formula*. Princeton, NJ: Princeton University Press, 2006.

Nasar, Sylvia. *A Beautiful Mind: The Life of Mathematical Genius and Nobel Laureate John Nash*. New York: Simon & Schuster, 2001.

Oliver, Dean. *Basketball on Paper: Rules and Tools for Performance Analysis*. Washington, DC: Brassey's, 2004.

Pullan, J. M. *The History of the Abacus*. New York: F. A. Praeger, 1969.

Rafiquzzaman, M. *Fundamentals of Digital Logic and Microcomputer Design*. Hoboken, NJ: Wiley, 2005.

Rudin, W. *Principles of Mathematical Analysis*. New York: McGraw-Hill, 1953.

Salem, Lionel, Frédéric Testard, and Coralie Salem. *The Most Beautiful Mathematical Formulas*. Hoboken, NJ: Wiley, 1992.

Schwarz, Alan. *The Numbers Game: Baseball's Lifelong Fascination with Statistics*. New York: St. Martin's Press, 2004.

Smith, D. E. *History of Mathematics*. Vol. 2. New York: Dover Publications, 1958.

Solow, Daniel. *How to Read and Do Proofs: An Introduction to Mathematical Thought Process*. Hoboken, NJ: Wiley, 1982.

Steen, Lynn A. *On the Shoulders of Giants: New Approaches to Numeracy*. Washington, DC: National Academy Press, 1990.

Sterrett, Andrew. *101 Careers in Mathematics*. Washington, DC: The Mathematical Association of America, 1996.

Suzuki, Jeff. *A History of Mathematics*. Upper Saddle River, NJ: Prentice Hall, 2002.

Taylor, Alan D. *Mathematics and Politics: Strategy, Voting Power, and Proof*. New York: Springer-Verlag, 1995.

van der Waerden, B. L. *Geometry and Algebra in Ancient Civilizations*. Berlin: Springer, 1983.

Venema, G.A. *The Foundations of Geometry*. Upper Saddle River, NJ: Pearson Prentice Hall, 2006.

Weygandt, Jerry J., Paul D. Kimmel, and Donald E. Keiso. *Managerial Accounting: Tools for Business Decision Making*. Hoboken, NJ: Wiley, 2008.

Winkler, Peter. *Mathematical Puzzles: A Connoisseur's Collection*. Natick, MA: AK Peters, 2004.

Wright, Tommy, and Joyce Farmer. *A Bibliography of Selected Statistical Methods and Development Related to Census 2000*. Washington, DC: U.S. Bureau of the Census, 2000.

Yeldham, F. A. *The Teaching of Arithmetic Through Four Hundred Years (1535–1935).* London: G. G. Harrap & Company, 1935.

Yong, L. L., and A. T. Se. *Fleeting Footsteps.* Singapore: Word Scientific Publications, 2004.

Zaslavsky, Claudia. *Africa Counts: Number and Pattern in African Culture.* Chicago: Lawrence Hill Books, 1999.

Zill, D. G. *Calculus with Analytic Geometry.* Boston: Prindle, Weber & Schmidt, 1985.

Journals and Magazines

The AMATYC Review
The American Mathematical Monthly
Association for Women in Mathematics Newsletter
Biometrics
Chance
The College Mathematics Journal
Experimental Mathematics
The Fibonacci Quarterly
Historia Mathematica
IMU-Net
Involve
Journal of Humanistic Mathematics
Journal of Integer Sequences
Journal of Recreational Mathematics
Journal of Statistics Education
Loci
MAA FOCUS
Math Horizons
Mathematics Magazine
Mathematics Teacher
NAM Newsletter
Notices of the American Mathematics Society
The Pentagon
Pi Mu Epsilon Journal
Plus Magazine
PRIMUS
Rose-Hulman Undergraduate Mathematics Journal
SIAM Review
Scholastic Math
Significance
Teaching Children Mathematics
Undergraduate Mathematics and Its Applications

Internet

American Institute of Mathematics
  www.aimath.org

The Algebra Project
  www.algebra.org
AMATYC
  www.amatyc.org
American Mathematical Society
  www.ams.org
American Statistical Association
  www.amstat.org
Association for Women in Mathematics
  www.awm-math.org
CryptoKids
  www.nsa.gov/kids
Datamath Calculator Museum
  www.datamath.org
Illuminations
  illuminations.nctm.org
MacTutor History of Mathematics
  www-history.mcs.st-and.ac.uk
Mathematical Fiction
  http://kasmana.people.cofc.edu/MATHFICT
Math for America
  www.mathforamerica.org
Math Forum
  www.mathforum.com
Math Fun Facts!
  www.math.hmc.edu/funfacts
MathDL
  mathdl.maa.org/mathDL
Mathematical Association of America
  www.maa.org
Mathematical Science Research Institute
  www.msri.org
The Museum of Mathematics
  www.momath.org
National Association of Mathematicians
  www.nam-math.org
National Council of Teachers of Mathematics
  www.nctm.org
RadicalMath
  www.radicalmath.org
Society for Industrial and Applied Mathematics
  www.siam.org
We Use Math
  www.weusemath.org
Wolfram MathWorld
  www.mathworld.wolfram.com

Glossary

absolute value

For a real number x, the absolute value of x, written $|x|$, is the "unsigned" version of the number. If x is non-negative, then $|x| = x$; if x is negative, then $|x| = -x$. For a complex number $x = a + bi$, the definition is slightly more complicated:

$$|x| = \sqrt{a^2 + b^2}\,.$$

In both cases, $|x|$ represents the magnitude of x, and $|x - y|$ represents the distance between x and y.

acute angle

A nonzero angle that is smaller than a right angle. The measure of an acute angle is between 0 and 90 degrees.

algebra

A branch of mathematics dealing with the formal properties and behavior of symbolic operations, relations, and structures. Mathematicians use the word "algebra" much, much more broadly than it is used in everyday usage; nonmathematicians often use the term specifically for what is taught in the secondary school curriculum under the heading "algebra," which is only the most elementary part of algebra.

algebraic number

A complex number is called algebraic if it is the root of some polynomial with integer coefficients. For example, all rational numbers are algebraic, as are $\sqrt{19}$, i,

$$\sqrt[3]{5 + 2\sqrt{5}}\,,$$

and all five roots of $x^5 - x + 1 = 0$. Though any number that can be expressed in terms of arithmetic operations and roots must be algebraic, not all algebraic numbers can be written in this way (in particular the roots of the given polynomial cannot be so written).

analysis

Elements of analysis are part of the curriculum under the name "calculus." The most elementary part of analysis is part of the curriculum under the name "calculus."

antiderivative

If $f(x)$ is the derivative of a function $F(x)$, then we say that $F(x)$ is an antiderivative (integral) of $f(x)$.

Arabic numerals

The familiar base-ten number system, with digits 0, 1, 2, 3, 4, 5, 6, 7, 8, 9 and a positional place value system based on powers of 10.

arithmetic mean

The arithmetic mean of a set of numbers $a_1, a_2, \ldots, a_n$ is

$$\frac{a_1 + a_2 + \cdots + a_n}{n}.$$

The arithmetic mean is often called the "average" or just the "mean."

arithmetic progression

A sequence (finite or infinite) such that the difference of any two consecutive terms is the same; equivalently, a sequence in which each term (except the first and last) is the arithmetic mean of the terms immediately preceding and following.

associativity

A binary operation is said to be associative, or (less frequently) to associate, if the (chronological) order in which it is evaluated does not affect the answer. In symbols, $\sim$ is associative if $(x \sim y) \sim z = x \sim (y \sim z)$ for all x, y, z for which the operation is defined.

base b

See *number bases*.

bijection

A function is a bijection (also called a "one-to-one correspondence") if it is both injective and surjective.

binary (notation system)

Base two. (See *number bases*.)

binomial coefficient

The binomial coefficient

$$\binom{n}{r}$$

(often read "n choose r"), defined for integers $n \geq r \geq 0$ by the formula

$$\frac{n!}{r!(n-r)!};$$

this binomial coefficient counts the number of combinations of r elements from a set of size n.

Binomial Theorem

A theorem of basic algebra which connects the binomial coefficients to the numbers appearing in the expansion of a binomial raised to a power (indeed this is the reason they are called binomial "coefficients"). The precise statement is

$$(x+y)^n = \sum_{k=0}^{n} \binom{n}{k} x^k y^{n-k}.$$

calculus

A branch of mathematics that focuses on (1) evaluating the rate at which a function changes and (2) evaluating the rate at which a function accumulates. If we plot a single-valued function $y = f(x)$, these correspond to (1) the slope of $f(x)$ and (2) the area under the curve $f(x)$.

Cartesian product

If A and B are any sets whatsoever, then the Cartesian product $A \times B$ is the set of all ordered pairs (a, b) with a in A and b in B.

circle

In a plane, the set of all points at a fixed distance (the "radius") from a given point (the "center").

claim

See *proposition*.

closed form

An expression or formula is in closed form if it is written explicitly in a way that could be directly evaluated. Recursions, summation (or product) notations, and ellipses to indicate omitted terms cannot be part of closed-form expressions.

codomain

The set where a function's values ("outputs") live. Specifying the codomain is part of defining a function.

combination

One of the fundamental enumeration problems. The combinations of r objects from a set of n objects are the "unordered" sets of r distinct objects from the set. There are

$$\frac{n!}{r(n-r)!}$$

combinations of r objects from a set of size n.

commutativity

A binary operation is said to be commutative, or to commute, if the left-to-right order of its arguments does not matter. That is, $\sim$ is commutative if $x \sim y = y \sim x$ for all x, y for which the operation makes sense. In the context of group theory, a commutative group operation is usually called "abelian."

complex conjugate

The (complex) conjugate of a complex number $z = x + iy$ is $\bar{z} = x - iy$; that is, complex conjugation changes the sign of the imaginary part. Complex conjugation preserves the structure of the complex number system; in particular $\overline{x + y} = \bar{x} + \bar{y}$ and $\overline{xy} = (\bar{x})(\bar{y})$.

complex numbers

The set of all numbers of the form $a + ib$, where a and b are real numbers and i is the imaginary unit, the square root of -1 (a is sometimes called the real part, and b the imaginary part). Complex numbers can be added, subtracted, multiplied, and divided (except by zero). Geometrically, the set of complex numbers can be visualized as a plane. The complex plane is the complex analogue of the number line. The set of complex numbers is traditionally denoted by a blackboard-bold $\mathbb{C}$.

composite

A positive integer n is composite if it can be written as a product $n = ab$, where a and b are positive integers greater than 1. (For positive integers *greater than* 1, "composite" means "not prime.")

congruent

In geometry, one object is congruent to another object of the same type if they are "the same size and shape." For line segments, this means they have the same length; for angles, this means they have the same angle measure. For triangles (and other figures, though the term is used most frequently to refer to triangles), congruence means that corresponding sides have the same length and corresponding angles have the same measure.

In modular arithmetic, the term "congruent" is often used as the analogue of "equal." For example 2 and 16 are congruent modulo 7. Equations in modular arithmetic are thus often called congruences.

conic section

Any geometric figure that can be realized as the intersection of a full (double) cone with a plane. Such a figure can be defined by an equation of the form $ax^2 + bxy + cy^2 + dx + ey + f = 0$. Setting aside degenerate cases, the conic sections have three types: "ellipses," "hyperbolas," and "parabolas."

conjecture

An assertion that is believed to be true (at least by some people) but that has not yet been rigorously demonstrated. The term "conjecture" is not generally applied to any wild guess; a conjecture is typically justified (but not proven) by a heuristic argument or experimental evidence.

continuous function

Intuitively, a function is continuous if whenever x is "near" y, then $f(x)$ is "near" $f(y)$. How exactly this is formalized depends on context, especially the domain and codomain of the function. For real-valued functions of a single real variable, it is often said that a function is continuous if its graph could be drawn "without picking up your pencil" (which may be a helpful heuristic, but is not, strictly, true). The formal definition in that case is that a function is continuous at x_0 if, for all $\varepsilon > 0$, there exists a $\delta > 0$ such that, $|x - x_0| < \delta$ implies $|f(x) - f(x_0)| < \varepsilon$. Continuity is conceptually very close to limits. Continuity is a central notion to analysis and to topology.

contrapositive

If a statement takes the form "If P, then Q," then the contrapositive is the statement "If not Q, then not P." For example, the contrapositive of "If I have meditated, then I am relaxed" is "If I am not relaxed, then I have not meditated." A statement and its contrapositive are either both true or both false. It is often more straightforward to prove the contrapositive form of a theorem than to prove the theorem directly.

coordinate geometry

A technique for studying Euclidean geometry by identifying points in the plane with ordered pairs of real numbers (or points in space with ordered triples, and generally points in n-dimensional space with ordered n-tuples) based on their position relative to a family of axes (sing. axis). This allows us to understand geometric objects in terms of numbers and equations, using algebra and arithmetic knowledge to solve geometric problems.

corollary

A proposition that is an easy consequence of a previous theorem.

countable (set)

A set is countable if it is finite or if it is "the smallest sort of infinite," if its elements can be listed $a_0, a_1, a_2, \ldots$. There are countably many integers, countably many rational numbers, and countably many polynomials with integer coefficients. However, the set of real numbers is uncountable.

decimal (notation system)

The familiar notation for writing real numbers, base ten. (See *number bases*.)

definite integral

The definite integral

$$\int_a^b f(x)\,dx$$

represents the "accumulation" of the function $f(x)$ over the interval $a \le x \le b$. Geometrically, this is the signed area between the graph of the function and the x-axis. ("Signed" indicates that area above the x-axis is weighted positively, and area below is weighted negatively.)

degrees

One of the two most commonly used units of angle measure, denoted by the symbol °. There are 360 degrees in one full circle, and 90 degrees in a right angle. The primary advantage of this unit (and the historical reason for its use) is that 360 has many factors, so that many important angles are a whole number of degrees.

derivative

See *differentiation*.

differential equations

Equations or systems of equations relating one or more functions to their derivatives. A wide array of special-

ized techniques have been developed to solve such equations exactly or to approximate solutions. Such equations have myriad uses in mathematics (pure and applied) and in science and engineering.

differentiation

The process of finding the "derivative" of a function, which is intuitively the "rate of change" of the value of a function with respect to the input.

discrete mathematics

A very wide branch of mathematics dealing with finite or countable objects and structures. This includes counting (enumeration) problems, partitions, graph theory, matroids, designs, and Ramsey theory.

domain

The set of valid arguments ("inputs") to a function. Specifying the domain is part of defining a function.

e (exponential or Euler's constant)

Arguably the most "natural" base for exponential and logarithmic functions. The constant e can be defined by

$$e = 1 + \frac{1}{1!} + \frac{1}{2!} + \frac{1}{3!} + \cdots$$

or by the formula

$$\lim_{n \to \infty} \left(1 + \frac{1}{n}\right)^n$$

(or numerous other formulae). It has the approximate value 2.71828. . . . See also *exponential function* and *logarithm*.

ellipse

The set of points at a fixed sum of distances from two given points. That is, if A and B are two points (called the *foci* – sing. *focus*) and r is any positive constant greater than the distance AB, then the set of all points P such that $AP + BP = r$ is an ellipse. After suitably rotating and translating the coordinate axes, any ellipse can be described by an equation of the form

$$\frac{x^2}{a^2} + \frac{y^2}{b^2} = 1.$$

enumeration

Counting and/or listing the objects or structures of a particular kind of type. Questions of the shapes "How many X have the properties Y?" or "What are all the X that have the properties Y?" lead to enumeration problems.

exponential function
For any positive base b, there is a base-b exponential function, traditionally written $f(x) = b^x$.

factorial
If n is a nonnegative integer, then $n!$ (read "n factorial") is defined by $0! = 1$, $1! = 1$, and generally $n! = 1 \times 2 \times 3 \times \cdots \times (n-1) \times n$. Factorials show up frequently throughout mathematics, especially in combinatorics or in answers to problems where combinatorics is "in the background." Using more advanced methods, it is possible to define $x!$ for some values of x that are not nonnegative integers. This generalization is called the "gamma function."

Fields Medal
Officially called the "International Medal for Outstanding Discoveries in Mathematics," the Fields Medal is widely considered the mathematician's analogue of the Nobel Prize. The prize is awarded by the International Mathematical Union once every four years to two, three, or four mathematicians no older than 40 years old.

finite (set)
A set is finite if its elements can be put in one-to-one correspondence with the elements of a set $\{1, 2, 3, \ldots, n\}$ for some n.

function
Formally, a function f has three parts: a domain set D, a codomain set C, and a rule which corresponds each domain element to a unique codomain element. More formally, there is a subset S_f of the product $D \times C$, such that each element of D is the first member of exactly one of the ordered pairs in the subset. If (x, y) is the unique element of S_f beginning with x, we say that $f(x) = y$. Informally, the domain is the set of inputs, and the codomain is the set of potential outputs. Functions are most often specified by an algebraic expression such as

$$f(x) = x^3 + x^2$$

but this is not necessary.

Fundamental Theorem of Algebra
Every polynomial

$$p(x) = a_n x^n + a_{n-1} x^{n-1} + \cdots + a_1 x + a_0$$

with coefficients in the complex numbers has at least one root in the complex numbers. Furthermore, if roots are counted with multiplicity, an nth-degree polynomial will always have exactly n roots.

Fundamental Theorem of Arithmetic
Every integer greater than 1 can be written as a product of primes; furthermore, the prime factorization is unique except for the order in which the factors are written.

Fundamental Theorem of Calculus
Any of several important theorems of elementary analysis relating differentiation and integration as (in some suitable sense) inverse processes. The most commonly given are the following two:

1. If a function F is an antiderivative of a function f on an interval $[a, b]$, then

$$\int_a^b f(x)\, dx = F(b) - F(a)$$

(We can easily compute definite integrals given an antiderivative.

2. If f is a continuous function on some interval containing a, then the function

$$F(x) = \int_a^x f(t)\, dt$$

is an antiderivative of f. (Definite integrals can be used to define antiderivatives.)

gamma function ($\tilde{A}(s)$)
A generalized version of the factorial function which makes sense for non-integral arguments, and even nonreal arguments. The gamma function is defined (for complex numbers with positive real part) by

$$\tilde{A}(s) = \int_0^\infty t^{s-1} e^{-t}\, dt$$

and satisfies $\tilde{A}(s) = (s-1)!$ and the factorial-like identity $\tilde{A}(s+1) = s\tilde{A}(s)$. (It is possible to extend this definition through complex analysis techniques to all complex numbers except nonpositive integers.)

geometric mean

The geometric mean of a set of numbers $a_1, a_2, \ldots a_n$ is

$$\sqrt[n]{a_1 a_2 \cdots a_n}.$$

This is the geometric mean of 2 and 8 is 4.

geometric progression

A sequence (finite or infinite) such that the ratio of any two consecutive terms is the same; equivalently, a sequence in which each term (except the first and last) is the geometric mean of the terms immediately preceding and following.

geometry

A branch of mathematics dealing with shapes, sizes, lengths, angles, areas, volumes, and so on. Because many contemporary mathematics curricula stress proofs and axiomatic reasoning for the first (and often last) time in secondary-school geometry class, geometry and axiomatic reasoning are sometimes conflated in the mind of the general public.

golden ratio

The number $\varphi = \dfrac{1+\sqrt{5}}{2} \approx 1.6180339887$, or roughly 8/5.

The golden ratio can be defined as the unique positive solution to the equation

$$\frac{1}{\varphi} = \varphi - 1.$$

This number appears throughout mathematics, nature, music, and art. The golden ratio also has the interesting continued fraction representation

$$\varphi = 1 + \cfrac{1}{1 + \cfrac{1}{1 + \cfrac{1}{1 + \cfrac{1}{1 + \cfrac{1}{\ddots}}}}}$$

golden rectangle

A rectangle for which the ratio of its side lengths is equal to the golden ratio. Such a rectangle has the notable property that it can be dissected into a square and a smaller rectangle which is similar to the original. Golden rectangles have been considered the most aesthetically pleasing rectangles, and they play a role in classical art and architecture.

gradians

A somewhat obscure unit of angle measure, still occasionally seen in certain texts and certain calculators. There are 400 gradians in a full circle.

harmonic mean

The harmonic mean of a set of positive numbers $a_1, a_2, \cdots, a_n$ is

$$\frac{n}{\dfrac{1}{a_1} + \dfrac{1}{a_2} + \cdots + \dfrac{1}{a_n}}.$$

harmonic progression

A sequence (finite or infinite) in which each term (except the first and last) is the harmonic mean of the terms immediately preceding and following. Equivalently, a sequence whose reciprocals form an arithmetic progression.

harmonic series

The series

$$\sum_{n=1}^{\infty} \frac{1}{n} = 1 + \frac{1}{2} + \frac{1}{3} + \frac{1}{4} + \frac{1}{5} + \cdots.$$

This series famously diverges to infinity, that is, has no finite value. Its partial sums

$$1 + \frac{1}{2} + \frac{1}{3} + \frac{1}{4} + \frac{1}{5} + \cdots + \frac{1}{N}$$

are sometimes called the "harmonic numbers" and are well-approximated by $\log N$.

hexadecimal (notation system)

Base sixteen (the digits used are traditionally 0, 1, 2, 3, 4, 5, 6, 7, 8, 9, A, B, C, D, E, F). (See *number bases*.)

homomorphism

A function from one mathematical structure to another mathematical structure that respects all the operations and properties of that type of structure. For example, a homomorphism of rings (structures in which addition and multiplication make sense) must preserve addition and multiplication. For example, the function that maps each polynomial to its constant term is a ring homomorphism from the set of polynomials to the set of real numbers.

hyperbola

The set of points at a fixed difference of distances from two given points. That is, if A and B are two points (called the *foci* – sing. *focus*) and r is any positive constant, then the set of all points P such that $|AP - BP| = r$ is a hyperbola. After suitably rotating and translating the coordinate axes, any hyperbola can be described by an equation of the form

$$\frac{x^2}{a^2} - \frac{y^2}{b^2} = 1.$$

iff

A commonly used mathematician's abbreviation for "if and only if," used to indicate that two statements logically imply one another.

image

The set of all values realized by a function. In symbols, if $f:D \rightarrow C$ is a function with domain D and codomain C, then the image of f is $\{y \in C : \exists x \in D \ s.t. \ f(x) = y\}$ ("$\exists$" stands for "there exists").

imaginary number

This term is used inconsistently. Some use it to refer to any complex number that is not real, others only for a *pure* imaginary number of the form bi, where b is real.

imaginary unit

The symbol i, whose defining property is that $i^2 = -1$.

indefinite integral

The indefinite integral $\int f(x)\,dx$ is the family of all antiderivatives of $f(x)$ (usually written in the form $F(x) + C$, where F is a particular antiderivative and C is a general constant).

induction

A proof technique used to prove that a statement or property holds in an infinite number of cases (e.g., to prove that $1 + 2 + 3 + \ldots + n = n(n+1)/2$ for all posi-

tive integers n). First, one checks the simplest cases individually (the "base case(s)"); then, one proves that, if the claim holds in the first n cases, it will also hold in the $(n+1)$-st case. In this way, it is possible to prove an infinite collection of statements with a finite proof. (Note that this is not the same as "inductive reasoning.")

infinite set

A set is infinite if its elements cannot be put in one-to-one correspondence with the elements of the set $\{1, 2, 3, \ldots n\}$ for any n. Equivalently, a set is infinite if its elements can be put in one-to-one correspondence with a proper subset of itself.

injective

A function $f:D \rightarrow C$ is injective (also called "one-to-one") if distinct inputs give distinct outputs. In symbols, $f(x) = f(y)$ implies $x = y$, that is, passes the horizontal line test—a test used to determine if a function is one-to-one. If no horizontal line intersects a function's graph more than once, the function is said to be "one-to-one."

integer

The set of integers includes the counting numbers 1, 2, 3, 4, 5, 6, … and their negatives, as well as 0. The set of integers is traditionally denoted by a blackboard-bold $\mathbb{Z}$ (from the German word *Zahl*, meaning "number").

integral

See *integration* for the broad concept. See *definite integral* and *indefinite integral* for particular uses common in classroom usage. (An adjective form of "integer.")

integration

Intuitively, the operation of "adding up" or "accumulating" the values of a function over part or all of its domain. More concretely but still informally, the integral of a real-valued single-valued function is the area under its graph. There are numerous types of integration, at varying degrees of complexity and technicality. The simplest is the Riemann integral, which is part of the standard calculus curriculum. The most commonly used in professional mathematics is arguably the Lebesgue integral.

irrational number

A number (usually the term is only used for real numbers) is called irrational if it is not rational, that is, if it cannot be expressed as the ratio p/q of two integers. The decimal expansion of an irrational number neither terminates nor repeats.

isomorphism

A homomorphism that is also a bijection. Two objects related by an isomorphism are called "isomorphic." Isomorphic objects are in a deep sense "the same." This notion is fundamental to modern mathematics.

lemma

A proposition which is proven primarily so that it can be used in the proof of another proposition, which is presumably considered more important.

limit

Intuitively, if whenever x is "near" a, then $f(x)$ is "near" L, we say $\lim f(x) = L$. How exactly this is formalized depends on context; there are even versions of the definition that allow variables to approach infinity or allow for infinite limit values. For real-valued functions of a single real variable, the formal definition is that $\lim f(x) = L$ if, for all $\varepsilon > 0$, there exists a $\delta > 0$ such that, $0 < |x - a| < \ddot{a}$ implies $|f(x) - L| < \varepsilon$. Derivatives and integrals are both defined in terms of limits. The limit is the foundational concept on which calculus and analysis are built.

linear algebra

In its more narrow meaning, linear algebra is the study of systems of linear equations, matrices, vectors, and their operations. More broadly and abstractly, linear algebra is the study of linear operators, functions which respect addition and multiplication by scalars.

logarithm

The inverse operation to exponentiation. For any positive base $b \neq 1$, we can define a continuous function $\log_b x$ for all $x > 0$ with the following formal properties:

$$\log_b (xy) = \log_b x + \log_b y$$
$$\log_b 1 = 0 \quad \log_b b = 1$$
$$\log_b \left(x^n \right) = n \log_b x.$$

We have $\log_b x = k$ if and only if $b^k = x$. If the base is not specified, then the base e is implied in standard mathematical usage; among nonmathematicians and in most secondary school textbooks, the default base is the less natural 10.

logic

The formal study of valid reasoning and inference. Logic is both a branch of mathematics (particularly symbolic logic) and part of its architecture.

matrix

A rectangular array of numbers or mathematical symbols. Two matrices of the same shape can be added or subtracted. In order to multiply two matrices, the number of columns of the first must match the number of rows of the second. The algebraic properties of these operations are a bit different from operations on numbers; for example, matrix multiplication is not commutative, and it is possible to multiply two non-zero matrices together and get zero. There are many applications of matrices and matrix algebra.

modular arithmetic

The arithmetic of congruences in number theory. When working *modulo n*, one considers two integers to be congruent if their difference is a multiple of n (equivalently, if they give the same remainder when divided by n). Working modulo 2 is just keeping track of whether numbers are even or odd (even + odd = odd, odd + odd = even, etc.). Working modulo 12 is so-called "clock arithmetic."

multinomial coefficient

The multinomial coefficient

$$\binom{n}{n_1, n_2, \ldots, n_k}$$

defined for nonnegative integers

$$n = n_1 + n_2 + n_3 + \cdots + n_k \text{ by the formula}$$

$$\frac{n!}{n_1! \; n_2! \; n_3! \cdots n_k!}.$$

This multinomial coefficient counts the number of ways to divide n objects into k piles so that the first pile contains n_1 objects, the second contains n_2, and so on.

Multinomial Theorem

A theorem of basic algebra which connects the multinomial coefficients to the numbers appearing in the expansion of a sum raised to a power (indeed this is the reason they are called multinomial "coefficients"). The precise statement is

$$\left(x_1 + x_2 + \cdots + x_k\right)^n =$$
$$\sum_{\substack{n_1, n_2, \cdots, n_k \geq 0 \\ n_1 + n_2 + \cdots + n_k = n}} \binom{n}{n_1, n_2, \cdots, n_k} x_1^{n_1} x_2^{n_2} \cdots x_k^{n_k}$$

natural number

A natural number is a "counting number": 1, 2, 3, 4, 5, ... Some sources include 0 as a natural number, while others do not. The set of natural numbers is traditionally denoted by a blackboard-bold $\mathbb{N}$.

number bases

For any integer $b > 1$, we can define a system for writing down real numbers using the digits $0, 1, 2, \ldots, (b-1)$. A string of digits $a_k \cdots a_2 a_1 a_0 . a_{-1} a_{-2} \ldots$ represents the number

$$a_k b^k + a_{k-1} b^{k-1} + \cdots + a_2 b^2 + a_1 b + a_0$$
$$+ a_{-1} b^{-1} + a_{-2} b^{-2} + \cdots.$$

For example, if $b = 7$, then 123.4 (base 7) represents the number

$$49 + 2 \times 7 + 3 + \frac{4}{7} = 66 \frac{4}{7}$$

(written in familiar base 10). This is called base b notation, and several of the more commonly-used bases have other names, such as "binary" for base 2.

number line

A commonly used device for visualizing the real number system in which every real number is represented by a point. Typically the number line is drawn so that the numbers increase toward the right.

number theory

A branch of mathematics concerned with the properties of number systems, especially integers. This includes modular arithmetic, prime numbers, Diophantine equations (an indeterminate polynomial equation that allows the variables to be integers only), and modular forms.

The concepts of "integer," "prime," and so on, can be generalized to much wider contexts than just the "ordinary" integers.

obtuse angle

An angle that is larger than a right angle but less than half a circle. The measure of an obtuse angle is between 90 and 180 degrees.

octal (notation system)

Base eight. (See *number bases*.)

one-to-one

See *injective*.

one-to-one correspondence

See *bijection*.

onto

See *surjective*.

open problem

An unsolved problem, an opportunity for mathematical research. A mathematical question is said to be "open" if it has not been answered in the existing mathematical literature. (compare *conjecture*)

opposite

The opposite of a number x is $-x$. Also called "additive inverse."

ordered pair

A pair of numbers or other mathematical objects, denoted (a, b), where the "order matters," so that (a, b) and (b, a) are different as ordered pairs (unless $a = b$).

ordered triple

A list of three numbers or other mathematical objects, denoted (a, b, c), where the "order matters," so that

$$(a, b, c), (a, c, b), (b, a, c), (b, c, a), (c, a, b), (c, b, a)$$

are different as ordered triples (if a, b, c are distinct).

parabola

The set of points that are at the same distance from a fixed point (the "focus") and a fixed line (the "directrix"). The graph of any quadratic function $f(x) = ax^2 + bx + c$ with a nonzero is a parabola, and every parabola can be described in this way after suitably rotating the coordinate axes.

parity

Whether a number is even or odd. (Considering numbers based only on parity is just another name for working modulo 2.)

perfect number

A positive integer is said to be perfect if it is equal to the sum of all its proper divisors. For example, 6 is perfect because $6 = 1 + 2 + 3$. The next smallest perfect number is $28 = 1 + 2 + 4 + 7 + 14$. It is not known whether there are infinitely many perfect numbers, nor whether there exists even one odd perfect number.

permutation (enumeration)

One of the fundamental enumeration problems. The permutations of r objects from a set of n objects are the "ordered" r-tuples of distinct objects from the set. There are

$$\frac{n!}{(n-r)!}$$

permutations of r objects from a set of size n.

permutation (function)

A bijection from a set to itself. Intuitively, a permutation is a rearrangement of a set of objects. There are $n!$ permutations of a set of n elements, including the trivial permutation, which leaves the order unchanged.

phi (φ). See *golden ratio*.

pi (π)

A constant defined as the ratio of the circumference of any circle to its diameter, or the number of radians in half a circle. The approximate value of π is 3.1415926536. The constant π is ubiquitous in mathematics and physics.

polygon

Traditionally a plane figure bounded by a closed path made up of a sequence of line segments. These line segments are called the "sides" of the polygon, and the places where one edge ends and the next starts are called "vertices" (sing. "vertex"). A polygon with three sides is usually called a "triangle." A four-sided polygon is a "quadrilateral," a five-sided polygon is a "pentagon," and so on. More generally an n-sided polygon is called an "n-gon." Depending on context, we may want to think of the boundary itself as the polygon, or we might prefer to consider the boundary and its interior together as the polygon.

polyhedron

The three-dimensional analogue of a polygon. The boundary of a polyhedron consists of a collection of polygons in space ("face") with common sides ("edges" of the polyhedron) and "vertices." The faces form a closed figure in space.

polynomial

A polynomial in variables $x_1, x_2, \ldots x_n$ is any expression that can be constructed from the variables and from constants by addition and multiplication. The expressions $2xy + z^3$ and $x^2 + y^2 + z^2 - 1$ are polynomials in x, y, z,

$$\text{but } x + y + \frac{1}{z} \text{ and } \sqrt{xyz} \text{ are not.}$$

polytope

The analogue of polygons and polyhedra in dimensions greater than three.

prime factorization

An expression of a positive integer as a product of primes (possibly raised to powers). For example, $42 = 2 \times 3 \times 7$ and $525 = 3 \times 5^2 \times 7$.

prime (number)

A positive integer is prime if it has exactly two factors, itself and 1. The first few primes are 2, 3, 5, 7, 11, 13, 17, ... (In particular, note that the number 1 is not considered prime.)

probability

A numerical measure taking values between 0 and 1 quantifying the likelihood of an event to occur (or beliefs about that likelihood). Also, the general theory for understanding and working with such measures of likelihood and expectation.

proof

Any rigorous demonstration of the validity of a proposition. Proofs may be written in formal or informal language and may consist of any proportion of words and symbols.

proof by contradiction

A proof technique in which one assumes that the intended conclusion of the theorem is false, and derives from this supposition and from the other hypotheses of the theorem and known facts some statement which contradicts something already known. This shows that one of our assumptions must have been wrong, and the only questionable one was the assumption that our theorem was false. This in turn shows that the conclusion of the theorem does in fact hold.

proposition

A proposition is a declarative statement. Depending on the importance of a statement and/or its role in a larger argument, a proposition may be called a "lemma," a "corollary," a "claim" or a "theorem," depending on its perceived importance.

Q.E.D.

Abbreviation for *quod erat demonstrandum* (Latin for "which was to be demonstrated"), classically used by mathematicians to indicate the conclusion of a proof, signifying that the claims may have now been fully justified. While Q.E.D. itself is less commonly used than it once was, it is still standard practice to include an "end-of-proof symbol" of some kind. The most commonly used end-of-proof symbol today is probably □, called the "Halmos tombstone" and named for mathematician Paul R. Halmos (1916–2006).

quadratic formula

The equation

$$x = \frac{-b \pm \sqrt{b^2 - 4ac}}{2a},$$

which expresses the solutions of the polynomial equation $ax^2 + bx + c = 0$ in terms of its coefficients. Analogous formulas exist for cubic equations

$$ax^3 + bx^2 + cx + d = 0$$

and quartic equations $ax^4 + bx^3 + cx^2 + dx + e = 0$, but they are less well-known because they are much more complicated. Notably, it is known that no such formulas exist for polynomials of degree five or larger.

radians

One of the two most commonly-used units for angle measure. Among mathematicians, radians are the standard unit of angle measure, and radian measure is considered implied unless some other unit is indicated. There are 2π radians in a circle. If a circle is drawn with center at the vertex of an angle, then the radian measure of the angle is the ratio of the length of the arc inside the angle to the radius of the circle. The trigonometric functions have the simplest properties (particularly in the context of calculus) if radians are used.

range

The set of values that a function $y = f(x)$ may take corresponding to the values of the input over the specified domain of x. Also called "codomain."

rational function

A rational function in variables $x_1, x_2, \ldots, x_n$ is any expression that can be constructed from the variables and from constants by addition, multiplication, and division. A rational function can always be written as a ratio of two polynomials.

rational number

The set of rational numbers (from the word "ratio") consists of those numbers which can be expressed as a fraction m/n, where m and n are integers with n nonzero. The set of rational numbers is traditionally denoted by a blackboard-bold $\mathbb{Q}$ (for "quotient").

real number

This is what is typically meant when one says "number" without further explanation, including all the integers, all the rational numbers, and many more numbers "between" the rational numbers. A real number can always be described by a (generally infinite) decimal expansion. Real numbers exclude imaginary numbers, complex numbers, and the square root of minus one. The set of real numbers is traditionally denoted by a blackboard-bold $\mathbb{R}$.

reciprocal

The reciprocal of a nonzero number x is $1/x$. Also called "multiplicative inverse."

recursion

A formula or equation relating each term of a sequence or object from a family to the previous terms or objects. This (usually together with one or more initial values) implicitly determines all the terms or objects, but it may be difficult or impossible to get a closed-form description of all the terms.

reductio ad absurdum

Latin for "reduction to absurdity." See *proof by contradiction*.

relation

Formally, a relation R on sets $X_1, X_2, \ldots X_n$ is a subset of the product $X_1 \times X_2 \times \cdots \times X_n$. If $(x_1, x_2, \ldots x_n)$ belongs to the subset we say that $R(x_1, x_2, \ldots x_n)$ is true or the relation holds. If not, the statement is false. Most often, $n = 2$ and we write the relation symbol between the arguments, as in "$x_1 < x_2$." Examples include "is less than," "is a factor of," and "does not equal." Examples with $n = 1$ include things like "is prime" or "is positive." Intuitively, a relation is a property that a combination of objects may or may not possess. (One can alternately think of a relation as a special kind of function whose output is "true" or "false.")

relatively prime

A set of integers is relatively prime if there is no number larger than 1 that is a common factor of all of them. For example 18 and 25 are relatively prime. So are 6, 10, and 15. A set of integers is "pairwise" relatively prime if every pair of number in the set is relatively prime. So 6, 10 and 15 are *not* pairwise relatively prime, but 69, 11, and 35 are.

right angle

An angle that is congruent to its own supplement; that is, a quarter of a full circle. The measure of a right angle is 90 degrees.

root

The nth roots of a number a, sometimes written

$$\sqrt[n]{a},$$

are the numbers x such that $x^n = a$. Also, a root of a polynomial

$$a_n x^n + a_{n-1} x^{n-1} + \cdots + a_1 x + a_0$$

is a solution of the equation

$$a_n x^n + a_{n-1} x^{n-1} + \cdots + a_1 x + a_0 = 0,$$

that is, a *zero* of the function

$$f(x) = a_n x^n + a_{n-1} x^{n-1} + \cdots + a_1 x + a_0.$$

scalar

A numerical quantity (as opposed to a vector or function) that may take one of several forms; for example, a real or complex number.

sequence

A finite or infinite list of terms (which can be numbers or any other type of object). A sequence can be given by listing its terms explicitly separated by commas or by giving a closed-form or recursive formula for the terms.

series

A sum of a finite or infinite (but more frequently the latter) list of terms (usually numbers or mathematical expressions). The terms of a series can be listed as the terms of a sequence can, but separated by plus signs, not commas. In the case of an infinite series, this goes beyond the ordinary concept of addition, and requires the notion of limit to make sense.

set theory

A branch of mathematics studying *sets*, which can be thought of as collections of objects. It also includes a vocabulary for making precise certain ideas about infinite sets and their relative sizes. Set theory sits close to the core of mathematical theory, and much of the architecture of mathematics is traditionally built on the foundations of set theory; set theory is also an object of study for its own sake.

surjective

A function $f: D \to C$ is surjective (also called "onto") if the image is the whole codomain. That is, for all y an element of C there exists some x an element of D such that $f(x) = y$.

symmetry

A symmetry of an object is a transformation or change of perspective after which the object is "the same as it was before." A figure has reflective symmetry if it looks the same when reflected across a certain line; the human face has at least approximate reflective symmetry. A figure has rotational symmetry if it looks the same when rotated around a certain point by a certain angle (traditional crossword puzzle grids have rotational symmetry). Symmetry is by no means a purely geometric concept; a unifying concept of much of modern mathematics is the problem of identifying and describing symmetries of all types.

tau (τ)

A circle constant defined as the ratio of the circumference of any circle to its radius, or the number of radians in a full circle. The approximate value of tau is 6.2831853072. While π is in much more common usage (for chiefly historical reasons), some consider τ to be the more mathematically significant constant.

theorem

A theorem is a mathematical statement that can be demonstrated to be true provided that the set of axioms and other theorems from which this theorem is derived is true. It is usually a general component of some larger theory. Its significance is often a subjective decision.

theory

A collection of related definitions and theorems on a particular topic, such as "number theory," "knot theory," or "graph theory"; an area of study or research in mathematics. Note that "theory" as used in mathematics means something like "formal study" and it definitely does not have the unproven, conjectural connotation present in scientific and everyday usage.

topology

A branch of pure mathematics dealing with those properties which are preserved by continuous deformations (stretching, twisting, enlarging, shrinking, and so on).

transcendental number

A complex number is called transcendental if it is not algebraic, that is, if it is not the root of any polynomial with integer coefficients. Famously transcendental numbers include π and e.

trichotomy

The principle that, if x, y are real numbers, then of the statements $x < y$, $x > y$, and $x = y$, "exactly" one is true.

Some other sets and order relations have trichotomy properties as well.

tuple

General term for "ordered pair," "ordered triple," "ordered quadruple," and "ordered n-tuple" for any larger n. The adjective "ordered" is usually regarded as implied from context and omitted.

variable

A symbol, often x (though just about any symbol can be used), which is meant to represent some unspecified object of a certain type (e.g., a real number). Sometimes a variable stands for a specific (unknown) number, as in "If $2x + 3 = 7$, then what is x?," while in other contexts variables are used formally without a particular value. A variable is to be contrasted with a "constant," which does not change within a given problem.

vector

Often defined as a quantity with a magnitude and a direction. Geometrically, a vector can be thought of as an arrow with a defined head and tail. More generally, the name "vector" is sometimes applied to any element of any vector space; the term is most commonly used for n-tuples of real numbers.

whole number

Some people use "whole number" as a synonym for "integer," others use it as a synonym for "positive integer," and still others as a synonym for "nonnegative integer."

zero (number)

A number that serves as the additive identity, characterized by the property that $x + 0 = x$ for all x. Many mathematical structures have additive identities (zero matrices, zero functions, and so on), and the term zero (and even the symbol 0) are often used for these. In contexts where multiplication makes sense, zero is also characterized by the property that $0 \times a = x$ for all a.

zero (of a function)

The zeroes of a function $f(x)$ are the domain values x_0 such that $f(x_0) = 0$.

MICHAEL "CAP" KHOURY
University of Michigan, Ann Arbor

Index

Index note: Text and page numbers in **boldface** refer to main topics. Page numbers in *italics* refer to photographs.

A

abacus, 8, 68, 71, 187, 713, 717–718
    See also soroban
Abbott, Edwin
    Flatland, 440, *557*, 557, 781, 859–860, 902
Abd al-Latif (Muhammad Taragay Ulughbek), 66
Abd al-Rahmân, 658
Abel, Niels Henrik, 34, 111, 362, 366, 786, 883, 1102
Abel Prize, 363
Aberdeen Proving Ground, 756
Able Danger program, 791
Aboriginal kinship system, 735–736
Aboriginal paintings, 736
Abstract Linking Electronically (ABLE), 141
Abu Abdallah Book of Addition and Subtraction by the Indian Method (al-Khwarizmi), 66–67
Abu al-Wafa Buzjani, 54, 882, 1094
    Those Parts of Geometry Needed by Craftsmen, 627
Abu Kamil Shuja ibn Aslam, 18, 54, 1094
Academy, The (Plato), 124
acceleration, 457
Accelerator Mass Spectrometry, 159
accident reconstruction, 1–2
accounting, 2–5
    assets/liabilities in, 3–5
    Benford's Law, 4
    as record keeping, *3*, 3–5
Achermann, Peter, 746
Achilles paradox, 552
ACNielsen Corporation, 320, 578
acoustics, 486–487
acrostics, word squares, and crosswords, 5–7
Acta Mathematica (journal), 1075

actuarial science, 164, 288–289, *605*, 605–606, 758
actuarial tables, 123, 546, 1100
Adams, John (composer)
    Dr. Atomic, 690
Adams, John Couch, 78, 345
Adams, John Quincy, 233, 234
Adams" Method, 233
addition and subtraction, 7–9
Adelard of Bath, 1094
Adelstein, Abraham Manie, 20
ADFGVX code, 1074
Adibi, Jafar, 509
adjustable rate mortgages, 484
Adleman, Leonard, 213
Adrain, Robert, 220, 707
Advance Circulation Model (ADCIRC), 492
Advanced Placement (AP) Exams, 151, 669, 946
Advanced Placement Calculus (AP Calculus), 143, 151, 1106
Advanced Research Projects Agency Network (ARPANET), 181, 518
Advances in Mathematics of Communications (journal), 224
Advancing HIV Prevention, 479
advertising, 10–12
Aerodynamic Proving Ground, 1073
aerodynamics, 29, 30, 64
    See also **airplanes/flight**
Africa, Central, 13–15
Africa, Eastern, 15–17
Africa, North, 19
Africa, Southern, 19–20
Africa, West, 20–22, 623
Africa Counts (Zaslavsky), 14, 16
African Fractals (Eglash), 14

African Institute for Mathematical Sciences, 20
African Mathematical Union (AMU), 22, 23
African Mathematics, 23–25
  development of, 23–24
  in Egypt, 24–25
  in sub-Saharan Africa, 25
African Mathematics Olympiads, 22
Afrika Matematica (journal), 22
Afrikaners, 19
Agassiz, Alexander, 250
An Agenda for Action (NCTM), 276–277
Agent Orange, 1037
agent-based models, 808
Agnesi, Maria Gaetana, 602
  Analytical Institutions, 143
  Instituzioni Analitiche, 1101
Agon, 120
Agora (movie), 682
Agreement Governing the Activities of States on the Moon
  and Other Celestial Bodies, 679
agricultural economics, 381–382
agriculture. *See* **farming**
aha calculations, 332
Ahlfors, Lars, 363
Ahmed ibn Yusuf, Abu Ja'far, 18
Ahmes (scribe), 1091
Ahmes papyrus. *See* Rhind papyrus
Ahnentafel, 418–419
AIDS. *See* **HIV/AIDS**
Aigner, Martin
  Proofs from THE BOOK, 625
air resistance, 390–391
air traffic control, 354
aircraft design, 25–28
  aircraft carriers, 28
  complex analysis and, 26
  helium airships, 1073
  Joukowski airfoil, 26
  nature-inspired algorithms, 26
  sonic booms, 27–28
airplanes/flight, 28–31
  flight speed, 30–31
  George Cayley and, *29*, 29
  lift and thrust, 30
  mathematical history of, 29–30
  principles of flight, 30
  World War II and, 1078–1079
Airy, George, 994
Akim, Efraim L., 679
al-. *See specific name following prefix*
Al Qaeda, 509–510, 791

Aladdin's saddlebag problem, 747
Alan M. Turing Award, 180
Albategnius, 1094
albatross, *47*
Alberti, Leone Battista, 248, 748
album cover art, 787
Alciphron, or the Minute Philosopher (Berkeley), 859
Aldiss, Brian, 900
Aldrin, Edwin "Buzz", 522
Aleksandrov, Pavel, 908
Alexander, James Waddell, II, 531
Alexander Polynomial, 531
Alfonso X of Castille
  Libro de los Juegos, 454, *454*
Algebra (al-Khwarizmi), 1094, 1095
Algebra (Bombelli), 1097
algebra and algebra education, 31–35
  Arabic/Islamic mathematics, 54
  Babylonian mathematics, 88
  as core subject, 39
  early history of, 31–32
  equation solving, 33
  European role in, 33
  Fermat's Last Theorem, 34–35
  al-Khwarizmi and, 32–33, 54, 66–67, 712, 1094, 1095
  later developments, 35
  modern period, 33–34
  theoretical mathematics and, 613, *616*, 619
algebra in society, 36–41
  "algebra for everyone", 35
  applications, 38–40
  early history of, 37–38
  as gateway, 36, 40–41
  usefulness, 36–37, 39–40
Algebraic Expressions in Handwoven Textiles (Dietz), 989
algebraic number theory, 720
algorithms
  genetic, 791
  Groups, Algorithms, Programming (GAP), 765
  Huffman, 382
  information theory and, 795
  iterative, 95
  for multiplication and division, 686–687
  nature-inspired, 26
  Stable Matching, 582
  traveling salesman problem and, 1009–1010
Alhambra Palace (Spain), 171, 781
Alhazen, 55, 111, 237, 1005, 1042
Alice in Wonderland (Carroll), 110, 464, 513, 557
Alice Through the Looking Glass (Carroll), 6
Alighieri, Dante, 860

alignment games. *See* **board games**

"al-jabr" ("algebra"), 33

Al-kitab al-muhtasar fi hisab al-jabr wa-l-muqabala
    (Compendium on Calculation by Completion and
    Reduction) (al-Khwarizmi), 33, 54

All Around the Moon (Verne), 521

"All is Number" (Pythagoras maxim), 729

All-American Midget Series (AAMS), 703

Almagest, The (Ptolemy), 18, 772, 1011, 1093, 1095

alternating current (AC), 341

Alvarez, Luis, 502, 503

Alvin, 295

Amazon, 303, *304*, 519

Amdahl, Gene, 753

Amend, Bill
    Foxtrot, 219

American City Planning Institute, 189

American Crossword Puzzle Tournament, 7

American Declaration of Independence, 847

American Express Corporation, 133, 253

American Institute of Physics, 592

American Invitational Mathematics Examination (AIME),
    227

American Journal of Math, 1104

American Journal of Physics, 1015

American Mathematical Monthly, 599, 810

American Mathematical Society (AMS), 119, 165, 170, 239,
    275, 710, 810, 1104

American Mathematical Society of Two Year Colleges
    (AMATYC), 811

American Mathematics Competition (AMC), 227

American Pension Corporation, 758

American Society for Communication of Mathematics, 222

American Society of Clinical Oncology, 183

American Sociological Association Section for
    Mathematical Sociology, 592

American Statistical Association (ASA), 809–810, 888, 944

American Statistical Society, 285

American system of units of length, 1021–1022

Ameritech, 175

Amitsur, Shimshon Avraham, 75

amortization, 483–484, 561–562

Ampere, Andre-Marie, 549, 837

amperes, 549

Ampere's law, 837

amplitude modulation (AM), 838–839, 1068–1069

Amsler, Jacob, 354

Amsler planimeters, 354

amyotrophic lateral sclerosis (ALS), 473

"Analysis situs" (Poincaré), 1105

Analyst, The (Berkeley), 859, 1088, 1101

analytic geometry. *See* **coordinate geometry**

analytic number theory, 720

Analytical Institutions (Agnesi), 143

Anastasi, Anne
    Psychological Testing, 816

Anaxagoras of Clazomenae, 1092

André, Désiré, 337

Andre, John, 869

Android open-source operating system, 977

anesthesia, 42–43
    mathematics applications, 42–43
    providers' education, 43

anesthesiologist assistants (AAs), 43

anesthesiologists, *42*, 42–43

Anh Le, 791

Anhgkor Wat, 70

Animalia kingdom, 44

animals, 43–49
    biological systematics, 44
    chimeras and hybrids, 48–49
    food webs, 47–48
    migration and, 46–47
    modeling based on, 44–45
    movement of, 45–46
    symmetry/fractals and, 49
    tissue structures, 44

animation and CGI, 49–51
    early devices, 50
    mathematics and, 50–51
    principles of, 50, 430, 1041

ankh, 857

Annuities upon Lives (de Moivre), 1100

annuity tables, 507, 1100

anthropometry, 200

Antikythera mechanism, 78

antinepotism rules, 589

Antiphon the Sophist, 1092, 1093

antiquarianism, 418

Anti-Submarine Warfare Operations Research Group, 1079

anxiety, math, 959–960

AP Calculus. *See* Advanced Placement Calculus

APBR metrics, 99

ape index, 200

Apgar, Virginia, 51, 52

Apgar scores, 51–53
    development and effectiveness, 52
    prediction of, 52

Apollo at Delos, oracle, 882

Apollo program, 276, 522, 678–679

Apollonius of Perga, 280, 1004
    Conic Sections, 237, 1092, 1093, 1094, 1100

Apollo's altar, 260

Apostol, Tom, 143

Appel, Kenneth, 1107

Appel–Haken, 1107

Appendix (Bolyai), 361

Apple Computer, Inc., 520, 767

Appleton, Edward, 838

Application de l'analyse à la géométrie (Monge), 1101

Applied Mathematics Panel, 1079

Approximatio ad summam terminorum binomii (a+b)n in seriem expansi (de Moivre), 707

Aquinas, Saint Thomas. *See* Thomas Aquinas, Saint

Arab Journal of Mathematics and Mathematical Sciences, 76

Arabic/Islamic mathematics, 53–55
  chronology, 1094
  combinatorics, 55
  decimal system, 53–54
  geometry, 54, 425
  numerical mathematics, 55
  religious tenets and, 627, 729
  trigonometry, 54–55

Arcadia (Stoppard), 774–775, 776, 777

arch bridges. *See* **bridges**

archery, 55–56

Archimedean solids, 783

Archimedes, 57–60
  circle measurement and, 552, 762
  contributions, *57*, 57–58, 67, 248, 280, 353, 544, 1092
  early life, 57
  Equilibrium of Planes, 544
  "Eureka", 60, 460, 652, 1025
  as father of integral calculus, 146; 460
  first law of exponents and, 370
  Greek mathematics and, 364
  King Hero II crown and, 1025
  legacy of, 60
  levers and, 544
  mathematics, *59*, 59–60
  Measurement of a Circle, 646, 1094, 1095
  The Method, 60
  palimpsest, 58
  On Plane Equilibriums, 1094
  The Sand Reckoner, 60
  siege of Syracuse and, 501–502
  simple machines and, 57, 67, 764
  On the Sphere and Cylinder, 1004, 1094
  volume of a sphere and, 261, 462, 653, 1004, 1094

Archimedes' screw/spiral, *57*, *57*, 67, 248, 353, 1092

architecture, *315*, 315–316
  careers in, 432
  Donald in Mathmagic Land (cartoon), 1084
  geometry and, 425–426, 428, 429, 430
  golden ratio in, 446–447
  houses of worship, 485–487
  symmetry and, 972
  The Ten Books on Architecture (Vitruvius), 1080
  tessellations in, 627
  trigonometry and, 1012
  See also Wren, Christopher; Wright, Frank Lloyd

Archiv der Mathematik und Physik (jounal), 1103

Archytas of Tarentum, 364, 1092

arcs and curves table, 1011

arenas, sports, 61–63

Arenstorf, Richard, 679

Arenstorf periodic orbits, 679

Arf, Cahit. *See* Cahit Arf

Arf invariant in algebraic topology, 75

Arf rings, 75

Arf semigroups, 75

Argand, Jean-Robert, 1102

Aristaeus the Elder, 1092

Aristarchus of Samos, 1092

Aristophanes
  Birds, 558

Aristotle, 305, *357*, 552, 1061
  Metaphysics, 1092
  Meteorologica, 1055

Aristotle's Wheel Paradox, 1061

arithmetic magic, 567

arithmetic puzzles, 820–821

Arithmetica (Diophantus), 1093, 1097
  Latin translation, 1098

Arithmetica infinitorum (Wallis), 1099

Arithmetica Integra (Stifel), 111, 858, 1096–1097

Arithmetica Logarithmica (Briggs), 371, 1098

Ark of the Covenant, 624

Armstrong, Lance, 109

Armstrong, Neil, 522

Army Corps of Engineers, 194

Army Signal Corps, 756

Arnold, Benedict, 869

Around the World in Eighty Days (Verne), 900

ARPANET, 910–911

Arrow, Kenneth, 217, 336, 1049

Arrow's Paradox, 217, 336

Ars Conjectandi (Bernoulli), 764, 801

Ars Magna (Cardano), 722

In artem analyticam (Viète), 1097

artificial intelligence, 165

artificial neural networks, 701–702

artillery, 63–66
  battlefield computers and, 63, 65–66

types of, 63, 191, 391
See also ballistics studies
Artin, Emil, 35
Artis analyticae praxis (Harriot), 765, 1098
ArXiv.org e-print archive, 220
Aryabhata the Elder, 73, 261, 424, 763, 1093
Aryabhatiya (Aryabhata), 73, 261, 424
Arzarchel, 664
Ascher, Marcia, 16, 736
Ascoli, Giulio, 364
Ashton, Frederick
   Scènes de Ballet, 90
Asia, Central and Northern (Russia), 66–67
Asia, eastern, 68–69
   China, 68–69
   educational philosophy, 68
   Hong Kong, 69
   Japan, 69, 298–299, *1033*
   Mongolia, 69
   North Korea, 69
   number system, 68
   South Korea, 69
   Taiwan, 69
Asia, southeastern, 70–72
   Brunei, Myanmar, and the Philippines, 72
   Cambodia, Laos, and Vietnam, 71–72
   early history of, 70–71
   Indonesia, 72
   Singapore and Malaysia, 71
   Thailand, 71
Asia, southern, 72–74
   history of, 72–73
   mathematics education, 73
   See also Indian mathematics
Asia, western, 74–76
   Babylon, 74–75
   Israel, 75–76
   Ottoman Empire and Turkey, 75
Asimov, Isaac, 791, 899, 901
   Foundation, 791
   The Realm of Algebra, 36
Aspect, Alain, 548
Aspin, Les, 767
Assessment Standards for School Mathematics (NCTM),
   277
assets, 3–5, 92–95
assisted reproductive therapy (ART), 796
Associated Press (AP), 242
Association for Computing Machinery, 180
Association for Professional Basketball Research, 99
Association for Symbolic logic, 811

Association for Women in Mathematics (AWM), 207, 811,
   1072, 1107
Association Mathématique Algérienne, 19
Assumption College (Bangkok), 71
astrodynamics, 521
astrolabe, 658
Astronaut Hall of Fame, 872
Astronomia Nova (Kepler), 365
astronomy, 76–78
   Almagest (Ptolemy) and, 1093
   astrophysics, 79
   Chinese and, 185
   early mural sextant, 67
   Greeks and, 77–78, 460–461
   heliocentric, 964, 1092
   Islamic Golden Age and, 18
   lunar calendars, 75, 77, 154–156
   Mayans and, 496
   measurements in, 1010–1011, 1022
   parallax measurements in, 78–79
   problems of representation and, 805
   during Renaissance, 78
   satellites and, 890–892
   See also Galileo (Galileo Galilei); **telescopes**
astrophysics. *See* **astronomy**
AT&T, 175, 976, 977
Atanasoff, John V., 1077
Atanasoff–Berry Computer (ABC), 1077
Atiyah, Michael F., 363
Atlas Recycled (Tsuchiya), *851*
atolls. *See* **coral reefs**
atomic bomb (Manhattan Project), 79–81
   authorization of, 79
   Cold War and, 214–215, 216
   energy-mass equivalence, 80–81
   moral questions and, **357–358**
   nuclear reactions, 80
   scientists and, 79–80, 216, 334–335
   World War II and, 79–80
atomic measurements, 1022
atonal music, 231
Auburn, David, 775
audio processing, 680–681, 788
Augustine, Saint (Aurelius Augustinius), 624
Austin, David, 610
Australia, 735–736
Australian Mathematical Society, 735
autism, 128
auto racing, 81–83
   car design, 83
   overview of, 82

race strategy, 83
race track design, 82–83
technology and safety, 83
Autographical Notes (Einstein), 335
automata, 1034
automobile manufacturers, 10–11
automobiles
city planning and, 189–190, 1081
design of, 148, 351
highway design, 476–477
purchasing, 225–226
radar guns and, 317
Segways versus, 906–907
smart cars, 922–923
speedometers, 242–243
thermostats, 990
traffic modeling, 808, 1000–1001
autoregressive integrated moving average (ARIMA), 401, 1020
Auto-Tune software, 788
Avempace, 664
Averroës (Ibn Rushd), 664
Avicenna, 664
axiom of choice (AC), 587
axiomatic systems, **84–85**, 426, 428, 586–587
See also Incompleteness theorem
Azerbaijan Journal of Mathematics, 76
Aztec civilization, 698

B
Babbage, Charles
Ada Lovelace and, 565–566
Bridgewater Treatises, 859
as father of computers, 718
mechanical computers and, 500, 768, 875, 1103
religious writings, 859
Uniform Penny Post and, 807
weaving systems and, 989
Babylonian mathematics, **87–89**
applied mathematics, 603
base-60 system, 725
BM 85200, 261
history of, 74–75
loan interest, 675
measurement, 646
and religion, 729
Bach, Johann Sebastian, *230*
Bachelier, Louis, 288
Bachelier Financial Society, 288
Back to Basics movement, 276, 542
background noise. *See* **cocktail party problem**

Bacon, Kevin, 915, *916*, 916
Bacon, Roger, 600–601, 619, 625, 664
Baha'i Faith symbols, 856
Baha'i House of Worship, 487
Ba-ila settlement, 16
Bain, Alexander, 382
Baire, René, 505
Baker, Alan, 363
Baker, Garth A., 167
Bakewell, Frederick, 382
Bakhshali manuscript, 73, 424, 1094
Balan, Radu, 211
Balinski, Michel, 232, 234
Ball State Project, 275
Ballanchine, George, 231
ballet, **90–91**, 231
Ballew v. Georgia, 117–118
ballistic pendulum, 64, 671
ballistics studies
artillery and, 63, 64–65, 671
computers and, 1077–1078
firearms and, 391
forensics, 255–256
military research, 1073, 1077–1078
underwater, 1079
ballot problem, *339*, 339–340
ballroom dancing, **91–92**
Ban Ki-moon, 289
Banach, Stefan, 95, 361, 587, 1030, 1106
Banach algebra theory, 360, 361
Banach–Tarski paradox, 587
Banchoff, Thomas, 486, 1042
BankAmericard, 253
bankruptcy, business, **92–94**
bankruptcy, personal, **94–96**
Bankruptcy Code of 1978, 94
al-Banna, al-Marrakushi ibn, 18, 764
Banneker, Benjamin, 668
Banu Musa, al-Hasan. *See* Ibn al-Haytham (Alhazen)
Banzhaf, John, 336, 338
Banzhaf Power Index, 338
bar codes, *96*, **96–97**
Baran, Paul, 517
Barazangi, Muawia, 774
Barbilian, Dan, 361
Bardeen, John, 768
Barkhausen effect, 489
Barnes, Ernest, 602
Barnes & Noble Nook, 303, 304
barometers, 344–345, 364, 1055, 1058, 1063
barometric pressure, 344–345

Barraquer, Jose, 1040–1041
Barricelli, Nils, 290
Barrow, Isaac, 407, 555
   Infinities, 776–777
   Lectiones opticae et geometricae, 1100
Barrow, John, 625, 776
Barton, James, 320
basal body temperature, 385
base-2 system, 9
base-10 system, 8, 67, 537, 711, 727–728, 1098
base-60 system, 712, 725
baseball, **97–99**, 378, 478
Baseball Abstract (James), 97, 378
basic number skills, 537
basketball, **99–102**, *100*
basketry, 16, **102–103**
bathyspheres, 295, 492
al-Battani, Muhammad ibn Jabir al-Harrani, 1094
battement tendu (ballet), 90
Batty, David, 511
Baudhayana Sulbasutra. *See* Sulbasutras
Baudot, Emile, 787
Baudot–Murray code, 1077, *1078*
Baumé scale, 246, 247
Bayes, Thomas, 701, 731
   See also Bayesian decision theory
Bayesian decision theory, 366, 796, 802, 843, 874, 936, 943
Bayt al-Hikma ("House of Wisdom"), 32–33, 74
Beal, Andrew, 600
Beal Conjecture, 600
Bean, Ronald, 1082
bearings, 1061
Beaufort, Francis, 212, 492, 1058, 1063–1064
Beaufort cipher, 212
Beaufort wind scale, 492, 1063–1064
Beaugris, Louis, 167
Beaujoyeulx, Balthazar
   Le Balet Comique de la Reine, 90
A Beautiful Mind (movie), 163, 681
A Beautiful Mind (Nasar) (novel), 559
Becher, Johann, 1028
Becquerel, Alexandre-Edmond, 930
bees, **103–105**
Beg, Ulugh, 1096
Begle, Edward G., 275
"Behold!" Proof, 814, 828
Bekenstein, Jacob D., 117
Bell, Alexander Graham, 976
Bell, John, 548
Bell, Robert J. T., 249

bell curve. *See* **normal distribution**
Bell Curve, The (Herrnstein and Murray), 708
Bell inequalities, 548
Bell Telephone Laboratories, 203, 307, 526, 831–832, 976, 977
Bell X-1 rocket-propelled airplane, 31
Bellaso, Giovan, 192–193
bells, *760*, 760, 761
Beltins Arena (Germany), 62
Benbow, Camilla, 1043
Benesh Movement Notation, 91
Benet Academy Math Team, *228*
Benford, Frank, 4
Benford's Law, 4, 498
Benjamin, Arthur, 569
Benjamin, Thomas, 995
Benjamin Banneker Association, 670
Benjamin–Lighthill conjecture, 995
Bentham, Jeremy, 356
Bentley, Wilson, 259
Bergman, C. A., 56
Berkeley, George, 122, 144
   Alciphron, or the Minute Philosopher, 859
   The Analyst, 859, 1088, 1101
Berlekamp, Elwyn, 415
Bernard, Serge, 167
Berners-Lee, Tim, 911
Bernoulli, Daniel
   fluid force studies and, 30
   Leonhard Euler and, 588, 671
   mathematical modeling and, 313, 314
   professional life, 359
   projectile trajectories and, 390, 391, 671
   trigonometric series and, 764
   vibrating string problem and, 410
Bernoulli, Jacob (Jacques, or James)
   Ars Conjectandi, 764, 801
   binomials and, 112
   curves and, 281
   exponentials/logarithms and, 370
   Law of Large Numbers and, 507, 707, 801
   normal distribution and, 707
   permutations and, 764
   polar coordinate system and, 354
   techniques of Leibniz and, 1100
   theoretical mathematics and, 617
Bernoulli, Johann (John, or Jean)
   equation of the catenary and, 281
   functions and, 410
   Leonhard Euler and, 588
   projectile trajectories and, 64, 390, 391

techniques of Leibniz and, 1100
theoretical mathematics and, 617
Bernoulli Society, 119
Bernoulli–Euler equation, 56
Bernoulli's principle, 30
Berry, Clifford, 1077
Bertalanffy, Ludwig von, 808
Berti, Gasparo, 1058
Bertrand, Joseph, 337
Bessarion, Cardinal, 364
Bessel, Friedrich, 78
Besson, Jacques, 862
Best Buy, 388
betting and fairness, 105–107, 368–369
Betz, Albert, 1067
Bezier, Pierre, 786
Bezier curves, 786
Bhagavad Gita, 731
Bhaskaracharya II
   Lilavati, 1095
   Siddhanta Siromani, 73
Bhaskara's dissection proof, 1095
Bhaumik, Mani Lal, 1041
bianzhong bells, *760*, *760*, 761
bible codes, 730–731
bicycles, 107–109
Biden, Jill, 540
Big Bang theory, 317, 474
Big Bang Theory (television show), 983
Bigollo, Leonardo Pisano. *See* Fibonacci, Leonardo
Bill James Baseball Abstract (James), 97, 378
billiards, 110–111
Billings, Evelyn, 385
Billings, John, 385
Billingsley, Henry, 1097
Binary Automatic Computer (BINAC), 768
binary states, 9
Binet, Alfred, 511
binomial theorem, 111–113, 1095
bin-packing problem, 912
biographies and memoirs, 559
bioinformatics, 164–165, 421–422
biological systematics, 44
biomathematics, 164–165, 606
biomechanics, 923–924
Biometrika (journal), 707, 942
biomimicry, 463, 973
biostatistics, 299–300, 606
bi-quaternion algebraic operator, 34
Birds (Aristophanes), 558
Birkoff, Garrett, 548

birthday problem, 113–115
al-Biruni, Abu Arrayhan, 54, 67, 248, 407, 425, 712
Bit Torrent, 388
bitmap graphics, 307–308
BITNET, 517
Black, Duncan
   The Theory of Committees and Elections, 335
Black, Fisher, 264
Black Flag, 787
black holes, 115–117, 441, 473–474, 837, 849, 934
Black Holes and Baby Universes and Other Essays (Hawking), 474
blackbody radiations, 393–394
Blackett, Patrick M. S., 1078
Blackmun, Harry A., 117–118
Blackwell, David, 118–119, 668, 943
Blake, John, 385
Bliss, Gilbert, 63, 671, 1073
Bloch, William Goldbloom, 557
blocks and tackles, 818
Blondel, Vincent, 174
blood oxygen level dependence (BOLD), 126
blood pressure measurement, 948
Blum, Lenore, 811
BM 85200 (Babylonian text), 261
BMI. *See* **body mass index**
board games, 119–122, *121*, *454*, 454–455
Bocaccio
   Decameron, 665
body clocks, 746
Body Mass Index, 122–124, 894, 1042
Boers, 19
Boethius, Anicius Manlius Severinus, 121, 364, 663–664, 1093
   Consolation of Philosophy, 663
Boethius's arithmetic, 121, 364
Bohr, Harald, 923
Bohr, Niels, 548, 775, 1027
Bohr Model of the Atom, 1027
Bohr radius, 1022
Boldyrev, Dmitry, 684
Bollettino dell'Unione Matematica Italiana (journal), 364
Boltzmann, Ludwig, 348, 988
Bolyai, János, 85, 438, 752, 1103
   Appendix, 361
Bolyai–Lobachevskian geometry, 438
   See also non-Euclidean geometry
Bolzano, Bernard, 553, 1103
Boma, A. N., 14
Bombelli, Rafael, 261–262
   Algebra, 1097
   L'Algebra, 722

Bonaventura, Saint (Giovanni de Fidanza), 624
Bonaventura Cavalieri, 1092
Bondi, Hermann, 350
bone movement, 526–528
Bonferroni, Carlo Emilio, 364
Bonferroni inequalities theory, 364
book ciphering, 869
Book of Addition and Subtraction by the Indian Method (al-Khwarizmi), 66–67
Book of Arithmetic About the Antichrist, A Revelation in the Revelation (Stifel), 858
Book of Signs, The (Theophrastus), 1055
Boole, George, 848, 977
    Investigation of the Laws of Thought, 362, 603, 1104
    The Mathematical Analysis of Logic, 362, 1103
    religion and, 602–603
    theoretical mathematics and, 617–618
Boolean algebra (Boolean logic), 212, 362, 617, 977, 1104
boosting, 291
Booth, Charles, 188–189
bootstrap aggregation (bagging), 291
Borda, Jean-Charles de, 337
Borda election method, 337, 1047–1048
Borel, Émile, 505, 841
Borges, Jorge Luis
    Library of Babel, 557
Borgia, Cesare, 364
Bose, Satyendra Nath, 114
Bosnian Mathematical Society, 364
Bougainville, Louis-Antoine de
    Traité du calcul-intégral, 869–870
Boulez, Pierre, 231
Bourbaki, Nicolas (collective pseud.), 366–367, 1106
Bourgainof, Jean, 367
Bourne, William, 294
Boussinesq, Joseph, 155
Box, George, 654
Box–Jenkins models, 654
    See also autoregressive integrated moving average (ARIMA)
Boyle, Willard, 307
Bradford–Hill criteria, 288
Bradwardine, Thomas, 1095
Bragg, Braxton, 193
Brahe, Tycho, 78, 365
Brahmagupta
    Brahmasphutasiddanta (*The Opening of the Universe*), 424, 908
    cyclic quadrilaterals and, 762, 1094
    Hindu-Arabic numerals and, 686
    negative numbers concept and, 695–696

    series and, 908
    zeros and, 1088
Brahmi numerals, 712
brain, 124–128
    composition and structure, 124–125
    cortical regions and mathematical ability, 598, *659*
    gender differences in, 1070
    imaging technologies and, 126–128, *659*, 1043
    nervous system and, 700–701
    nutrition and, 733–734
    optical illusions and, 739–741
    plasticity, 513, 541
    right versus left brain learning, 541–542
brain teasers. *See* **mathematical puzzles**
brainbow, 126
Bramer, Benjamin, 171
Branch, Stefan, 112
Brandenburg, Karl-Heinz, 684
Brattain, Walter, 768
Bravais, August, 259
Bravais lattice, 259
Breaking the Code (Whitmore), 777
Breakthrough, 120
Brecht, Bertold
    The Life of Galileo, 775
Brenner, Sydney, 124
bridges, 129–130
Bridgewater Treatises (Babbage), 859
A Brief History of Time (Hawking), 473, 474
Briggs, Henry
    Arithmetica Logarithmica, 371, 1098
Brimberg, Jack, 481
Bringing Down the House (Mezrich), 682
Brink, Chris, 19
Brisson, Barnabé, 155
Brix scale, 246
Broca, Paul, 128
Broca's (frontal lobe), 128
Brodetsky, Selig, 30
Brouncker, William, 1099
Brouwer, Luitzen Egbertus Jan "L. E. J.", 1075, 1105
Brown, Earl, 167
Brown, Robert, 288, 696
Browne, Marjorie, 668
Brownian motion, 696
Brun, Viggo, 721
Brunelleschi, 861
Brush, Charles, 1064
Bryant, Gridley J. F., 897
Buckingham, Edgar, 915
Buckland, Jonny, 787

Buckmire, Ron, 167
Budd, Chris, 255
budgeting, 130–132
Buell, Don Carlos, 193
Buffet, Jimmy, 787
Buffon, Comte de, 1101
Buffon, Georges, 770–771
Buildings, Antennae, Spans, and Earth (BASE) jumping, 374
Bukkapatnam, T. S., 134
Bulgarian Competition in Mathematics and Informatics, 361
Bulletin of the American Mathematical Society , 1104
Bunsen, Robert, 79
Bureau International de l'Heure (The International Time Bureau), 204
Bureau of Labor Statistics, 1019
Burger, Dionys
    Sphereland, 902
Burger, Warren E., 117
Burgeson, John, 377–378
Bürgi, Jobst, 1098
Burj Khalifa, 920
Burning Index (BI), 403
Burns, Ursula, 132–133
Burundi, 16
bus scheduling, 133–135
Busa Xaba, Abraham, 20
Bush, George W., 450, 509, 1047, 1049
business, economics, and marketing
    agricultural, 381–382
    bankruptcy, 92–96
    budgeting, 130–132
    business-to-business marketing (B2B), 579
    business-to-consumer marketing (B2C), 579
    credit, 224, 253–255, *254*, 387, 482–483
    debt, 482–483, 695–697
    economic order quantity (EOQ), 523
    European Economic Community (EEC), 337–338
    forecasting in, 400
    gross domestic product (GDP), 466–468, 696, 697
    interest, 95, 483–484, 561–562, 675, 1018
    Internet and, 519–520
    inventory models, 523
    investments and, 676–677
    loans, 482–484, *561*, 561–563, 675–676
    market research and, 578–580
    mutual funds, 691–692
    The New Economics (Edwards), 299
    stocks/stock market, 93, 288, 400, 691, 692, 696, 949–951
    See also **data analysis and probability in society**
business forecasting, 400
business-to-business marketing (B2B), 579

business-to-consumer marketing (B2C), 579
Buston, Dr., 249
Buteo, Johann (Jean Borell)
    Logistica, 764
Butler, C. Allen, 166
Butokukai, 582
buxiban (cram schools), 69
Buxton, Dr., 455, 658
Buys-Ballot, Christophorus, 1002
Byram, George, 403
Byron, Annabella Milbanke, 565
Byron, George Gordon Byron (Lord Byron), 565
Byzantine culture, 664

C
C. elegans, 124, 125
cabbage, goat, wolf problem, 594
Cabri Geometry, 427
Cabri II Plus, 929
Cadogan, Charles C., 167
Caenorhabditis elegans, 124, 125
Caesar, Julius, 154, 192, 212, 563
Cahit Arf, 75
Calandri, Filippo, 1096
Calculated Industries (CI), 141, 142
On the Calculation with Hindu Numerals (al-Khwarizmi), 712
calculators in classrooms, 137–139, 273, 718–719
calculators in society, 139–142
    early history of, 139–140, 887, 1106
    graphing calculators, 141–142
    pocket calculators, 1107
    special-purpose, 140–141
Calculus (Djerassi), 776
Calculus, Concepts, Computers and Cooperative Learning (C4L), 152
calculus and calculus education, 142–147
    contradictions of infinitesimals and, 505–506
    differential, 143–144
    equation of tangent, 144–145
    higher order derivatives, 145
    history of, 33, 143, 148, 405
    integral, 145–147
    Weierstrass Definition, 144
Calculus and Mathematica (C&M), 152
Calculus Consortium at Harvard (CCH), 152
Calculus for a New Century, 151
calculus in society, 148–153
    applications of, 148–150
    mathematics curriculum reform, 150–151
    measurement and, 650–651
    traditional versus reformed calculus, 151–152

Calculus of Extension, The (Grassmann), 1029
calculus theorem, 1100
Calcuus (Doxiadis), 143
Calder, Alexander, 545
    L'empennage, 544
calendars, 153–155
    ancient Mesoamerica and, 179, 240
    farming, 23, 799
    Julian and Gregorian, 154, 879–880, 1097
    lunar, 75, 77, 154–155
    Mayan, 155, 494–495, 1087
    menstrual, 19
    number "12" and, 729
    religious festivals and, 24, 74–75, 240
    solar, 155
    zeros and, 1087, 1088
    See also timekeeping
Cal-Tech (calculator), 140
cameras. *See* **digital cameras; movies, making of**
Campanus, Johannes, 1095
Campbell, Jack "Johnny", 182
Campbell, Lucy Jean, 167
Campbell, Merville O'Neale, 166–167
    See also soroban
Canada, mathematics in, 709
    See also **North America**
Canadian Mathematical Society (CMS), 709
canals, 155–156
cancer, 183–184
cannons, 671
Canon of Samos, 280
canons, 230–231
Canterbury Tales (Chaucer), 665
Cantor, Georg, 504–505, 618, 626, 1104
Capek, Karel, 875
Caplan, Seth, 440
capsid geometry, 1038–1039
car purchases, 225–226
car racing. *See* **auto racing**
Caratheodori, Constantin, 146
carbon dating, 157–159
carbon dioxide emissions, 199
carbon footprints, 159–162
    calculation of, 160, *161*, 463–464
    economy/policy and, 160
    marginal abatement curve and, 160–161
    of people, 159–160
    per country, 161–162
carbon-14 dating. *See* **carbon dating**
card and dice magic, 567–568
card games, 415–417, *416*

Cardan grill, 212
Cardano, Gerolamo
    Ars Magna, 722
    cubic equations and, 261–262, 786, 862, 1097
    dice games and, 302
    puzzles and games, 820
Cardan(o)-Tartaglia formula, 261, 262
cardinal numbers, 1106
careers, 162–166
    in actuarial science, 164, 288–289, *605*, 605–606
    analytical thinking and, 162–163
    applied mathematics and, 163, *605*, 605–607
    in architecture, 432
    in communications, 221
    emerging fields, 164–165, 299–300
    employers of mathematicians, 39, 163–164, 214
    financial mathematics and, 164
    medical and pharmaceutical, 42–43, 162–166, 288, 299–300
    online job listings, 165–166
    in statistical science, 284–287
    using algebra in, 36, 39, 40, 41
    See also **data analysis and probability in society**
Carey, Mariah, 787
car-following traffic modeling, 1000
Caribbean America, 166–167
Caribbean Journal of Mathematical and Computing Sciences, The, 166, 167
Caritat, Marie Jean Antoine Nicolas de. Marquis de Condorcet, 337
Carleson, Lennart, 363
Carley, Kathleen, 509
Carlyle, Thomas, 1102
Carnegie Corporation, 133
Carnot, Lazare Nicolas Marguerite, 501, 1102
Carolingian Renascence, 663–664
carpentry, 167–169
Carraher, David, 226
Carroll, Lewis (Charles Dodgson)
    Alice in Wonderland, 110, 464, 513, 557
    Alice Through the Looking Glass, 6
    The Game of Logic, 122
    word puzzles and, 820
Carruth, Shane, 1082
Cartan, Élie, 35
Cartan, Henri, 367
Carte Blanche, 253
Carter, Jimmy, 767
Cartesian plane, 33, 34, 247, 353, 894, 1042
    See also **coordinate geometry**; Descartes, René
cartography, 502, 570, 572–574, 738

Casazza, Peter, 211

casinos, 105, 106

Casio, 141–142

Cassini, Giovanni, 890

Castaneda v. Partida, 117

Castillo-Chávez, Carlos, 169–171

castles, 171–172

CAT scans, 306–307

Catalan, Eugène, 337

Cataldi, Pietro Antonio, 1097

catastrophe theory, 365

Catherine II (Catherine the Great), 359

Cathode-Ray Tube Amusement Device, 1034

Cauchy, Augustin Louis
    complex analysis and, 723
    complex function theory and, 146, 147, 366
    convergence and, 650–651
    Cours d'analyse, 553
    Green's theorem and, 1030, 1102
    permutations and, 765–766
    probability and, 803
    Résumé of Lessons of Infinitesimal Calculus,
      144

Cauchy–Riemann equations, 723

Cauchy's theorem, 766

Cavalieri, Bonaventura
    Archimedean spiral and, 353
    criticism of, 652
    Geometria indivisilibus continuorum, 353
    John Wallis and, 1099
    spiral curves and, 248
    theory of indivisibles and, 1092, 1098

Cavalieri's Principle, 186

caves and caverns, 172–174

Cayley, Arthur, 500, 766, 966, 1005, 1104

Cayley, George, 29, *29*

CCD chips, 305, 307–308

CD drives, 769

CDs, 309–310

Cedar Point Amusement Park (Ohio), 877, 878

celestial mechanics, 520–521

cell phone networks, 174–176

cell phone towers, 174–175, *175*

cell phones, 142, 977

Celsius, Anders, 988

Celsius scale, 988, 1058

census, 176–178

Census Act of 1800, 177

Cent Mille Milliards de poèmes (One Hundred Thousand Billion Poems) (Queneau), 779

Center for Bioinformatics, 183

Center for Promotion of Mathematical Research (Thailand), 71

center of gravity (CoG), 457

Centers for Disease Control and Prevention (CDC), 479

centigrade scale, 988

Central African Republic. *See* **Africa, Central**

Central America, 178–180

Central Limit theorem (CLT), 369, 507–508, 802

central processing units (CPUs), 752–753, 768

centripetal force, 878

Cerf, Sigrid, 180

Cerf, Vinton, *180*, **180–181**, 911

CERN httpd (W3C httpd), 911

Certified Registered Nurse Anesthetists (CRNAs), 43

Cetti, Franceso, 364

Ceva, Giovanni, 364

Cézanne, Paul, 749

CGI. *See* **animation and CGI**

Chadwick, Edwin, 500

chai (life), 730

Chain-Links (Karinthy), 915

Challenger (space shuttle), 870–871

Chamberlain, Nira, 28

Chandrasekhar, Subrahmanyan (Chandra), 115–116, 837

Chang, Joseph, 419

Change the Equation, 133

chaos theories, 803, 891
    See also ergodic theory

Chapter 7 bankruptcy (liquidation), 94–95

Chapter 13 bankruptcy (reorganization), 94, 95

Charlemagne, 663

Chartier, Timothy, 51

charts, 456

Chasles, Michel, 249, 1103

Châtelet, Gabrielle Émilie Le Tonnelier de Breteuil. Marquise du, 777, 1101

Chaucer
    Canterbury Tales, 665

Chebyshev, Pafnuty, 500, 803, 875

cheerleading, 181–183

Cheever, John
    The Geometry of Love, 558

chemical half-lives, 319

chemotaxis, 45

chemotherapy, 183–184

Cheney, Dick, 510

Cheng Dawei
    Suanfa Tongzong (General Source of Computational Methods), 187

Chern, Shiing-Shen, 68

Cherokee Mixed-Use Lofts, *462*

Cherry, Colin, 210
chess, 6, 120
Chiang
  Division by Zero, 901
  Story of Your Life, 901
Chiang, Ted, 901
Chiao Wei-Yo, 156
Child-Pugh score, 52
children's programming (television), 982
Children's Television Workshop (CTW), 982
Ch'in Chu-shao, 1095
China. *See* **Asia, eastern**
Chinese mathematics, 184–188
  East meets West period, 187
  foundation period, 185–186
  golden period, 185, 186–187
  number system, 8, 1091
  religion and, 626–627
Chinese Remainder theorem, 186, 187, 720
Chisholm, Grace. *See* Young, Grace Chisholm
Chitika, 12
Chokwe people, 14
cholera outbreak, 188, 1042, 1052
*Chouren Zhuan (Biographies of Astronomers and
  Mathematicians)* (ed. Ruan Yuan), 187
Christian religious tenets, 624–626, 664, 729–730, 731, *856*,
  857
Chronbach, Lee, 817
Chronicle of Higher Education, 166
chronometers, 502
  See also **measuring time**
Chrystal, George, 455
Chu Shih-chieh, 1095
Chung, Fan, 589
Chuquet, Nicolas
  Triparty en la science des nombres, 1096
Church, Albert, 193
Church, Alonso, 768
Church of Latter-day Saints (LDS), 418
Church of San Lorenzo, 861
Churchill, Winston, 508
ciphers, 191, 192
  See also **coding and encryption**
cipher-text, 212–214
circles, 88, 514–515, 552, 1092, 1093, 1096
  See also Measurement of a Circle (Archimede); **Pi**
Citrabhanu, 73
city planning, 188–190, *189*, 462, *462*
Civil War, U.S., 191–194
  cryptography in, 191–193
  impact on education, 193–194

Clairaut, Alexis Claude, 1005
  Elements de géometrie, 658
  Théorie de la figure de la Terre, 1101
Clark, Arthur C., 899
  2001: A Space Odyssey, 901
Clarke, Somers, 455
Class Phenotype Probability, 114
  See also **birthday problem**
classic mathematical problems
  Aladdin's saddlebag, 747
  ballot problem, *339*, 339–340
  bin-packing problem, 912
  birthday problem, 113–115
  cabbage, goat, wolf problem, 594
  cocktail party problem, 210–211
  Correlation Problem, 495–496
  coupon collector problem, 204, 253
  dining philosophers problem, 211
  double bubble problem, 653
  duplication of the cube, 260–261, 1091, 1092
  Get Off the Earth puzzle, 820
  heat conduction problem, 410, 987–988
  packing probem, 746–747
  parallel climbers puzzle, 201
  party problems, 210–211
  rabbit problem, 1095
  river-crossing puzzle, 16
  ruler and compass construction, 882
  stable marriage problem, 582
  three construction problems, 459
  three construction problems and, 459
  Tower of Hanoi puzzle, 412, 593–594
  traveling salesman problem (TSP), 912, 1009–1010
  two container problem, 594
  vibrating string problem, 410
  word problems, 1094
classical test theory, 816
"Classification of Countable Torison-Free Abelian Groups"
  (Campbell), 166–167
Classification theorem for Finite Simple Groups, 587
Clausius, Rudolf, 349
Clavis mathematicae (Oughtred), 1098
Clavius, Christopher, 812, 1097
Clay Mathematics Institute, 110
Clean Water Act, 1054
Clifford, William Kingdon, 34, 349, 965
Clifford torus (or donut), 965
climate change, 194–200
  as a distribution, 195–196
  global warming, 196–199, 442
  sunspots and, 964

climatology, 1057

climbing, 200–201

Clinton, Bill, 526

clock arithmetic, 201, 719–720

clocks, 201–204, 655–656

closed-box collecting, 204–206

Closs, Michael, 698

clouds, 206–207

clubs and honor societies, 207–208

CMOS chips, 305

Coandă, Henri, 30

Coandă effect, 30

Coasta, Celso, 966

Cobb, Paul, 222

cobweb theorem, 379

Coca-Cola Company, 252

cochlear implants, 209–210

Cochran, William, 889

cocktail party problem, 210–211

code breaking, 191

    See also **coding and encryption**

Code of Hammurabi, *506*, 507

code talking, 839, 1074

codebooks, 192, 1076

Codex Atlanticus (Leonardo da Vinci), 918

Codex Vigilanus (compilation), 712

coding and encryption, 212–214

    bible codes, 730–731

    Civil War and, 191–193

    code talking, 839, 1074

    credit card encryption, 224

    fax machines, 382–384, *383*

    Fermat's little theorem and, 113

    number theory and, 165

    public key cryptosystems, 282, 786

    Revolutionary War and, 869

    World War I and, 1074

    World War II and, 756–757, 1076, 1077

    See also military code; Morse code

coffeehouses, 596

cognitive epidemiology, 513

cognitive psychology, 741

Cohen, Bram, 388

Cohen, Paul J., 1106

coil springs, 636

Cold War, 214–218

    arms race, 214–216, 517

    competition during, 217–218, 360, 365, 709

    intelligence/counterintelligence and, 508

Coldplay, 787

collectibles. *See* **closed-box collecting**

Collins, Brent, 904–905

Colojoara, Ion, 361

color wheels, 515

Columbia (space shuttle), 871

Columbine High School shootings, 897

Columbus, Christopher, 576, 862, 932, 1011

Columbus Dispatch (newspaper), 40–41

combinations. *See* **permutations and combinations**

Combinatorial Game Theory, 122

combinatorics, 55, 444

comic strips, 218–219

Commandino, Federigo, 1097

Commedia (Dante), 665

Commentary on Euclid, Book I (Proclus), 1093

Commercial and Political Atlas, The (Playfair), 456

Commission on History of Mathematics in Africa, 23

Commissions on Mathematics Education in Africa, 22

Committee for Statistical Studies, 944

Committee on Data for Science and Technology, 1026

Committee on the Undergraduate Program in Mathematics (CUPM), vi

Common Core State Standards (CCSS), 275, 279

common school movement, 274

communative property, 8

communication in society, 219–225

    mathematical applications/technologies and, 224

    mediums for, 220–221

    proofs and, 222–224

community colleges. *See* two-year colleges

community-supported agriculture (CSA), 382

compact discs (CDs). *See* CDs

comparable transaction method, 93–94

comparison shopping, 225–227

compasses, 9

competitions and contests, 227–229

Complete Mancala Games Book, The (Russ), 13, 16

complex analysis, 26, 723–724

Complicite Theater Company, 776

composing, 229–231

compound interest, 95

compound pulley, 57–58

computable functions. *See* **functions, recursive**

computation of Pi. *See* **Pi**

computational aids, 713

computational geometry, 429

Computed Axial Tomography (CAT) scans, 1013

    See also functional MRI (fMRI); imaging technologies; magnetic resonance imaging (MRI); medical imaging

computer aided design (CAD), *352*, 353, 514, 990

computer aided looms (CAL), 990

computer algebra systems (CAS), 273

computer graphics, 165, 786

computer numerical control (CNC), 430

computer programming, 412, 565, 566

computer software. *See* **software, mathematics**

Computer Speaks, The (Khalifa), 730

computer visions, 165

computer-aided design (CAD), 168, 430

computer-generated imagery (CGI). *See* **animation and CGI**

computerized axial tomography (CAT scanning), 306–307

computers
    applied mathematics and, 603, 607
    battlefield, 65–66
    early history of, 517, 1106, 1107
    hacking, 192, 224
    mechanical, 875, 1103
    number/operations and, 718–719
    steam powered, 500
    super, 1107
    theoretical mathematics and, 617
    World War II and, 1077–1078
    zeros and, 1088–1089
    See also **calculators in classrooms; calculators in society; personal computers**

"Comrade Deuch" (Gaing Kek Ieu), 71

conception. *See* **fertility**

conditional probability tables (CPTs), 874

Condorcet criterion, 217

Condorcet Winner, 337, 1048–1049

Conférence Général des Poids et Mesures, 547

congressional representation, 231–235

conic sections, 235–238, 1092, 1093, 1094, 1100

Conic Sections (Apollonius), 237, 1092, 1093, 1094, 1100

connections in society, 238–243
    interconnected curriculum, 238–239
    mathematics as universal language, 240
    nutrition labeling and mathematics, 240–241
    sports and mathematics, 239, 241–242

Connectivity in Geometry and Physics Conference, 240

Connes, Alain, 343, 367

Consolation of Philosophy (Boethius), 663

Consortium for Mathematics and Its Applications, 229

Construction Master Pro software, 141, 142

constructivism, 542

Contact (movie), 683

Contact (Sagan), 901–902

contact lenses, 1040

contaminants, 1053–1054

contests. *See* **competitions and contests**

continuous quality improvement, 832

continuum modeling, 1001

contra and square dancing, 243–244

convergence, 650–651

Conversations on Mathematics With a Visitor from Outer Space (Reulle), 901

conversion rates, 11–12

Conway, John H., 122, 415

cooking, 244–247

Coolidge, Julian Lowell, 353

Coonce, Harry B., 628–629

Cooper, Lionel, 20

Cooper, Paul, 1015

coordinate geometry, 33, 34, **247–249**, 353, 727, 1041

Copenhagen (Frayn), 775–776

Copenhagen Telephone Company, 977

Copernicus, Nicolaus, 78, 365, 801
    De Revolutionibus, 772, 1096
    heliocentric astronomy and, 862, 964

coral reefs, 250, **250–251**

Coriolis, Gaspard-Gustave de, 110, 111, 995

Coriolis effect, 110, 998

Coriolis–Stokes force, 995

Correlation Problem, 495–496

Corson, William, 1037

Cos, Gertrude, 943

cost per click (CPC), 11

cost per mile (CPM), 11

cost-benefit optimization, 873–874

Costley, Charles Gladstone, 167

Coulomb, Charles-Augustin de, 340–341, 770

Coulomb's Law, 340–341, 770

Count (Martin and Craig), 776

counterintelligence. *See* **intelligence and counterintelligence**

counting numbers. *See* natural numbers

counting skills, 537

coupon collector problem, 204, 253

coupons and rebates, 252–253

Cournot, Antoine, 802

Cours d'analyse (Cauchy), 553

covering problems, 747

cowry shells, 23, *24*

Cox, David, 617, 1006

Cox, Elbert, 668

Cox, Richard, 803

Coxeter, Harold Scott MacDonald "Donald", 355, 781, 783

Cox's theorem, 803

Craig, Timothy
    Count, 776

Cramer, Kathryn
    Forbidden Knowledge, 901

crane, origami, *742*

credit bureaus, 387
credit cards, 253–255
    credit bureaus and, 254
    data mining and, 12, 254
    encryption, 224
    fraud detection and, 253, *254*
credit ratings, 482–483
Crelle's Journal, 1104
Cremona, Luigi, 1005
Cremona transformations, 1005
Crichton, Michael, 901
Crick, Francis, 674
crime scene investigation (CSI), 255–257
Crippen, Robert L., 871
Crispin, Mark, 518
criterion-referenced tests (CRTs), 326–327
Critical Infrastructure Protection (Lewis), 509
Critique of Practical Reason (Kant), 626
Critique of Pure Reason (Kant), 626
Croatian Mathematical Society, 364
crochet and knitting, 257–258
Crome, Augustus, 894
crosses, *856, 857*
crossmultiplication, 686
crosswords. *See* **acrostics, word squares, and crosswords**
Crother, Stephen, 115
Crozet, Claudius, 193, 429
CRT televisions, 986–987
Cruickshank, Steven
    Mathematics and Statistics in Anaesthesia, 43
cryptarithms, 820–821
cryptography, 719–720, 763
    See also **coding and encryption**; military code
cryptology, 224
crystallography, 259–260, 547
Csicsery, George, 227
Cuba, 167
    See also **Caribbean America**
Cube (movie), 683
cubes and cube roots, 260–262
cubic equations, 785–786, 1095
cubists, 748–749
Cummings, Alexander, 997
currency exchange, 262–264
Current Employment Statistics (U.S. Bureau of Labor Statistics), 286
Current Population Survey, The (U.S. Census Bureau), 286, 1019
curricula, international, 264–267
curriculum, college, 267–273
    general education mathematics requirements, 268–269

mathematical sciences concentration, 270–271
mathematics and partner disciplines, 269–270
two-year colleges, 272–273
undergraduate research, 271–272
curriculum, K–12, 274–279
    Back to Basics movement, 276
    Common Core State Standards (CCSS), 275, 279
    common school movement, 274
    measurement systems, 644–645, 658
    national standards-based, 277–278
    New Math movement, 275–276
    No Child Left Behind (NCLB) Act and, 278–279
    number and operations, 710–711
    Progressive movement, 274–275
    Race to the Top (RTTT) program, 279
    racial/minority disparities, 669
Curriculum and Evaluation Standards for School Mathematics (NCTM), 277
    See also Principles and Standards for School Mathematics (NCTM)
Curriculum Focal Points for Prekindergarten through Grade 8 Mathematics (NCTM), 277–278
curves, 280–282, 650, 1011, 1092
    See also Conic Sections (Apollonius)
Cusanus, Nicholas. *See* Nicholas of Cusa
Custer, George Armstrong, 193
cyan, magenta, yellow, key black (CMYK) subtractive model, 515
Cybernetics (Weiner), 875
cyberpunk, 902
cyclic redundancy check (CRC), 310
cycloid, 280
cyclones. *See* **hurricanes and tornadoes**
cyclotron, 358
"Cylinder Measures" (Millington), 167
Cyprus Mathematical Olympiad, 76
Cyprus Mathematical Society, 76, 364
Czech–Polish–Slovak Match, 361

D
da Vinci. *See* Leonardo da Vinci
Dai Fujiwara, 431
d'Alembert, Jean Le Rond, 410, 553
Dalgarno, George, 1028
D'Ambrosio, Ubiritan, 933
dams, 283–284
dancing
    ballet, 90–91
    ballroom, 91–92
    bee choreography, 105
    contra and square, 243–244

step and tap, 946–947
in West Africa, 22
Dancing With the Stars (television show), 983
Dante
Commedia, 665
Dantzig, George, 218, 952
Darboux, Jean-Gaston, 249, 1030
Darcy–Weisbach equation, 30
dark energy, 350
Darken, Joanne, 811
Darwin, Charles
Structure and Distribution of Coral Reefs , 250
data analysis and probability in society, 284–290
entertainment/gambling and, 289–290
finance/insurance and, 288–289
government and, 285–287
industry/manufacturing and, 287–288
medicine/pharmacy and, 288
problem of distributions and, 806
professional education, 285
data compression, 684
data mining, 290–292, 509
data rot, 310–311
Daubechies, Ingrid, 292–294
David, Florence Nightingale, 843
Davies, Charles, 193
Davies, Donald, 517
Davis, Jefferson, 193
Davis, Philip, 357
Day, Jeremiah
An Introduction to Algebra, 193
De Algebra Tractatus (Wallace), 112
De Arte Combinatoria (Leibniz), 1028
de Cari, Gioia, 1082–1083
Truth Values, 777, 1083
De Divina Proportione (About Divine Proportion) (Pacioli), 886
de Finetti, Bruno, 106–107
de Gusmão, Bartolomeu, 932
De lattitudinibus formarum (attrib. d'Oresme), 455
de Moivre, Abraham, 112, 507
Annuities upon Lives, 1100
Approximatio ad summam terminorum binomii (a+b)n in seriem expansi, 707
Doctrine of Chances, 801–802, 1100
Miscellanea analytica, 1100
See also Central Limit theorem (CLT)
de Moivre–Laplace theorem, 707
See also Central Limit theorem (CLT)
de Morgan, Augustus, 565, 770, 957
De Ratiociniis in Lundo Aleae (Huygens), 801

De Revolutionibus (Copernicus), 772, 1096
de Veaux, Richard, 969
De Ventula (anon.), 800
De Viribus Quantitatis (Pacioli), 821
Dean's Method, 233
Deary, Ian, 511
debt. *See* loans; mortgages; **national debt**
debt-to-income ratios, 482–483
Debussy, Claude, 894
Decameron (Bocaccio), 665
Dechales, Claude, 393
decimal system, 53–54, 69, 185, 727–728, 1095, 1097–1098
deciphering. *See* ciphers
decryption. *See* **coding and encryption**
Dedekind, Richard, 35, 726, 727, 1104
Dedekind cuts, 727
deductive logic, 459–460
Dee, John, 1097
deep submergence vehicles, 294–295
Defense Advanced Research Projects Agency (DARPA), 181
deforestation, 295–298
deism, 626
del Ferro, Scipione, 261, 786, 1096
Delamain, Richard, 1098
Delambre, Jean, 890
Delaunay, Charles-Eugene, 909
Deligne, Pierre, 367
DeLoache, Judy, 325
Demidovich, Boris Pavlovich, 360
Deming, W. Edwards, 287, 298–299, 944
The New Economics, 299
Out of the Crisis, 299
Statistical Methods from the Viewpoint of Quality Control, 832
Democritus of Abdera, 342, 652, 1092
Dennison, William, 192
deoxyribonucleic acid (DNA). *See* DNA
Derbyshire, John
Prime Obsession, 560
DeRose, Tony, 50, 679, 966
DeSalvo, Stephen, 963
Desargues, Gérard, 237, 1099
Descartes, René
algebra and, *32*
analytic geometry and, 365
animal/machine connections and, 45
Cartesian plane and, 33, 34, 247, 353, 894, 1042
conic sections and, 237
Discourse on a Method, 248–249, 365–366, 847, 859
imaginary numbers and, 722, 939
La Géométrie, 1098–1099

Marin Mersenne and, 588
mechanics and, 349
Meditations on First Philosophy, 365, 859
notation systems and, 785
as philosopher, 365–366
polyhedral formula and, 783
Principles of Philosophy, 366
religious writings, 859
spiral curves and, 354
tides/waves and, 994
volcanos and, 1044
Descent into the Maelstrom (Poe), 558
descriptive geometry, 429
design principles, 514, 533
Dessouky, Maged, 134
Devanagari numerals, 712
Devol, George, 876
Dewey, John, 274
DeWitt, Simeon, 870
diagnostic testing, **299–301**
Dialogues Concerning the Two Chief World Systems (Galileo), 858–859
Dias, Bartholomeo, 862
diatonic scales, 893
Diaz, Joaquin, 668
Dicaearchus of Messana, 248
dice games, 301–303
Dickens, Charles
    Hard Times, 560
dictatorships, 1048, 1049
Dido of Tyre (queen), 103
Die Coss (Rudolff), 1096
Die Theorie der Parallellinien (Lambert), 1101
diets, 733, *733*
Dietz, Ada
    Algebraic Expressions in Handwoven Textiles, 989
Dieudonne, Jean, 366–367
differential calculus, 143–144, 148, 149–150, 1101
differential geometry, 429
differential GPS (DGPS), 452
diffusion tensor imaging (DTI), 126
digital book readers, 303–304
digital cameras, 304–306
digital elevation models, 345
digital images, 306–308
Digital Opportunity Index (DOI), 769
digital storage, 308–311
digital video discs (DVDs). *See* DVDs
digital video recording devices (DVRs). *See* **DVR devices**
dikes, 155
Diners Club, 253

dining philosophers problem, 211
Dinostratus, 1092
Diocles of Carystus, 280
Dionysius
    Parthenopaeus, 6
Diophantus of Alexandria, 32, 461
    Arithmetica, 1093, 1097, 1098
direct acyclic graphs (DAGs), 874
direct current (DC), 341
direct proofs, 813
Dirichlet, lejeaune, 410
Dirichlet, Peter Gustav Lejeune, 1103
A Disappearing Number (McBurney and Complicite Theater Co.), 776
Discourse on a Method (Descartes), 248–249, 365–366, 847, 859
Discourses on Livy (Machiavelli), 861
Discover (magazine), 525
Discovering Geometry: An Investigative Approach (Serra), 780
discrete finite state automata (DFA), 1034
discrete geometry, 429
Discworld (Pratchet), 902
disease survival rates, 311–312
diseases, tracking infectious, 170–171, **312–314**, 513
displace games. *See* **board games**
Disquisitiones Arithmeticae (Gauss), 712, 939–940, 1102
dissections, 828
Dissertation Abstracts International, 629
Distance Measuring Equipment (DME) system, 1079
distribution-free tests, 844
divergence theorem, 1030
divination, *623*
divine proportion. *See* **golden ratio**
division. *See* **multiplication and division**
Division by Zero (Chiang), 901
Djerassi, Carl
    Calculus, 776
DNA
    analysis, 256, *420*, 421–422
    bioinformatics and, 164–165
    as blueprint of creation, 886
    electron microscopes and, 547
    mapping and sequencing, 606
    math gene, 584–585
    molecular structure of, 532, *673*, 673–674
    topology, 614
    viruses and, 1038
    See also **genetics**
Doctrine of Chances (de Moivre), 801–802, 1100
dodecahedron, 782
Dodge, H. F., 832

Dodgson, Charles (pseud. Lewis Carroll). *See* Carroll, Lewis (Charles Dodgson)

Doll, Richard, 362

Domain Name System (DNS) servers, 518

domains, 723–724

domes, *315*, **315–316**, 1045

Domesday Book, 362, 1020

Donald in Mathmagic Land (cartoon), 111

Donaldson, Simon, 363

Doppler, Christian, 316, 756

Doppler radar, **316–317**, 1002

Dore, Richard, 778

d'Oresme, Nicole
  De lattitudinibus formarum (attrib.), 455

Dorman, James, 774

Dorsey, Noah, 547

dose-response curve, 535

dot-com bubble, 519–520

Dots & Boxes, 122

double bubble problem, 653

Doubleday, Abner, 193

Douglas, Jess, 966

Dow Jones Industrial Average (DJIA), 288, 692, 949

Doxiadis, Apostolos
  Calcuus, 143
  Logicomix, 219
  Seventeenth Night, 776

Doyle, Arthur Conan
  The Final Problem, 558

Dr. Atomic (Adams), 690

Dresden Codex, 496

Drexler, K. Eric, 693

drinking water, *1053*, 1053–1054

Droid, 977

Drosnin, Michael, 731

drug dosing, **317 319**

drums, 761

dry adiabatic lapse rate, 207

Dryden, John, 596

Dudeney, Henry, 821

Duncan, Arne, 275

Dunham, William, 610

Dunnett, Charles, 319

Dupain-Triel, Jean-Louis, 345

duplication of the cube problem, 260–261, 1091, 1092

Durenmatt, Friedrich
  The Physicists, 775

Dürer, Albrecht, 783, 1096

DVD drives, 769

DVDs, 309–310, 705

DVR devices, **319–321**, 705

dynamical systems, 128

DynaTAC8000x, 977

dyscalculia, 128, 714

Dzhumadidayev, Askar, 67

E

"E=MC2 " album (Carey), 787

Early Childhood Longitudinal Survey (ECLS), 669

earth, measurement of, 762–763, 1092

From the Earth to the Moon (Verne), 521, 900

earthquakes, **323–324**

Eastern Bloc. *See* **Europe, eastern**

Eastman, George, 305

Eastman Kodak Company, 305

Eberlein, Ernst, 684

Eckert, John Presper, 768, 1078

eclipses, *77*, *78*, 677–678

École Polytechnique, 366, 501

ecological design. *See* **green design**

economic order quantity (EOQ), 523

economics. *See* business, economics, and marketing

ecosystems, 48

ecumene, *573*

Eddington, Arthur, 116, 855

Edgeworth, Francis, 369

Edidin, Dan, 211

Edison, Thomas, 182, 549, 986

Educate to Innovate campaign, 133, 1035

Educational Amendments Act, 1071

educational manipulatives, **324–326**

Educational Recovery Act, 279

educational testing, **326–329**

Edward I (King), 172, 648

EEG/EKG, 42–43, 126, 128, **329–330**

Egan, Greg
  The Infinite Assassin, 901
  Schild's Ladder, 559

Eglash
  African Fractals, 14

Egorov, Dmitry, 505

Egyptian mathematics, 32, **330–333**, 603, *638*, 638–639, 711

Ehlinger, Ladd, Jr., 440

Ehrenfest, Paul, 837

Eiffel, Gustave, 919

Eiffel Tower, 919

Eight Ball, 110

Ein Sof, 626

Einstein, Albert, **333–335**
  Autographical Notes, 335
  conservation of mass/energy and, 334–335
  "E=MC2" album (Carey), 848

Einstein on the Beach (Glass), 775
elegant proofs and, 611–612
field equations, 770
field equations and, 775
gravity/weightless and, 1059
Insignificance (Johnson), 775
light speed and, 547–548
mathematics/reality link and, 1027
personal and professional life of, 333–334, *335*, 335, 598, *598*, 887, 959
photoelectric effect and, 930
quantum physics and, 974
Riemannian geometry and, 438, 555
Theory of Everything, 441
theory of relativity and, 38, 78, 115, 429, 474, 848, 854–855, 980, 1074, 1105
universal constants and, 1026
See also **atomic bomb (Manhattan Project)**; **relativity**
Einstein Institute of Mathematics, 75
Einstein on the Beach (Glass), 690
Einstein's field equations, 775
Einthoven, Willem, 329
Eisenhart, Luther, 232
Eisenstein, Gotthold, 370
EKG. *See* **EEG/EKG**
elections, 335–340
ballot problem, *338*, 339–340
exit polling, 336–337, 339
types of, *336*, 337
U.S. Electoral College, 336, 338–339
weighted voting, 336, 337–338
Electrical Numerical Integrator and Calculator (ENIAC), 768, 769, 1077–1078, 1079
electricity, 340–342
electrocardiogram (ECG). *See* **EEG/EKG**
electrodynamics, 549
electroencephalogram (EEG). *See* **EEG/EKG**
electromagnetic radiation (EMR), 662, 837
electromagnetic wave equation, 547
electron spin resonance (ESR) dating, 941
electronic ink, 303–304
electronic passwords, 517–518
Elementary Mathematics from an Advanced Standpoint (Klein), 616
elementary particles, 342–344
Elements (Euclid)
Abraham Lincoln and, 193
deductive logic and, 459–460
editing/revision of, 1093, 1099
Golden Ratio and, 1116
metaphysical reasoning and, 624

origination of, 1092
parallel postulate and, 750–751, 752
ruler/compass construction and, 881–882
solids and, 782
terminology, 53, 84
Theaetetus of Athens and, 1092
translations, 18, 187, 1094–1095, 1097
See also **Euclid of Alexandria**; **Euclidean geometry**
Elements de géometrie (Clairaut), 658
Élements de géométrie (Legendre), 1102
Elements of Physical Biology (Lotka), 789
elevation, 344–346
elevators, 346–348
Elkies, Noam, 227
Elliott, Ralph, 400
Elliott waves, 400
ellipses, 235
elliptical curves. *See* **curves**
elliptical orbits, 437–438, 457, 772
Ellis, George, 115
Ellis, Robert, 802
Else-Quest, Nicole, 1070
Emilie (Gunderson), 777
en l'air (ballet), 90
Enclosure Acts, 500
encryption. *See* **coding and encryption**
energy, 348–350
energy, geothermal. *See* **geothermal energy**
Energy Conscious Scheduling (ECS), 463
Energy Explained Web site. *See* **U.S. Federal Bureau of Investigation (FBI)**
Energy Information Administration (EIA), 287
Energy Kids Web site, 287
Energy Policy Act, 998
Engelbach, Reginald, 455
Engelberger, Joseph, 876
Engels, Friedrich, 500
engineering design, 351–353
Engle, Robert, 466
English language. *See* **acrostics, word squares, and crosswords**
Enigma code, 212, 763, 1076–1077
Eniwetok Atoll, 250
Enneper, Alfred, 966
Enron Corporation, 509
entanglement, 548
enterprise value (EV), 93
Epic of Gigamesh, The (Mesopotamian epic poem), 899–900
epidemiology, 479, 513, 942
See also **diseases, tracking infectious**

epistemology, 540

Epitome of Copernican Astronomy, The (Kepler), 365

equal tuning, 824–825

equations, polar, 353–354

equiangular spiral, *762*, 763

Equilibrium of Planes (Archimedes), 544

equilibrium theory, 519

Eratosthenes of Cyrene, 18, 77–78, 460–461, 762–763, 1092

Erdös, Paul, 361, 518, 589, 596

Erdös number, 589, 916

    See also **Six Degrees of Kevin Bacon**

Erdös–Rényi graphs, 453, 518

e-readers. *See* **digital book readers**

ergodic theory, 490

    See also chaos theories

Erie Canal, 156

Eritrea, 16

Erlang, Agner, 977

Erlangen Program, 438, 965, 971, 1104

Error-Correcting Code (ECC) memory, 911

Escher, M.C., 354–356

    as amateur mathematician, 599

    mandalas and, 515

    Metamorphosis III, 355–356

    Regular Division of the Plane with Asymmetric Congruent Polygons, 355, 765

    symmetry and, 748, 970

    tessellations and, 781

Eskin, Alex, 110

Essai Philosophique sur les Probabilités (Laplace), 368

ethics, 356–358

ethnomathematics, 22, 458, 485, 515, 670, 933

Euclid Freed of Every Flaw (Saccheri), 1101

Euclid of Alexandria

    Abraham Lincoln and, 193

    Arabic/Islamic mathematics and, 53 54

    axiomatic geometry and, 424–425

    binomial theorem and, 111

    Euclidean philosophers and, 357, 427

    golden ratio and, 446

    as library leader, 459

    musical pitches and, 433

    optics and, 1042

    pinhold camera and, 305

    postulates of, 426, 750–751, 752

    ruler/compass construction and, 881–882

    superposition and, 1004

    theoretical mathematics and, 615

    Theory of Intervals, 433

    See also **Elements** (Euclid); Euclidean geometry; Greek mathematics; non-Euclidean geometry

Euclidean geometry, 84–85, 424–425, 427, 428–429

    See also **Elements** (Euclid); Euclid of Alexandria

Eudemian Summary (Proclus), 1092

Eudemus of Rhodes, 1092

Eudoxus of Cnidus, 146, 652, 727, 772, 1092, 1093

Eukaryota domain, 43, 44

Euler, Leonhard

    algebra and, 35, 38

    contributions of, 1101

    exponential functions and, 1012

    functions and, 144, 146, 366, 370

    geodesics and, 281

    Introductio in analysin infinitorium, 410

    Pi and, 771

    polyhedra and, 783

    projectile trajectories and, 64

    Seven Bridges of Königsburg and, 129, 130, 366, 454–455, 755, 821

    at St. Petersburg Academy, 359

    transformations and, 1005

Euler characteristic, 965, 966

Eulerian Graphs, 14

Eulerian–Lagrangian-Agent Method (ELAM), 1054

Euler–Lotka equations, 385

Euler's method, 1101

Eupalinian aqueduct, 1014, 1015

Eupalinos of Megara, 1015

"Eureka" (Archimedes), 60, 460, 652, 1025

Europe, eastern, 358–361

Europe, northern, 361–363

Europe, southern, 363–365

Europe, western, 365–367

European Economic Community (EEC), 337–338

European Organization for Nuclear Research (CERN), 911

European Women in Mathematics, 811, 1072

Eutocius of Ascalon, 1094

event horizons, 115, 116

Expected Benefit (EB), 873–874

expected values, 368–369

expenditure method (GDP), 467

Exploring Our Solar System (Ride), 871

Exploring Small Groups (ESG), 273

exponentials and logarithms, 370–372, 1011

Exposition du Systeme du Monde (Laplace), 115

extinction, 372–373

Extracts (Stobaeus), 619

Extraordinary Hotel, The (Lem), 901

extreme sports, 374–375, *375*

eyeglasses, 1040

Eytzinger, Michael, 418

F

fabrics. *See* **textiles**

face recognition, 165

Facebook, 926

factorials, 765

Fahrenheit, Gabriel, 988

Fahrenheit temperature scale, 988, 1058

Fair Credit Reporting Act (FCRA), 387

Fair Isaac Corporation (FICO). *See* **FICO score**

fair market value (FMV), 93

Faltings, Gerd, 367

family trees. *See* **genealogy**

Fan Chung. *See* **Graham, Fan Chung**

Fannie Mae, 254

Fanning, Shawn, 388

Fantasy Football Index (magazine and Web site), 378

fantasy sports leagues, 377–379

Faraday, Michael, 837

Faraday's law of induction, 837

al-Farisi, 764

Farmer, John,, 418

farming, 23, 286, **379–382**, *380*

Farr, William, 288, 313, 942

Farrell, Edward, 167

fashion design, 431–432

Fast Fourier Transforms. *See* Fourier Transforms

fatigue, 955

fax machines, 382–384; *383*

Fechner, Gustav, 447, 653–654

Federal Aid Highway Acts, 476

Federal Bureau of Investigation (FBI). *See* U.S. Federal Bureau of Investigation (FBI)

Federal Communications Commission, 175

Federal Deposit Insurance Corporation (FDIC), 691

Federal Reserve System, 675

federal tax tables, 497–498

Fedorov, E. S. "Yevgraf", 781

Feingold, Graham, 206

feng shui, 534

Fennema, Elizabeth, 961

Fennema–Sherman mathematics attitudes scale, 961

Ferguson, Claire, 904

Ferguson, Helaman, 904

Fermat, Pierre de
    algebra and, 34–35
    analytic geometry and, 1098
    Arcadia (Stoppard), 1083
    Blaise Pascal and, 105–106, 1099
    conic sections and, 237
    coordinate geometry and, 248
    dice games and, 302

Fermat's Enigma (Singh), 560

Fermat's Last Tango and, 689–690

integral formula of arc length, 763

Marin Mersenne and, 588

probability and, 105–106, 112–113, 366, 806, 841

squares/square roots and, 883, 939

See also **Wiles, Andrew**

Fermat primes, 883, 939

Fermat's Enigma (Singh), 560

Fermat's Last Tango (Lessner & Rosenblaum), 689–690

Fermat's Last Theorem, 34, 112–113, 282, 362, 366, 720, 829, 1107

See also **Wiles, Andrew**

Fermat's Room (movie), 683

Fermi, Enrico, 79

Ferrari, Lodovico, 261, 786

Ferrel, William, 1063

fertility, 384–386

Feynman, Richard, 79, 228, 349, 548
    Surely You're Joking, Mr. Feynman, 682
    What Do You Care What Other People Think, 682

Fibonacci, Leonardo, 18, 54, 412, 886, 1088, 1094
    Liber Abaci, 412, 819, 939, 1088, 1095

Fibonacci sequence, 62, 364, 412, 886, 908

Fibonacci tuning. *See* **Pythagorean and Fibonacci tuning**

FICO score, 254, 288, **386–387**, 482–483

FidoNet, 517

Fields Medal, 69, 363, 367

Fignon, Laurent, 109

file downloading and sharing, 387–388

Final Problem, The (Doyle), 558

fingerprints, 388–390

"Finite Simple Group (of Order Two)" (Klein Four Group), 787

finite state machines, 1034

Finkel, Benjamin, 810

Finley, John, 490

Fiore, Antonio Maria, 1096

Fiqh al-Hisab (Ibn Mun'im), 55

firearms, 390–392

fireflies, 973

fireworks, 392–394, *393*

First, Outside, Inside, Last (FOIL), 41

First Circle, The (Solzhenitsyn), 558

FIRST Robotics, 133

First-fit algorithm, 912

first-generation (1G) cell technology, 175

Fischer, Carl, 139, 141

Fischer, Gwen, 834

fish schooling, 973

Fisher, Ronald, 379, 803, 842, 943, 944

Fisher–Neyman–Pearson inferential methods, 943

fishing. *See* **data mining**

fishing (aquatic), **394–395**

Fiss, Andrew, 193

Fitzgerald, George, 547

Five Hysterical Girls theorem, 776

fixed rated mortgages, 483

Fizeau, Hippolyte, 316

flash memory, 309–310

Flatland (Abbott), 440, 557, *557*, 781, 859, 902

Flatland the Movie (movie), 440, 781, 1083, 1084, 1085, 1086

Flatterland (Stewart), 902

flavonoids, 732

Fleischmann, Martin, 350

Fletcher, Thomas, 926

flight, animals, *44*, 46–47, *47*

Flintusehel, Eliot, 901

floods, **395–398**

FLOW-MATIC, 754

Flynn, Morris, 488

Focus in High School Mathematics (NCTM), 278

folded normal, 708

FONE F3, 303

Fons, W. R., 402

Fontana, Niccolò. *See* Tartaglia, Niccolò

Food and Drug Administration (FDA), 319

food webs, 47–48

football, **398–400**, 529–531

Forbidden Knowledge (Cramer), 901

Ford Motor Co., 10, 523

forecasting, **400–402**

foreign exchange (FX) market, 263–264

forensic ballistics. *See* ballistics studies

forest fires, **402–404**

formal concept analysis (FCA), 509

Foster, Donald, 958

Foundation (Asimov), 791

Four-Color theorem, 587, 611, 814, 1107

four-dimensional geometry, 548

Fourier, Jean Baptiste Joseph
  contributions of, 1102
    heat conduction and, 410, 987–988, 1083
    Napoléon Bonaparte and, 501
    On the Propagation of Heat in Solid Bodies, 987
    squares/square roots and, 939

Fourier analysis. *See* Fourier transforms

Fourier series, 410, 1040

Fourier transforms, 165, 209–210, 211, 360, 547, 684, 769

Fourier's heat equation, 410, 987–988, 1083

fourth-generation (4G) cell technology, 176

Foxtrot (Amend), 219

fractals
  in African societies, 13–14
  in animal kingdom, 49
  coastlines and, 570–571
  coral reefs and, 251
  in education, 426
  houses of worship and, 486
  lightning patterns and, 551–552
  in village design, 515, 534
  visualization and, 1042
  See also patterns

fractional exponents, 1096

fractional linear transformations, 1005

fractions, continued, 1097

Fraenkel, Abraham, 587, 1106

Framingham Heart Study, 123

Francesca, Piero della, 248

Frank Lloyd Wright School of Architecture, *1080*

Frankenstein (Shelley), 877, 900

Franklin, Benjamin, 274, 869, 886, 913

fraud detection
  accounting and, 4–5
  in communication technologies, 224
  credit card, 253, *254*
  data mining and, 292
  neural networks and, 125
  probability theory and, 5
  Social Security and, 759
  taxes, 498

Fraunhofer, Joseph von, 79

Frayn, Michael, 775
  Copenhagen, 775–776

Fréchet, Maurice, 651, 1105

Freddie Mac, 254

Freeman, Greydon, 517

Frege, Gottlob, 219

Frémont, John Charles, 344

French Academy, 1099

Frenet, Jean Frédéric, 249, 281

Frenet–Serret Formulas, 249, 281

frequency modulation (FM), 838–839, 1068–1069

frequentist approach, 802

Fresnel, Augustin-Jean, 547

Frézier, Amédée-François, 393

Friedman, William, 756, 1074

Frisch, Karl von, 105

Fritts, Charles, 930

Fröbel, Friedrich, 324

Fröbel Gifts, 324

"From Fish to Infinity" (Strogatz), 610

Frost, Wade Hampton, 314

Fuchs, Ira, 517

fuel consumption, 404–405

Fujita, Tetsuya Theodore, 492

Fujita–Pearson scale, 492

Fujiwara, Masahiko

    An Introduction to the World's Most Elegant Mathematics, 612

Fulke, William, 122

Fuller, Buckminster "Bucky", 62, 674

Fuller, Thomas, 599, 668

fullerenes, 674, 782

function rate of change, 405–408

functional MRI (fMRI), 126, 127

    See also magnetic resonance imaging (MRI); medical imaging

functions, 408–410

functions, recursive, 410–412

Fundamental theorem of Algebra, 722, 785

Fundamental theorem of Calculus, 149

Fürer algorithm, 688

Fusaro, Marc, 675

fusion, 349–350

Futurama (television show), 829, 903, 1082, 1083, 1084

Future Shock (Toffler), 402

futures market, 510

fuzzy logic/sets, 290, 803, 843

FX market, 263–264

fx-7000G, 141–142

G

Gabriel's Horn, 505

"Gadget" (nuclear test bomb), 80

Gaing Kek Ieu "Comrade Deuch", 71

Gale, David, 122, 582

Galerkin, Boris, 283

Galileo (Galileo Galilei)

    animal/machine connections and, 45

    contributions of, 1098

    Dialogues Concerning the Two Chief World Systems, 858–859

    infinite sets and, 505

    influence of, 64, 364

    invention of thermometer, 1055

    normal distribution and, 707

    pendulum clocks and, *202*, 202–203

    principle of relativity and, 853

    proofs and, 812

    religious writings, 858–859

    square/square roots and, 939

    Star Messengers (Zimet and Maddow), 690

    telescopes and, 978

    theory of gravity and, 64, 364

    Two New Sciences, 545

Galileo Galilei (Glass), 690

Galileo's principle of relativity, 853

Gallup, George, 888

Gallup Polls, 889

Galois, Évariste, 34, 362, 366, 786, 883, 1103

    algebra and, *32*

Galois Theory, 34

Galton, Francis, 389, 803, 895, 942, 1042

    Regression Towards Mediocrity in Hereditary Stature, 895

Galton distribution, 708

Gama, Vasco da, 862

gambling. *See* **betting and fairness; dice games**

Game of Logic, The (Carroll), 122

Game of Pistols, 415

game shows, 982–983

game theory, 216, **413–415**

    in baseball, 99, 378

    in basketball, 101

    Cold War and, 214, 1037–1038

    David Blackwell and, 119

    in football, 399

    strategy and tactics in, 952–954

    topology and, 119

    See also **board games**

games. *See* **board games; video games**

games, board. *See* **board games**

Gamow, George, 853

Gardner, Howard, 543

Gardner, Martin, 569, 595, 823, 901

    The Island of Five Colors, 902

    Mathematics, Magic and Mystery, 569

Garfield, James, 828

Garfield, Richard, 415–417

Garibaldi, Skip, 200

gas volume, 1025–1026

Gates, Bill, 768

Gatun Locks, *156*

Gauguin, Paul, 737

Gauss, Carl F.

    abstract groups and, 35

    arithmetic sequence and, 908–909

    axiomatic systems and, 85

    complex numbers and, 722

    contributions of, 239, 249, 366, 574

    curved space and, 854

    Disquisitiones Arithmeticae, 712, 939–940, 1102

    error modeling and, 369

    hyperbolic geometry and, 1103

    laws for electricity and magnetism, 837

linear equation simplification and, 634–635

linear transformations and, 1005

non-Euclidean geometry and, 848

normal distribution and, 802

parallel processing and, 752

polygons and, 781

ruler/compass construction and, 883

Theory of Celestial Movement, 707

theory of curves and, 281

thermostats and, 991

Gauss, Karl Friedrich, 220

Gauss–Bonnet theorem, 965

Gaussian curvature, *965*, 965

Gaussian distribution, 366, 706, 802

See also **normal distribution**

Gaussian elimination, 634–635

Gaussian thermostat, 991

Gauss–Jordan elimination, 634–635

Gauss's divergence theorem, 1030

Gauss's laws for electricity and magnetism, 837

Gavin, M. Katherine, 539–540

Gawrych, Billy, 374

gay-related immune deficiency (GRID), 478–479

GDP. *See* **gross domestic product (GDP)**

Geary, David, 537

Geber (Jabir ibn Aflah), 664, 1095

GEE system, 1079

Gelfand, Israel Moiseevich, 360

Gelfand representation, 360

gelosia, 686

Gemini program, 522

Gender Bias Elimination Act, 1072

gender schemas, 1071

Genealogical Data Communication (GEDCOM), 418

genealogical numbering systems (GNS), 418

Genealogical Society of Utah, 418

genealogy, 417–419, 454

General Conference on Weights and Measures, 640

General Electric Company (GE), 288, 832

General Motors Company (GM), 10

General Theory of Employment, Interest and Money, The (Keynes), 1018

genetic algorithms, 791

genetic engineering, 422

genetic variability, 421

genetically modified foods, 733

genetics, 419–422

Geneva score, 52

Geni (Web site), 419

genotype, 420–421

GeoEye, 891

GeoGebra, 929

Geographia (Ptolemy), 1011

geographic information systems (GIS), 430, 570

Geometer's Sketchpad (GSP), 427, 929

Geometria indivisilibus continuorum (Cavalieri), 353

geometric art, 736–737

See also patterns

geometric magis, 568

geometry and geometry education, 422–427

Arabic/Islamic, 54

Babylonian, 88–89

of castle defense, 171–172

computational, 429

dance as, 90, 92

differential, 429

discrete, 429

early history of, 422–425, 798–799

hyperbolic, 1103

plane and spherical, 54

prehistorical, 798–799

recent developments in, 426–427

sacred, 885–886

software, 929

synthetic projective, 1099

theoretical mathematics and, 613–614, 615–617, 619

See also **coordinate geometry**; *Elements* (Euclid); Euclidean geometry; non-Euclidean geometry

Geometry Center for the Computation and Visualization of Geometric Structures, 426, 1042

Geometry Forum, 220, 426

Geometry from Africa (Gerdes), 14

geometry in society, 427–433, *431*

design/manufacturing and, 430, *431*

early history of, 428–429

fashion design and, 431–432

graphics/visualization and, 430

information systems and, 430–431

occupational connections and, 431–433

types of, 429–430

Geometry of Love, The (Cheever), 558

geometry of music, 433–436

geometry of the universe, 436–441

dimensionality and, 440

Euclidean geometry, 436–437

global geometry and, 439–440

triangles and, 437–438

Georgian National Mathematical Committee, 76

geosynchronous satellites, 891

geothermal electricity, 442–443

geothermal energy, 441–443

geothermal heating, 442

Gerard of Cremona, 1094

Geraschenko, Anton, 778

Gerbert d'Aurillac (Pope Sylvester II), 1094

Gerdes, Paulus
Geometry from Africa, 14

Gerhard of Cremona
algebra and, 33

Germain, Sophie, 1102

German ciphers, 212

Gerry, Elbridge, 443

gerrymandering, 443–445

Gershgorin, Semyon Aranovich, 635

Gessen, Masha
Perfect Rigor, 559

Gestalt psychology, 741

Get Off the Earth puzzle, 820

Gfarm Grid File System, 290, 388

g-forces, 878

Giant's Causeway (Ireland), 781

Gibbard, Allan, 1050

Gibbs, Josiah Willard, 909, 1029–1030, 1104

giftedness, *538*, 539–540

Gilbert, W. S.
Pirates of Penzance, 690

Gill, John, 201

Gilliver's Travels (Swift), 559

Gini, Corrado, 769

Giotto di Bodone, 748, 860

Girard, Albert, 1098

Girls' Guide to Fantasy Football (Web site), 378–379

Glass, Philip
Einstein on the Beach, 690
Galileo Galilei, 690

Glauert, Hermann, 31

Gleason Grading system, 52

global geometry, 439–440

global positioning systems (GPS). *See* **GPS**

global warming. *See* **climate change**

globalization, 733

gnomonics, 202, 534

Gnutella, 388

God. *See* **mathematics and religion; numbers and God**

Gödel, Kurt
contributions of, 1106
Incompleteness theorem, 85, 360, 366, 625, 776, 814, 901
Seventeenth Night (Apostolos), 776
ZF set theory and, 587, 1106

Gold Bug, The (Poe), 558

Goldbach, Christian, 359, 721

Goldbach's Conjecture, 683, 721

Goldberg, Ian, 842

Goldberg Extension, 114
See also **birthday problem**

Golden Ratio, 445–448, 729, *730*, 748, 886, 986

Golden Rectangles, 729, 748

golden spirals, 446, 448

Goldin, Gerald, 863

Goldman, David, 128

Goldman equation, 128

Goldsmith, Thomas, 1034

Goldstein, Raymond, 941

Goldstine, Herman, 1079

Goldwasser, Eric, 320

Goldwasser, Romi, 320

Golenischev papyrus. *See* Moscow papyrus

"Gompertzian growth" model, 184

Good Will Hunting (movie), 163, 681–682

Google, 12, 180, 518–519

Gore, Al, 1047, 1049–1050

Gorgas, Josiah, 193

Gossett, William (pseud. Student), 708, 943

Gottman, John, 793

Gougu theorem, 185, 423

"Governable Parachute" (Cayley), *29*

government and state legislation, 448–450

Gowers, Timothy, 363

Gowers, William, 763

GPS, 450–453
satellites, 774, 855, 891
smart cars and, 922–923
trilateration and, *451*, 451–452, 1012–1013

Graham, Fan Chung, 453–454

Graham, Ronald, 453, 589

Grand Design, The (Hawking), 625

Granger, Clive, 466

Grant, Ulysses S., 193

Granville, Evelyn Boyd, 668, 671, 679

graph paper, 249, 455, 658

graph theory, 454, 455, 519, *534*, 701–702, 963

graphical user interfaces (GUIs), 291

graphing calculators, 138, 141–142, 273, 427, 456

graphs, 454–456, 465, 509, 696, 963, 1042

On graphs not containing independent circuits (Lovász), 361

graph-theoretic tournaments, 999

Grassmann, Hermann
The Calculus of Extension, 1029

Grassmann, Hermann Günther, 34, 249, 1103

Graunt, John, 546

gravitational time dilation, 854

gravity, 456–458, 1026, 1059

Gray, Elisha, 976

Gray, Mary, 811
Great Fire of London, 1100
Greater Cleveland Mathematics Program, 275
Greek Anthology (Metrodorus), 1093
Greek gods and godesses, 729
Greek mathematics, 458–461
    applied mathematics and, 603–604
    Archimedes and, 364
    astronomy and, 77–78, 460–461
    decline of, 424–425
    deductive logic and, 459–460
    early mathematicians, 458–461
    Golden Ratio and, 446
    measurement and, 446, 646, 653
    written history of, 1092
    See also Elements (Euclid); Euclid of Alexandria
Green, Ben, 975–976
Green, George, 155, 653, 1030, 1067, 1102
Green, Judy
    Pioneering Women in American Mathematics, 1072
Green Card Lottery program, 563
green design, 461–463
green mathematics, 463–466
Green Monster (Fenway Park), *62*
greenhouse gases (GHGs), 442–443, 1063
Green's theorem, 653, 1030, 1102
Greenwald, Sarah, 829
Greenwaldian theorem, 829
Greenwich Time, 577
Gregorian calendar, 495–496
Gregory, James, 111, 112, 354, 1100
Gregory of Rimini
    Lectures, 858
Gregory XIII (pope), 154
Gribeauval, Jean-Baptiste Vaquette de, 287
Grienberger, Christopher, 354
Grill, Bernhard, 684
Groff, Rinne, 776
groma, 879
Gromov, Mikhail, 367
Gross, Mark, 877
gross domestic product (GDP), 466–468, 696, 697
Grosseteste, Robert, 664
Grossman, Alex, 293
Grossman, Sharon, 420
Grothendieck, Alexander, 1037
ground resonance effect, 475
group theory, 35, 362, 674, 971
Groups, Algorithms, Programming (GAP), 765
Grover, Lov, 548
growth charts, 468–469

Grundlagen der Geometrie (The Foundation of Geometry)
    (Hilbert), 85, 1105
Gudermann, Christof, 249
Guggenheim Museum, 1081
Gunderson, Lauren
    Emilie, 777
    Leap, 776
Gundisalvo, Domingo, 1094
guns. *See* **firearms**
Gunter, Edmund, 371
Guo Shoujing (Kuo Shou-ching), 156, 186
Gupta numerals, 712
Guthrie, Francis, 19
gymnastics, 469–470

H
hacking, computer, 192, 224
Hadamard, J., 1105
    The Psychology of Invention in the Mathematical Field, 560
Hadamard, Jacques, 560
Hagia Sophia dome, *315*, 315
hagwons (academies), 69
Hahn, Hans, 1030
Haken, Wolfgang, 1107
Hales, Thomas, 104–105, 746
Halley, Edmund, 288, 344, 546, 1100
    See also **data analysis and probability in society**
Halley's Comet, 288–289, 1063
Halma, 120
Halmos, Paul Richard, 361
Hamilton, Alexander, 232
Hamilton, William Rowan, 34, 122, 249, 362, 722,
    1029–1030, 1103
Hamiltonean Graph, 122
Hamilton's Method, 232, 233, 234
Hamilton-type circuits, 11
Hammerstein, Oscar, 1082
Han dynasty, 185, 423
handicapping. *See* **sport handicapping**
Hankel, Hermann, 1030
Hankins, Thomas, 455
Hansen, Morris, 890
Hanson, Howard, 231
hard disk drives (HDDs), 309
Hard Problems (movie), 1086
Hard Times (Dickens), 560
Hardy, Godfrey (G. H.)
    contributions of, 73, 1074, 1075, 1105
    A Disappearing Number (Complicite Co.), 776
    A Mathematician's Apology, 559, 597, 608, 776
    Srinivasa Ramanujan and, 73, 599

Hare system, 1048

Harmonices Mundi (Kepler), 365

harmonics, 211, 434–435, **471–473**, 475, 1066

Harnack, Carl Gustav Axel, 146

Harrell, Marvin, 781

Harrington, John, 997

Harriot, Thomas, 964

    Artis analyticae praxis, 765, 1098

Harris Interactive, 889

Harrison, John, 203

Hart, George, 904

Hatori, Koshiro, 742

Hauck, Frederick H., 871

Hauptman, Ira

    Partition, 776

Haussmann, Baron, 189

Hawk, Tony, 374

Hawking, Stephen, 115, 116, 117, 362, **473–474**, *474*, 625

    Black Holes and Baby Universes and Other Essays, 474

    A Brief History of Time, 473, 474

    The Universe in a Nutshell, 474

Hawking radiation, 116, 117

Hawthorne effect, 889

Haynes, Martha Euphemia, 668

Headley, Velmer, 167

heap (Egyptian mathematics), 332

hearing. *See* **cochlear implants**

Hearst, Patty, 958

heat conduction problem, 410, 987–988

Heath, Thomas, 60

Heaviside, Oliver, 838, 977, 1029–1030, 1104

Heaviside Layer, 838

Heeding the Call for Change (MAA), 944–945

Heegner, Kurt, 599

Hein, Piet, 122

Heinlein, Robert A., 899, 902

Heisenberg, Werner, 766, 770, 775–776, 1027

Heisenberg uncertainty principle, 770, 1027

helicopters, **475**

heliocentric hypothesis, 1092

Helmholtz, Herman, 740

Hemachandra, Acharya, 764

hemoglobin, 672–673

Henlein, Peter, 202

Hennessey, Andrew, 990

Henry, Leighton, 167

Henry, Warren, 1037

Henry I, King, 639

Heraclides, 6

Hérigone, Pierre, 765

Hermann grid, 739

Hermite, Charles, 1104

Heron of Alexandria (Hero), 762, 1014, 1015, 1093

    Metrica, 646

Heron's formula, 1093

Herrnstein, Richard

    The Bell Curve, 708

Herschel, John, 305, 1063

Hertz (Hz), 1017, 1068–1069

Hertz, Heinrich, 341, 1068

Herzberg, Agnes, 963

Hess, Harry, 773–774

Hewlett, Bill, 140

Hewlett-Packard (HP), *140*, 141, 142, 165

Hex, 120, 122

Hexagrams, 185

HEXI, 211

Hickman, C. N., 56

Hideaki Tomoyori, 771

hieroglyphics, 493

Hieron II, King, 57–58, 60

high occupancy toll (HOT) lanes, 489

high occupancy vehicle (HOV) lanes. *See* **HOV lane management**

higher math. *See* number theory

highly optimized tolerance (HOT), 551

highways, **476–477**, 488–489, 1000–1001

Higson, Nigel, 904

Hilbert, David

    axiomatic systems and, 587

    contributions of, 366, 504, 1074–1075

    Grundlagen der Geometrie (The Foundation of Geometry), 85, 1105

Hilbert spaces, 361, 366, 504

Hildebrand, Harold "Dr. Andy", 788

Hill, Austin Bradford, 288, 362, 944

Hilton, Conrad, 254

Hindu-Arabic numerals

    addition/subtraction and, 8

    number and operations, 712–713

    place-value structures and, 686

    widespread usage, 54, 55, 862, 932, 1088, 1094, 1095

    zeros and, 494, 1088

    See also Indian mathematics

Hines, Gregory, 946–947

Hipparchus of Rhodes, 79, 782, 1011, 1093

Hippias of Elis, 1092

Hippocrates of Chios, 236, 353, 1091

Hironaka, Heisuke, 69

Hiroshima, Japan, 79–80

Hirzebruch, Friedrich, 367

Hispalensis, Isidorus, 364

hitting a home run, *477*, 477–478

HIV/AIDS, 170, **478–480**

Hobbes, Thomas, *357*, 652

hockey, 480–482

hockey stick graph, 482

Hodgkin, Alan, 128, 700

Hodgkin–Huxley equations, 700

Hohmann, Walter, 521

Hohmann Transfer Orbit, 521

Holland, Clifford, 1014

Holland, John, 791

Hollerith, Herman, 718

Holmegaard bows, 55

Holmes, Arthur, 773

holomorphic functions, 723

Holzmann, William (Xylander), 1097

home buying, 482–485

Home Insurance Building, 919

home runs. *See* **hitting a home run**

Homes, Oliver Wendell, 849

homological algebra, 531

honor societies. *See* **clubs and honor societies**

Hood, John Bell, 193

Hooker, Joseph, 193

Hopfield neural networks, 702

Hopper, Grace Murray, 754

Hörmander, Lars, 147, 363

Horner, William George, 50, 1102

Horner–Ruffini method, 186, 187

horologium, 202

horse racing, 937–938

horsepower, 499

Horton, Joseph Warren, 203

Hot X (McKellar), 35, 1082

Hotelling, Harold, 944

House Bill 246, 449

House of Representatives. *See* **congressional representation**

House of Wisdom, 32–33, 74

House Resolution 224, 450

houses of worship, 485–487

houses purchases, 225–226

HOV lane management, 488–489

hovercraft, 1037

"How Google Finds Your Needle in the Web's Haystack" (Austin), 610

"How Long is the Coast of Britain? Statistical Self-Similarity and Fractional Dimension", 570

Howard, Ebenezer, 190

howitzers, *64*, 191

   See also **artillery**

Hubbard, Mont, 375

Hubble, Edwin, 317

Hubble Space Telescope, 317, 441, 891

Hudde, Johann, 1100

Huffman, David, 382, 684, 977

Huffman coding, 382, 684

Huizanga, Johan, 119–120

Hull, Robert "Bobby", 481

Human Connectome Project, 125

Human Genome Project, 421

human immunodeficiency virus (HIV). *See* **HIV/AIDS**

Hunt, Fern, 167, **489–490**

Huntington, Edward, 232

Huntington–Hill Method, 232, 234

Hurley, William, 481

Hurricane Katrina, *491*, *994*

hurricanes and tornadoes, 490–492, *491*, *994*

Huth, Andreas, 973

Hutlee'/Umyuarchdelee program, *272*

Hutton, Charles, 129

Huxley, Aldous, 559

Huxley, Andrew, 128, 700

   Young Archimedes, 559

Huygens, Christiaan, 281, 370, 547, 745, 1100

   De Ratiociniis in Lundo Aleae, 801

Huzita, Humiaki, 742

Huzita–Hatori axioms, 742–743

hydraulics, 155, 347

hydrodynamic modeling, 1001

hydroelectric power, 283–284

hydrometers, 246

hydrostatics, 1092

Hyman, Albert, 745

Hypatia of Alexandria, 18, 461, 880, 961, 1093

Hypatia Scholarship program, 961

Hyperbolic Crochet Coral Reel, 258, 905

hyperbolic geometry, 236, 437–438, 1005, 1043, 1103

   See also non-Euclidean geometry

hypersonic aircraft, 31

Hypsicles of Alexandria, 782

I

Iacob, Caius, 361

IBM Corporation, 12, 96, 718, 767, 977, 1106

Ibn al-Haytham (Alhazen), 55, 111, 237, 1005, 1042

Ibn Baija (Avempace), 664

Ibn Hayyan. *See* Jabir ibn Aflah (Geber)

Ibn Ibrahim. *See* Yusuf ibn Ibrahim

Ibn Mun'im

   Fiqh al-Hisab, 55

Ibn Rushd (Averroës), 664

Ibn Sina, 764
Ibn Turk, Abd al-Hamid, 73
Ibn Yunus ibn Abd al-Rahman, 18
iconometry, 485
ID3 data blocks, 685
Identification Friend or Foe (IFF), *502*, 503
Identity theorem, 723
Illuminations Web site, 221
illusions, optical, 739–741
imaginary numbers, 34, 722, 723, 939, 1012, 1097
imaging technologies, 126–128, 643–644, *644*, 985–986, 1043
   See also Computed Axial Tomography (CAT) scans; magnetic resonance imaging (MRI); medical imaging
immunology, 480, 732
Imposter syndrome, 584
Incan and Mayan mathematics, 493–496
   astronomy, 496
   base-10 system, 698
   calendars, 155, 494–496
   Incan civilization, 931–932
   quipus, 493–494
   zeros and, 493, 494
income method (GDP), 467
income tax, 496–499
incommensurables, 1092
Incompleteness theorem, 85, 360, 366, 625, 776, 814, 901
   See also **axiomatic systems**; Gödel, Kurt
independence of irrelevant alternatives method, 1049
indeterminate equations, 1095
Indian mathematics
   arithmetic and, 1095
   astronomy and, 763
   decimal system and, 53, 67
   measurement and, 638
   negative numbers and, 73, 1088
   number systems, 712
   place-value structures and, 686, 712
   religion and, 626–627, 729
   sine function and, 54–55, 1011
   square/cube roots and, 939, 1093
   Sulbasutras and, 424
   translations of, 1094
   zeros and, 73, 1088
   See also **Asia, southern**; Hindu-Arabic numerals; **Vedic mathematics**
Indianapolis 500, 82
indirect proofs, 813–814
individual retirement accounts (IRAs). *See* **pensions, IRAs, and social security**
indivisibles, theory of, 1098

Industrial Revolution, 499–501
   accounting and, 2
   catalysts, 1041, 1063, 1115
   employment and, 379, 500, 1018
   mass production and, 287, 380–381, 466, 896, 990
   mathematics and, 500–501
   operations research (OR) and, 806–807
   steam engines and, 499, 500, 1063
infantry (aerial and ground movements), 501–503
infertility. *See* **fertility**
Infinite Assassin, The (Egan), 901
infinitesimal calculus, 552–553
infinitesimals, 60, 144
Infinities (Barrow), 776–777
infinity, 504–506
   calculus and, 148
   curves and, 281
   Euclid's fifth postulate and, 460–461
   literature and, 557
   measurement and, 650, 652
   normal distribution and, 802
   orders of, 1104
   reasoning/proof and, 845
   religious tenets and, 624–625, 627, 858–860
   universe and, 439
Infinity (movie), 682
inflation, 696
information systems, 180, 181, 430–431
information theory, 212, 795, 1068
innerspring mattresses, 636
Insignificance (Johnson), 775
Institute for Figuring, 258
Institute for Operations Research and the Management Sciences (INFORMS), 229, 592
Institute for Strengthening the Understanding of Mathematics, 170
Institute of Electrical and Electronic Engineers (IEEE), 181
Institute of Mathematical Sciences (University of Malaya), 71
Institute of Mathematical Statistics, 119
Institute of Mathematics of the National Academy of Sciences (Republic of Armenia), 76
Instituzioni Analitiche (Agnesi), 1101
instructional technology, 273
   See also **calculators in classrooms**
insurance, 506–508
integers classification, 1093, 1096
integral calculus, 145–146, 148, 149–150, 460
integrated moving average (ARIMA), 401
Intel Corporation, 769
intelligence and counterintelligence, 508–511

intelligence quotients (IQ), 511–513
  bell curves and, 708
  content of tests, *512*, 512–513
  correlative variables and, 512
  development of, 511–512
  learning exceptionalities and, 537, 539–540
  link to health, 513
interdisciplinary mathematics research. *See* **mathematics research, interdisciplinary**
interest rates, 483–484, 561–562
Intergovernmental Panel on Climate Change (IPCC), 194
interior design, 514–515
International Association for Statistics in Physical Sciences, 119
International Atomic Time (TAI), 656
International Baccalaureate Calculus (IB Calculus), 151
International Baccalaureate (IB) Programme, 267
International Bureau of Weights and Measures (BIPM), 204
International Business Machines Corporation. *See* IBM Corporation
International Congress of Mathematicians, 1037
international curricula. *See* **curricula, international**
international debt, 696–697
International Earth Rotation and Reference Systems Service (IERS), 204
International Linear Algebra Society, 811
International Mathematical Olympiad (IMO), 19, 20, 67, 227–228, 664
  Azerbaijan, 76
  China and, 68
  eastern Europe, 361
  Israel, 75
  Kuwait, 76
  Malaysia, 71
  Mongolia, 69
  northern Europe, 363, 365
  Republic of Armenia, 76
  South Korea, 69
  southern Asia, 73
  Turkey, 75
  Vietnam, 71
  western Europe, 367
International Mathematical Union, 1105
International Mathematics and Design Association, 514
International Mobile Telecommunications, 977–978
International Space Station, *876*
International Study Group on Ethnomathematics, 670
International Sun/Earth Explorer 3 (ISEE--3), 891
International System of Units (SI), 349, 640, 643, 1021–1022, 1024
International System of Units (SI units), 349

International Telecommunications Union, 382
Internet, 515–520
  advertising, 11–12
  Cerf Vinton and, 180–181
  data searches, x
  economics and, 519–520
  Grigori Perelman and, 1107
  history of, 516
  mathematical problems and, 516–518, *517*
  mathematical sciences codevelopment and, 516
  MP3 players and, 684
  networks and, 518–519
Internet Assigned Numbers Authority, 518
Internet Corporation for Assigned Names and Numbers, 181
Internet Mathematics (journal), 840
Internet Message Access Protocol (IMAP), 518
Internet Protocol (IP), 181, 911
interplanetary travel, 520–522
interprocess communication (IPC), 911
Introductio in analysin infinitorium (Euler), 410
Introduction to Algebra, An (Day), 193
Introduction to Arithmetic (Nicomachus), 1093
Introduction to Computational Studies (Suanxue Qimeng) (Zhu Shijie), 187
Introduction to the World's Most Elegant Mathematics, An (Fujiwara and Ogawa), 612
Invalides dome, 315
inventory models, 523–524
Investigation of the Laws of Thought (Boole), 362, 603, 1104
investments, 676–677
  See also **mutual funds**
Invisible Man, The (Wells), 900
iPhone, 977
IQ. *See* **intelligence quotients (IQ)**
I.Q. (movie), 682
IRAs. *See* **pensions, IRAs, and Social Security**
irrational numbers. *See* **numbers, rational and irrational**
irrigation, 1051–1052
IRS. *See* U.S. Internal Revenue Service (IRS)
Isaac, Earl, 254
Ishango bone, 657, 686, 1091
Islamic mathematics. *See* **Arabic/Islamic mathematics**
Islamic religious traditions, 729
Island of Five Colors, The (Gardner), 902
isothermal coordinates, 249
Israel. *See* **Asia, western**
Israel Journal of Mathematics , 75
Israel Mathematical Union, 75
Italian Mathematical Union, 364

item-response theory (IRT), 328, 816–817
iterative algorithms, 95
Ito, Kiyoshi, 69
iTunes, 684, 788
Iwasawa theory, 690

J
Jabir ibn Aflah (Geber), 664, 1095
Jabir ibn Hayyan. *See* Jabir ibn Aflah (Geber)
Jackson, Shirley Ann, 525–526, *526*
Jackson, Thomas Jonathan "Stonewall," 193
Jacobi, Carl Gustav, 1102
Jacquard, Joseph, 875, 989
Jacquard loom, 875, 989
Jacques, Cassini, 890
Jahresbericht (journal), 1105
James, George William "Bill", 290
  Bill James Baseball Abstract, 97, 378
James, Lancelot F., 167
James Webb Space Telescope, 441
jamitons, 488
Jansky, Karl Guthe, 838
Japanese paper folding. *See* **origami**
Japanese Railway Ministry, 582
Japanese Technology Board, 756
Jean-Michel, Jean-Michelet, 167
Jeans, James, 825
Jefferson, Thomas, 232, 847
Jefferson's Method, 232, 233, 234
Jeffrey, Harold, 323
Jenkins, Gwilym, 654
Jenney, William Le Baron, 919
Jennings, Thomas, 517
Jeopardy (mathematics), 208
Jesus, 731
Jewish religious tenets, 625, 730
Jia Xian, 186, 187
Jia Xian Triangle, 186, 187
Jigme Khesar Namgyel Wangchuck, King, 73
jigsaw puzzles, 822–823
Jigu Suanjing (*Continuation of Ancient Mathematics*)
  (Wang Xiao-tong), 186
Jim Crow Laws, 668
jiu jitsu, 583
Jiuzhang Suan Shu (Chinese text). *See Nine Chapters on the Mathematical Art* (Chinese text)
Job Related Almanac (Krantz), 162
John of Seville, 1094
Johnson, Art, 781
Johnson, Brown, 982
  Insignificance, 775

Johnson, Dana, 539–540
Johnson, Neil, 792
Johnson, Terry, 775
Johnston, Albert Sidney, 193
Johnston, Joseph, 193
joint loading, 527–528
joints, 526–528, *528*
Jonas, David, 749
Jones, Vaughan, 735
Jones, William, 1100
Jordan, Camille, 35
Jordan, Michael, 163
Jordan, Wilhelm, 635
Jordanus de Nemore, 1095
Joukowski airfoil, 26
joule, 349
Joule, James, 348, 349
joule-seconds, 1027
Journal de l'École Polytechnique , 1101
Journal of Mathematical Chemistry, The , 592
Journal of Mathematical Physics , 592
Journey Through Genius (Dunham), 610
A Journey to the Center of the Earth (Verne), 900
Judeo-Christian religious tenets. *See* Christian religious tenets
juku schools, 69
Juran, Joseph M., 832
Jurassic Park (movie), 683
Jyesthadeva, 73

K
kabala, 730
Kaczynski, Ted
  Unabomb Manifesto, 958
Kadison, Richard, 211
Kagan, Normal
  The Mathenauts, 901
Kahn, Robert E., 180, 911
Kai Fang Shu, 185–187
Kaigun Ango-Sho D (JN-25B), 756
Kaku, Michio, 441
Kakutani, Shizuo, 919
Kamil Shuja ibn Aslam, Abu. *See* Abu Kamil Shuja ibn Aslam
Kandinsky, Wassily, 749
Kangaroo Mathematics Contest, 361
Kant, Immanuel, 425, 626
  Critique of Practical Reason, 626
  Critique of Pure Reason, 626
Kaplan, Edward, 1006
Kappa Mu Epsilon, 208

kappa test, 1076
Kaput, James, 863, 864
Karaji, 54
al-Karaji, Abu Bakr, 1094
Karinthy, Frigyes
 Chain-Links, 915
Karlin, Samuel, 523
Kashani, see al–Kashi, Jamshid
al-Kashi, Jamshid, 53–54, 514, 726
 The Key to Arithmetic, 73
 Miftah al-Hisab (Calculator's Key), 55
 Treatise on the Circumference, 73
Kasimov, Aslan, 488
Katyayana, 72
 Sulbasutra, 762
Kavli Institute for Theoretical Physics, 777
Kedlaya, Kiran, 7
Kelly, Larry, Jr., 107
Kelly criterion, 107
Kelvin, Lord. *See* Thomson, William (Lord Kelvin)
Kelvin scale, 988
Kempe, Alfred, 500
Kennelly, Arthur Edwin, 838
Kennelly–Heaviside Layer, 838
Kenschaft, Patricia, 838
kente cloth, 21–22, 875
Kepler, Johannes
 Astronomia Nova, 365
 The Epitome of Copernican Astronomy, 365
 Harmonices Mundi, 365
 laws of planetary motion, 40, 78, 237, 362, 365, *366*, 772, 890, 1098
 measurement of volume and, 647, 652
 Nova Stereometria Doliorum Vinarorum, 652
 polyhedra and, 782
 solar system model, 366
 Somnium, 900
Kepler curve, 1060
Kepler–Poinsot polyhedra, 782, 783
Kepler's Laws, 40, 78, 237, 362, 365, *366*, 772, 890, 1098
Kerala, 73
Kermack, W. O., 1039
Kerr, Roy, 115
Key to Arithmetic, The (al-Kashi), 73
Keynes, John, 466
 The General Theory of Employment, Interest and Money, 1018
Keynes, John Maynard, 803, 1018
Khalifa, Rashad
 The Computer Speaks, 730

Khayyam, Omar, 54, 73, 111, 261, 425, 1005, 1095
 Treatise on Demonstration of Problems of Algebra, 237
Khovanov, Mikhail, 532
al-Khujandi, Abu Mahmud, 67
al-Khwarizmi, 248, 425, 664
 Abu Abdallah Book of Addition and Subtraction by the Indian Method, 66–67
 Algebra, 1094, 1095
 algebra and, *32*
 Al-kitab al-muhtasar fi hisab al-jabr wa-l-muqabala (Compendium on Calculation by Completion and Reduction), 32–33, 53–54
 Book of Addition and Subtraction by the Indian Method, 67
 On the Calculation with Hindu Numerals, 712
kicking a field goal, 529–531
Kilby, Jack, 139
kilowatts (kWh), 341
Kim, Scott, 823
Kim II-Sung University, 69
al-Kindi, Abu Yusuf Ya'qub ibn Ishaq
 On the Use of the Indian Numerals, 712
Kindle (Amazon), 303, *304*
kinematic redundancy, 527
kinetic body data, 733
King, Ada (Countess of Lovelace). *See* **Lovelace, Ada**
King, Anne Isabella, 565
King, Byron, 565
King, Martin Luther, Jr., 849
King, Ralph, 565
King, William, 565
kinship systems, 735–736
Kircher, Athanasius, 1028
Kirchhoff, Gustav, 79
Kish, Leslie, 384, 888
Kiss My Math (McKellar), 1082
Kittinger, Joseph, 1060
Klein, Felix
 abstract groups and, 35
 Aerodynamic Proving Ground and, 1073
 curves and, 280
 Elementary Mathematics from an Advanced Standpoint, 616
 Erlangen program, 426, 438, 971, 1104
 geometric spaces and, 429
 geometry education and, 426
 Klein 4-group, 637
 Primary Colors , 958
 surfaces and, 964, 965
Klein, Joe, 958
Klein 4-group, 637
Klein bottle, 616, 964, 965

Klein Four Group
   "Finite Simple Group (of Order Two)," 787
Kleinrock, Leonard, 181, 517
Klemen, Michael, 1063
Klotz, Eugene, 427
Klugel, G. S., 751
Knaster, Bronislaw, 95
"Knee Deep in the Big Muddy" (Straw), 1038
knitting. *See* **crochet and knitting**
knots, 531–532, *534*
Knowing and Teaching Elementary Mathematics, 265
knowledge discovery in databases (KDD), 291
Knowlton, Nancy, 920
Knuth, Donald, 221
Koch snowflake, 281, 763
Kodaira, Kunihiko, 69
Kodak. *See* Eastman Kodak Company
Kolmogorov, A. N., 106, 360, 803
Kolpas, Sidney, 780–781
Konane, 122
Königsburg bridges. *See* Seven Bridges of Königsburg
Koopmans, Tjalling, 913
Korean Women in Mathematics, 1072
Korotkoff, Nilolai, 948
Korotkoff sounds, 948
Kovalevsky, Sonja, 588–589, 596
   *On the Problem of the Rotation of a Solid Body About a
      Fixed Point*, 1104
Kramer, Briton, 684
Kramp, Christian, 765
Krantz, Les, 162
Krbalek, Milan, 134
Kronecker, Leopold, 812
Kubrick, Stanley, 903
Kummer, Eduard, 34
Kunz, Hanspeter, 746
Kuo Shou-ching (Guo Shoujing), 156, 186
Kuratowski, Kazimierz, 146
Kurten, Bernd, 684
kurtosis, 369
Kuse, Allan, 1043

L
La Géométrie (Descartes), 1098–1099
La Hire, Philippe de, 1063
La nova scientia (Tartaglia), 390
La Vallée-Poussin, Charles De, 146
Laban, Rudolf, 947
Labanotation, 91
Laborde, Jean-Marie, 427
labyrinths, *534*

Ladd, Harry, 250
Ladd-Franklin, Christine, 1070
LaDuke, Jeanne
   Pioneering Women in American Mathematics, 1072
Laennec, René, 948
Lagrange, Joseph Louis, 34, 366, 574, 652–653
   Mécanique analytique, 1101
   Réflexsur la résolution algébrique des équations, 765
Lagrangian–Eulerian flow models, 441
Lake Ponchartrain Causeway (Louisiana), 130
L'Algebra (Bombelli), 722
Lambda calculus, 768
Lambert, Johann Heinrich, 574, 909
   Die Theorie der Parallellinien, 1101
Lamé, Gabriel, 249
Lanchester, Frederick, 214, 1073
Lanchester model of warfare, 214–215
Lanczos, Cornelius, 360
land measurement. *See* **units of area**
Landau, H. G., 999
Landsberger, Henry, 889
landscape design, 533–534
Langdell, Christopher, 849
Langley Laboratory, 1073
language of math, 538–539
Lao People's Democratic Republic (Laos), 71
Laplace, Pierre de, 112, 366, 507
   Essai Philosophique sur les Probabilities, 368
   Exposition du Systeme du Monde, 115
   Théorie Analytique des Probaés, 802
   Traité de mécanique céleste, 1103
Laplace distribution. *See* **normal distribution**
Larmor, Joseph, 547
laser-assisted in situ keratomileusis (LASIK), 1040–1041,
   1041
lasers, 973–974, 977
LASIK, 1040–1041, *1041*
LaTeX, 221
Latini, Brunetto, 860
lattice multiplication, 686, 713
lattice theory, 509–510
lava domes, 1045
Lavrentev, Mikhail, 155
law of exponential growth, 95
Law of Large Numbers (LLN), 507, 707, 801
laws of planetary motion, 40, 78, 237, 362, 772, 890, 1098
Lax, Peter, 30
Lay, Kenneth, 509
LCD televisions, 987
LD50/median lethal dose, 535–536
Le Balet Comique de la Reine (Beaujoyeulx), 90

Le Système International d'Unités (SI), 349
Leadership in Energy and Environmental Design (LEED), 462, *462*
Lean Six Sigma, 288
Leap (Gunderson), 776
Learned Ignorance (Nicholas of Cusa), 858
learning exceptionalities, 536–540
  giftedness, *538*, 539–540
  mathematical disabilities, 536–553
learning models and trajectories, 540–544
learning styles. *See* **learning exceptionalities; learning models and trajectories**
Lebesgue, Henri Léon, 146, 360, 505, 651, 1105
Lebesgue integral, 146, 360, 651
Lebombo bone, 19, 173
Leclerc, Georges Louis, 1101
Lecroix, Sylvestre-Francois, 501
Lectiones opticae et geometricae (Barrow), 1100
Lectures (Gregory of Rimini), 858
Lee, Robert E., 193
left-brain learning, 541–542
Legendre, Adrien-Marie, 429, 501, 970
  Élements de géométrie, 1102
legislation. *See* **government and state legislation**
Lego Group, 877
Lehrer, Thomas, 787
Leibniz, Gottfried Wilhelm
  biographical information, 149
  calculus and, 144, 362, 365, 407–408, 552, 1028, 1100
  combinations and, 764
  De Arte Combinatoria, 1028
  energy and, 349
  equation of the catenary and, 281
  functions and, 409
  limits/continuity and, 552–553
  mathematics education and, 359
  plagiarism accusations, 149
  rate of change and, 407–408
  religion and, 601–602
  seconds pendulum and, 639
  symbolic language and, 847–848, 1028
Lem, Stanislaw
  The Extraordinary Hotel, 901
LeMond, Greg, 109
L'empennage (Calder), *544*
Lenin, Vladimir, 214
Lenormand, Louis-Sébastien, 918
Lenstra, Hendrik, 355
Leo XIII, Pope, 626
Leonardo da Vinci
  Codex Atlanticus, 918

flight studies, 29–30
  golden ratio and, 446–447
  humidity measurements and, 1055
  paintings, 748, 861
  sculptures, 904
  siege machines and, 364
  Vitruvian Man, 200, *200*, 650, 886
Leonardo de Pisa. *See* Fibonacci, Leonardo
Leontief, Wassily, 218
Leray, Jean, 367
Leslie, Joshua, 167
Leslie, Nandi, 297
Lesser, Lawrence, 787
Lesser, Mary, 748
Lessner, Sydney
  Fermat's Last Tango and, 689–690
lethality, 535
Leverrier, Urbain, 78
levers, 544–545
Lévy, Paul, 803
Lewis, Ted
  Critical Infrastructure Protection, 509
l'Hospital, Marquis de, 1100
l'Hospital rule, 1100
Li Chunfeng, 186
Li Zhi (Li Yeh), 186, 1095
  Sea Mirror of the Circle Measurements (*Ce Yuan Hai Jing*), 187
  Yi Gu Yan Duan (*New Steps in Computation*), 187
Libby, Willard, 157
Liber Abaci (Fibonacci), 54, 412, 819, 939, 1088, 1095
Library of Alexandria, 18, 66, 459, 460
Library of Babel (Borges), 557
Librié, 303
Libro de los Juegos (Alfonso X), 454, *454*
Lichtman, Jeff, 126
Lidwell, Mark, 745
Lie, Sophus, 35, 344
Lie groups, 344, 360, 844
Life Adjustment schools, 274–275
life expectancy, 545–547
Life of Galileo, The (Brecht), 775
Life of Pythagoras (Porphyry), 826
light, 547–549
light bulbs, 549–550
light-emitting diodes (LEDs), 695
Lighthill, Michael, 890, 995
Lighthill–Whitham–Richards (LWR) traffic theory, 1001
lighting effects, 486–487
lightning, 550–552
light-years, 1022

Lilavati (Bhaskara II), 1095
limits and continuity, **552–553**, 728
Lincoln, Abraham, 192, 193
Lind, James, 288, 732
Lindemann, Ferdinand, 883, 1104
Lindenstrauss, Elon, 75
linear algebra, 555
linear concepts, **553–556**
linear equations, 554–555
linear optimization, 555
linear programming problems, 218, 555, 592, 951–952
linear transformations. *See* **transformations**
linkages, 545
links, 531–532
Linn, Marcia, 1069
Linnaeus, Carolus, 988
Linnaeus, Charles, 862
Lions, Pierre-Louis, 367
Liouville, Joseph, 724
Liouville's theorem, 724
Lippershey, Hans, 978
literature, **556–561**
    fictional mathematicians, 557–559
    genres, 559–560
    mathematical connections/commonalities, 556–557, 560
    mathematical imagery, *557*, 557
    mathematical problems and, 559–560
    in mathematics education, 558
Littlewood, Andrew, 618
Littlewood, John, 671, 1073
Liu Hui, 552
    Nine Chapters on the Mathematical Art, 186, 652
    The Sea Island Mathematical Manual (Haidao Suan Jing), 186
    Six Arts, 185
Livio, Mario, 885
"Lo Shu" magic square, 568
loans, 482–484, *561*, **561–563**, 675–676
Lobachevskian geometry, 1103
Lobachevsky, Nicolai Ivanovitch, 85, 359, 438, 848, 1103
locations systems, 672
Lockheed Corporation, 31, 880–881, 1037
locomotives, 1002
logarithmic spiral, *762*, 763
logarithms. *See* **exponentials and logarithms**
Logicomix (Doxiadis), 219, 776
Logistica (Buteo), 764
lognormal distribution, 708
London Mathematical Society, 1104
Long Count Days, 495–496
Long Range Navigation (LORAN), 238, 1079

Longstreet, James, 193
Lorch, Lee, 356
Lorentz, Hendrik, 547, 1005
Lorentz transformations, 547, 853–854, 1005
Lorenz, Max, 769
Lotka, Alfred
    Elements of Physical Biology, 789
Lotka–Volterra system, 789
lotteries, 106, **563–565**, 665–666, 667
Lotus Temple, 487
Louis XIV (king), 177
Lovász, László, 361
Lovelace, Ada, 500, **565–566**
Lovell, James, 869
Loyd, Samuel, 820
Lu Chao, 771
Lucas, Edouard (N.Claus), 412, 593, 821
Lucas, Henry, 1099
Lucasfilm LTD, 51
Lucasian professorship, 1099, 1100
Lucent Technologies, 976, 977
Lucien of Samosata
    True Histories (True Tales), 900
Ludu Algebraicus, 122
Ludus Astronomorum, 122
Ludus Latrunculorum, 120
luminous efficacy, 549
Luoshu, 185
Luotonen, Ari, 911

M
Ma, Liping, 265
McBurney, Simon, 776
McCall, David, 982
McCalla, Clement, 167
McClellan, George, 193
McCleskey v. Kemp, 118
McCready, Mike, 788
McCulloch, Warren, 701
McCulloch–Pitts Theory of Formal Neural Networks, 701
McCurty, Kevin, 519
Mach, Ernst, 27
Mach bands, 739
Mach Number *(M)*, 27, 31
Machiavelli, Niccolò
    Discourses on Livy, 861
    Prince, 861
Machin's formula, 722, 723
MCI Digital Information Services, 180, 181
MCI Mail, 181
Macintosh computers, 767

McKellar, Danica, 35, 163
  Hot X, 35, 1082
  Kiss My Math, 1082
  Math Doesn't Suck, 1082
McKendrick, A. G., 1039
McKenna, P. Joseph, 129
Maclaurin, Colin, 35, 38, 994, 1100
McLean, Malcolm, 913
McMichaels, Robert, 489
McNamara, Frank, 253
McNamara, Robert, 414, 1037–1038
MacTutor History of Mathematics Archive, 221, 1072
Maddow, Ellen
  Star Messengers, 690
Madison, Ann, 119
Madow, William, 890
Maggi, Girolamo, 364
al-Maghribi, Samu'il (al-Samaw'al), 54, 55, 111
magic, 567–569
MAGIC code, 757
magic squares, cubes and circles, 568–569
Magic: The Gathering, 415–417, *416*
magnetic disk drives, 769
magnetic resonance imaging (MRI), 126, 165, 541, 560, 1013
  See also Computed Axial Tomography (CAT) scans; functional MRI (fMRI); imaging technologies; medical imaging
magnetic tunnels, 769
Magnetoencephalography (MEG), 127, 128
magnetoperception, 45
magnitudes. *See* irrational numbers
magnitudes (as line segments), 726
Magnus, Albertus, 664
Magnus, Heinrich, 671
Magnus Effect, 671
al-Mahani, 726
Mahavira, 762, 1094
Mahoney, Michael Sean, 248
Maimonides, Moses, 94
Mair, Bernard, 167
Makridakis, Spyros, 401
Malaysian Mathematical Sciences Society, 71
Malliavin calculus, 147
malnutrition, 733
Malthus, Thomas, 177
management science, 807
managerial accounting. *See* **accounting**
Mancala, 13, 15–16, *16*, 120
mandalas, 515, 885–886, *886*
Mandelbrot, Benoit, 361, 551, 570, 722, 763, 949, 1042

Mandelbrot set, 361, 722
Manhattan Project. *See* **atomic bomb (Manhattan Project)**
manifolds, 555
Manin, Yu, 222
Mann, Estle Ray, 1034
Mann, Horace, 274
Mannheim, Amédée, 1103
Mansa Musa of Mali, 24
Mansur, Abu Nasr, 67
Mantle, Mickey, 477
manufacturing design, 430, *431*
Maori culture, 736, *736*
Maple (software), 273, 372, 768
mapping coastlines, 570–571
maps, 571–574
  See also cartography
Marar, K. M., 73
Marcellus, General, 59
Marchetti, Alessandro, 364
Marconi, Guglielmo, 838
Marianas Trench, 295
Marin, Mario, 850
marine navigation, 574–577, 738
Maris, Roger, 477
market research, 578–580
marketing. *See* business, economics, and marketing
market-value-weighted stock indices, 950
Markham, Beryl, 31
Markopoulou, Athina, 791
Markov, Andrei, 803, 1007
Markov chains, 134, 398
Markov decision process model, 787, 1007
Markowitz, Harry, 691–692
al-Marrakushi ibn al-Banna. *See* al-Banna, al-Marrakushi ibn
marriage, 580–582
Marrison, Warren, 203
Mars Climate Orbiter, 639
Marshall Islands, 737, 738
martial arts, 582–584, *583*
Martin, Artemas, 599
Martin, David, 920
  Count, 776
Martin, John, 776
martingale stochastic (random) processes, 803
Marx, Karl, 214
Mascheroni, Lorenzo, 883, 1101
Maslow, Abraham, 795
Maslow's hierarchy, 795
Massey, William, 977
M A S S I V E, 787

"math castle," 171

Math Doesn't Suck (McKellar), 1082

Math Forum, 220, 426

Math Fun Facts, 221

math gene, 584–586

Math Kangaroo, 361

math rock, 230, 787, 788

Math Standards (NCTM), 151

Mathcad software, 61

mathcore, 230, 787

MATHCOUNTS, 227

Mathematica (software), 142, 152, 273, 372

mathematical ability, 584–585

Mathematical Analysis of Logic, The (Boole), 362, 1103

Mathematical Association of America (MAA), vi, ix, 151, 166, 207, 221, *709*, 709–710, 810, 1105

mathematical certainty, 586–588

Mathematical Collection (Pappus), 1093

Mathematical Contest in Modeling (MCM), 228–229

Mathematical Correspondent, The (journal), 870

mathematical disabilities, 536–538

mathematical engines, 875, 1103

mathematical epidemiology, 170–171, 797, *797*

mathematical expectation, 1100

mathematical friendships and romances, 588–589

Mathematical Games (column), 595

mathematical giftedness, *538*, 539–540

mathematical magic. *See* **magic**

Mathematical Markup Language (MathML), 221

mathematical modeling, 589–593

    for accident reconstruction, 1–2

    in accounting, 5

    for animals in nature, 44–45, 463, 789–790, 973

    archery bows, 56

    of auditory processing, 684

    for Barkhausen effect, 489

    bus scheduling and, 134–135

    for climbing, 201

    combat modeling, 503

    comparison shopping and, 226

    cycling equipment and, 109

    data mining and, 291–292

    for dynamic systems, *250*, 251, 297–298

    for economic order quantity (EOQ), 523

    energy sustainability and, 463

    for firefly activity, 973

    flood predictions and, 396–397

    forecasting and, 401, 654

    forensic ballistics and, 63, 64–65, 255–256

    for forest fires, 402–404

    for geothermal processes, 441

    global warming and, 464, 592

    for helicopter flight, 475

    highway design and, 476

    history of, 590–592

    hockey and, 481

    for hurricanes and tornadoes, 490–491, 607

    for hydrostatics, 591

    for infectious diseases, 313, 314, 1038–1039

    of kinship systems, 735

    linear programming models, 218, 555, 592, 951–952

    Mathematical Contest in Modeling (MCM), 228–229

    for motion and gravity, 591, 592

    for motion of moon, 678

    for moving fluids, 591

    for nutrition, 732–733

    for population growth, 591

    for predicting attacks, 790–792

    for probability for survival, 535

    process, 590

    recycling and, 852

    search protocols and, 518–519

    SIR model for, 1039

    for surgery, 966–968

    for swimming, 969

    for taxes, 498

    telecommunications and, 977

    for traffic, 488, 1000–1001

    train timetables and, 1002

    for tunnels, 1015

    for volcanos, 1044, 1045

    for water supplies, 1052, 1054

    for wind, 1064

Mathematical Olympiad Summer Program, 227

mathematical problems, classic. *See* classic mathematical problems

"Mathematical Psychology of War, The" (Richardson), 1075

mathematical puzzles, 593–595

Mathematical Reviews (journal), 220, 1106

mathematical sciences majors, 270–271

Mathematical Society of Serbia, 365

Mathematical Treatise in Nine Sections (Ch'in), 1095

mathematician defined, 595–597

A Mathematician Plays the Stock Market (Paulos), 400, 949

mathematicians, amateur, 597–600

mathematicians, religious, 600–603

Mathematicians Action Group (MAG), 811

A Mathematician's Apology (Hardy), 559, 608

Mathematicians of the African Diaspora, 221, 297, 1072

Mathematico-Physical Journal, 360

Mathematics (record label), 787

mathematics, applied, 603–607
    actuarial science and, *605*, 605–606
    biomathematics, 606
    biostatistics, 606
    careers in, 163, *605*, 605–607
    historical context of, 603–604
    for mathematical modeling, 607
    operations research (OR) and, 606
mathematics, Arabic/Islamic. *See* **Arabic/Islamic**
    mathematics
mathematics, Babylonian. *See* **Babylonian Mathematics**
mathematics, Chinese. *See* **Chinese Mathematics**
mathematics, defined, 608–610
mathematics, Egyptian. *See* **Egyptian mathematics**
mathematics, elegant, 610–612
mathematics, Greek. *See* **Greek mathematics**
mathematics, green. *See* **green mathematics**
mathematics, Incan and Mayan. *See* **Incan and Mayan**
    Mathematics
Mathematics, Magic and Mystery (Gardner), 569
mathematics, Native American. *See* **Native American**
    Mathematics
Mathematics, Physics, and Astronomy Society of Slovenia,
    365
mathematics, Roman. *See* **Roman mathematics**
mathematics, theoretical, 613–618
    algebra, 613
    analysis, 614
    geometry and topology, 613–614
    mathematicians and, 617–618
    number theory, 614
    training and education, 615–617, *616*, 619
mathematics, utility of, 618–620
mathematics, Vedic. *See* **Vedic mathematics**
mathematics: discovery or invention, 620–622
mathematics and religion, 622–627
    Chinese religious tenets, 626–627
    Christian religious tenets, 624–626
    divination, 623, *623*
    Indian religious tenets, 626–627
    Islamic religious tenets, 627
    pattern drawing, *623*, 623–624
Mathematics and Statistics in Anaesthesia (Cruickshank), 43
Mathematics Anxiety Rating Scale (MARS), 960
Mathematics Awareness Month, 448, 450
Mathematics Genealogy Project, 628–629
Mathematics Genealogy Project Website, 629
Mathematics in Africa (Int'l Mathematical Union), 15
mathematics literacy and civil rights, 630–632
Mathematics of Investment (Rider and Fischer), 139
Mathematics of Marriage, The (Gottman, Murray, et al.), 581

mathematics research, interdisciplinary, 632–633
mathematics software. *See* **software, mathematics**
Mathenauts, The (Kagan), 901
Math-Jobs Web site, 166
MathSciNet, 220
MathTrek (Peterson), 221
MATLAB, 273
matrices, 512, **634–636**, 635, 686
matrix multiplication, 686
matrix theory, 635
Mattangs, 737, 738
mattresses, *636*, **636–637**
Mauchly, John, 1078
Maxwell, James, 305, 547, 662, 837, 838, 988, 1005, 1030
    Treatise on Electricity and Magnetism, 1029
Maxwell v. Bishop, 118
Maxwell–Boltzmann kinetic theory of gases, 988
Maxwell's equations, 1005
Mayan mathematics. *See* **Incan and Mayan mathematics**
Mayer, J. C. A., 389
Mayer, Tobias, 1042
mazes, *534*
Meade, George G., 193
meal planning, 733
mean, 654–655
Mean Girls (movie), 683
Measure Master, 141
measurement, systems of, 637–640, *638*, 643
measurement in society, 640–645
    accuracy and precision, 641–642
    everyday applications, 642–644, *644*
    Pre- K–12 curricula and, 644–645
    systems, 640–641
Measurement of a Circle (Archimede), 1094, 1095
measurements, area, 645–647
measurements, length, 647–651
measurements, volume, 651–653
measures of a center, 653–655
measuring time, 202, *202*, **655–657**
measuring tools, 657–658
Mécanique analytique (Lagrange), 1101
Mecca, 24, 74–75, *75*
mechanical clocks, 202
mechanics, 1098, 1101
Meddos, 737, 738
median, 654
median lethal dose. *See* **LD50/median lethal dose**
medical imaging, 126–128, *527*, **659–660**, 1017–1018
    See also Computed Axial Tomography (CAT) scans;
        functional MRI (fMRI); imaging technologies;
        magnetic resonance imaging (MRI)

medical simulations, 660–661

Meditations on First Philosophy (Descartes), 365, 859

Meehl, Gerald, 196

Meet the Press (news program), 510

Mega Millions, 563

Mehmed-II, Sultan, 75

Mei Juecheng, 187

Meier, Paul, 1006

Meister, A. L. F., 781

Melanchthon, Philip, 862

Melchizedek, Drunvalo, 885

Melinda Gates Foundation, 133

Melville, Herman
    Moby Dick, 737

memory foam, 636

memory latency, 769

Menabrea, Luigi, 565–566

Menaechmus, 236–237, 1092

Mendel, Gregor, 420, 421

Menelaus of Alexandria
    Sphaerica, 425, 1011, 1093

Meno (Plato), 858

Mercator, Gerardus, 577

Mercator, Nicolaus, 1099

Mercator chart of the world (1569), *576*

Mercury program, 522

Méré, Chevalier de, 801

Meril, Alex, 167

meromorphic function, 724

Merriam, Thomas, 938

Merrill, Winifred Edgerton, 1070

Mersenne, Marin, 588

Merton, Robert C., 264

Mesoamerica, 424

Message Found in a Copy of Flatland (Rucker), 902

Message Passing Interface (MPI), 753

Messiaen, Olivier, 894

metalcore, 230

Metamorphosis III (Escher), 355–356

metaphysics, 624

Metaphysics (Aristotle), 1092

Metastasis (Xenakis), 231

Meteorologica (Aristotle), 1055

meteorology, 317

meter, 777–778, 995–996

Method, The (Archimedes), 60

method of equilibrium, 1092

method of exhaustion, 1092

method of fluxions, 144

Method of Fluxions (Newton), 354

method of indivisibles, 1092

Method of Largest Remainders, 232

metric spaces, 651

metric system, 639, 1021–1022, 1024, 1025

Metrica (Heron), 646

Metrodorus, 1093

Metromachia, 122

Meyer, Yves, 293

Meyerhoff Scholoars Program, 961

Méziriac, Bachet de, 1098

Mezrich, Ben
    Bringing Down the House, 682

Michelangelo Buonarotti, 861

Michell, John, 115

Michell, Keith, 166

Michelson, Albert, 547

Michelson–Morley experiment, 853

Micolich, Adam, 749

microgravity, 1059–1060

microphones, 354

microscopes, *548*, 674, 1098

microscopic modeling, 1000

Microsoft Corporation, 768

microwave ovens, 661–663

microwave technology, 661–663, *662*

Middle Ages, 663–665

Miftah al-Hisab (Calculator's Key) (al-Kashi), 55

Milgram, Stanley, 388, 915

military code, **see also coding and encryption**
    code talking, 839, 1074
    Enigma code, 212, 763, 1076–1077
    Morse code, 192, 838, 1055, 1074
    superencypherment, 1076, 1077
    trench codes, 1074

military draft, 665–667, *666*

military research in mathematics, 756, 1073–1074, 1075

Millau Bridge (France), 130

millimeter wave scanners, 643–644, *644*

Millington, Hugh G. R., 167

Milnor, John, 228

Milton Bradley Company, 324

Mindstorms NXT, 877

minerals, 732

Ming Antu, 69

Ming dynasty, 187, 423

Minié, Claude, 191

Minié ball ammunition, 191

minimax theorem, 413

Minkowski, Hermann, 548, 954–955, 1005

Minkowski space, 548

minorities, 170–171, **667–670**, 984–985

Miraflores Locks, *156*

Mirifici Logarithmorum Canonis Descriptio (Napier), 713
Mirollo, Renato, 973
Miscellanea analytica (de Moivre), 1100
Mises, Richard von , 113, 803
missiles, *671*, **671–672**
Mitofsky, Warren, 336
Mittag,Leffler, Gosta, 1075
Miura, Koryo, 891
Mo Jing, 185
Mo Ti, 305
mobiles, *544*, 545
Möbius, August, 249, 901
Möbius band, 965, *965*
Mobius transformations, 1005
Moby Dick (Melville), 737
mode, 653–654, 655
modeling. *See* mathematical modeling
modes, musical, 893
A Modest Proposal (Swift), 559
MODFLOW, 442
Modified Mercalli Intensity Scale, 323, 324
modular arithmetic, 201, 719–720
Mohr, Georg, 882–883, 1100
Mohr–Mascheroni theorem, 883
Moisil, Grigore C., 361
molecular structure, **672–674**
Molien, Theodor, 35
Mondrian, Piet, 748
money, **674–677**
Monge, Gaspard, 281, 425, 455, 501, 1005
    Application de l'analyse à la géométrie, 1101
Mongkut, King, 71
monotonicity criterion, 1049
Monte Carlo simulation, 28, 419, 837, 842
Montenegro Mathematical Society, 364–365
Montessori, Maria, 324, 325
moon, **677–679**
    eclipses, 77, 78, 677–678
    human exploration of, 276, *678*, 678–679
    lunar calendars, 75, 77, 154–156
moonlanding, *678*, 678–679
Moore, Eliakim, 239, 249, 658
Moore, Gordon, 769
Moore, Robert Lee, 268
Moore Method, 268
Moore's Law, 769
More, Thomas
    Utopia, 900
Morgan, Augustus de, 722
Morgan, J. P., 163
Morgan, Ryan, 929

Morgan's theorem, 929
Mori, Shigefumi, 69
Morpheus Laboratory, 26
Morse, Marston, 232
Morse, Samuel F.B., 1055
Morse code, 192, 838, 1055, 1074
    See also **coding and encryption**
mortality as dose response, 535, 536
mortality tables, 123, 546, 1100
mortgages, 482–484
Morton, A. Q., 957
Moscow papyrus, 554, 646, 651, 1091
    See also Rhind papyrus
Moser, Jurgen, 367
Moses, Robert, 631, *631*
Moss, Jamal, 787
Most Beautiful Mathematical Formulas, The (Salem, Testard, Salem), 612
most recent common ancestor (MRCA), 419
Mosteller, Frederick, 944
Motion Pictures Experts Group (MPEG), 320
Motorola, Inc., 287–288, 832
Mouchot, Augustin, 930
Mouhe Fanggai (double vault), 186
Moulton, Forest Ray, 1073
Mount St. Helens, 1044, *1045*
Mouton, Gabriel, 639
movies, making of, **679–681**
movies, mathematics in, **681–684**
Moving On 2000, 543
Mozart, 447
MP3 players (MPEG Audio Layer III), 320, **684–685**, *685*
Mr. Gasket Hot Rod Calc, 141
Muhammad ibn Muhammad, 24, 111, 785
Mulcahy, Ann, 132
Müller, Johann, 1096
Multi-Angle Imaging SpectroRadiometer, 206
multinomial distribution, 112
    See also **binomial theorem**
"Multi-Objective and Large-Scale Linear Programming" (Osei-Bryson), 167
Multiple Agent Simulation System in Virtual Environment (MASSIVE), 289–290
multiple intelligences, 543
multiplication and division, **685–689**
    checking results, 687–688
    computational speed, 688
    division algorithms, 687
    generalizing, 688–689
    history of algorithms for, 686–687
    multiplication by addition, 688

"A Multi-Variate EWMA Approach to Monitor Process Dispersion" (Bernard), 167
multivariate probability inequalities, 10–11
Mumford, David, 228, 363
Mumford, Lewis, 190
Munk, Max, 756
Munroe, Randall, 219
    xkcd, 219
Münster, Sebastian, 573
muqarnas, 514
Murray, Charles
    The Bell Curve, 708
Murray, H. J. R., 120
Murty, M. Ram, 963
music, 231, 786–788, 823–824, 893, 956–957
    golden ratio in, 446–447
music, geometry of. See geometry of music
music, popular. See popular music
Music Intelligence Solutions, 788
Music IP, 788
musical theater, 689–691
mutual funds, 691–692
    See also investments; stocks/stock market
Mutually Assured Destruction (MAD), 214–215, 216, 414, 1037–1038
myriad-grouping system, 68
Mystery of Mars, The (Ride), 871

N
Nabokov, Vladimir, 6
"nach Adam Riese" (according to Adam Riese), 1096
Nader, Ralph, 1047, 1049–1050
Nadir, Mehmet, 600
Nagari numerals, 712
Nagasaki, Japan, 80
Name Worshipping, 505
nanocars, 694
Nano-robots, 875
nanotechnology, 674, 693–695, 694, 769, 875
nanotubes, 674, 769
Napier, John, 370, 371, 1011, 1098
    Mirifici Logarithmorum Canonis Descriptio, 713
Napier's anologies, 1098
Napier's bones, 713
Napier's rods, 1098
Napoleon (howitzer), 191
Napoleon Bonaparte (Napoleon III), 156, 189, 191, 392, 501
Napoleon's theorem, 501
Napster, 388
Nasar
    A Beautiful Mind, 559

NASCAR, 703–704
NASDAQ, 691
Nash, John, 122, 216, 414, 559, 681
Nash Equilibria, 217, 414
Nasîr al-Din Shah, 1095
A Nation at Risk (NCEE), 150, 276, 277
National Academy Foundation, 133
National Academy (French Indonesia), 71
National Academy of Science, 119, 228
National Aeronautics and Space Administration (NASA)
    Apollo program, 276, 522, 678–679
    data collection/analysis, 195, 206, 286
    elementary particles and, 343
    establishment of, 1106
    interplanetary communication and, 181
    Planetary Flight Handbook, 881
    Sally Ride and, 870–871
    space elevators, 347, 348
    weightless research, 1060
National Agricultural Statistics Service, 286–287
National Arbor Day Foundation, 196
National Assessment of Educational Progress (NAEP), 328
National Association for Stock Car Auto Racing (NASCAR), 82
National Association of Mathematicians (NAM), 810–811, 1107
National Association of Securities Dealers Automated Quotations (NASDAQ). See NASDAQ
National Association of Stock Car Drivers (NASCAR). See NASCAR
National BankAmericard (NBI), 253
National Basketball Association (NBA) Draft Lottery, 563
National Cancer Institute, 183
National Center for Atmospheric Research (NCAR), 196
National Center for Education Statistics, 151, 287
National Center for Electron Microscopy, 548
National Commission on Excellence in Education, 150–151
National Conference on City Planning, 189
National Council of Supervisors of Mathematics (NCSM), 539
National Council of Teachers of Mathematics (NCTM)
    algebra education and, 36, 40
    establishment of, 810
    Illuminations Web site, 221
    interconnected curriculum and, ix, 238, 239
    International Conference on Teaching Statistics (ICOTS), 946
    New Math and, 1106
    No Child Left Behind (NCLB) Act and, 710
    reform recommendations, 151, 276–278, 279
    roles of proofs in education and, 222

National Council of Teachers of Mathematics Principles and Standards for School Mathematics, 946
national debt, 695–697
National Defense Education Act (NDEA), 275, 449
National Elevation Dataset, 346
National Fire-Danger Rating System (NFDRS), 403
National Football League, 398
National Hockey League (NHL), 481
National Institute of Biomedical Imaging and Bioengineering (NIBIB), *527*
National Institute of Standards and Technology, 203, 489–490, 843
National Institutes of Health, 125, 170
National Mathematical Olympiad (Malaysia), 71
National Mathematics Advisory Panel, 325, 450
National Oceanic and Atmospheric Administration (NOAA), 195, *994*
National Popular Vote Compact, 339
National Renewable Energy Laboratory, 1065
National Research Council (NRC), 151
National Science Foundation, 238, *272*, 272, 449–450
National Security Agency (NSA), 39, 214, 229, 508, 509
national standards-based mathematics curriculum, 277–278
National Women's Hall of Fame, 525, 871
Native American mathematics, 697–699
    code talking and, 839, 1074
    ethnomathematics and, 668
    Mary G. Ross and, 880–881
    measurement and, 649
    minorities in mathematics and science, 170–171, 178, 669–670, 811
    sweat lodge design, 486, 487
    weather forecasting and, 1055
Natural Magic (Porta), 862
natural numbers
    Giuseppe Peano and, 1105
    logic tools and, 1105
    orders of infinity and, 1104
    series, 908–909
    zero and, 1087–1088
Navaratna, Channa, 746
Navaratna, Menaka, 746
Nave, Jean-Christophe, 488
Navier, Claude-Louis, 30, 129, 591, 995, 1052
Navier–Stokes equations, 30, 591, 995, 1052
navigation systems, 238, 429, 458, 658, 672
    See also **maps; marine navigation**
navigational clocks, 202–203
Nazca lines, 932
Nazis, 857
Nechunya ben Hakanah, 731

negative numbers, 722
    acceptance of, 261
    addition/subtraction and, 8–9
    black/red ink custom, 1095
    Chinese mathematics and, 187
    complex numbers and, 262
    Indian mathematics and, 73, 1088
    multiplication and, 938
    topographic maps and, 345
Neilsen, Arthur C., 578
Neonativist theories, 741
Neoplatonism, 862
Neptune, discovery of, 78
nerds, 981, 983–985
Nernst, Walther, 128
Nernst equation, 128
nervous system, 700–701
    See also neurons
Netflix, 12, 520
Netscape Navigator Web browser, 842
network science, 519
networks, Internet, 517, 518–519
Neugebauer, Otto, 87, 89
Neumann, John von, 79, 216, 218, 232, 360, 922, 988
neural networks, 701–703
neurobiology, 541–542
neurochip, 701
NEURON computer simulation system, 700
neurons, 124–128, *127*
    See also **nervous system**
neurophysiology, 741
neuroscience, 700–701
    See also **brain**
Neutral Buoyancy Simulator, 1060
New and Complete System of Arithmetick, The (Pike), 870
New Economics, The (Deming), 299
New England Historic Genealogical Society (NEHGS), 418
New Math, 275–276, 1037, 1106
New Principles of Gunnery (Robins), 391
New Testament, 731
new urbanism, 462
New York Stock Exchange, 691
New Zealand, 735, 736
New Zealand Mathematical Society, 735
Newcombe, Nora, 1069
Newcomen engine, 500
    See also steam engines
Newman, Krissie, 704
Newman, Ryan, 703–704, *704*
Newton, Paul, 963

Newton, Sir Isaac
    aerodynamics and, 64
    animal/machine connections and, 45
    binomial theorem and, 111, 112
    biographical information, 148–149, 154
    birthday observance, 30
    coordinate systems and, 354
    imaginary numbers and, 722
    invention of calculus, 362, 405, 1011–1012, 1100
    Kepler's third law and, 40
    laws of motion/laws of gravity and, 115, 406–407, 591, 772, 847, 1029, 1055
    light and, 547
    limits/continuity and, 552–553
    linear concepts and, 555
    Method of Fluxions, 354
    method of fluxions and, 144
    normal distribution and, 707
    Philosophiae Naturalis Principia Mathematica, 362, 602
    Principia, 847, 1029, 1101
    religion and, 601–602
    telescopes and, *362*, 981
Newtonian, 981
Newton's laws, 30, 40
Next 50 Years, The (Stewart), 607
Neyman, Jerzy, 889, 943
Ngo Bao Chau, 367
Nicholas of Cusa, 280, 601, 625, 1096
    Learned Ignorance, 858
Nickelodeon, 982
Nicomachus of Gerasa
    Introduction to Arithmetic, 1093
Nicomedes, 280, 1092
Nielsen, Arthur, 704–705
Nielsen, Henrik Frystyk, 911
Nielsen Media Research, 705
Nielsen ratings, 320, 578, **704–706**
Nielson, Arthur, 288
Nightingale, Florence, 288, 942, 1042
Nigrini, Mark, 498
Nim, 122
Nine Chapters on the Mathematical Art (Chinese text), 155, 185–187, 261, 634, 652, 887, 914, 1093, 1095
Nine Men Morris, 120, 992–993
1984 (Orwell), 849
Nipkow, Paul, 986
Nipkow disk, 986
Nixon, Richard M., 117
No Child Left Behind (NCLB) Act, 278–279, 328, 710
Noether, Amalie Emmy, 35, 1105
Noether, Emmy, 1030, 1072

Noether, Max, 1105
Noetherian Ring groups, 1072
nominal GDP, 467
Nomos Alpha (Xenakis), 787
non-Euclidean geometry, 79, 85, 425–426, 783, 848–849, 1101
    parallel postulate and, 586–587
    See also Euclid of Alexandria; hyperbolic geometry
non-Euclidean polyhedra, 783
nonparametric tests, 844
Nook, 303, 304
normal distribution, 220, **706–708**, 802–803
    See also Gaussian distribution
norm-referenced tests (NRTs), 326–327
Norse Greenland society, 198
North America, 708–710
North Atlantic Treaty Organization (NATO), 215, 767
Norton, Larry, 183
Norton–Simon hypothesis, 184
Nouvelles annales de mathématiques (journal), 1103
Nova Stereometria Doliorum Vinarorum (Kepler), 652
NP-Complete problems, 912
NP-hard, 1010
nuclear bombs. *See* **atomic bomb (Manhattan Project)**
nuclear fission, 80
NUMB3RS (television show), 163, 984, 1083, 1084, 1085
number and operations, 710–714
    computational aids, 713
    early number systems, 711
    Hindu-Arabic numerals, 712–713
    Indian or Hindu numerals, 712
    Roman numerals, 8, 418, 711–712, 862, 879
number and operations in society, 714–719
    calculation tools, 717–718
    computers and, 718–719
    economics/demographics and, 716
    estimation, 717
    measurement, 716
    mental arithmetic, 717
    operations, 715
    tally marks, 714–715
    types of numbers, 715–716
"number blindness" (dyscalculia), 128
number colors, black and red, 1095
number theory, 614, 617, **719–721**, 975
numbers, complex, 721–724
numbers, rational and irrational, 724–726
numbers, real, 727–728
numbers and God, 729–731
    "7," 729–730
    "12," 729

"19," 730
bible codes, 730–731
Golden Ratio, 729
infinity and, 460–461
Pythagoras maxim, 729
resurrection of Jesus, 731
numerical weather prediction, 1055–1056
Nunes, Pedro, 576
Nunes, Terezinha, 226
"Nuremberg eggs", 202
nutrition, 240–241, **731–734**
nutrition labeling, 240–241

O

Obama, Barack, 133, 228, 289, 526, 1035
ocean tides and waves. *See* **tides and waves**
Oceania, Australia, and New Zealand, 735–737
Oceania, Pacific Islands, 737–739, 774
Ockeghem, Johannes, 230
Ocneanu, Adrian, 904
O'Connor, John, 221
"Octatube" (Ocneanu), 904
October Revolution (Russia), 154
Official Guide to Japan (Japanese Railway Ministry), 582
Ogawa, Yoko
    An Introduction to the World's Most Elegant Mathematics, 612
ogive graph, 654
Ohl, Russell, 930
Ohm, Georg, 341
Ohm's Law, 341
Okrent, Daniel, 378
Oliver, Dean, 99
Olson, Steve, 419
Olympic Games, 91, 374
Omicescu, Octav, 361
"On Computable Numbers" (Kleinrock), 517
"On Operations on Abstract Sets and their Application to Integral Equations" (Banach), 361
Once Were Warriors (movie), 737
One Laptop Per Child, 769
Opana Point, 756
Open Handset Alliance, 978
operations. *See* **number and operations; number and operations in society**
operations research (OR), 165, 218, 481, 606, 807, 1078, 1079
Operator Algebras, 548
Oppenheim, Slexander, 71
Oppenheimer, J. Robert, 79, 80, 116
Optical Character Recognition (OCR), 913

optical illusions, 739–741
optical scanners, 96
optics, 978–979, 1040, 1100
    See also visualization
Opus Geometricum (Saint-Vincent), 354
oracle bones, 1091
orbifolds, 435
orbits, planetary. *See* **planetary orbits**
Oresme, Nicole, 409, 1095–1096
organ transplants, *1006*, 1006–1007
origami, 741–743, *742*
origami technology, 741
"Origin of Polar Coordinates" (Coolidge), 353
Orthello, 120
orthonormal polynomials, 1040
Orwell, George
    1984, 849
Oscar II of Sweden, King, 909
Osei-Bryson, Kweku-Muata Agyei, 167
Oughtred, William, 371
    Clavis mathematicae, 1098
Ouranomachia, 122
Ouspensky, Peter
    Tertium Organum, 627
Out of the Crisis (Deming), 299
Outer Space Treaty, 679
Outside In (video), 1042
Ouvroir de Littérature Potentielle (Workshop of Potential Literature), 779
overblowing, 1066
overtone series, 472–473

P

pacemakers, 745–746
    body clocks/jet lag and, 746
    heart rhythms and, 745
Pacific Islands. *See* **Oceania, Pacific Islands**
Pacific Ring of Fire, 774
Pacioli, Luca
    algebra and, *32*, 33
    De Divina Proportione (About Divine Proportion), 886
    De Viribus Quantitatis, 821
    Summa de arithmetica, Geometria, Proportioni et Proportionalita, 2, 289, 754, 800–801, 1096, 1099
Packe, Christopher, 344
Packer, Claude, 167
packing problems, 746–747
Page, Lawrence, 518–519
PageRank (Google), 518–519, 610
painting, 736, **748–750**, 861
Paisano, Edna Lee, 178

palimpsest, 58
Palin, Bristol, 983
Panama Canal, *156*, 156
Pangea, 773
Panini, 72–73
paper folding. *See* **origami**
Pappus of Alexandria, 762, 829
    Synagoge (The Collection), 425, 1093
parabolic flight, 1060, *1060*
parabolic segments, area of, 1092
paradoxes, 505
paradoxical preferences, 794–795
parallax measurements, 78–79
parallel climbers puzzle, 201
parallel postulate, 586–587, **750–752**, 1101
parallel processing, 752–753
Parallelogram Law of Vector Addition, 1030
parametric sensitivity analysis, 874
Parent, Antoine, 1100
Pareto condition, 1049
Park, Bletchley, 1077
Parlett, David, 120
Parthenon, 280, 729–730, *730*
Parthenopaeus (Dionysius), 6
particle physics. *See* **elementary particles**
"Particle Zoo, The", 342
Partition (Hauptman), 776
party problems, 113–115, 210–211
Pascal, Blaise
    barometric pressure/elevation and, 344–345
    binomial theorem and, 111, 112
    combinations and, 764
    contributions of, 1100
    dice games and, 302
    invention of Pascaline, 887
    Marin Mersenne and, 588
    Pensées, 860
    Pierre de Fermat and, 105–106, 1099
    probability theory and, 366
    Provincial Letters, 860
    religion and, 601
    Traité du Triangle Arithmétique, 112, 1099
Pascal, Étienne, 887
Pascal's Pyramid, 112
Pascal's Simplex, 112
Pascal's Triangle, 111, 112, 1095
Passarola, 932
patterns
    caves and caverns, 173–174
    decorative, 13–15, 698–699, *736*, 736–737
    drawing, 623–624

figure skating, 917
    geometric, 16, 21–22, 102, 1091
    recognition of, 702
    step and tap dancing, 947
    See also fractals; tessellations; tilings
Pauling, Linus, 334
Paulos, John Allen
    A Mathematician Plays the Stock Market, 400, 949
payroll, 754–755
PCs. *See* **personal computers**
Peacock, George, 1103
    Treatise on Algebra, 1103
Peano, Giuseppe, 281, 1030, 1105
Pearl Harbor, attack on, 755–757
Pearson, Egon, 943
Pearson, Karl, 369, 654, 706–707, 894, 942
Peart, Paul, 167
Peaucellier, Charles-Nicolas, 500
Peaucellier cell, 545
Peirce, Charles, 552
Pell, John, 1099
Pell equation, 1099
Penrose, Lionel, 336, 355
Penrose, Roger, 115, 116, 355, 886
Penrose tilings, 515, 903
Penrose–Banzhaf Power Index, 336
Pensées (Pascal), 860
pensions, IRAs, and Social Security, 757–760
Penske, Roger, 703
pentagrams, 886
Pentominoies, 120
perceptrons, 701–702
percussion instruments, 760–761
Perelman, Grigori "Grisha", 360, 559, 1107
Perelman, Yakov Isidorovich, 360
Perfect Rigor (Gessen), 559
perimeter and circumference, 761–763
permutations and combinations, 113, **763–766**
Perry, William J., 766–767
"person of color." *See* **minorities**
personal computers, 767–770
    See also computers
Persons, Jan, 17
Péter, Rózsa, 360, 411
Peter the Great, 359
Petersen, Anne, 1069
Peterson, Ivars, 221
Peterson, Julius, 455
Peterson, W. Wesley, 310
Petrarca, Francesco, 860
Petteia, 120

Peurbach, Georg von, 1096
Ph.D. programs, 1070–1071, 1072
Philadelphia Storage Battery Company (Philco), 96
philosophers
    dining philosophers problem, 211
    Euclidean, 357
    mathematical reasoning and, 847
    on nature/meaning of mathematics, 608–609
    Neoplatonists, 862
    René Descartes, 365–366
    Rithmomachia (Philosopher's Game), 121–122
    See also Aristotle; Plato of Tivoli; Socrates
Philosopher's Game (Rithmomachia), 121–122
Philosophiae Naturalis Principia Mathematica (Newton), 362, 602
Philosophical Dictionary (Voltaire), 847
Philosophy of Composition (Poe), 560
Phoenix Mathematics Inc., 509
photovoltaic cells, 930
phugoids, 30
Physicists, The (Durenmatt), 775
Pi, 55, 59, 450, 723, **770–771**, 1092, 1093, 1095, 1100
    See also circles; *Measurement of a Circle* (Archimede)
Pi (movie), 682
Pi Mu Epsilon, 208
Piaget, Jean, 540–541, 1043
Piaget's theory, 540–541, 1043
Picard, Charles, 724
Picard's little theorem, 724
Picasso, Pablo, 748
Piccard, Auguste, 295
Piccard, Jacques, 295
Pickett, George E., 193
Pickover, Clifford, 512, 730
Picone, Mauro, 1073
pie chart, 456
Pie Day, 450
Pierce, Benjamin, 35
Pierce, John, 890
Pierce, R. C., Jr., 371
Piero della Francesca, 861
Pike, Nicholas, 870
    The New and Complete System of Arithmetick, 870
Pincherle, Salvatore, 364
pioems, 779
Pioneering Women in American Mathematics (Green and LaDuke), 1072
Pirates of Penzance (Gilbert & Sullivan), 690
Pithoprakta (Xenakis), 231
Pitiscus, Bartholomaus, 1011, 1098
    Trigometria, 1011

Pixar Animation Studios, 50, 51
pixels, 660, 1013
Place of Mathematics in Modern Education, The (NCTM), 40
placeholders, 785
place-value structures, 154, 686, 712, 785
plain-text, 212–214
Planck, Max, 548, 837, 1027
Planck length, 1022
On Plane Equilibriums (Archimede), 1094
Planetary Flight Handbook (NASA), 881
planetary orbits, **771–773**, 890, 891
    See also Kepler's Laws
planimeters, 657–658
Plankalkül, 1077
planning departments. *See* **city planning**
plasma televisions, 987
plate tectonics, **773–774**
Plateau, Joseph, 966
Platinum Blue, 788
Plato of Tivoli
    Academy of, 424, 619, 1092
    Aristotle and, *357*
    Meno, 858
    oracle of Apollo at Delos and, 882
    religious writings, 858, 886
    The Republic, 173, 356–357, 858
    The Timaeus, 858, 886
Platonic solids, 782
Platonists, 621
Plato's Academy, 424, 619, 1092
Platzer, Maximillian, 1065
Playfair, William, 288, 696, 894–895
    The Commercial and Political Atlas, 456
Playfair Square, 212
Playfair's Postulate, 751
plays, **774–777**
Plimpton 322 Tablet, 828–829, 1091
Plücker, Julius, 249, 1103
plurality elections, 1047
plus (+) and minus (−) signs, 1096
Plutarch, 59, 763
plutonium bombs, 79–80, *80*, 81
pocket calculators, 1107
Poe, Edgar Allan
    Descent into the Maelstrom, 558
    The Gold Bug, 558
    Philosophy of Composition, 560
poetry, 596–597, **777–779**
Poincaré, Jules Henri, 360, 521, 531, 547, 803, 1005, 1042–1043
    "Analysis situs," 1105

Poincaré conjecture, 360, 1107
Poincaré disc model, 1043
Poinsot, Louis, 782
Poisson, Siméon-Denis, 707, 802, 1103
Pol Pot (Saloth Sar), 71
polar coordinate system, 353–354
polarized light, 547
Pollock, Jackson, 749–750
Pólya, George, 355, 360, 369, 707, 766
polygons, 779–782
polyhedra, 782–784
polynomials, 784–786
Pompeiu, Dimitrie, 361
Poncelet, Jean-Victor, 249, 281, 883, 1102
Pons, Stanley, 350
Ponte Vecchio (Italy), 130
Pontryagin, Lev, 404
Pontryagin's Maximum Principle, 404
Ponzi schemes, 759
pool. *See* **billiards**
popular music, 786–788
population growth, 385, 408
population paradox, 234
Porphyry
    Life of Pythagoras, 826
Porta, Giambattista della
    Natural Magic, 862
portable document format (PDF), 303
Portuguese Society of Mathematics, 365
Posidonius of Rhodes, 1063
positive rational numbers, 725
positron-emission tomography (PET), 126, 127
Post, Charles, 252
Post Cereal, 252
Post Office Protocol (POP), 518
Postel, Jonathan, 518
Potts, Renfrey, 1014
power centrality, 926
power laws, 791
Powerball, 563, 564
Powers, Kerns H., 986
Prairie style architecture, 1080, *1080*
Prandtl, Ludwig, 31
Prandtl–Glauert, 31
Pratchet, Terry
    Discworld, 902
prayer wheels, 1066
pre-calculus, 1098, 1099
Prechter, Robert, Jr., 599
Precious Mirror of the Four Elements (*Si Yuan Yujian*)
    (Zhu Shijie), 187

precolonial Africa, 23, 24
predator–prey models, 788–790
predicting attacks, 790–792
    National Security Agency and, 508
    National Security Agency (NSA), 39, 214, 229, 509
predicting divorce, 792–793
predicting preferences, 793–796
preference ballots, 1047–1048
pregnancy, 796–797
prehistory, 798–799
premiums, insurance, 506, 507–508
Preparing Mathematicians to Educate Teachers (PMET), 270
Presidential Medal of Freedom, 180
President's Council of Advisors, 526
pressurization, 26
Preyer, Lunsford Richardson, 287
Price, G. Baley, 1079
Price, Richard, 802
Primary Colors (Klein), 958
primary sampling units (PSUs), 1019
Prime Number formula, 722
prime numbers, 719, 720–721, 722, 883, 939
Prime Obsession (Derbyshire), 560
Primer (movie), 1082
Prince (Machiavelli), 861
Principia (Newton), 847, 1029, 1101
Principia Mathematica (Russell and Whitehead), 362,
    1074–1075, 1105
Principles and Standards for School Mathematics (NCTM),
    151, 222, 277, 867
Principles of Philosophy (Descartes), 366
printing, 1096–1097
prions, 673
Prisoner's Dilemma, 414, 983
private mortgage insurance (PMI), 483
probability, 800–803
    in baseball, 98–99
    in basketball, 99–100
    betting and, 105–107
    birthday problem, 113–115
    fraud detection and, 5
    Native Americans and, 699
    normal distribution theory and, 802–803
    objective and subjective approaches, 802
    study of, 800–802
    subjective, 106–107
    of survival, 535
    See also probability theory
Probability of Precipitation (PoP), 1056–1057
probability theory, 5, 112, 368–369, 507–508
    See also **probability**

On the Problem of the Rotation of a Solid Body about a Fixed Point (Kovalevsky), 1104

problem solving in society, 733–734, **804–808**

Proceedings of the London Mathematical Society (journal), 1104

Process Standards (NCTM), 151

Proclus Diadochus
    Commentary on Euclid, Book I, 1093
    Eudemian Summary, 1092

product method (GDP), 466

product-limit estimator, 1006

professional associations, 809–811

Professional Standards for Teaching Mathematics (NCTM), 277

Programme for International Student Assessment (PISA), 265, 278, 328, 1070

Progressive Education movement, 274–275

Project 8 (video game), 374

Project Gueledon, 172

Project Gutenberg, 303

ProjectCalc series calculators, 141

proof, 222–224, 611–612, **812–815**

Proof (movie), 163, 682, 1085

Proof (play), 775, 776

Proofs from THE BOOK (Ziegler and Aigner), 625

On the Propagation of Heat in Solid Bodies (Fourier), 987

protractors, 658

PROVERB computer program, 7

Provincial Letters (Pascal), 860

psychological testing, 815–817

Psychological Testing (Anastasi), 816

Psychology of Invention in Mathematical Field, The (Hadamard), 560

psychometrics, *816*, 816–817

psychophysics, 653–654

Ptorygou, Ouon, and Pelekys (Wine, Egges, and Hatchet) (Simias), 779

Ptolemy, Claudius
    The Almagest, 18, 18, 772, 1011, 1093, 1095
    arcs and curves table, 1011
    coordinate geometry and, 248
    Earth centered universe and, 78, 437–438, 460
    ecumene description, *573*
    Geographia, 1011
    table of chords, 54

Ptolemy I Soter, 1092

Ptolemy II Philadelphus, 1092

public key cryptosystems, 282, 786

Pujols, Albert, *477*

pulleys, 818–819

Purkinje, Johannes, 389

Putin, Vladimir, 385

puzzles, 819–823
    See also **mathematical puzzles**

pyramids, 423, 447

pyrotechnics. *See* **fireworks**

Pythagoras' Cave, 173

Pythagoras of Samos, 38, 364, *458*, 458–459, 727, 824, 825–826, 1091

Pythagorean and Fibonacci tuning, 823–825

Pythagorean numbers, 34, 121

Pythagorean School, 586, 624, **825–827**

Pythagorean theorem, 827–829
    algebra and, 38
    building structures and, 649–650
    coordinate geometry and, 248
    dissection proof of, 1094, 1095
    Greek mathematics and, 459
    irrational numbers and, 726, 727
    in Sulbausutras, 424

Pythagorean Triple, 828–829, 1091

Pythagorean tuning, 67

Q

Qin Jiushao
    Shushu Jiuzhang (Mathematical Treatise in Nine Sections), 186–187

Qin Shi Huang (emperor), 185, 1093

quadratic equations, 40, 785–786, 939–940, 1095, 1101

quadratrix, 1092

quadrature of the circle problem. *See* **circles**

quality control, 831–833

quanta, 836

quantitative literacy, 714

quantum computing, 548

quantum field theory (QFT), 548
    See also quantum mechanics

quantum groups, 532

quantum mechanics, 80, 548, 673, 775–776, 855

quarterback ratings, 398

quartic equations, 785–786, 1097

quartz crystal clocks, 203

quasi-empirism, 622

quaternions, 724, 1029–1030

Queneau, Raymond
    Cent Mille Milliards de poèmes (One Hundred Thousand Billion Poems), 779

Questi et inventioni diverse (Tartaglia), 390

Quetelet, Adolphe, 500, 707, 803, 894, 942, 1042
    A Treatise on Man and the Development of His Faculties, 653

Quetelet index, 122–123

queuing theory, 134–135, 211, 509, 1000
Quillen, Daniel, 228
quilting, 833–834
Quincy School, 897
Quine, Willard Van Orman, 914
Quipus, 493–494, 931–932
Quota Rule, 233, 234
Qur'an, 730
QWERTY keyboard calculators, 142

R
R (software), 291
R peak, 43
rabbit problem, 1095
Race to the Top (RTTT) program, 279
Racine, Father, 73
racquet games, 835–836
radar, *502, 503*
   See also **doppler radar**
radiation, 836–838
radio, 838–840, *839*
Radio Corporation of America (RCA), 96
radio-frequency identification (RFID), 577, 913
Radó, Ferenc, 361
Radon, Johann, 146
Rafaello Sanzio (Raphael). *See* Raphael
Raffles, (Thomas) Stamford, 71
Raffles Institute, 71
Raghavan, Prabhakar, 840–841
Rahn, J. H., 1099
railroads. *See* **trains**
Rajagopal, C. T., 73
Ramanujan, Srinivasa
   as amateur mathematician, 598, 599
   mathematics education and, 73
   number theory and, 262, 1074, 1105
   religion and, 600
Rameau, Jean-Philippe
   Treatise on Harmony, 229
Ramelli, Agostino, 862
"ramiform" pattern, 174
Ramsay, Michael, 320
Ramsey, Frank, 211
Ramsey theory, 211, 453–454
RAND Corporation, 36, 41, 119, 215, 414, 842
random access memory (RAM), 769
randomness, 571, 800, **841–843**
rankings, 843–845
Rao–Blackwell theorem, 119
Raphael
   School of Athens, 861

rate of change. *See* **function rate of change**
Raven's Matrices, 512
Rawls, John, 356
reactive transport (RT) models, 1054
read-only memory (ROM), 140, 769
Reagan, Ronald, 150
reality, measuring, 1027–1028
Realm of Algebra, The (Asimov), 36
reasoning and proof in society, 845–850
   abstractions/symbolism and, 847–848, 863
   Euclidean Logic and, 848–849
   legal arguments and, 849
   origins of mathematical proofs, 845–846
rebates. *See* **coupons and rebates**
Rebbelibs, 737, 738
Reber, Grote, 838
recipes. *See* **cooking**
Record system, 418
Recorde, Robert, 1097
recreational mathematics, 595
recursive functions. *See* **functions, recursive**
recycling, 850–853
red, green, blue (RGB) additive model, 515
redistricting, 443–445
redshift, 317
reduction to absurdity, 559, 1101
Reed, Lowell, 314, 320
Reed–Frost epidemic model, 314
Reed–Solomon codes, 309, 320, 784
Rees, Mina, 1079
Reeve, W.D., 40
reflexive theory, 509
Réflexsur la résolution algébrique des équations (Lagrange), 765
refraction, 547
Regelous, Stephen, 289
regimento das léguas (regiment of the leagues), 576
Register System, 418
Regression Towards Mediocrity in Hereditary Stature (Galton), 895
Regular Division of the Plane with Asymmetric Congruent Polygons (Escher), 355, 765
Reidemeister, Kurt, 531
Reidemeister moves, 531
Rejewski, Marian, 1077
relativity, 853–855, 1005
   See also Einstein, Albert; theory of relativity
religious symbolism, 855–857
religious writings, 857–860
Renaissance, 860–862
Rényi, Alfréd, 361, 518, 596

representation theory, 863
representations in society, 863–868
   in 21st century, 866
   internal/external structures, 863, *864*
   mathematics as language and, 867–868
   multiple approaches to, 863–865
   problem solving and, 866–867
   translational skills and, 865–866
Republic, The (Plato), 173, 356–357, 858
Research and Development Corporation (RAND).
   See RAND Corporation
Research Experiences for Undergraduates (REUs), 272
Resin Identification Codes, 852
resource intensity, 381
Résumé of Lessons of Infinitesimal Calculus (Cauchy), 144
resurrection of Jesus, 731
retirement planning. *See* **pensions, IRAs, and Social**
   Security
Reulle, David
   Conversations on Mathematics With a Visitor from Outer
     Space, 901
Revenue Act (1926), 757
Revere, Paul, 212, 869
Revolutionary War, U.S., 868–870
Rey, José-Manuel, 581
Reynolds, Osborne, 30, 1052
Reynolds, Simon, 787
Reynolds number, 30, 1052
Rheticus, George Joachim, 1097
Rhind papyrus
   construction and, 423
   discovery of, 146
   doubling/halving numbers and, 332
   estate description, 411
   linear equations and, 554, *554*
   origination of, 1091
   puzzles and, 819
   recreational mathematics and, 333
   See also Moscow papyrus
rhythms, 996–997
ribonucleic acid (RNA), 421
Riccati, Jacopo, 155
Ricci, Matteo, 187
Rice, Marjorie, 599
Richard the Lion Hearted (king), 648
Richards, Donald St. P., 167
Richardson, Lewis Fry, 791, 1055, 1064, 1075
Richman, Hal, 377
Richter, Charles, 370
Richter scale, 323–324
Rich-Twinn Octagon House, *780*

Ride, Sally, 870–872
   Exploring Our Solar System, 871
   The Mystery of Mars, 871
   To Space and Back, 871
   The Third Planet, 871
   Voyager, 871
Rider, Paul, 139, 141
Riemann, Bernhard
   contributions of, 201, 560, 724, 1104
   geometric formulations and, 334, 429, 438, 555, 848, 854
   Green's theorem and, 1030
   limits and, 144
   number theory problems and, 721
   Prime Obsession (Derbyshire), 560
Riemann, Hugo, 434
Riemann hypothesis, 201, 560, 721, 724
Riemann integral, 146
Riemannnanian geometry, 438
Riese, Adam, 1096
Riesz, Frigyes, 146, 360
right-brain learning, 541–542
Rind, Alexander Henry, 146
Ring, Douglas, 977
Ring of Fire, *442*
Rio Riot 20GB, *685*
Rior, Antonio Maria, 261
risk management, 872–874
risk pooling, 506–507
risk transfer, 506–507
risk-return relationship, 677
Rissanen, Jorma, 977
Rithmomachia (Philosopher's Game), 121–122
Rittenhouse, David, 674
river-crossing puzzle, 16
Rivest, Ronald, 213
RNA, 421
Robbins, Benjamin, 63, 64
Robert of Chester, 33, 1094
Roberts, Louis, 476
Robertson, Edmund, 221
Robertson, Malcolm S., 628–629
Roberval, Gilles Personne de, 354, 1099
Robins, Benjamin, 391–392, 671
   New Principles of Gunnery, 391
Robinson, Abraham, 144, 553
Robinson, David, 162–163
Robinson, Karl, 167
Robinson, Michael, 101
roBlocks construction system, 877
Robonaut 2, *876*
robots, 522, 678–679, **874–877**

Rockefeller, John D., 131
Roe v. Wade, 117
Roger, Everett, 378
Rohde, Douglas, 419
Rolle, Michel, 1100
roller coasters, 877–878
ROM. *See* read-only memory (ROM)
Roman mathematics, 8, 711–712, **878–880**
Roman numerals, 8, 418, 711–712, 862, 879
Romanian Master of Sciences, 361
Romanian National Olympiad, 361
Romanowski, Miroslaw, 708
Romig, H. G., 832
rond de jambe à terre (ballet), 90, 91
Röntgen, Wilhelm Conrad, 659, 837
Roosevelt, Franklin D., 79, 334, 755, 888
Rosa, Edward, 547
Rosales, Rodolfo, 488
Rosenblaum, Joshua
    Fermat's Last Tango and, 689–690
Rosencrantz and Guildenstern are Dead (Stoppard), 775
Rosnaugh, Linda, 781
Ross, Mary G., 880–881
Ross, Ronald, 314
rostro, 879
Rota, Gian-Carlo, 597
Roth, Klaus, 363
Rothermel, Richard C., 402
Rothschild, Linda, 811
Rotisserie League Baseball, 378
row operations, 634
Royal Air Force (RAF) Coastal Command, 1078
Royal Society of London, 115, 149, 1099
Royal Spanish Mathematical Society, 365
Royal Statistical Society, 119
Rozycki, Zerzy, 1077
RSA public key system, 213–214
Ruan Yuan, 187
"Rubaiyat" (Khayyam), 1095
Rubik's Cube, 765, 822
Rubin, Andrew, 977
Rucker, Rudy, 902–903
    Message Found in a Copy of Flatland, 902
    Spaceland, 902
Ruddock, Graham, 265–266
Rudin, Mary Ellen, 589
Rudin, Walter, 589
Rudolff, Christoff
    Die Coss, 1096
Ruffini, Paolo, 765, 786
"rule of circular parts," 1098

ruler and compass construction, 881–883
    classical problems of, 882
    Euclid and, 881–882
    proofs and, 883
    tool variations, 882–883
Rumsfeld, Donald, 510
Russ, Laurence
    The Complete Mancala Games Book, 13, 16
Russell, Bertrand, 219, 335
    Principia Mathematica, 362, 1074–1075, 1105
    Why I Am Not a Christian, 860
Russian Peasant algorithm, 686
Ruth, George Herman "Babe", 477

S
S&P 500, 692
sabermetrics, 97–99
Saccheri, Girolamo, 751–752
    Euclid Freed of Every Flaw, 1101
Sacred Cubit, 638–639
sacred geometry, 515, 534, 799, **885–886**
Saffir, Herbert, 492
Saffir–Simpson scale, 492, 1059
Sagan, Carl, 900
    Contact, 901–902
Saint Petersburg Academy, 359
Saint-Venant, Jean Claude, 156
Saint-Venant equations, 156
Saint-Vincent, Grégoire de, 1099
    Opus Geometricum, 354
Salamis stone tablet, 8
Salem, Coralie
    The Most Beautiful Mathematical Formulas, 612
Salem, Lionel
    The Most Beautiful Mathematical Formulas, 612
sales tax and shipping fees, 886–888
Sally Ride Science Academy, 871
Salmon, George, 602, 966
Samarkand Observatory, 66
al-Samaw'al, 54, 55, 111
Samphan, Khieu, 71
sample surveys, 888–890
Sampling Methods for Censuses and Surveys (U.S. Census Bureau), 176
Sanborn, John B., 117
Sand Reckoner, The (Archimedes), 60
Sandburg, Carl, 218
sangaku, 485
Santa Maria del Fiore, 315
Sar, Saloth (Pol Pot), 71
Sarigul-Klijn, Nesrin, 1065

Sarrus, Frederic, 500
Sarukkai, Sundar, 1028
SAT. *See* Scholastic Aptitude Test (SAT or SAT-M)
satellites, 890–892
    coastline mapping and, 571
    GPS and, 450–451, 452
    marine navigation and, 577
    military uses of, 503
    moon as, 677
satires, 559, 560
Sato, Mikio, 69
Satterthwaite, Mark, 1050
Saudi Association for Mathematical Sciences, 76
Savile, Henry, 1098
Savilian professorships, 1098, 1099, 1100
scaffolding, 542–543
scales, 823–824, **892–894**
scanning, *383*, 383
scanning tunneling microscope, 674
scatter diagram, 894
scatterplots, 894–895
Scènes de Ballet (Ashton), 90
Schattschneider, Doris, 427
scheduling, 896–897
Scheiner, Christoph, 964
Scheinman, Victor, 876
Scherk, Heinrich, 966
Scheubel, Johannes, 1097
Schild's Ladder (Egan), 559
Schliemann, Analucia, 226
Schmidt, Stefan, 510
Schmitt, Harrison H., *678*
Schneider, Robert, 230
Schoenberg, Arnold, 231
Schoenberg, Frederic, 403–404
Schoenberg, Isaac Jacob, 361
Schofield, Keith, 787
Scholastic Aptitude Test (SAT or SAT-M), 326, 328, 844,
    1069–1070
Scholasticism, 624, 625, 664
Scholes, Myron, 264
Schönhage, Arnold, 688
Schönhage–Strassen algorithm, 688
School of Athens (Raphael), 861
School of Mathematical and Navigational Sciences
    (Moscow), 358–359
Schoolhouse Rock! (television show), 982
schools, 897–899
    buxiban (cram schools), 69
    Columbine High School shootings, 897
    Frank Lloyd Wright School of Architecture, *1080*

    juku schools, 69
    Life Adjustment schools, 274–275
    Pythagorean School, 586, 624, 825–827
    Quincy School, 897
Schooten, Frans van (the Younger), 1099
Schor, Peter, 548
Schramm, Oded, 75
Schultheis, Michael, 748, 750
Schwabe, Heinrich, 964
Schwartz, Laurent, 367
Schwarz, Hermann, 653
Schwarz, Stefan, 360
Schwarzschild, Karl, 115, 853
Schweikardt, Eric, 877
Science, Technology, Engineering, and Mathematics
    Education (STEM), 133, *222*, 897, 958–959
    National Science Foundation and, 448, 449, 450
science, technology, engineering, and mathematics (STEM),
    540
science fiction, 899–903
    authors, 902–903
    literary, 559, 899–903
    mathematical, 900–902
    space ships and, 935
    visual media and, 903
Science Friday (radio program), 838
Science News (journal), 221
Science of Discworld, 902
Science of the Better, 807
Scientific American (magazine), 595
Scientific and Technological Research Council (Turkey), 75
scintigraphy, 659–660
Scipione del Ferro, 33
Scotus, Duns, 664
Scudder, Kathyrn, 325
sculpture, 903–905
    mobiles, *544*, 545
scurvy, 732
Sea Island Mathematical Manual, The (Haidao Suan Jing)
    (Liu Hui), 186
Sea Mirror of the Circle Measurements (Ce Yuan Hai Jing) (Li
    Zhi), 187
Seaborg, Glenn T., 358
search engines, 905–906
Search for Extraterrestrial Intelligence (SETI), 1028
Sears, Roebuck and Co., 141
Sears Tower, 920
seasonal extension (SARIMA), 1020
SeaWiFS Data Analysis System, *286*
Seba, Petr, 134
second, 656

secondary sampling units (SSUs), 1019
second-generation (2G) cell technology, 175
seconds pendulum, 639
security body scanners, *644*
Sefer Yetzira (Book of Creation), 730
Segway, 906–907
Seibold, Benjamin, 488
seismic tomography, 173
Seki Takakazu, 143
Selberg, Atle, 363
Seldon, Hari, 607
Selective Service, 665
semiconductor systems, 525
sequences and series, 908–910
Sequential Pairwise elections, 337
Séquin, Carlo, 904–905
Séquin–Collins Sculpture Generator, 904–905
Serra, Michael, 780, 781
Serre, Jean-Pierre, 367, 908
    Discovering Geometry: An Investigative Approach, 780
Serret, Joseph, 249, 281
servers, 910–912
Sesame Street (television show), 982
Sesostris, King, 428
Set Theory, 587, 1104, 1106
Seurat, Georges, 749, *749*
Seven Bridges of Königsburg, 129, 130, 366, 454–455, 775, 821
Seventeenth Night (Doxiadis), 776
severe acute respiratory syndrome (SARS), 510
sexagesimal notation, 87–88, 89
sextant, 658
Shafarevich, Igor, 360
Shafer, Alice, 811
Shakespeare, William, 957, 958
Shamir, Adi, 213
Shanks, William, 909
Shannon, Claude, 212, 387–388, 768, 803, 843, 977, 1068
Shapely, Lloyd, 94, 582
Shattuck, Lemuel, 285, 288, 942
Shelah, Saharon, 75
Shelley, Mary, 877, 900
    Frankenstein, 877, 900
shells and mortars, 393
Sherman, Julia, 961
Sherman, William Tecumseh, 193
Shewhart, Walter, 287, 298
    Statistical Methods from the Viewpoint of Quality Control, 832
Shewhart, Walter E., 831–832
Shewhart control charts, 832

Shimura, Goro, 1062
Shipman, Barbara, 105
shipping, 912–914
Shisima, 16
Shockley, William Bradford, 768
Shor, Peter, 769
Shrapnel, Henry, 63
shuffle function, 685
Shuli Jingyun (ed. Mei Juecheng), 187
Shushu Jiuzhang (Mathematical Treatise in Nine Sections) (Qin Jiushao), 186–187
SI units, 349, 640, 643, 1021–1022, 1024
SIAM News (journal), 467
Siddhanta Siromani (Bhaskaracharya II), 73
Siegel, Carl, 367
Sielbee, J. Lyman, 1080
Sierpinski, Waclaw, 113, 1074
Sierpinski's Triangle, 113
sieve, 1092
Sieve Of Eratosthenes, 18
siffusion spectrum imaging, 126
Signal Intelligence Service, 756
signal processing, 209–210, 211, 680–681, 909
Silver, Bernard, 96
Simias of Rhodes
    Pteryges, Ooon, and Pelekys (Wine, Egges, and Hatchet), 779
similarity, 914–915
Simon Personal Communicator, 977
Simple Mail Transfer Protocol (SMTP), 518
Simplicius, 1093
Simpson, George, 1064
Simpson, Thomas, 369, 492, 654
Sinclair, John
    Statistical Accounts of Scotland, 285, 942
"Singapore Math Method", 71
Singapore Mathematical Olympiad, 71
Singapore Mathematics Project Festival, 71
Singer, Isadore "Iz", 211
Singh, Simon
    Fermat's Enigma, 560
Single Photon Emission Computed tomography (SPECT), 127–128
sinoatrial node (SA node), 745
Sino-Korean number system, 69
SIR model, 1039
Six Arts (Liu Yi), 185
Six Degrees of Kevin Bacon, 915–916, 926
    See also Erdös number
Six Sigma Black Belt, 285, 287–288, 633, 916
SixDegrees.org, 916

skating, figure, **916–917**
Sketchpad, 273, 427
Skewes, Stanley, 19
Skewes number, 19
skewness, 369, 655
skydiving, 918–919
skyscrapers, 919–920
Slint
    Spiderland, 787
Sluze, René François Walter de, 1099
"small world" phenomenon, 915
smallpox, 313
SMART board, 920–921
smart cars, 922–923
Smarter Planet campaign, 12
smartphones, 977–978
Smith, David, 385
Smith, Edmund Kirby, 193
Smith, George, 307
smoothbore artillery. *See* **artillery**
Snedecor, George, 943
Snell's Law, 547
Snow, John, 188, 288, 313, 1042
snowflakes, 259–260
Sobolev, Sergei Lvovich, 360
soccer, 923–924, 999
Social Choice and Individual Values (Arrow), 336
social choice theory, 232, 510, 1047
Social Constructivism, 621–622
social network analysis (SNA), 508–509
social networks, 508–509, **924–926**
Social Security. *See* **pensions, IRAs, and Social Security**
socialism, 217–218
Sociedad Matemática Mexicana (Mexican Mathematical
    Society), 179
Société des Sciences Naturelles et Physiques du Maroc, 19
Society for American Baseball Research (SABR), 97
Society for Industrial and Applied Mathematics (SIAM),
    163, 170, 207, 229, 604–605, 810
Society for Mathematical Biology, 592
Society for Mathematical Psychology, 592
Society for Technical Communication, 221
Society for the Advancement of Chicanos and Native
    Americans in the Sciences (SACNAS), 170, 811
Society of Actuaries, 285
Society of Women Engineers, 881
Socrates, 268, 1092
Socratic Method, 268
software, mathematics, 273, **927–929**
solar panels, *930*, **930–931**
Soldo, Fabio, 791

solid-state drivers (SSD), 309
Solomon, Gustave, 309, 320
Solzhenitsyn, Aleksandr
    The First Circle, 558
Somayagi, Nilakantha, 345
"Some Properties of Markoff Chains" (Blackwell), 119
"Some Results Related to the Generators of Cyclic Codes
    Over Zm" (Beaugris), 167
Somerville, Mary Fairfax, 565, 1103
Somnium (Kepler), 900
Sophocles, 6
Sorby, Sheryl, 1069
soroban, *68*, 69
    See also abacus
sound telescopes, 980
South African Mathematics Olympiad, 20
South America, 931–933
Southern Africa Mathematical Sciences Association, 20
Soviet Union, 67, 214, 217, 275, 1106
    See also **Asia, Central and Northern (Russia); Europe,
    eastern**
To Space and Back (Ride), 871
space elevators, 347, 348
Spaceland (Rucker), 902
spaceships, 522, 678–679, 870–871, **933–935**
space-time geometry, 115, 116, 1005
spam filters, 935–936
Spartan ciphers, 212
speech recognition, 165
speedometers, 242–243
speleology, 172
spelunking, 172
Spencer (rifle), 191
sperm count, 385–386
Sperry, Roger, 1043
Sphaerica (Menelaus), 425, 1011, 1093
Sphaerica (Theodosius), 1094
On the Sphere and Cylinder (Archimede), 1094
On the Sphere and Cylinder (Archimedes), 1004
Sphereland (Burger), 902
spherical geometry, 54, 1011
spherical trigonometry, 1093, 1094–1095
sphygmometer, 948
Spiderland (Slint), 787
Spinoza, Baruch, *357*
spiral of Archimedes. *See* Archimedes' screw/spiral
spontaneous order. *See* **synchrony and spontaneous
    order**
Sporer, Thomas, 684
sport handicapping, 936–938
sports, 97–99, 239, 241–242

sports arenas. *See* **arenas, sports**

sports engineering/equipment, 375

Sports Illustrated (magazine), 704

Sputnik, 275, 276, *359*, 359, 449, 709, 812, 1106

square dancing. *See* **contra and square dancing**

square wheel bike, *109*

squares and square roots, 243, **938–940**

squaring a double-digit number, 594–595

squaring of the circle problem. *See* circles

SR-52, 140

stable marriage problem, 582

Stable Matching Algorithm, 582

Stager, Anton, 192

stained glass windows, 486, 487

stalactites and stalagmites, 940–941

Stam, Jos, 966

Stancel, Valentin, 932

Stand and Deliver (movie), 683

Standard and Poor's 500 (S&P 500). *See* S&P 500

"A Standard City Planning Enabling Act" (U.S. Dept. of Commerce), 190

Standard Model of Particle Physics, 342

"A Standard State Zoning Enabling Act" (U.S. Dept. of Commerce), 190

standardized measurements, 648

Stanford arm, 876

Stanford–Binet test, 511

Stanley, Julian, 1043

Stanton, Elizabeth Cady, 849

star compasses, 575

star maps, 575

Star Messengers (Zimet and Maddow), 690

Star of David, 856

star polygons, 1095

Star Trek (television show), 877, 903, 984

Star Wars (movies), 51, 899, 903

Stark, Harold, 599

Stark–Heegner theorem, 599

stars, collapse of. *See* **black holes**

stars, symbolism of, 855–856, *856*, 886

START II treaty, 767

static budget, 130–131

Statistical Accounts of Scotland (Sinclair), 285, 942

Statistical Averages (Zizek), 653

Statistical Methods from the Viewpoint of Quality Control (Shewhart and Deming), 832

statistical process control (SPC), 831–832

statisticians, 284–287

statistics
    problem solving and, 805–806
    Rao–Blackwell theorem, 119

sports, 97–99, 241–242
    survival analysis, 1006
    See also **data analysis and probability in society**

statistics education, 941–946

Staudt, Karl Georg Christian von, 1103

Steadman, Robert, 1058

steam engines, 460, 499, *499*, 500, 1063

Steel and Foam Energy Reduction (SAFER) barrier, 83

Stefan, Josef, 988

Steiner, Jakob, 762, 883, 1103

Steinhaus, Hugo, 95

STEM. *See* Science, Technology, Engineering, and Mathematics Education (STEM)

Ste-One, 51

step and tap dancing, 946–947

stereoblindness, 1043

stereotypes in mathematics, 630–631, 960–961, 981, 983–985, 1043

stethoscopes, 947–949

Stevin, Simon, 726, 1097–1098

Stewart, Ian, 607, 902

stick charts, 737, 738

Stifel, Michael
    Arithmetica Integra, 111, 858, 1096–1097
    Book of Arithmetic About the Antichrist, A Revelation in the Revelation, 858

Stobaeus, Joannes
    Extracts, 619

Stochastic calculus, 147, 231, 442, 696

stock market indices, 949–951

Stockhausen, Karlheinz, 231

stocks/stock market, 93, 288, 400, 691, 692, 696, 949–951
    See also **mutual funds**

Stoilow, Simion, 361

Stokes, George, 30, 591, 995, 1030, 1052

Stokes' theorem, 1030

Stonehenge, 767

stonemasons arithmetic, 8

Stoppard, Tom
    Arcadia, 774–775, 776, 777
    Rosencrantz and Guildenstern are Dead, 775

Story of Your Life (Chiang), 901

Strassen, Volker, 688

Strategic Air Command (SAC), 215

strategic voting, 1049–1050

strategy and tactics, 951–954

Strat-O-Matic, 377

stratus clouds, 207

Stravinsky, Igor, 90

Straw, Barry, 1038

street maintenance, 954–956

Strengthening Underrepresented Minority Mathematics Achievement (SUMMA) program, 669, 670

string instruments, 956–957

Strode, Thomas, 764

Strogatz, Steven, 388, 609–610, 973

Structure and Distribution of Coral Reefs (Darwin), 250

Stuart, J.E.B., 193

Student (William Gossett). *See* Gossett, William (pseud. Student)

Studies in Conflict and Terrorism (journal), 509

Study, Eduard, 35

Stylianides, Andreas, 223

stylometry, 957–958

Su, Francis, 221

Suan Shu Shu. See Nine Chapters on the Mathematical Art (Chinese text)

Suanfa Tongzong (General Source of Computational Methods) (Cheng Dawei), 187

subitizing, 714

submarines. *See* **deep submergence vehicles**

subsonic speeds, 27

subtraction. *See* **addition and subtraction**

succeeding in mathematics, 958–962

    achievement measurements and, 959

    failure/anxiety and, 959–960

    organizations and, 961–962

    research on, 960–961

    STEM and, 958–959

    stereotype threat and, 960–961

sudoku, 962–963

sukkah, 487

Sulbasutras, 424, 646, 762

    See also **Vedic mathematics**

Sullivan, Arthur

    Pirates of Penzance, 690

Summa de arithmetica, Geometria, Proportioni et Proportionalita (Pacioli), 2, 289, 754, 800–801, 1096, 1099

sunspots, 963–964

supercomputers, 1107

superconducting quantum interference devices (SQUIDS), 127

superencypherment, 1076, 1077

superposition, 1004

Supply of and Demand for Statisticians , 944

support vector machines (SVMs), 791

Supreme Court decisions, 117–118

Surely You're Joking, Mr. Feynman (Feynman), 682

surfaces, 964–966

surgery, 966–968

Surplus Phoenix missiles, *671*

Surveyor's formula, 781

survival analysis, 1006

survival rates. *See* **disease survival rates**

Sutras, 1031–1033

Swanson, Irena, 834

swastika, 857

sweat lodges, 486, 487

Swerling, Peter, 671

Swift, Jonathan

    Gilliver's Travels, 559

    A Modest Proposal, 559

swimming, 969–970

Swinburne, Richard, 731

Sylow, Peter

    Théorèmes sur les groupes de substitutions, 766

Sylow theorems, 766

Sylvester, James Joseph, 455, 500, 1005, 1104

symbolism

    abstractions and, 847–848

    logic and, 811

    religious, 855–857, *856*

    representations and, 863, 867–868

    of "zero", 186, 493, 494

symmetry, 435–436, 970–972

Synagoge (The Collection) (Pappus), 425

synchrony and spontaneous order, 972–974

Syntaxis mathematica. See Almagest, The (Ptolemy)

Synthesis Report, 2007 (IPCC), 194

synthetic projective geometry, 1099

Syracuse, 58, 60

Systeme Internal d'Unites (SI), 349, 640, 643, 1021–1022, 1024

T

Taichi, 185

Taimina, Daina, 258, 905

Tait, Peter, 1029–1030

Taliesin West, *1080*

tally marks, 714–715, 798–799

tallying systems, 1091

Talon, Jean, 177

Tan, Tony, 71

tangents, 144–145

Taniyama, Yutaka, 1062

Taniyama–Shimura Conjecture, 1062

    Fermat's Last Tango and, 689–690

Tanner, C. Kenneth, 898

Tao, Terence, 735, 975–976

tap dancing. *See* **step and tap dancing**

Tapia, Richard, 669

Tarski, Alfred, 587

Tartaglia, Niccolò, 33, 261, 671, 786, 820, 1096, 1097
    La nova scientia, 390
    Questi et inventioni diverse, 390
tartan setts, 989–990
Taschenuhr, 202
Taussky-Todd, Olga, 635
taxes. *See* **income tax**
taxonomies, 44
Taylor, Brook, 784, 1100
Taylor, Richard, 749, 1062
Taylor polynomials, 784
TCP/IP protocols, 180
telegraphy, 191, 192, 1055
telephones, 976–978
telescopes, 238, **978–981**, 1098
television, mathematics in, 981–985
    children's programming, 982
    reality/game shows, 982–983
    stereotypic nerds and, 981, 983–985
televisions , 985–987
Teller, Edward, 79
temperature, 987–988
    See also **thermostats**
tempered scale, 433–434
Temple, Blake, 350
temples, 485–486
Ten Books on Architecture, The (Vitruvius), 886
Ten Computational Canons (ed. Li Chunfeng), 186
10-10-80 principle, *131*, 131
tephra, 1044–1045
Tepping, Benjamin, 890
terrorist attacks. *See* **predicting attacks**
terrorist cells, 509–510
Tertium Organum (Ouspensky), 627
Tesla, Nikola, 838
tessellations
    in applied design, 513, 514–515, 833–834, 970
    in Islamic architecture, 627
    M.C. Escher and, 781
    pattern blocks and, 103–104
    in stained glass windows, 428
    See also patterns
Testard, Frédéric
    The Most Beautiful Mathematical Formulas, 612
TeX, 221
Texas Instruments, 140, 141, 273
text conversion. *See* **coding and encryption**
textile industry, 990
textiles, 989–990
Thabit ibn Qurra, 1005, 1094
Thales of Miletus, 323, 364, 424, 458, *458*, 914, 1091

"That's Mathematics" (Lehrer), 787
Theaetetus of Athens, 782, 1092
"theaters of machines" (Besson and Ramelli), 862
Theodorius of Cyrene, 1092
    Sphaerica, 1094
Theon of Alexandria, 1093
Theophrastus
    The Book of Signs, 1055
theorema Egregium of Gauss, 965
Théorèmes sur les groupes de substitutions (Sylow),
    766
Théorie Analytique des Probaés (Laplace), 802
Théorie de la figure de la Terre (Clairaut), 1101
Theory of Celestial Movement (Gauss), 707
Theory of Committees and Elections, The (Black), 335
Theory of Everything, 440, 441
Theory of Intervals (Euclid), 433
theory of linkages, 500
theory of proportion, 1092
theory of relativity, 38, 78, 80–81, 115, 429, 474, 848,
    854–855, 980, 1074, 1105
Theory of Sets of Points, The (Young and Young), 1105
thermodynamics, 116
thermometers, 988, 1055
thermoscopes, 988
thermostats, 990–991
    See also **temperature**
"thinking fluid", 62
Third International Mathematics and Science Study
    (TIMSS), 328
Third Planet, The (Ride), 871
third-generation (3G) cell technology, 175
Thom, René, 365, 367
Thomas Aquinas, Saint, 664, 957
Thomson, William (Lord Kelvin), 988, 994, 1030
Those Parts of Geometry Needed by Craftsmen (Abu'l Wafa),
    627
three construction problems, 459
three-body problem, 909
3-D graphing calculators, 142
3-D televisions, 986
Thurston, William, 431–432, 1042
Tian Yuan Shu (Method of Coefficient Array), 186, 187
Tiao, George, 401
Tic-Tac-Toe, 992–993
tides and waves, 993–995
Tiedt, Iris, 779
tilings, 103–104, 699, 833–834, 886
    See also patterns
Timaeus (Plato), 858
Time Machine, The (Wells), 900, 902

time series data, 1020

time signatures, 995–997

time zones, 1002

Timeas, The (Plato), 886

timekeeping, 24, 201–204, 655–656, 746

   See also calendars

timescales, *655*, 655–656

time-series procedures, 807

timetables (train), 1002

time-to-pregnancy (TTP), 796

Tippet, Leonard, 842

Tirthaji, Sri Bharati Krsna

   Vedic Mathematics, 1032

Tissot, Nicolas Auguste, 574

Titeica, Gheorghe, 361

Title IX, 1071–1072

Tits, Jacques, 367

TiVo, 320

Tlatelolco massacre, 170

Toba volcano, 1045

Tobacco Masaic Virus (TMV), 1038

Toffler, Alvin, 402

toilets, 997–998

toleta de marteloio, 575–576

Tolstoy, Leo

   War and Peace, 557

tonal harmonies, 434–435

Tonnetz (Tonal Network), 434–435

tools, measuring. *See* **measuring tools**

Topics in Mathematical Modeling (Tung), 196

topographic maps, 345–346

topoisomerases, 532

topological puzzles, 821

topological quantum field theory, 531

topology, 119, 531, *534*, 613

tornadoes. *See* **hurricanes and tornadoes**

Torricelli, Evangelista, 364, 763, 1055, 1058, 1063, 1099

   Gabriel's Horn and, 505

total quality management (TQM), 832

Tour de France, 108, 109

tournaments, 7, 998–999

tower clocks, 202

Tower of Hanoi puzzle, 412, 593–594

town planning. *See* **city planning**

toxicity, 535–536

traditional versus reformed calculus, 151–152

traffic, 1000–1001

trains, 1001–1004

Traité de mécanique céleste (Laplace), 1103

Traité du calcul-intégral (Bougainville), 869–870

Traité du Triangle Arithmétique (Pascal), 112, 1099

trajectories

   firearms and, 390–391, *391*

   fuel consumption and, 404

   learning, 542–543

   See also **learning models and trajectories**

trampolining, 470

transformations, 1004–1005

   compositional, 230–231

   linear, 1004–1005

   Lorentz, 853–854

Transmission Control Protocol (TCP), 181, 911

transplantation, *1006*, **1006–1007**

transport planners, *1008*, 1008–1009

Transportation Security Administration (TSA), 643

Transvaal, 19

transverse flow effect, 475

travel planning, 1007–1009

traveling salesman problem (TSP), 912, 1009–1010

Traveller's Dodecahedron, 122

Travers, Jeffrey, 915

Treatise on Algebra (Peacock), 1103

Treatise on Demonstration of Problems of Algebra (Khayyam), 237

Treatise on Electricity and Magnetism (Maxwell), 1029

Treatise on Harmony (Rameau), 229

A Treatise on Man and the Development of His Faculties (Quetelet), 653

Treatise on the Circumference (al-Kashi), 73

Treatise on the Quadrilateral (al-Tusi), 54

Treaty of the Meter, 639–640

Tremain, Janet, 211

trench codes, 1074

   See also cryptography

Trends in International Mathematics and Science Study (TIMSS), 76, 133, *265*, 265, 266

Trends in Mathematics and Science Survey, 1070

Triangle Law, 1030

triangles, measuring. *See* **Pythagorean theorem;**

   trigonometry

triangulation, 1012–1013

Trieste, 295

Trigometria (Pitiscus), 1011

trigonometric functions, 723, 1097, 1098

trigonometric series, 764, 909

trigonometry, 1010–1013

   Arabic/Islamic mathematics and, 54–55

   functions, 723, 1097, 1098

   series, 764, 909

   spherical trigonometry, 1093, 1094–1095

   See also **Pythagorean theorem**

Trigrams, 185

trilateration, *451*, 451–452, 1012–1013
"Trinity" (nuclear bomb test), 79, *80*
Triparty en la science des nombres (Chuquet), 1096
trisection problem, 1092
tropical forests. *See* **deforestation**
True Histories (True Tales) (Lucien), 900
Truman, Harry, 449
Truth Values (De Cari), 777, 1083
Tsiolkovsky, Konstantin, 347
Tsoro Yematatu, 16
Tsu Ch'ung Chih, 1093
Tsuchiya, Tom
   Atlas Recycled, 851
tube resonators, 1065–1066
Tucker, Alan, 201
Tukey, John, 944
Tung, Ka-Kit
   Topics in Mathematical Modeling, 196
Tunisian Mathematical Society, 19
Tunnel of Samos, 1014, 1015
tunnels, *1014*, **1014–1015**
Turing, Alan, 508, 517, 777, 1078
Turkish Mathematical Society, 75
Turner, Claude, 35
Turner, Joseph, 749
al-Tusi, Nasir, 73
   Treatise on the Quadrilateral, 54
Twenty Thousand Leagues Under the Sea (Verne), 900
21 (movie), 682–683
Twersky, Victor, 837
Twin Prime Conjecture, 721
twining, 102
twistor theory, 618
Twixt, 122
two container problem, 594
Two New Sciences (Galileo), 545
2001: A Space Odyssey (Clark) (novel), 901
2001: A Space Odyssey (movie), 903
two-year colleges, 272–273, 540, 811

U
UK Census, 177
Ulam, Stanislaw, 79, 80
Ulrish, Richard, 905
ultrasound, **1017–1018**
Ulughbek, Muhammad Taragay (Abd al-Latif), 66
Unabomb Manifesto (Kaczynski), 958
unemployment, estimating, **1018–1020**
Uniform Penny Post, 807
Unimaginable Mathematics of Borges' Library of Babel, The (Bloch), 557

Union Canal Company, 155
Union of Concerned Scientists, 891
United Midget Auto Racing Association (UMARA), 703
United Nations Commission on Statistical Sampling (U.S. Census Bureau), 176
United Nations Framework convention on climate Change, 295
United Nations World Population Prospects, 546
United States Auto Club (USAC), 703
units of area, **1020–1021**
units of length, **1021–1023**
units of mass, **1023–1024**
units of volume, **1024–1026**
Universal Automatic Computer (UNIVAC), 768
universal constants, **1026–1027**
universal language, **1027–1028**
Universal Product Code (UPC), 96
Universal Time (UT), 656
Universe in a Nutshell, The (Hawking), 474
University of Alexandria, 1092
University of Maryland Mathematics Project, 275–276
University of Stellenbosch, 20
Unreasonable Effectiveness of Mathematics in the Natural Sciences, The (Wigner), 609, 612, 619, 620
al-Uqlidisi, Abu'l-Hasan, 712
uranium bombs, 79–80, 81
US News and World Report, 843
U.S. Army Signal Corps, 191, 490
U.S. Bureau of Justice Statistics, 287
U.S. Bureau of Labor Statistics, 40, 286
U.S. Census Bureau
   data collection, 286, 718, 890
   history of, 177–178, 889
   unemployment estimates and, 1018, 1019
   See also **census**
U.S. Centers for Disease Control and Prevention (CDC), 288, 314
U.S. Coast Guard, 452
U.S. customary system of units of length, 1021–1022
U.S. Department of Agriculture, 286
U.S. Department of Commerce, 190, 466
U.S. Department of Defense, 181
U.S. Department of Education, 449, 898
U.S. Department of Energy, 350
U.S. Department of Health and Human Services, 288
U.S. Department of the Navy, 131
U.S. Electoral College, 336, 338–339
U.S. Environmental Protection Agency (EPA), 1053–1054
U.S. Federal Bureau of Investigation (FBI), 287, 293, 898
U.S. Geological Survey (USGS), 346
U.S. Internal Revenue Service (IRS), 287

U.S. Joint Space Operations Center, 891
U.S. Military, 665, *666*
U.S. Military Academy, 193, 268, 429, 501
U.S. Mint, 674
U.S. Missile Defense Agency, 672
U.S. National Bureau of Standards, 203
U.S. National Center for Health Statistics, 287, 469
U.S. National Collegiate Mathematics Championship, 222
U.S. National Medal of Technology, 180
U.S. National Oceanic and Atmospheric Administration (NOAA), 206
U.S. Naval Academy, 194
U.S. Nuclear Regulatory Commission (NRC), 526
U.S. Postal Service, 912, 913
U.S. Safe Drinking Water Act, 1054
U.S. Space and Rocket Center, 208
U.S. Supreme Court, 117–118, 178, 190
USA Junior Mathematical Olympiad (USAJMO), 227
USA Mathematical Olympiad, 7
USA Mathematical Talent Search (USAMTS), 227
USA Today (newspaper), 242, 510, 871
On the Use of the Indian Numerals (al-Kindi), 712
Usonian design, 1081
Utopia (More), 900
Uttal, David, 325
Uzelac, Tomislav, 684

V
vaccination campaigns, 1039
Vaidyanathaswamy, Ramaswamy, 73
Valian, Virginia, 1071
Vallée Poussin, C. J. de la, 1105
Van Allen belts, 522, 837, 891
Van Allen, James, 522, 837
Van de Hoop, Maarten, 773
Van der Hilst, Robert, 773
van Gogh, Vincent, 749
van Heuraet, Hendrik, 763
van Hiele model, 541, 1043
van Hiele, Pierre, 541, 1043
van Hiele-Geldof, Dina, 541, 1043
Vandenburg, steven, 1043
Varahamihira, 1093
Varignon, Pierre, 354
Vasco da Gama Bridge (Portugal), 130
Vassiliev, Victor, 531
VCRs, 320
Veblen, Oswald, 1073
vectors, 1029–1031
Vedic mathematics, 72, 626, 646, 778, **1031–1033**
   See also Indian mathematics; Sulbasutras

Vedic Mathematics (Tirthaji), 1032
Vedic scholars, 72
Velez-Rodriguez, Argelia, 167
velocity of light, 1026–1027
vending machines, 1033–1034
Venn, John, 113, 455, 802, 1042
Venn diagrams, 113, 455, 1042
VENONA, 1076
Verdu, Sergio, 1068
Verizon Communications, 181
Verne, Jules
   All Around the Moon, 521
   Around the World in Eighty Days, 900
   From the Earth to the Moon, 521, 900
   A Journey to the Center of the Earth, 900
   Twenty Thousand Leagues Under the Sea, 900
vertical curves, 476–477
vibrating string problem, 410
Vico, Giambattista, 542
Victoria Martin (Wallet), 777
video games, 1034–1037
   geometry in design, 1035
   programming, 1035–1037, *1036*
videocassette recorders (VCRs). *See* VCRs
Vidinli Hüseyin Tevfik Pasha, 75
Viète
   In artem analyticam, 1097
Viète, François, 111, 1096–1097
Viète, FrançoisIn, 1097
Vietnam War, 667, **1037–1038**
Vietnamese Mathematical Society, 71
Vigenère, Blaise de, 192–193
Vigenère cipher, 192–193, 212
vigesimal and duodevigesimal number system, 493
Village Power Optimization Model for Renewables (ViPOR), 1065
Villani, Cedric, 367
Vinge, Vernor, 769
Vinton, Samuel, 232
Vinton's Method, 232
Virginia Tech (VT) shootings, 897, 898
virtual reality, 741
virtual simulations. *See* **medical simulations**
viruses, 1038–1039
Visa, 253–254
vision correction, 1039–1041
visual perception, 739–741
visual symmetry, 970, *971*
visualization, 515, **1041–1044**
   See also optics
visual-spatial skills, 539

vitamins, 732, 734
Vitruvian Man (Leonardo da Vinci), 200, *200*, 650, 886
Vitruvius (Marcus Vitruvius Pollio), 514, 1004
  The Ten Books on Architecture, 886
Viviani, Vincenzo, *202*, 1099
vocal cords, 1065
voice, human, 1065
volcanos, 1044–1045
volleyball, 1046–1047
Volta, Alessandro, 209, 341
Voltaire
  Philosophical Dictionary, 847
Volterra, Vito, 789, 1073
volts (v), 341
von Hippel, Paul, 655
von Koch, Helge, 281, 763
von Neumann, John, 413, 548, 671, 768, 1078, 1107
Voronoi diagrams, 430–431
voting. *See* **elections**
voting methods, 1047–1050
Voyager (Ride), 871
Vranceanu, Gheorghe, 361

W
al-Wafa Buzjani, Abu. *See* Abu al-Wafa Buzjani
Wagner, David, 842
Wainer, H., 969
Wake, W. C., 957
Walden, Byron, 7
Waldo, c. A., 449
Walker, Augustus, 329
Wall Street crash, 696
Wall Street Journal (newspaper), 690
Wallace, David, 401
Wallace, Henry, 943
Wallace, John, 111
  De Algebra Tractatus, 112
Wallet, Kathryn
  Victoria Martin, 777
Wallis, John, 601
  Arithmetica infinitorum, 1099
Walt Disney Studios, 50
Walter Reed Army Medical Center, *1006*
Wang Xiao-tong, 186, 1094
  Jigu Suanjing (*Continuation of Ancient Mathematics*), 186
Wantzel, Pierre Laurent, 883
War and Peace (Tolstoy), 557
War Between the States. *See* **Civil War, U.S.**
war contracts, 1079
War of Independence. *See* **Revolutionary War, U.S.**
War of the Worlds (Wells), 521

"war on terror," 508–510
War Policy Committee (War Preparedness Committee), 755
Waring, Edward, 262, 1005
Waring–Goldbach, 262
Warlpiri kinship system, 736
Warnock, John, 303
Warring States period (China), 8, 185
Warschawski, Stefan E., 360
Washington, George, 234, 563, 870
water distribution, 1051–1053
water footprints, 1052
water quality, *1053*, **1053–1054**
Water Works Bureau, 156
waterwheels, 380
Watson, James D., 674
Watson, Thomas, 718
Watt, James, 341, *499*, 500, 1063
Watts, Duncan, 388
Wave Model (WAM), 492
wave theory of light, 547
wavelet analysis, 293–294, 389–390
waves. *See* **tides and waves**
Weak Law of Large Numbers, 803
weather forecasting, 492, **1054–1057**
Weather Prediction by Numerical Process, 1064
weather scales, 492, **1057–1059**
Weaver, Warren, 1079
weaving, 21–22, 875, 989–990
Web based communication, 220–221
Web comics, 219
Web crawlers, 906
Web search engines. *See* **search engines**
Web servers. *See* **servers**
Web sites, 11–12, 226
  See also specific Web sites
Webern, Anton, 231
Webster, Daniel, 233, 234
Webster's Method, 233, 234
Wechsler, David, 511
Wechsler Adult Intelligence Scale, 511
Wechsler Intelligence Scale for Children, 327, 511
Wedgwood, Josiah, 2
Wegener, Alfred, 773, 774
Weibull, (Ernst) Waloddi, 803, 1064–1065
Weibull distribution, 1064–1065
Weierstrass, Karl, 35, 144, 588–589, 596, 762, 1103
Weierstrass Definition, 144
Weight Watchers Online program, 733
weighted voting, 1047
weightless flight, 1059–1060, *1060*
Weil, André, 366–367

Weiner, Norbert
    Cybernetics, 875
WEKA, 291
Weldon, Walter, 707
Weldon, William, 942
Wells, Herbert George (H. G.), 942
    The Invisible Man, 900
    The Time Machine, 900, 902
    War of the Worlds, 521
Wells, Katrina, 781
Wenninger, Magnus, 783
Werner, Wendelin, 367, 808
Wernicke, Carl, 128
Wernicke's (temporal lobe), 128
Wessel, Caspar, 781, 1101
West African Examinations Council, 22
Westat, Inc., 890
Westbrook, Jeff, 1082, 1083
Westphal, Heinrich, 636
Weyl, Hermann, 39
What Do You Care What Other People Think (Feynmann),
    682
"What's New" blog, 976
wheel, 1060–1061
Whish, Charles, 73
White, Shaun, 374
"White City" (World's Columbian Expo. of 1893 in
    Chicago), 189, *189*
"white matter," 125
Whitehead, Alfred North, 40
    Principia Mathematica, 362, 1074–1075, 1105
Whitmore
    Breaking the Code, 777
Whitney, Eli, 287
whole-tone scales, 894
Why I Am Not a Christian (Russell), 860
Whyte, Frederick, 1002
Whyte system, 1002
wide augmentation system (WAAS), 452
Widman, Johann, 1096
Wiebe, Edward, 325
Wien, Wilhelm, 349, 836–837
Wiener, Norbert, 147, 768, 1030, 1073
Wigner, Eugene
    *The Unreasonable Effectiveness of Mathematics in the
        Natural Sciences*, 609, 612, 619, 620
Wilbraham, H., 909
Wiles, Andrew, 262, 362, 363, 366, 720, 829, **1061–1063**, 1107
    Fermat's Last Tango (Rosenblum), 689–690
    See also Fermat's Last Theorem
Wilhelm Karl Theodor, 144

Wilkins, J. Ernest, 350, 837
Wilkinson Microwave Anisotropy Probe (WMAP), 439
Wilks, Samuel, 112, 708, 944
William Lowell Putnam Mathematical Competition, 228
William the Conqueror, 177
Williams, Scott, 221
Willis Tower, 920
Wilson, Alexander, 964
Wilson, Kenneth G., 228
Wilson, Snoo, 777
wind and wind power, 1063–1065, 1066–1067
wind instruments, 1065–1066
wind tunnels, 1064
wind turbines, 1064, 1066–1067
windmills, 1066–1067
Winer, David, 304
Winkler, Johann, 738
wireless communication, 1068–1069
Wise, Michael, 767
Wishart, John, 803, 944
Wissel, Christian, 973
Witten, Edward, 344, 531
Wittgenstein, Ludwig, 219, 621
Wohlstetter, Albert, 215
Wohlstetter, Roberta, 215
Wolf, Rudolph, 964
Wolf Prize in Mathematics, 68, 69, 363, 367
Wolfram MathWorld, 221
women, 1069–1073
    mathematical aptitude and, 1069–1070
    mathematics education and, 668–669, 1069–1070
    professional employment and, 1070–1072, 1079
Women in Mathematics in Africa, 22
The Wonder Years (television show), 163
Wood, Melanie, 227
wood carvings, 736–737
Woodland, Norman J., 96
word problems, 1094
word squares. *See* **acrostics, word squares, and crosswords**
World Association of Veteran Athletes (WAVA), 937
World Cup Finals, 239
World Health Organization (WHO), 123, 314
World Squash Federation, 835
World Statistics Day, 289
World War I, 1073–1075
World War II, 1075–1080
    aircraft carriers, 28
    atomic bomb (Manhattan Project) and, 79–80
    codes/cyphers and, 1076–1077
    computers and, 1077–1078
    intelligence/counterintelligence and, 508

mathematicians and, 502–503, 1079
navigation and, 1078–1079
operations research (OR) and, 1078
See also military code
World's Columbian Exposition of 1893 (Chicago), *189*, 189
wormholes, 116
Wren, Christopher, 763, 1100
Wright, Benjamin, 156
Wright, Frank Lloyd, 190, **1080–1081**
writers, producers, and actors, **1081–1086**
    Danica McKellar, 1082
    goals/impact of, 1083
    mathematical consultants and, 1083–1085
    portrayals of mathematicians, 1085
    production similarities to mathematics, 1082
Wu, Sijue, 995
Wynne, Arthur, 6

X

X & Y album (Coldplay), 787
X games, 374
Xenakis, Iannis
    Metastasis, 231
    Nomos Alpha, 787
    Pithoprakta, 231
Xenocrates of Chalcedon, 763
Xerox Corporation, 132, 133
Xian, Jia, 111
Xiangjie Jiuzhang Suanfa (Yang Hui), 187
xkcd (Munroe), 219
X-rays, 308, 659, 674
Xu Guangqi, 187
Xylander. *See* Holzmann, William (Xylander)

Y

Yackel, Carolyn, 257
Yackel, Erna, 222
Yahoo, 12, 180
Yahoo! Research Labs, 840
Yamamoto Isoroku, 755
Yang Hui, 186, 1095
    Xiangjie Jiuzhang Suanfa, 187
Yang–Baxter equations, 532
Yang–Mills gauge theory, 525
Yates, Frank, 176
Yau, Shing-Tung, 68
Yeager, Charles, 31
Yellowstone caldera (supervolcano), 1044
Yellowstone Volcano Observatory, 1044
Yerkes Observatory, 980
Yi Gu Yan Duan (*New Steps in Computation*) (Li Zhi), 187

Yi Jing (I-ching I or *Book of Changes)*, 185
yield, crop, 381
Ying Yang, 185
Yoccoz, Jean-Christophe, 367
Yost, Charles, 636
Young, Grace Chisholm
    The Theory of Sets of Points, 1105
Young, James, 525
Young, Peyton, 232, 234
Young, Thomas, 547
Young, W. Rae, 977
Young, William Henry, 1105
Young Archimedes (Huxley), 559
Young Choon Lee, 463
Yule, Udny, 957
yupana, 494
Yupaporn Kemprasit, 71
Yusuf ibn Ibrahim, 18, 425, 1005

Z

Z3 computer, 1077
Zadeh, Lotfi, 290, 803
al-Zarqali (Arzarchel), 664
Zaslavsky, Claudia, 13
    Africa Counts, 14, 16
Zeno of Elea, 144, 505, 552
Zenodorus, 653, 762
Zeno's Paradox, 505, 775, 1091
Zentralblatt für Mathematik (journal), 220
Zentralblatt MATH, 220
Zermelo, Ernst, 587, 1106
Zermelo-Fraenkel (ZF) set theory. *See* Set Theory
zero, 186, 493, 494, 711, 901, **1087–1089**
Zero Gravity Research Facility, 1060
Zero-G, 1059–1060
zero-point energy, 1088
Zhang Heng, 323
Zhangjiashan's tomb, 185
Zhao, Jiamin, 134
Zhou dynasty, 185
Zhoubi suanjing (Anon.), 185, 423
Zhu Shijie, 111, 186
    Introduction to Computational Studies, 187
    Precious Mirror of the Four Elements (*Si Yuan Yujian*), 187
Zhui Shu (Zu Chongzhi), 186
Ziegler, Günter
    Proofs from THE BOOK, 625
ziggurats, 485, *486*
Zimbabwe, 16
Zimet, Paul
    Star Messengers, 690

Zizek, Frank
   Statistical Averages, 653
zodiac, 729
Zoetrope, 50, *50*
Zoltan, Kecskemeti B., 51

Zomaya, Albert, 463
Zu Chongzhi, 186
Zu Geng, 186
Zuse, Konrad, 1077
Zygalski, Henyrk, 1077

Photo Credits

Photos.com: 3, 24, 42, 44, 45, 47, 56, 75, 77, 90, 100, 103, 110, 131, 154, 161, 168, 173, 226, 239, 250, 254, 256, 263, 315, 446 bottom, 447 center, 451, 481, 495, 517, 534, 538, 551, 561, 563, 583, 605, 636, 638, 649, 659, 675, 715, 730, 733, 842, 864, 891, 899, 923, 986, 1003, 1008, 1014, 1053, 1065; iStockphoto: 16, 68, 98, 127, 182, 216, 246, 265, 296, 302, 307, 327, 339, 345, 352, 372, 375, 380, 383, 386, 389, 391, 393, 420, 628, 655, 834, 907, 983, 1033, 1036; Library of Congress: 189, 192, 334; Arthur's Clip Art: 57; Abel Prize: 293; National Archives: 81; National Snow and Ice Data Center: 197; National Aeronautics and Space Administration: 27, 116, 286, 334, 347, 359, 491, 678, 839, 871, 876, 1060; Centers for Disease Control and Prevention: 313 (James Gathany); Department of Energy: 662; Transportation Security Administration: 640; National Institutes of Health: 209, 528, 673; U.S. Army: 666 (Elizabeth M. Lorge), 704, 1006 (Laura Owen); U.S. Air Force: 82 (Larry McTighe), 272 (Todd Paris), 502 (Michele G. Misiano), 930; U.S. Navy: 223, 396 (Gina Wollman), 1041 (Brien Aho); National Center for Electron Microscopy: 548 (John Turner); USAID: 336 (Maureen Taft-Morales); NOAA Aviation Weather Center: 994, 1056; World Economic Forum: 526; Sandia National Laboratories: 37 (Randy Montoya); Rice University: 694 (Yasuhiro Shirai); U.S. Geological Services: 442, 773, 1045 (Austin Post); Cornell University: 431; Darrah Chavey: 13, 14; Wikimedia Commons: 18, 21, 29, 32, 50, 59, 62, 64, 80, 108, 119 (Konrad Jacobs), 121, 129 (Kevin Madden), 140 (Seth Morabito), 156 (Stan Shebs), 171 (Yves Remedios), 175, 180, 200, 202, 203, 215 (Tony Peters), 228, 230, 256, 258, 282, 294, 304, 310, 331, 357, 362, 366, 416, 423, 428, 446 top, 447 bottom, 454, 458, 462, 474, 477, 486, 530, 544, 554, 557, 565, 573, 588, 593, 598, 616, 623, 631, 668, 671, 685 (Steve Jurvetson), 696, 699, 709, 736, 738, 742, 756, 760, 762, 768, 780 (Doug Kerr), 783, 786, 851 (Tom Tsuchiya), 856, 861 (Stefan Bauer), 869, 886, 895, 904, 916 (Joan Garvin), 921, 957, 965 (David Benbennick), 971 (Sam Oth), 1039, 1078 (Ricardo Ferreira de Oliveira), 1080 (Greg O'Beirne); Wellesley Archives: 1070; Yorck Project: 749; Map on pg. 576 courtesy Wilhem Kruecken: http://www.wilhelmkruecken.de. Thank you to Stan Wagon for providing the photo on page 109 of Wayne Roberts at Macalester College.